P9-ECR-037

PLANT POPULATION GENETICS, BREEDING, AND GENETIC RESOURCES

PLANT POPULATION GENETICS, BREEDING, AND GENETIC RESOURCES

Edited by Anthony H. D. Brown
CSIRO, Canberra, Australia
Michael T. Clegg
University of California, Riverside
Alex L. Kahler
Biogenetic Services Inc., Brookings, S.D.
Bruce S. Weir
North Carolina State University, Raleigh

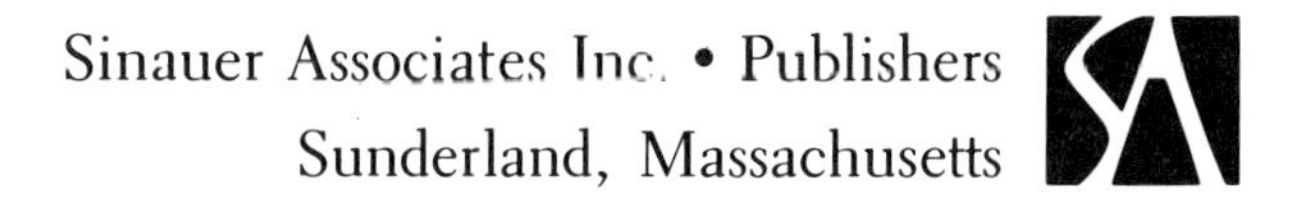
Sinauer Associates Inc. • Publishers
Sunderland, Massachusetts

Cover photograph by Major M. Goodman,
North Carolina State University

PLANT POPULATION GENETICS, BREEDING, AND GENETIC RESOURCES

This book evolved from the International Symposium on Population Genetics and Germplasm Resources in Crop Improvement, held August 11–13, 1988 at the University of California, Davis. For information address Sinauer Associates, Inc., Sunderland, MA 01375 U.S.A.

Library of Congress Cataloging-in-Publication Data

Plant population genetics, breeding, and genetic resources / edited by Anthony H. D. Brown . . . [et al.].

p. cm.

"Evolved from the International Symposium on Population Genetics and Germplasm Resources in Crop Improvement, held August 11–13, 1988, at the University of California, Davis"—T.p. verso.

Includes bibliographical references.

ISBN 0–87893–116–3 : $50.00 (est.). — ISBN 0–87893–117–1 (pbk.) : $29.50 (est.)

1. Plant population genetics—Congresses. 2. Germplasm resources, Plant—Congresses. 3. Plant Breeding—Congresses. 4. Allard, R. W. (Robert Wayne), 1919– —Congresses. I. Brown, Anthony Hugh Dean, 1941– II. International Symposium on Population Genetics and Germplasm Resources in Crop Improvement (1988 : University of California, Davis)

SB123.P564 1989

582'.015—dc20 89–37761

CIP

Printed in U.S.A.

3 2 1

To Robert W. Allard

Contents

Preface

Plant population genetics today stands on the threshold of a new period of scientific advance. The traditional theoretical and empirical methods of genetics and population genetics have melded with the new and powerful tools of molecular biology to provide an unparalleled view of genetic diversity and of the evolutionary mechanisms that shape genetic diversity. It is important now to take stock of our present knowledge and consider new interpretations of old problems as well as challenges for the future.

Population genetics sits at the nexus of several major areas of biology. It connects molecular biology to population and evolutionary biology and provides the basis for understanding environmental adaptations and the theoretical framework for crop improvement. No previous book has dealt with plant population genetics as a scientific enterprise. Our goal is to fill this void and, in so doing, to provide a guide to current issues, unsolved problems, and research opportunities in the field of plant population genetics. This book should be of interest and value to a wide audience, including students and research workers in population genetics, molecular evolution, evolutionary biology, ecological genetics, and crop improvement. We hope to stimulate students of genetics to enter plant population genetics and to mine the empirical lode that is indicated in these chapters.

This volume grew out of a symposium organized at Davis, California in August 1988 to review the status and future prospects of plant population genetics. A further purpose of the symposium was to honor Professor Robert W. Allard, who founded experimental plant population genetics as a scientific discipline.

These chapters bring together the contributions of research workers in genetics, population genetics, plant breeding, and genetic resources. We begin with an introductory chapter by R. W. Allard that points to new opportunities and unsolved problems. Following this introduction, the remainder of the book is divided into three broad areas. The first section considers the kinds of genetic diversity found in plant species and the statistical theory for evaluating variation. The second section is devoted to the structure of genetic variation and the microevolutionary processes that shape genetic diversity. The third section considers the applications of plant population genetics in foresty, crop improvement, and the conservation and use of crop genetic resources. The Literature Cited provides a comprehensive entrée into the current scientific literature.

Many individuals contributed to the symposium and to the inception of this volume. Particular credit goes to Drs. Calvin Qualset and Patrick

Macguire and to Mrs. Roberta Hooker of the Genetic Resources Conservation Program of the University of California. In addition, we thank the organizations that sponsored the original symposium, including the National Science Foundation; the California Crop Improvement Association; Calgene, Inc.; Cornnuts, Inc.; Monsanto Company; Pioneer Hi-Bred International, Inc.; the Rockefeller Foundation; the California Dry Bean Advisory Board; Campbell Soup Company; Eschegen Corporation; Northrup King Company; and Sigco Company.

A. H. D. BROWN
M. T. CLEGG
A. L. KAHLER
B. S. WEIR

The Contributors

ROBERT W. ALLARD, Department of Agronomy and Range Science, University of California, Davis, California 95616 U.S.A.

SPENCER C. H. BARRETT, Department of Botany, University of Toronto, Toronto, Ontario M5S 1A1 Canada

FREDRICK A. BLISS, Department of Pomology, University of California, Davis, California 95616 U.S.A.

ANTHONY H. D. BROWN, CSIRO, Division of Plant Industry, GPO 1600, Canberra, A.C.T. 2601 Australia

J. J. BURDON, CSIRO, Division of Plant Industry, GPO 1600, Canberra, A.C.T. 2601 Australia

MICHAEL T. CLEGG, Department of Botany and Plant Sciences, University of California, Riverside, California 92621 U.S.A.

JAN DVOŘÁK, Department of Agronomy and Range Science, University of California, Davis, California 95616 U.S.A.

R. A. ENNOS, Department of Forestry and Natural Resources, University of Edinburgh, Mayfield Road, Edinburgh EH9 3JU Scotland

B. K. EPPERSON, Department of Botany and Plant Sciences, University of California, Riverside, California 92621 U.S.A.

PAUL GEPTS, Department of Agronomy and Range Science, University of California, Davis, California 95616 U.S.A.

M. J. GODT, Department of Genetics, University of Georgia, Athens, Georgia 30602 U.S.A.

JAMES L. HAMRICK, Department of Botany, University of Georgia, Athens, Georgia 30602 U.S.A.

ALAN HASTINGS, Division of Environmental Studies, University of California, Davis, California 95616

BRIAN C. HUSBAND, Department of Botany, University of Toronto, Toronto, Ontario M5S 1A1 Canada

A. M. JAROSZ, CSIRO, Division of Plant Industry, GPO 1600, Canberra, A.C.T. 2601 Australia

D. R. MARSHALL, Waite Agricultural Research Institute, The University of Adelaide, Glen Osmond, S.A. 5064 Australia

O. MAYO, Biometry Section, Waite Agricultural Research Institute, The University of Adelaide, Glen Osmond, S.A. 5064 Australia

OUTI MUONA, Department of Genetics, University of Oulu, SF-90570 Oulu Finland

MASATOSHI NEI, Center for Demographic and Population Genetics, University of Texas Graduate School of Biomedical Sciences, Box 20334, Houston, Texas 77225 U.S.A.

OLIVER E. NELSON, Laboratory of Genetics, University of Wisconsin, Madison, Wisconsin 53706 U.S.A.

C. O. QUALSET, Department of Agronomy and Range Science, University of California, Davis, California 95616 U.S.A.

KERMIT RITLAND, Department of Botany, University of Toronto, Toronto, Ontario M5S 1A1 Canada

CHARLES W. STUBER, U.S. Department of Agriculture, Agricultural Research Service, North Carolina State University, Raleigh, North Carolina 27595 U.S.A.

W. E. WEBER, Institute for Applied Genetics, University of Hannover, Herrenhauserstrasse 2, D-3000 Hannover 21 Federal Republic of Germany

BRUCE S. WEIR, Department of Statistics, North Carolina State University, Box 8203, Raleigh, North Carolina 27695 U.S.A.

G. WRICKE, Institute for Applied Genetics, University of Hannover, Herrenhauserstrasse 2, D-3000 Hannover 21 Federal Republic of Germany

CHAPTER 1

FUTURE DIRECTIONS IN PLANT POPULATION GENETICS, EVOLUTION, AND BREEDING

R. W. Allard

I believe it will be useful to start by quoting from the brochure that announced the symposium on which this volume is based:

> The discipline of plant population genetics lies at the core of modern approaches to plant population biology, evolution, plant breeding, and genetic resource conservation . . . the overall aim of this symposium is to increase understanding of the role of plant population genetics in guiding evolutionary theory and plant improvement.

The first inkling I had that a symposium with this goal might be in the works came three or four years ago when I began to hear comments from various colleagues that plant population genetics had changed a great deal since I joined the staff at Davis in 1946 and that the time might be approaching to hold a symposium to assess the present state of plant population biology and to attempt a look into the future. I thought this was a good idea and I want to express my appreciation to those who came up with the idea, and to those, especially Cal Qualset, who persevered in bringing this symposium about. I am most pleased with the program the organizing committee has arranged and also with their decision to dedicate the symposium to me. It gives me a warm feeling to find that my efforts have been noticed.

When I learned that the symposium was indeed going to take place I rejected offers to be one of the speakers; however, over the past couple of years one thing led to another and ultimately I found myself confronted with the title "Future directions in plant population genetics, evolution, and breeding." This title is quite intimidating, consequently, rather than address it I will consider a safer title, something on the order of, "Some directions I hope plant population genetics will take in the future."

I believe it is fair to say that, as least as far as textbooks were concerned, plant population genetics had not been invented when I arrived at Davis in 1946. There was no population genetics textbook at the time and, in fact, the term "population genetics" occurred neither in the general genetics textbooks nor in the plant breeding or animal breeding textbooks of the day. There was usually an opening statement in the genetics texts to the effect that "A knowledge of the chromosome basis of heredity is essential to evolutionists" and, in the breeding texts, that such knowledge is essential to breeders. But what followed were descriptions of one-locus and two-locus transmission genetics in segregating families, linkage maps of various organisms, procedures for genetic analysis of pedigrees and quantitative characters, tables of chromosome numbers, discussions of techniques such as polyploidy breeding, interspecific hybridization, mutation breeding, and the like. So far as I am aware, the first clear statement of the importance of population genetics was in Dobzhansky's *Genetics and the Origin of Species*, first published in 1937. In that volume Dobzhansky recognized three levels at which *genetics* has an impact on *evolutionary processes*. The first level was that of *mechanisms of transmission of hereditary characteristics from parents to offspring*. Dobzhansky's comment was that "The signal successes of genetics to date have been at this first level . . . types possessing a desired set of characteristics may, within limits, be synthesized at will, and the schemes of such 'syntheses', as worked out in theory, are almost always realized in detail in actual experiments." The second level was that of *gene action*, which Dobzhansky considered to be "far from being solved." The third subdivision "has as its province the processes taking place in groups of individuals—in populations—and therefore is called the *genetics of populations*. The rules governing the genetic structure of populations are distinct from these governing the genetics of individuals . . . although in fact they are only integrated forms of the latter." Dobzhansky gave Hardy's formula and mentioned that Haldane, Fisher, and others had treated effects of nonrandom mating and selection on one-locus allelic and genotypic frequencies mathematically. He concluded, however, "that it is necessary to proceed by inference because the theory has not been adequately tested in nature." He went on to say, "Nonetheless, a growing body of observational and experimental evidence gives at least a promise that an adequate analysis of evolutionary dynamics will be possible in the not too distant future." Dobzhansky was aware of Sewall Wright's 1931 paper "Evolution in Mendelian Populations." He commented: "The importance of this work can hardly be overestimated. The experimental work itself is, however, still a task for the future."

What Dobzhansky seemed to be saying was that, even though our ignorance greatly exceeded our knowledge, we did have at least some notion of the future of evolutionary genetics. This leads me to a question: could plant oriented persons of the late 1940s have predicted the breadth and the depth of the remarkable series of advances in plant population biology that are indicated by the program of this symposium? Some of the research that was going on in the 1940s, e.g., that of Harry Harlan and Coit Suneson with inbreeding species and that of Merle Jenkins and George Sprague with corn, leads me to believe that there were some who might have identified, at least in broad outline, the general course that plant population biology was to take. These workers were keen observers of genetic diversity in their experimental populations and they were aware that numerous biological factors, such as mating systems, variability in selective values, fluctuations in populations of pathogens, and complex interactions among such factors, all play a role in determining the course of genetic change. Their vision, which flew in the face of the single-factor approaches that many favor to this day, was that frequent cycles of intermating, segregation, and recombination produced numerous novel genotypes and that selection, both natural and artificial, reduced the frequency of the "weaker sorts," as Harlan put it, and increased the frequency of the more favorable genotypes in their populations. This perspective led them to adopt population-type breeding programs that foreshadowed the recurrent selection procedures that have been applied so successfully in recent years.

I will now attempt to assess the situation as it is at present to set the stage for a few remarks on the directions I hope plant population genetics, evolution, and breeding will take in the near future. Fortunately, the titles of the papers of this symposium have already done most of the stage setting for me because in themselves they give a good idea of the broad range of issues that has been addressed and the richness of the advances that have been made since the 1940s. However, I think it will be useful to introduce some additional information, still largely unpublished, from several ongoing population studies that have been the core of my project for the past 40 years; these studies are the only ones in my project that I can claim as singularly my own. Although I have not yet fully analyzed the data of these experiments, some of the results that have emerged have badly shaken various concepts that I had held firmly over the years as established doctrine, whereas other results have supported a number of concepts I had arrived at largely on intuitive grounds.

The studies I refer to grew out of two sets of observations I made in the 1930s. The first observations were made in the period 1931–1937 when I planted, cared for, and harvested experimental populations of corn and ryegrass (both outbreeders), and common beans, lima beans, garbanzos, and fava beans (all moderate to heavy inbreeders) that W. W. Mackie of the Berkeley campus grew on my father's farm. The second set of observations was made here at Davis, when, as an undergraduate student, I worked for Coit Suneson on his composite crosses of barley and his natural populations of *Avena fatua*. After my undergraduate work I went to the University of Wisconsin as a graduate student. There, working under forced draft, I tried to complete a Ph.D. before the Navy called me to World War II duty, and I did not have time to think about my experiences with plant populations. However, during World War II the Navy provided me with frequent slack periods during which I pondered the discussions I had had with Bill Mackie, Coit Suneson, Harry Harlan, Gus Wiebe, and some of the corn breeders, and, by the time I returned to Davis in the spring of 1946, I had in mind some experiments I wanted to do. Mackie and Suneson were very encouraging and they generously provided me with reserve seed of a succession of generations of their populations. The main question asked in these experiments was, "What do alternative alleles of variable loci *do* in populations, and, in particular, how do specific alleles affect *adaptedness* at the population level?" This, it seemed to me, was the "bottom line" question of population genetics. The structure of these experiments was very simple: it involved determining the frequency of specific alleles of a large number of discretely-inherited single-locus characters while simultaneously determining the expression of a number of quantitative traits over many successive generations in experimental and natural populations of several species. The populations were monitored over generations in two ways: first, for those discretely inherited characters and for those quantitative traits that can be classified or measured conveniently on an entire population basis in natural populations or in agricultural populations grown at commercial rates of planting; second, the populations were monitored for those traits that can be classified or measured only on a single plant basis in spaced plantings. Quantitative characters that were monitored on an entire population basis included characters such as flowering time, height, and seed size. Seed yield was also monitored on an entire population basis in the cultivated cereal and cereal legume species; however, yield testing in replicated yield trials repeated in three to six seasons is labor intensive and it was consequently done only at 5 to 10 generation intervals. The original set of discretely inherited characters monitored on an entire population basis included several easily classified morphological variants such as two-row versus six-row spikes and rough versus smooth awn in barley. However, in the 1950s a second category was added, namely, resistance versus susceptibility to specific pathotypes of fungal disease organisms, which was determined by inoculating seedlings with single pathotypes in the greenhouse and counting the numbers of individual seedlings that were resistant or susceptible. A third category, a number of moderately to highly polymorphic isozyme variants detectable by starch gel electrophoresis, was added in the 1960s and a fourth category, a number of DNA restriction fragment loci, was added in the late 1970s. The number of loci was adequate to blanket all chromosomes and hence give an idea of the sweep of selection over the entire genome. The restriction fragment loci turned out to be the most informative loci once we had worked out rapid assay methods.

The second way in which the populations were monitored was by isolating random individuals in a succession of generations of the populations, selfing these individuals (or making biparental matings among them), and then classifying individuals within the resulting families for the discretely inherited characters and also measuring the same individuals for a dozen or so quantitative characters. Thus progeny-test data were taken in such a way that the individual founder plants that had been isolated from the population could be categorized as homozygous or heterozygous for each single-locus marker loci and also as to phenotype for the quantitative characters. The family progeny tests were repeated in two or three seasons. This practice was adopted because it had become apparent early in these studies that estimates of quantitative trait expression made on individual plants varied so much with the local environment as to be of little use in estimating parameters of quantitative variability and that even family progeny-test data varied a great deal from season to season.

I emphasize that large populations were grown each generation with the goal of reducing the effects of genetic drift to triviality. Actual results indicated that the effects of directed processes, such as selection and departures from random mating, were overwhelmingly larger than random processes in these studies, in fact, usually more than two orders of magnitude larger. Similarly, the samples taken from the various generations were large enough so that standard errors of estimates were small for both the enumeration data and the quantitative data.

I am certain you do not want me to give details of the various kinds of evidence provided by the massive data sets that accumulated over the years. Consequently, I will proceed directly to an abbreviated summary of the main conclusions that emerged from studies of the several species that were investigated. In this necessarily terse overview I will make 10 points in total, proceeding from three points that emerged from studies of the *quantitative* characters, to two points from studies of the discretely inherited, *single-locus* characters considered one by one, to five points from *multilocus analyses* of the discretely inherited characters.

First, three points concerning quantitative characters:

Point 1. Reproductive capacity manifested as higher seed yields, larger numbers of seeds/plant, greater spike weight, and other traits directly related to reproductive capacity, increased in all populations. In highly variable populations of the cereal grains and the cereal legumes that had been synthesized from samples of germplasm from diverse ecogeographical regions, reproductive capacity increased during the first 15–20 generations of natural selection from about 60% to about 95% of that of modern cultivars included as checks. Reproductive capacity continued to increase into the latest generations, for example, it continued to increase for more than 50 generations in Barley Composite Cross II; the rate of increase, although slower in the late generations, kept performance within about 95% of that of the latest adapted cultivars. Selection was clearly very strong and there could be no doubt that highly favorable evolutionary changes occurred in all populations.

Point 2. Changes in quantitative traits directly related to reproductive capacity, such as seed yield and numbers of seeds/plant, were positively and significantly correlated with each other in most generations. However, such changes in reproductive capacity were usually *not* significantly correlated, or they were only weakly correlated, with the other quantitative traits measured such as time to maturity, height, vegetative biomass, and similar favorite characters in ecogenetic studies.

Point 3. Data from the quantitative characters indicated that the underlying genetic systems were complex and much affected by environment. Quantitative data unfortunately provided, at the very most, distressingly meager information about what was going on genetically within the populations.

Now to two points reached from analyses of data from discretely inherited characters:

Point 4. Every discretely inherited marker locus, including the morphological variants, disease resistance loci, isozyme loci, and the restriction fragment loci, had *large* effects on reproductive capacity as well as on several and often all of the other quantitative characters that were measured. Persistent attempts to disassociate quantitative trait effects from specific alleles of the marker loci by segregation were consistently unsuccessful. This indicates that the marker loci studied were major genes not only for the discrete alternative characters for which they are named, but also for many quantitative characters. Quantitative-trait differences between marker-locus homozygotes were usually large, which is to say that single-locus additive effects were large. However, single-locus dominance effects were nearly always small.

Point 5. Allelic frequency changes at all marker loci studied were strongly correlated with changes in reproductive capacity. High grain yield and high seed number/plant were consistently associated positively with the *predominant* allele of each marker locus. Interestingly the *majority* of alleles that confer disease resistance had negative effects on yield and reproductive capacity in the absence of disease and often also in the presence of disease. No consistent patterns were detected for other quantitative characters, e.g., early versus late flowering, tall versus short stature, and high versus low vegetative biomass were associated about equally frequently with marker-locus alleles with *low* reproductive capacity and *low* survival value as with marker-locus alleles with *high* reproductive capacity and *high* survival value. Thus, predominant alleles were about equally likely to increase as to decrease the mean expression of quantitative traits (other than reproductive capacity) as the populations evolved.

Now to five points from *multilocus analyses* of the marker loci.

Point 6. Twenty or so marker loci were scored over many generations for most populations in most species so that very large numbers of pairwise as well as three-locus, four-locus, and higher order multilocus comparisons were ultimately available for analysis. Rapid build-up of highly significant associations between specific alleles at pairs of loci occurred in early generations in all inbreeding populations. However, the extent of associations between pairs of loci frequently changed significantly in amount, and even in direction, in ways that indicated that associations between pairs of loci are much affected not only by year-to-year fluctuations in the environment, but also by evolutionary changes in the background multilocus genotype of the population. These changes occurred in bewildering patterns such that I was unable to deduce higher order multilocus genetic structure from the two-locus structures. Another finding was that additive × additive (A × A) types of interactions were large at the two-locus, and more particularly at higher multilocus levels, whereas additive × dominance (A × D) and dominance × dominance (D × D) type interactions were small at all levels.

Point 7. Log-linear, canonical correlation and cluster analyses uncovered features of microevolutionary change that were not apparent from the two-locus analyses. The polymorphic loci monitored in various inbreeding populations usually became significantly clustered into groups of three or four loci each by the third or fourth generation after synthesis of the population. A series of breakdowns of associations and complex repatternings and amalgamations then started that led to clusters of six to eight or more loci in the mid and later generations. The picture of adaptive change that emerges is that increases in adaptedness in a given environment are correlated with the development of clusters of associated alleles at different loci and gradual amalgamation of clusters into large synergistically interacting complexes. Far less genetic structure developed within *outbreeding* than within *inbreeding* populations. Adaptedness manifested as superior ability to live and reproduce in specific environments also appeared to develop more slowly within outbreeding populations than within inbreeding populations. Ecogenetic differentiation among outbreeding populations was much smaller than ecogenetic differentiation among inbreeding populations.

Point 8. None of the populations of either inbreeders or outbreeders had become genetically uniform even after as many as 50 generations of propagation in closed isolated populations grown continuously in a single area. Barley CC II, for example, still responded dramatically to artificial selection in both plus and minus directions for quantitative traits, including traits relevant to agricultural suitability, such as time to maturity, superior standing ability, and disease resistance, after more than 50 generations of propagation without conscious selection. Barley CC II also remained highly variable for quantitative characters and it responded rapidly to selection for such characters in late generation. It also remained polymorphic for marker loci and complexes of marker loci regardless of the environment in which it was grown, e.g., whether it had been grown in the Mediterranean climate of Davis or the continental climate of Montana.

Point 9. All of the composite crosses of barley developed essentially the same multilocus structure when they were grown in ecologically similar environments, e.g., in various facies of Mediterranean climates. However, they developed very different multilocus structures in each major climatic region in which they were grown, e.g., the structure that developed in Mediterranean climates was very different from that which developed in continental climates. In cultivated barley there appear to be only about a half dozen major genetic structures worldwide. But there are also a number of subtypes—relatively minor variations on the major themes—within each major type. This leads to my final point—the evolution of differing genetic organizations among populations of a species, which I will illustrate with *Avena barbata*, the Slender Wild Oat.

Point 10. Avena barbata, which was accidentally introduced to California originally from Spain during the explorer and mission period, has become a major component of grassland and grass oak savanna habitats in California where it now occurs in populations of millions and millions of individuals. Surveys of about 20 loci, mostly isozyme and restriction fragment loci, indicate that virtually all alleles, excepting a few rare alleles, present in the Spanish gene pool are also present, and at similar frequencies, in California. Genetic identity measures show that the Spanish and California gene pools are very similar to each other in allelic content. The Spanish and California gene pools are, however, both highly structured on a multilocus basis and this structuring is highly correlated with various factors of the environment, particularly with moisture and temperature differences. Although many of the three- or four-locus associations found in Spain occur in California, not a single one of the half dozen or so very large multilocus complexes found in California has been found in Spain. It is therefore apparent that within a period of 150 years or so, the allelic ingredients of the Spanish gene pool have become structured under the environmental conditions of California into a limited number of *novel* multilocus genotypes, each of which is very closely associated with a specific distinct habitat. Natural selection has clearly done a good job of developing populations of this desirable range forage species that are uniquely adapted to the main ecological niches of California. The process closely mimics the highly favorable evolutionary processes that have occurred in the experimental populations, such as the barley composite crosses, which were grown in agricultural environments, and the evolutionary processes, which have occurred during the domestication and improvement of barley by man.

Each of the above 10 points has implications, and I believe they are important implications, concerning one or more aspects of plant population genetics, evolution, and/or breeding, e.g., implications concerning the generality of the multiple factor hypotheses, host-pathogen interactions at the population level, the evolutionary consequences of mating systems, genetic changes that occur during domestication and breeding, and genetic resource conservation. In particular these results tell us that highly useful evolutionary processes go on in genetically variable populations grown under agricultural conditions and they also tell us why it is that modern breeders of major crops rely so heavily on elite cultivars and on various kinds of genetically enhanced populations for their genetic diversity, rather than on traditional collections that emphasize obsolete cultivars, primitive landraces, and wild relatives stored in static state. However, each of the 10 points I have made is a story in itself which cannot be pursued further here.

So much for attempts to assess the state of our knowledge of plant population genetics, evolution, and breeding as it is at present. This assessment gives me the distinct feeling that I arrived on the scene 40 years too early. I very much wish the issues had been as well defined 40 years ago as they are now; I would not have stumbled about to the extent that I did. I wish even more the tools we have now, biochemical and especially DNA markers, ways of measuring selection, ways of estimating mating system parameters and parameters specifying genetic organization from actual populations, and procedures for evaluating the usefulness of specific alleles in various genetic backgrounds under various natural and agricultural conditions, had been available then. I see the stage as now being set to improve the knowledge/ignorance ratio at rates that could hardly have been imagined even a decade ago. But the rate of improvement that is now possible is unlikely to be realized unless we maintain the balance among disciplines that we have had in our programs in recent years. My concern is that we are in the process of losing that balance by overemphasizing "high-tech" methodologies to the detriment of the basic underlying genetic, ecological, statistical, and evolutionary components that have been essential to the progress we report in this symposium. There is a great deal yet to learn about what really goes on genetically within and among different populations of a species and my hope for the future is that we continue a comprehensive and balanced attack on those things that are most important to know about populations living in natural and agricultural environments.

Finally, several persons have asked me to comment on the impact of molecular biology on future progress in practical plant breeding. With respect to "down in the trenches" plant breeding, my guess is that, at least for the dozen or so main crops that feed and clothe the world, major progress is going to come much in the same way as it has in the past, i.e., through the efforts of "honest-to-goodness" plant breeders, as one of my midwestern colleagues put it recently, "patiently" putting together favorable complexes of genes much in the same way they are doing it now. I am apprehensive that the production of general plant breeders will continue its drastically downward course in our universities to the detriment of progress in plant breeding and efficient conservation of the germplasm resources that are needed for long-term breeding progress.

PUBLICATIONS OF R. W. ALLARD

1946

Ennis, W. B., Jr., C. P. Swanson, R. W. Allard and F. T. Boyd. Effects of certain growth-regulating compounds on Irish potatoes. Bot. Gaz. 107: 568–574.

Allard, R. W., H. R. DeRose and C. P. Swanson. Some effects of plant growth-regulators on seed germination and seedling development. Bot. Gaz. 107: 575–583.
Allard, R. W., W. B. Ennis, Jr., H. R. DeRose and R. J. Weaver. The action of isopropylphenyl-carbamate on plants. Bot. Gaz. 107: 589–596.

1948

Allard, R. W. Root-knot nematode. Calif. Agric. 2: 9.

1949

Allard, R. W. *Manual of California Field Crops*. Associated Students Store, Davis, California. 167 pp.
Allard, R. W. A cytogenetic study dealing with the transfer of genes from *Triticum timopheevi* to common wheat by backcrossing. J. Agric. Res. 78: 33–64.
Allard, R. W. Plant breeding disease resistance genes of nonagricultural wheat transferred to commercial bread wheat. Calif. Agric. 3: 6.

1952

Allard, R. W. The precision of lattice designs with a small number of entries in lima bean yield trials. Agron. J. 44: 200–202.
Kendrick, J. B., Jr., and R. W. Allard. A root rot tolerant lima bean. Phytopathology 4: 515.
Allard, R. W. Inheritance of hypocotyl color in lima beans. Proc. Am. Soc. Hortic. Sci. 60: 387–390.

1953

Allard, R. W. Production of dry-edible lima beans in California. California Agricultural Experiment Station Extension Service Circular No. 423. 26 pp.
Briggs, F. N., and R. W. Allard. The current status of the backcross method of plant breeding. Agron. J. 45: 131–138.
Allard, R. W. Inheritance of some seed-coat colors and patterns in lima beans. Hilgardia 22: 167–177.
Allard, R. W. A gene in lima beans pleiotropically affecting male-sterility and seedling abnormality. Proc. Am. Soc. Hortic. Sci. 61: 467–471.
Allard, R. W. The inheritance of four morphological characters in lima beans. Hilgardia 22: 383–389.

1954

Allard, R. W. Sources of root-knot nematode resistance in lima beans. Phytopathology 44: 1–4.
Allard, R. W., and F. L. Smith. Production of dry-edible beans in California. California Agricultural Experiment Station Circular No. 436. 38 pp.
Allard, R. W., and J. Hills. The Mackie standard lima bean variety. Calif. Agric. 8: 5.
Allard, R. W., and R. G. Shands. Inheritance of resistance to stem rust and powdery mildew in cytologically stable spring wheats derived from *Triticum timopheevi*. Phytopathology 44: 266–274.
Allard, R. W. Frijol que tolera el calor. Hacienda 49: 47.
Allard, R. W. Natural hybridization in lima beans in California. Proc. Am. Soc. Hortic. Scie. 64: 410–416.

1955

Kendrick, J. B., Jr., and R. W. Allard. A lima bean tolerant to stem rot. Calif. Agric. 9: 8.

1956

Allard, R. W. Formulas and tables to facilitate the calculation of recombination values in heredity. Hilgardia 24: 235–278.
Allard, R. W. The analysis of genic-environmental interactions by means of diallel crosses. Genetics 41: 305–318.
Allard, R. W. Genes modifying the Cc and Rr loci in lima beans. Proc. Am. Soc. Hortic. Sci. 68: 386–391.
Allard, R. W. Biometrical approach to plant breeding. Brookhaven Symp. Biol. 9: 69–88.
Allard, R. W. Estimation of prepotency from lima bean diallel cross data. Agron. J. 48: 537–543.
Allard, R. W., D. G. Smeltzer and D. S. Mikkelsen. *Manual of California Field Crops*, Second Edition. Associated Students Store, Davis, California. 183 pp.

1957

Allard, R. W. Effect of genotypic-environmental interactions on the prediction of genetic advance under selection. Biometrics 13: 550.

1958

McGuire, D. C., and R. W. Allard. Testing nematode resistance in the field. Plant Dis. Rep. 42: 1169–1172.

1959

Allard, R. W., and W. M. Clement. Linkage in lima beans. J. Hered. 50: 63–67.
Parsons, P. A., and R. W. Allard. Seasonal variation in lima bean seed size: An example of genotypic-environmental interaction. Heredity 14: 115–123.

1960

Crumpacker, D. W., and R. W. Allard. A diallel cross analysis of heading date in wheat. Genetics 45: 982–983.
Allard, R. W. *Principles of Plant Breeding*. Wiley, New York. 485 pp.
Jain, S. K., and R. W. Allard. Population studies in predominantly self-pollinated species. I. Evidence for heterozygote advantage in a closed population of barley. Proc. Natl. Acad. Sci. U.S.A. 46: 1371–1377.
Allard, R. W., and H. L. Alder. The effect of incomplete penetrance on the estimation of recombination values. Heredity 14: 263–282.
Allard, R. W. Stabilizing yields of self-pollinated crops. Calif. Agric. 14: 7.
Allard, R. W. Plant exploring in Latin America. Report Fourth Annual Dry Bean Research Conf. 4: 17–23.

1961

Allard, R. W. Relationship between genetic diversity and consistency of performance in different environments. Crop Sci. 1: 127–133.
Yermanos, D. M., and R. W. Allard. Detection of epistatic gene action in flax. Crop Sci. 1: 307–310.
McGuire, D. C., R. W. Allard and J. A. Harding. Inheritance of root-knot nematode resistance in lima beans. Proc. Am. Soc. Hortic. Sci. 78: 302–307.

1962

Crumpacker, D. W., and R. W. Allard. A diallel cross analysis of neading date in wheat. Hilgardia 32: 275–318.

Allard, R. W., and S. K. Jain. Population studies in predominantly self-pollinated species. II. Analysis of quantitative genetic changes in a bulk-hybrid population of barley. Evolution 16: 90–101.

Pattimore, E. D., and R. W. Allard. Host-parasite interactions between lima bean strains and four species of root-knot nematodes. Proc. Am. Soc. Hortic. Sci. 81: 299–303.

Fasoulas, A. C., and R. W. Allard. Nonallelic gene interactions in the inheritance of quantitative characters in barley. Genetics 47: 899–907.

Workman, P. L., and R. W. Allard. Population studies in predominantly self-pollinated species. III. A matrix model for mixed selfing and random outcrossing. Proc. Natl. Acad. Sci. U.S.A. 48: 1318–1325.

1963

Allard, R. W., and P. L. Workman. Population studies in predominantly self-pollinated species. IV. Seasonal fluctuations in estimated values of genetic parameters in lima bean populations. Evolution 17: 470–480.

Allard, R. W. Current dry bean production research. Annual Dry Bean Research Conference 6: 37–38.

Edwards, K. J. R., and R. W. Allard. The influence of light intensity on competitive ability. Am. Nat. 97: 243–248.

Allard, R. W. An additional gametophytic factor in the lima bean. Der Zuchter 33: 212–216.

Allard, R. W., and J. Harding. Early generation analysis and prediction of gain under selection in a wheat hybrid. Crop Sci. 3: 454–456.

Allard, R. W. Evidence for genetic restriction of recombination in the lima bean, a predominantly self-pollinated species. Genetics 48: 1389–1395.

Allard, R. W., and P. E. Hansche. Population and biometrical genetics in plant breeding. Proc. XI Int. Congr. Genet. 3: 665–679.

Allard, R. W. Some population parameters and the utilization of exotic variability in predominantly self-pollinated crop plants. 1963 Annual Rept. W–1 Tech. Committee. pp. 5–37.

Jain, S. K., and R. W. Allard. A theoretical study of population changes under stabilizing selection. In "Genetics Today," Proc. XI Int. Congr. Genet. 1: 148–149.

Allard, R. W., G. A. Baker, Jr. and J. Christy. Stochastic processes and genotypic frequencies under mixed selfing and random mating. Ann. Math. Stat. 34: 687–688.

Allard, R. W., G. A. Baker, Jr. and J. Christy. Analysis of genetic change in finite populations composed of mixtures of pure lines. Ann. Math. Stat. 34: 688.

1964

Allard, R. W., and C. Wehrhahn. A theory which predicts stable equilibrium for inversion polymorphisms in the grasshopper, *Moraba scurra*. Evolution 18: 129–130.

Hartmann, R. W., and R. W. Allard. The effect of nutrient and moisture levels on competitive ability in barley. Crop Sci. 4: 424–426.

Workman, P. L., and R. W. Allard. Population studies in predominantly self-pollinated species. V. Analysis of differential and random viabilities in mixtures of competing pure lines. Heredity 19: 181–189.

Allard, R. W., and P. E. Hansche. Some parameters of population variability and their implications in plant breeding. Adv. Agron. 16: 281–325.

Weil, J., and R. W. Allard. The mating system and genetic variability in natural populations of *Collinsia heterophylla*. Evolution 18: 515–525.

Allard, R. W., and A. D. Bradshaw. The implications of genotype-environmental interactions in applied plant breeding. Crop Sci. 4: 503–508.

Kannenberg, L. W., and R. W. Allard. An association between pigment and lignin formation in the seed coat of the lima bean. Crop Sci. 4: 621–623.

1965

Allard, R. W. Genetic systems associated with colonizing ability in predominantly self-pollinated species. pp. 49–76. In: *Genetics of Colonizing Species*, Edited by H. E. Baker, Academic Press, New York.

Harding, J., and R. W. Allard. Genetic variability in highly inbred isogenic lines of the lima bean. Crop Sci. 5: 203–206.

Imam, A. G., and R. W. Allard. Population studies in predominantly self-pollinated species. VI. Genetic variability between and within natural populations of wild oats from differing habitats in California. Genetics 51: 49–62.

Wehrhahn, C., and R. W. Allard. The detection and measurement of the effects of individual genes involved in the inheritance of a quantitative character in wheat. Genetics 51: 109–119.

1966

Allard, R. W., J. Harding and C. Wehrhahn. The estimation of selective values and their use in predicting population change. Heredity 21: 547–564.

Jain, S. K., and R. W. Allard. The effects of linkage, epistasis and inbreeding on population changes under selection. Genetics 53: 633–659.

Harding, J., R. W. Allard and D. G. Smeltzer. Population studies in predominantly self-pollinated species. IX. Frequency dependent selection in *Phaseolus lunatus*. Proc. Natl. Acad. Sci. U.S.A. 56: 99–104.

Hansche, P. E., S. K. Jain and R. W. Allard. The effects of epistasis and gametic unbalance on genetic loads under inbreeding. Genetics 54: 1027–1040.

Allard, R. W. Population structure and performance in crop plants. Ciencia E. Cultura 19: 145–150.

1967

Kannenberg, L. W., and R. W. Allard. Population studies in predominantly self-pollinated species. VIII. Genetic variability in the *Festuca microstachys* complex. Evolution 21: 227–240.

1968

Allard, R. W., S. K. Jain, and P. L. Workman. The genetics of inbreeding populations. Adv. Genet. 14: 55–131.

Allard, R. W., and L. W. Kannenberg. Population studies in predominantly self-pollinated species. XI. Genetic divergence among the members of the *Festuca microstachys* complex. Evolution 22: 517–528.

Allard, R. W. The application of population genetics to plant breeding. Proc. XIII Int. Congr. Genet. (Tokyo) 2: 12.

Brown, A. H. D., and R. W. Allard. Inheritance of isozyme differences among the inbred parents of a reciprocal recurrent selection population of corn. Crop Sci. 9: 72–75.

1969

Allard, R. W., and J. Adams. The role of intergenotypic interactions in plant breeding. Proc. XII Int. Congr. Genet. 3: 349–370.

Marshall, D. R., and R. W. Allard. The genetics of electrophoretic variants in *Avena*. I. The esterase E_4, E_9, E_{10}, phosphatase P_5 and anodal peroxidase APX_5 loci in A. *barbata*. J. Hered. 60: 17–19.

Allard, R. W. Some observations of breeding for drought resistance in plants. pp. 459–471. In: *Man, Food and Agriculture in the Middle East*. Centennial Symposium, American University of Beirut, Beirut, Lebanon.

Kikuchi, F., R. W. Allard and S. R. Chapman. The effect of natural selection on phenotypic plasticity and genetic polymorphism in a barley population. pp. 517–

526. In: *Man, Food and Agriculture in the Middle East*. Centennial Symposium, American University of Beirut, Beirut, Lebanon.

Harding, J., and R. W. Allard. Population studies in predominantly self-pollinated species. XII. Interactions between loci affecting fitness in a population of *Phaseolus lunatus*. Genetics 61: 721–736.

Clay, R. E., and R. W. Allard. A comparison of the performance of homogeneous and heterogeneous barley populations. Crop Sci. 9: 407–412.

Chapman, S. R., R. W. Allard and J. Adams. Effect of planting rate and genotypic frequency on yield and seed size in mixtures of two wheat varieties. Crop Sci. 9: 575–576.

Brown, A. H. D., and R. W. Allard. Further isozyme differences among the inbred parents of a reciprocal recurrent selection population of maize. Crop Sci. 9: 643–644.

Allard, R. W., and J. Adams. Population studies in predominantly self-pollinated species. XIII. Intergenotypic competition and population structure in barley and wheat. Am. Nat. 103: 621–645.

1970

Allard, R. W. Population structure and sampling methods. pp. 97–107. In: *Plant Exploration and Conservation of Genetic Resources in Plants*. Int. Biol. Program Symp.

Allard, R. W. Problems of maintenance of genetic variability. pp. 491–494. In: *Plant Exploration and Conservation of Genetic Resources in Plants*. Int. Biol. Program Symp.

Kahler, A. L., and R. W. Allard. The genetics of isozyme variants in barley. I. Esterases. Crop Sci. 10: 444–448.

Allard, R. W., A. L. Kahler and B. S. Weir. Isozyme polymorphisms in barley populations. Proc. Second Int. Barley Genet. Symp. pp. 1–13.

Marshall, D. R., and R. W. Allard. Isozyme polymorphisms in natural populations of *Avena fatua* and A. barbata. Heredity 25: 373–382.

Brown, A. H. D., and R. W. Allard. Estimation of the mating system in open-polinated maize populations using isozyme polymorphisms. Genetics 66: 133–145.

Marshall, D. R., and R. W. Allard. Maintenance of isozyme polymorphisms in natural populations of *Avena barbata*. Genetics 66: 393–399.

Allard, R. W. Plant breeding. Encyclopedia Britannica. 15: 496–500.

1971

Allard, R. W., and A. L. Kahler. Allozyme polymorphisms in plant populations. Stadler Genet. Symp. 3: 9–24.

Brown, A. H. D., and R. W. Allard. The effect of reciprocal recurrent selection for yield on isozyme polymorphisms in maize. Crop Sci. 11: 888–893.

1972

Clegg, M. T., and R. W. Allard. Patterns of genetic differentiation in the slender wild oat species *Avena barbata*. Proc. Natl. Acad. Sci. U.S.A. 69: 1820–1824.

Hamrick, J. L., and R. W. Allard. Microgeographical variation in allozyme frequencies in *Avena barbata*. Proc. Natl. Acad. Sci. U.S.A. 69: 2000–2004.

Clegg, M. T., R. W. Allard and A. L. Kahler. Is the gene the unit of selection?: Evidence from two experimental plant populations. Proc. Natl. Acad. Sci. U.S.A. 69: 2474–2478.

Allard, R. W., G. R. Babbel, M. T. Clegg and A. L. Kahler. Evidence for coadaptation in *Avena barbata*. Proc. Natl. Acad. Sci. U.S.A. 69: 3043–3048.

Allard, R. W., A. L. Kahler and B. S. Weir. The effect of selection on esterase allozymes in a barley population. Genetics 72: 489–503.

Weir, B. S., R. W. Allard and A. L. Kahler. Analysis of complex allozyme polymorphisms in a barley population. Genetics 72: 505–523.

Allard, R. W., and A. L. Kahler. Patterns of molecular variation in plant populations. Proc. Sixth Berkeley Symp. Math. Stat. Prob. 5: 237–254.

Allard, R. W., et al. Chap. V. Food. In: Contributions of the Biological Sciences to Human Welfare, Federation of American Society of Experimental Biology Proceedings 31(6) Part II: 81–90.

1973

Clegg, M. T., and R. W. Allard. The genetics of electrophoretic variants in *Avena*. II. The esterase E_1, E_2, E_5, E_6 and anodal peroxidase APX_4 loci in *A. fatua*. J. Hered. 64: 2–6.

Clegg, M. T., and R. W. Allard. Viability versus fecundity selection in the slender wild oat, *Avena barbata* L. Science 181: 667–668.

Allard, R. W., and A. L. Kahler. Multilocus organization and morphogenesis. Brookhaven Symp. Biol. 25: 329–343.

1974

Marshall, D. R., and R. W. Allard. Performance and stability of mixtures of grain sorghum. I. Relationship between level of genetic diversity and performance. Theor. Appl. Genet. 44: 145–152.

Allard, R. W., and M. M. Green. Biological control of populations. Science 185: 96–97.

Weir, B. S., R. W. Allard and A. L. Kahler. Further analysis of complex allozyme polymorphisms in a barley population. Genetics 78: 911–919.

1975

Kahler, A. L., M. T. Clegg and R. W. Allard. Evolutionary changes in the mating system of an experimental population of barley (*Hordeum vulgare* L.). Proc. Natl. Acad. Sci. U.S.A. 72: 943–946.

Allard, R. W., A. L. Kahler and M. T. Clegg. Isozymes in plant population genetics. pp. 261–171. In: *Isozymes: Genetics and Evolution IV*. Academic Press, New York.

Allard, R. W. The mating system and microevolution. Genetics 79: 115–126.

Hamrick, J. L., and R. W. Allard. Correlations between quantitative characters and enzyme genotypes in *Avena barbata*. Evolution 29: 438–442.

1976

Zali, A. A., and R. W. Allard. The effect of level of heterozygosity on the performance of hybrids between isogenic lines of barley. Genetics 84: 765–775.

1977

Allard, R. W. Coadaptation in plant populations. pp. 223–231. In: *The Genetic Control of Diversity in Plants*, Edited by Amir Muhammed, R. Aksel, and R. C. von Borstel. Plenum, New York.

Adams, W. T., and R. W. Allard. The effect of polyploidy on phosphoglucose isomerase diversity in *Festuca microstachys*. Proc. Natl. Acad. Sci. U.S.A. 74: 1652–1656.

Allard, R. W., A. L. Kahler and M. T. Clegg. Estimation of mating cycle components of selection in plants. pp. 1–19. In: *Measuring Selection in Natural Populations*. Edited by F. B. Christiansen and T. M. Fenchel. Springer Verlag, Heidelberg.

1978

Jackson, L. F., A. L. Kahler, R. K. Webster and R. W. Allard. Conservation of scald resistance in barley composite cross populations. Phytopathology 68: 645–650.

Clegg, M. T., A. L. Kahler and R. W. Allard. Estimation of life cycle components of selection in an experimental plant population. Genetics 89: 765–792.

Allard, R. W., R. D. Miller and A. L. Kahler. The relationship between degree of environmental heterogeneity and genetic polymorphism. pp. 49–73. In: *The Structure and Functioning of Plant Populations*. Edited by A. H. J. Freysen and J. W. Woldendorp. North Holland Publishing Company, Amsterdam, Oxford, New York.

Clegg, M. T., A. L. Kahler and R. W. Allard. Genetic demography of plant populations. pp. 173–188. In: *Ecological Genetics: The Interface*, Edited by P. F. Brussard. Springer-Verlag, New York.

1979

Perez de la Vega, M., R. W. Allard and A. L. Kahler. Determination of several parameters in rye populations using electrophoresis techniques. J. Genet. Luso-Espanolas 15: 99–101.

Kahler, A. L., M. Krzakowa and R. W. Allard. Isozyme phenotypes in five species of the Eubromus Section. Proc. Polish Acad. Sci. 168: 45–52.

1980

Kahler, A. L., R. W. Allard, M. Krzakowa, C. F. Wehrhahn and E. Nevo. Associations between enzyme phenotypes and environment in the slender wild oat (*Avena fatua*) in Israel. Theor. Appl. Genet. 56: 31–47.

Shaw, D. V., and R. W. Allard. Analysis of mating system parameters and population structure in Douglas fir using single-locus and multilocus methods. pp. 18–22. In: *Isozymes of Forest Trees and Insects*, Edited by M. T. Conkle.

Garber, M., and R. W. Allard. Mid-arm recombination fractions. J. Hered. 71: 211–213.

1981

Kahler, A. L., and R. W. Allard. Worldwide patterns of genetic variation among four esterase loci in barley (*Hordeum vulgare* L.). Theor. Appl. Genet. 59: 101–111.

Shaw, D. V., A. L. Kahler and R. W. Allard. A multilocus estimator of mating system parameters in plant populations. Proc. Natl. Acad. Sci. U.S.A. 78: 1298–1302.

Kahler, A. L., S. Heath-Pagliuso and R. W. Allard. Genetics of isozyme variants in barley. II. 6-Phosphogluconate dehydrogenase, glutamate oxalate transaminase, and acid phosphatase. Crop Sci. 21: 536–540.

Kahler, A. L., M. I. Morris and R. W. Allard. Gene triplication and fixed heterozygosity in diploid wild barley. J. Hered. 72: 374–376.

1982

Muona, O., R. W. Allard and R. K. Webster. Evolution of resistance to *Rhynchosporium secalis* (Oud.) Davis in barley Composite Cross II. Theor. Appl. Genet. 61: 209–214.

Shumaker, K. M., R. W. Allard and A. L. Kahler. Cryptic variability at enzyme loci in three plant species, *Avena barbata*, *Hordeum vulgare* and *Zea mays*. J. Hered. 73: 86–90.

Adams, W. T., and R. W. Allard. Mating system variation in *Festuca microstachys*. Evolution 36: 591–595.

Shaw, D. V., and R. W. Allard. Isozyme heterozygosity in adult and open-pollinated embryo samples of Douglas fir. Silva Fenn. 16: 115–121.

Jackson, L. F., R. K. Webster, R. W. Allard and A. L. Kahler. Genetic analyses of changes in scald resistance in barley composite cross V. Phytopathology 72: 1069–1072.

Shaw, D. V., and R. W. Allard. Estimation of outcrossing rates in Douglas fir using allozyme markers. Theor. Appl. Genet. 62: 113–120.

1983

Price, S. C., J. Hill and R. W. Allard. Genetic variability for herbicide reaction in plant populations. Weed Sci. 31: 652–657.

Saghai-Maroof, M. A., R. K. Webster and R. W. Allard. Evolution of resistance to scald, powdery mildew and net blotch in barley composite cross II populations. Theor. Appl. Genet. 66: 279–283.

Hutchinson, E. S., A. Hakim-Elahi, R. D. Miller and R. W. Allard. The genetics of the diploidized tetraploid *Avena barbata:* acid phosphatase, esterase, leucine aminopeptidase, peroxidase, and 6-phosphogluconate dehydrogenase loci. J. Hered. 74: 325–330.

Hakim-Elahi, A., and R. W. Allard. Distribution of homoeoalleles at two loci in a diploidized tetraploid: leucine aminopeptidase loci in *Avena barbata*. J. Hered. 74: 379–380.

Hutchinson, E. S., S. C. Price, A. L. Kahler, M. I. Morris and R. W. Allard. An experimental verification of segregation theory in a diploidized tetraploid: Esterase loci in *Avena barbata*. J. Hered. 74: 381–383.

1984

Kahler, A. L., C. O. Gardner and R. W. Allard. Non-random mating in experimental populations of maize (*Zea mays* L.) Crop Sci. 24: 350–354.

Kahler, A. L., R. W. Allard and R. D. Miller. Mutation rates for enzyme and morphological loci in barley (*Hordeum vulgare* L.). Genetics 106: 729–734.

Price, S. C., K. M. Shumaker, A. L. Kahler, R. W. Allard and J. E. Hill. Estimates of population differentiation obtained from enzyme polymorphisms and quantitative characters. J. Hered. 75: 141–142.

Epperson, B. K., and R. W. Allard. Allozyme analysis of the mating system in lodgepole pine populations. J. Hered. 75: 212–214.

Perez de la Vega, M., and R. W. Allard. Mating system and genetic polymorphism in populations of *Secale cereale* and *S. vavilovii*. Can. J. Genet. Cytol. 26: 308–317.

Heath-Pagliuso, S., R. C. Huffaker and R. W. Allard. Inheritance of nitrogen reductase and regulation of nitrate reductase, nitrite reductase, and glutamine synthetase isozymes. Plant Physiol. 76: 353–358.

Saghai-Maroof, M. A., R. A. Jorgensen, K. Soliman and R. W. Allard. Ribosomal DNA (rDNA) spacer-length (sl) variation in barley: Mendelian inheritance, chromosomal location, and population dynamics. Proc. Natl. Acad. Sci. U.S.A. 81: 8014–1018.

Muona, Outi, R. W. Allard and R. K. Webster. Evolution of disease resistance and quantitative characters in barley Composite Cross II: Independent or correlated? Hereditas 101: 143–148.

1985

Polans, N. O., and R. W. Allard. The inheritance of electrophoretically-detectable variants in ryegrass. J. Hered. 76: 61–62.

Price, Steven C., Robert W. Allard, James E. Hill and James Naylor. Associations between discrete genetic loci and genetic variability for herbicide reaction in plant populations. Weed Sci. 33: 650–653.

1986

Biljsma, R., R. W. Allard and A. L. Kahler. Nonrandom mating in an open-pollinated maize population. Genetics 112: 669–680.

Zhang, Qifa, and R. W. Allard. The sampling variance of the genetic diversity index. J. Hered. 77: 54–55.

Goodwin, S. B., M. A. Saghai-Maroof, R. W. Allard and R. K. Webster. Isozymes in *Rhynochosporium secalis*. Phytopathology 76: 843.

Webster, R. K., M. A. Saghai-Maroof and R. W. Allard. Evolutionary response of barley Composite Cross II to *Rhynchosporium secalis* analyzed by pathogenic complexity and by gene-by-race relationships. Phytopathology 76: 661–668.

Goodwin, S. B., M. A. Saghai-Maroof, R. W. Allard and R. K. Webster. Isozyme and restriction fragment length polymorphisms in *Rhynochosporium secalis*. Phytopathology 76: 1102.

Neale, D. B., N. C. Wheeler and R. W. Allard. Paternal inheritance of chloroplast DNA in *Pseudotsuga menziesii* (Mirb.) Franco. Can. J. For. Res. 16: 1152–1154.

1987

Cluster, P. D., O. Marinkovic, R. W. Allard and F. J. Ayala. Correlations between development rates, enzyme activities, ribosomal DNA spacer-length phenotypes and adaptation in *Drosophila melanogaster*. Proc. Natl. Acad. Sci. U.S.A. 84: 610–614.

Epperson, B. K., and R. W. Allard. Linkage disequilibrium between allozymes in natural populations of lodgepole pine. Genetics 115: 341–352.

Wagner, D. B., G. K. Furnier, M. A. Saghai-Maroof, S. M. Williams, B. P. Dancik and R. W. Allard. Chloroplast DNA polymorphisms in lodgepole and jack pines and their hybrids. Proc. Natl. Acad. Sci. U.S.A. 84: 2097–2100.

Zhang, Q., R. K. Webster and R. W. Allard. Geographical distribution and associations between resistance to four races of *Rhynchosporium secalis*. Phytopathology 77: 352–357.

Goodwin, S. B., R. W. Allard and R. K. Webster. Isozyme analysis reveals geographical differentiation of *Rhynchosporium* populations. Phytopathology 77: 6.

Soliman, K., G. Fedak and R. W. Allard. Inheritance of organelle DNA in barley and *Hordeum* × *Secale* intergeneric hybrids. Genome 29: 867–872.

1988

Price, Steven C., James E. Hill and R. W. Allard. The morphological and physiological response of the slender wild oat (*Avena barbata*) to the herbicides Barbon and Difenzoquat. Weed Sci. 36: 60–69.

McDonald, B. A., R. W. Allard and R. K. Webster. The responses of two-, three-, and four component barley mixtures to a variable pathogen population. Crop Sci. 28: 447–452.

Allard, R. W. Genetic changes associated with the evolution of adaptedness in cultivated plants and their wild progenitors. J. Hered. 79: 225–238.

Neale, D. B., M. A. Saghai-Maroof, R. W. Allard, Q. Zhang and R. A. Jorgensen. Chloroplast DNA diversity in populations of wild and cultivated barley. Genetics 120: 1105–1110.

1989

Epperson, B. K., and R. W. Allard. Spatial autocorrelation analysis of the distribution of genotypes within populations of lodgepole pine. Genetics 121: 369–377.

McDonald, B. A., J. M. McDermott, R. W. Allard and R. K. Webster. Coevolution of host and pathogen populations in the *Hordeum vulgare-Rhynchosporium secalis* pathosystem. Proc. Natl. Acad. Sci. U.S.A. 86: 3924–3927.

Allard, R. W. Future directions in plant population genetics, evolution and breeding. pp. 1–19. In: *Plant Population Genetics, Breeding and Genetic Resources*. Edited by A. H. D. Brown, M. T. Clegg, A. L. Kahler and B. S. Weir. Sinauer, Sunderland, Massachusetts.

Kahler, A. L., D. V. Shaw and R. W. Allard. Nonrandom mating on tasseled and detasseled plants in a open pollinated population of Maize. Maydica 34: 15–21.

Soliman, K. M. and R. W. Allard. Genetic control of the triplicate esterase 4 locus in diploid *Hordeum sponteneum* × *H. vulgare* crosses. J. Hered. 80: 70–81.

In Press

Soliman, K. M., and R. W. Allard. Chromosome locations of additional barley enzyme loci identified using wheat-barley addition lines. Zeit. Pflanzenzucktung 101:

Polans, Neil O. and R. W. Allard. An experimental evaluation of the recovery potential of ryegrass populations from genetic stress resulting from restriction of population size. Evolution 43:

McDermott, J. M., B. A. McDonald, R. W. Allard and R. K. Webster. Genetic diversity for pathogenicity, isozyme, ribosomal DNA and colony color variants in populations of *Rhynchosporium secalis*. Genetics 121:

McDonald, B. A., J. M. McDermott, S. B. Goodwin and R. W. Allard. The population biology of host-pathogen interactions. Annu. Rev. Phytopathol. 27:

Garcia, P., F. J. Vences, M. Perez de la Vega and R. W. Allard. Comparison of the Spanish and Californian Gene Pools of *Avena barbata:* Evolutionary implications. Genetics 121:

SECTION 1

GENETIC DIVERSITY: KINDS AND AMOUNTS

Genetic diversity is ubiquitous throughout nature. The relentless process of mutation guarantees a continuous input of new variants, while the equally relentless processes of environmental adaptation and random genetic drift shape the distribution of genetic diversity in time and space. The extent to which populations or species can adapt to environmental challenges is determined by the store of genetic variation contained within local populations and shared among networks of populations. Similarly, the applied geneticist still depends on genetic variants available within the gene pool of domesticated species and their close relatives as a resource for crop improvement. This section discusses the subject of genetic diversity in plant populations from several viewpoints. Emphasis is placed on our increasingly detailed knowledge of the types and patterns of genetic diversity observed at the biochemical and molecular levels. Major attention is also given to the mutational and recombinational processes that serve to generate genetic novelties.

Genetic diversity is a statistical concept because it is a property of collections of individuals. Usually the collection of interest is connected in the sense that the individuals share breeding relationships. In the first chapter of Section 1 Weir develops the statistical concepts needed to make evolutionary inferences about populations from diversity statistics. Particular attention is given to the partition of diversity statistics within individuals, between individuals within populations, and between populations. Diversity statistics that incorporate information from several loci are also developed. The calculation of variances for these measures is presented for several models of population structure. Finally, appropriate estimators for variance components are presented. This work sets the stage for the consideration of empirical results.

Hamrick and Godt apply diversity measures to the analysis of allozyme data from plant species. The allozyme data set is now quite large (more than 450 species are included in Hamrick and Godt's analysis) and it provides a very general picture of genetic diversity in plant species. Most plant species contain large stores of genetic variation and, as might be expected, mating system and species size are major determinants of relative diversity. Gepts considers a specific class of protein-coding loci, those coding for storage proteins in seeds, to develop the theme of genetic diversity with respect to biochemical traits of major agronomic interest.

The remaining four chapters of Section 1 develop various facets of molecular diversity in plant species. Dvořák considers the evolution of multigene families and particularly the mechanisms that shape the coordinated evolution of the ribosomal RNA loci. Clegg reviews the present state of our knowledge of molecular diversity within plant species. Current information on diversity with respect to single-copy genes in plant species is rudimentary and it is necessary to consider animal systems (particularly *Drosophila*) to anticipate the kinds of empirical questions that will arise. Nelson provides a detailed discussion of the properties of plant transposable elements with a particular focus on their role in the generation of genetic novelties. Nei concludes the section with a discussion of the adaptive nature of some DNA polymorphisms. Here again it is necessary to turn to animal research (in this case taken from the human genetics literature) to find examples of the action of selection at the DNA sequence level. Taken together these chapters provide a guide to our present knowledge about genetic diversity in plant species and they indicate where future research is needed.

CHAPTER 2

SAMPLING PROPERTIES OF GENE DIVERSITY

B. S. Weir

ABSTRACT

A convenient means of measuring genetic variation is given by the gene diversity, defined as one minus the sum of squares of allelic frequencies at a locus. Using this statistic to make inferences about a population requires knowledge of its sampling properties. This chapter presents results for the mean and variance of diversity, whether expectations are taken over samples from a single population, or over all replicate populations.

INTRODUCTION

Knowledge of genetic variation is necessary for determining the course of evolution in a species, and it can be quantified with several different statistics. These may be as simple as the number of different alleles at a locus or as complicated as functions of genotypic frequencies at several loci. Here the focus is on gene diversity.

Gene diversity, defined at a single locus as one minus the sum of squared allelic frequencies, is a more appropriate measure of variation than heterozygosity in predominantly selfing species of plants. These species may contain a great deal of variation, but very few heterozygous individuals. For random-mating species, gene diversity and heterozygosity have very similar values. The quantity was introduced, under the name "polymorphic index" by Marshall and Allard (1970), and as "gene diversity" by Nei (1973).

Most of the discussion concerns the sampling variance of estimated diversities, and will show the effects of reference frame, sample size, mating system, and gene frequencies on the variance. Methods for estimating the variance will be given, and the problems with using the variance between loci will be explained.

This work serves to illustrate three features in the analysis of population genetic data that are often overlooked. First, it is necessary to take account of the sampling unit. When individuals are sampled, results must involve genotypic frequencies rather than gene frequencies. Second, the structure of the population must be taken into account. Quite different results are appropriate for mixed self and random mating infinite populations than for finite populations mating at random. Third, the sampling variances attached to parameter estimates must recognize the scope of inference to be made. If conclusions are to be limited to the one population sampled, then within-population variances that refer to all samples from that population are appropriate. For inferences about a collection of populations, of which the one sampled is but a representative, however, variances must include between-population variation. Only the within-population variance of gene diversity was considered by Brown and Weir (1983).

SAMPLE GENE FREQUENCIES

As some of the notions of within- and between-population variance are still not generally accepted, the notation and methodology will be introduced for gene frequencies. An extension to gene diversity is then straightforward.

Within populations

Suppose the frequency of allele A at locus **A** is p_A in some population, and that a sample of n individuals is taken at random from that population. The $2n$ genes in the sample can be described by a variable x that indicates whether or not each gene is allele A. For the kth gene ($k = 1,2$) in individual j ($j = 1, \cdots, n$) define

$$x_{jk} = \begin{cases} 1 & \text{if gene is A} \\ 0 & \text{otherwise} \end{cases}$$

The sample allelic frequency of A is then

$$\tilde{p}_A = \frac{1}{2n} \sum_{j=1}^{n} \sum_{k=1}^{2} x_{jk}$$

Over repeated samples from within (W) the same population, the indicator variable x has moments

$$\mathscr{E}_W x_{jk} = p_A$$
$$\mathscr{E}_W x_{jk}^2 = p_A$$

$$\mathscr{E}_W x_{jk} x_{jk'} = P_{AA}, \qquad k \neq k'$$
$$\mathscr{E}_W x_{jk} x_{j'k'} = p_A^2, \qquad j \neq j'$$

where P_{AA} is the population frequency of AA homozygotes and the last result follows from the fact that different individuals are sampled independently from the population. These expectations provide

$$\mathscr{E}_W \tilde{p}_A = p_A$$
$$\mathscr{E}_W \tilde{p}_A^2 = p_A^2 + \frac{1}{2n}(p_A + P_{AA} - 2p_A^2)$$

so that the within-population sampling variance of a sample gene frequency is

$$Var_W(\tilde{p}_A) = \frac{1}{2n} p_A(1 - p_A)(1 + f) \tag{1}$$

using the inbreeding coefficient f within populations (Wright's F_{IS}) defined from

$$P_{AA} = p_A^2 + p_A(1 - p_A)f$$

Only for random-mating populations, when $f = 0$, does the variance reduce to what it would be if sample allele numbers followed a binomial distribution. In other cases, the variance follows from the assumed multinomial distribution for genotypic numbers. For the mixed-mating model, which supposes a constant fraction s of selfing in an infinite population that otherwise mates at random, $f = s/(2 - s)$ at equilibrium and then

$$Var_W(\tilde{p}_A) = \frac{1}{n(2 - s)} p_A(1 - p_A)$$

As the selfing fraction approaches 1, the variance approaches the binomial result for a sample of size n, rather than $2n$. This reflects the fact that the population will be composed of homozygotes and the sample is essentially composed of n independent pairs of identical genes.

Between populations

An alternative reference frame takes account of the variation between replicate populations. When the genetical sampling between generations is considered, populations may differ even though they have been subjected to the same forces. Taking expectations over samples, and over replicate populations, gives the total (T) expectations

$$\mathscr{E}_T x_{jk} = p_A$$
$$\mathscr{E}_T x_{jk}^2 = p_A$$

$$\mathscr{E}_T x_{jk} x_{jk'} = P_{AA}, \qquad k \neq k'$$
$$\mathscr{E}_T x_{jk} x_{j'k'} = P_{A|A}, \qquad j \neq j'$$

where frequencies p_A and P_{AA} now refer to the average over replicate populations, and $P_{A|A}$ is the joint frequency of A genes in two different individuals. Differences between populations are reflected in correlations between individuals within populations. In biological terms, genes in different individuals are dependent in the sense that they may both be the same allele by virtue of identity by descent.

With differences among populations being allowed for, it is necessary to employ the total inbreeding coefficient F (Wright's F_{IT}) and the coancestry θ (Wright's F_{ST}). These measures allow the frequencies of gene combinations, over all populations, to be written as

$$P_{AA} = p_A^2 + p_A(1 - p_A)F$$
$$P_{A|A} = p_A^2 + p_A(1 - p_A)\theta$$

The total variance of a sample gene frequency becomes

$$\begin{aligned} \mathrm{Var}_T(\tilde{p}_A) &= p_A(1 - p_A)\left[\theta + \frac{1}{n}(F - \theta) + \frac{1}{2n}(1 - F)\right] \\ &= \sigma_a^2 + \frac{1}{n}\sigma_b^2 + \frac{1}{2n}\sigma_w^2 \end{aligned} \tag{2}$$

where the three components reflect variation among populations, σ_a^2, between individuals within populations, σ_b^2, and between genes within individuals, σ_w^2, respectively.

For populations with random union of gametes, pairs of genes are equally related whether they are located in one or two individuals, $F = \theta$ (and $f = 0$), so that $\sigma_b^2 = 0$ and

$$\mathrm{Var}_T(\tilde{p}_A) = p_A(1 - p_A)\left[\theta + \frac{1}{2n}(1 - \theta)\right]$$

For the mixed self and random mating model, different individuals are unrelated, $\theta = 0$, so that $\sigma_a^2 = 0$ and

$$\mathrm{Var}_T(\tilde{p}_A) = \frac{1}{2n} p_A (1 - p_A)(1 + F)$$

At equilibrium for mixed mating then,

$$\begin{aligned} \mathrm{Var}_T(\tilde{p}_A) &= \frac{1}{n(2 - s)} p_A(1 - p_A) \\ &= \mathrm{Var}_W(\tilde{p}_A) \end{aligned}$$

For this situation of infinite populations, there is no variation between populations and the total variance is just the variance within populations.

Estimation

Equations 1 and 2 are theoretical expressions for variances of sample gene frequencies. They are appropriate for predicting the variance in future samples, and may be useful in establishing sample sizes to achieve estimates with desired variances. As they depend on unknown parameters, however, they do not give numerical values to attach to estimated frequencies.

The sample variance of $\tilde{p}_A$ within populations can be estimated simply by substituting observed values into the variance formula

$$\text{Var}_W(\tilde{p}_A) \hat{=} \frac{1}{2n}(\tilde{p}_A + \tilde{P}_{AA} - \tilde{p}_A^2) \tag{3}$$

and, ignoring terms in n^{-2}, n^{-3} etc., this will be unbiased.

The total variance cannot be estimated from data from a single population. Methods for estimating the three variance components necessary have been discussed previously by Cockerham (1969, 1973) and follow from setting out various sums of squares of xs in an analysis of variance framework. For data from several populations, the indicator variable x needs an additional subscript, and x_{ijk} refers to gene k in individual j from population i. Genes from different populations are assumed to be independent, so that

$$\mathscr{E}_T x_{ijk} x_{i'j'k'} = p_A^2, \qquad i' \neq i$$

Under the assumption that the populations sampled have a common ancestral population, they have the same expected gene and genotypic frequencies. Additional work would be needed to remove the assumption that these populations are maintained under the same conditions.

The analysis of variance framework uses the following sums of squares (SS) and means squares (MS) for data on equal-sized samples from r populations:

Between genes within individuals

$$\begin{aligned} \text{SSG} &= \sum_{i=1}^{r}\sum_{j=1}^{n}\sum_{k=1}^{2} x_{ijk}^2 - \frac{1}{2}\sum_{i=1}^{r}\sum_{j=1}^{n}\left(\sum_{k=1}^{2} x_{ijk}\right)^2 \\ &= n\sum_{i=1}^{r}(\tilde{p}_{A_i} - \tilde{P}_{AA_i}) \\ \text{MSG} &= \frac{1}{nr}\text{SSG} \end{aligned}$$

where $\tilde{p}_{A_i}$ *and* $\tilde{P}_{AA_i}$ are the sample gene and homozygote frequencies for A in the ith population.

Between individuals within populations

$$\mathrm{SSI} = \frac{1}{2}\sum_{i=1}^{r}\sum_{j=1}^{n}\left(\sum_{k=1}^{2} x_{ijk}\right)^2 - \frac{1}{2n}\sum_{i=1}^{r}\left(\sum_{j=1}^{n}\sum_{k=1}^{2} x_{ijk}\right)^2$$

$$= 2n\sum_{i=1}^{r} \tilde{p}_{A_i}(1 - \tilde{p}_{A_i}) - \mathrm{SSG}$$

$$\mathrm{MSI} = \frac{1}{r(n-1)}\mathrm{SSI}$$

Between populations

$$\mathrm{SSP} = \frac{1}{2n}\sum_{i=1}^{r}\left(\sum_{j=1}^{n}\sum_{k=1}^{2} x_{ijk}\right)^2 - \frac{1}{2nr}\left(\sum_{i=1}^{r}\sum_{j=1}^{n}\sum_{k=1}^{2} x_{ijk}\right)^2$$

$$= 2n\sum_{i=1}^{r} (\tilde{p}_{A_i} - \bar{p}_A)^2$$

$$\mathrm{MSP} = \frac{1}{r-1}\mathrm{SSP}$$

where $\bar{p}_A$ is the average sample frequency for A over all r populations.

The three mean squares have the following total expectations

$$\mathscr{E}_T(\mathrm{MSG}) = \sigma_w^2$$
$$\mathscr{E}_T(\mathrm{MSI}) = \sigma_w^2 + 2\sigma_b^2$$
$$\mathscr{E}_T(\mathrm{MSP}) = \sigma_w^2 + 2\sigma_b^2 + 2n\sigma_a^2$$

and these lead to estimates for the three components of variance

$$\hat{\sigma}_w^2 = \mathrm{MSG}$$
$$\hat{\sigma}_b^2 = \frac{1}{2}(\mathrm{MSI} - \mathrm{MSG})$$
$$\hat{\sigma}_a^2 = \frac{1}{2n}(\mathrm{MSP} - \mathrm{MSI})$$

The total variance of a sample gene frequency from one of the populations can therefore be estimated as

$$\mathrm{Var}_T(\tilde{p}_{A_i}) \triangleq \hat{\sigma}_a^2 + \frac{1}{n}\hat{\sigma}_b^2 + \frac{1}{2n}\hat{\sigma}_w^2 = \frac{1}{2n}\mathrm{MSP} \tag{4}$$

which is just the sample variance among populations. Although this derivation used the terminology of analysis of variance, there is no assumption that the underlying indicator variables are normally distributed and so the sampling distributions of the mean squares cannot be given.

DIVERSITY AT ONE LOCUS

At locus **A**, the gene diversity is defined as

$$d_{\mathbf{A}} = 1 - \sum_{\mathrm{A}} p_{\mathrm{A}}^2$$

where the sum is over all alleles A at the locus. The sample value is simply

$$\tilde{d}_{\mathbf{A}} = 1 - \sum_{\mathrm{A}} \tilde{p}_{\mathrm{A}}^2$$

From the previous expression for the expected value of a sample squared allelic frequency, the expected value of sample diversity, within populations, is

$$\mathscr{E}_{\mathrm{W}}\tilde{d}_{\mathbf{A}} = d_{\mathbf{A}}\left[1 - \frac{1}{2n}(1 + f)\right]$$

while the total expectation is

$$\mathscr{E}_{\mathrm{T}}\tilde{d}_{\mathbf{A}} = d_{\mathbf{A}}\left[(1 - \theta) - \frac{1}{2n}(1 + F - 2\theta)\right]$$

or

$$\mathscr{E}_{\mathrm{T}}\tilde{d}_{\mathbf{A}} = (1 - \theta)\, d_{\mathbf{A}}\left[1 - \frac{1}{2n}(1 + f)\right]$$

While the estimates are biased, the bias within populations will be small for large sample sizes. The total bias depends on the descent measures F and θ. Cockerham (1967) defined a group coancestry coefficient θ_{L} for a group of n individuals as

$$\theta_{\mathrm{L}} = \frac{n - 1}{n}\theta + \frac{1}{2n}(1 + F)$$

and this allows the simplification of the expectations of sample diversity

$$\mathscr{E}_{\mathrm{T}}\tilde{d}_{\mathbf{A}} = (1 - \theta_{\mathrm{L}})d_{\mathbf{A}}$$

It is therefore the group coancestry coefficient that determines the total bias in sample gene diversity.

Within populations

The most direct way to find the variance of $\tilde{d}_{\mathrm{A}}$ is to determine the variances and covariances of squared sample gene frequencies. If only terms of order n^{-1} are needed, the within-population variances and covariances follow directly from Fisher's formula (see Appendix 2 of Bailey 1961) for functions of multinomial counts. More general expressions, allowing both within-population and total values, however, can be found with the indicator variables x. It is necessary to find the expected value of $\tilde{p}_A^4$ and $\tilde{p}_A^2\tilde{p}_{A'}^2$, as detailed in Appendix 1.

Among samples from one population,

$$\mathscr{E}_W\tilde{p}_A^4 = p_A^4 + \frac{3}{n}(p_A^3 + p_A^2P_{AA} - 2p_A^4)$$

$$\mathscr{E}_W\tilde{p}_A^2\tilde{p}_{A'}^2 = p_A^2p_{A'}^2 + \frac{1}{2n}\Big[p_Ap_{A'}(p_A + p_{A'}) + 4p_Ap_{A'}P_{AA'} + p_A^2P_{A'A'} + p_{A'}^2P_{AA} - 12p_A^2p_{A'}^2\Big], \qquad A \neq A'$$

which provide

$$\text{Var}_W(\tilde{p}_A^2) = \frac{2}{n}(p_A^3 + p_A^2P_{AA} - 2p_A^4)$$

$$\text{Cov}_W(\tilde{p}_A^2, \tilde{p}_{A'}^2) = \frac{2}{n}p_Ap_{A'}(P_{AA'} - 4p_Ap_{A'}), \qquad A \neq A'$$

Using the within-population inbreeding coefficient,

$$\text{Var}_W(\tilde{p}_A^2) = \frac{2}{n}p_A^3(1 - p_A)(1 + f)$$

$$\text{Cov}_W(\tilde{p}_A^2, \tilde{p}_{A'}^2) = -\frac{2}{n}p_A^2p_{A'}^2(1 + f), \qquad A \neq A'$$

$$\text{Var}_W(\tilde{d}_A) = \frac{2}{n}(1 + f)\left[\sum_A p_A^3 - \left(\sum_A p_A^2\right)^2\right]$$

For a locus with only two alleles, this last expression reduces further to

$$\text{Var}_W(\tilde{d}_A) = \frac{2}{n}(1 + f)p_A(1 - p_A)(1 - 2p_A)^2$$

where p_A now refers to the frequency of one of the two alleles. There is the striking result that, when high-order powers of n^{-1} are ignored, gene frequencies of one-half give sample gene diversities of one-half with zero variance.

Between populations

For the total variance, frequencies of three or four genes are needed, analogous to $P_{A|A}$ for a pair of genes. The notational device of separating genes in different individuals by a vertical rule is continued, so that $P_{A|A|A}$ refers to three genes in three distinct individuals. Similarly $P_{AA'|A|A}$ refers to different alleles in one individual and two further copies of one of these alleles in two other separate individuals. The necessary moments are (Appendix 1)

$$\mathscr{E}_T\tilde{p}_A^4 = P_{A|A|A|A} + \frac{3}{n}(P_{A|A|A} + P_{AA|A|A} - 2P_{A|A|A|A})$$

$$\mathscr{E}_T\tilde{p}_A^2\tilde{p}_{A'}^2 = P_{AIAIA'IA'} + \frac{1}{2n}\left[P_{AIA'IA'} + P_{AIAIA'} + 4P_{AA'IAIA'} + P_{AAIA'IA'} + P_{A'A'IAIA} - 12P_{AIAIA'IA'}\right], \qquad A \neq A'$$

A fairly cumbersome expression will result for the total variance of gene diversity, but evidently it is a function of the joint frequencies of genes two, three and four at a time. Note that these frequencies arise because the data are collected at the genotypic level. For the record, the total variance formula is given, where the double summations now include the cases when the alleles A and A′ are the same or different:

$$\begin{aligned}\mathrm{Var}_T(\tilde{d}_A) = {} & \left[\sum_A \sum_{A'} P_{AIAIA'IA'} - \left(\sum_A P_{AIA}\right)^2\right] \\ & + \frac{1}{n}\left[2\sum_A P_{AIAIA} - \sum_A P_{AA} \sum_A P_{AIA} + 2\left(\sum_A P_{AIA}\right)^2\right. \\ & \left. + \sum_A \sum_{A'} (2P_{AA'IAIA'} + P_{AAIA'IA'} - 6P_{AIAIA'IA'})\right]\end{aligned} \tag{5}$$

For the mixed-mating system, the total variance reduces to the within-population value with F substituted for f. At equilibrium

$$\mathrm{Var}_T(\tilde{d}_A) = \frac{4}{n(2-s)}\left[\sum_A p_A^3 - (\sum_A p_A^2)^2\right] \tag{6}$$

For random-mating populations, some simplicity of presentation is afforded by the use of further descent measures that allow joint frequencies to be written in terms of gene frequencies. In addition to $F = \theta$, for the probability of identity of a pair of genes, there is need for γ, δ, and Δ, which are the probabilities for identity by descent of three genes, four genes, or two pairs of genes, respectively (Cockerham 1971). Employing these measures leads to

$$\begin{aligned}\mathrm{Var}_T(\tilde{d}_A) = {} & \left[(\Delta - \theta^2) + 2(\theta^2 - \delta)\sum_A p_A^2\right. \\ & + 4(\theta - 2\gamma - \Delta + 2\delta)\sum_A p_A^3 \\ & \left. - (4\theta - 8\gamma - 3\Delta + 6\delta + \theta^2)(\sum_A p_A^2)^2\right] \\ & + \frac{1}{n}\left[(2\gamma - 3\Delta + \theta^2) + 2(\theta - 3\gamma + 3\delta - \theta^2)\sum_A p_A^2\right.\end{aligned}$$

$$+ 2(1 - 9\theta + 14\gamma + 6\Delta - 12\delta)\sum_{A} p_A^3$$
$$- (2 - 16\theta + 24\gamma + 9\Delta - 18\delta - \theta^2)\,(\sum_{A} p_A^2)^2\Big] \tag{7}$$

Evaluation of the variance is sometimes easier in terms of the joint frequencies, however.

DIVERSITY AT SEVERAL LOCI

It has been common in the literature to present estimates of diversity averaged over several loci. Indexing loci by ℓ now, the average measure is

$$D = \frac{1}{m}\sum_{\ell=1}^{m} d_\ell$$

and the estimate is just the average of single-locus estimates.

This average diversity is biased to the same extent as the component parts:

$$\mathscr{E}_W\tilde{D} = D\left[1 - \frac{1}{2n}(1 + f)\right]$$

within populations, and

$$\mathscr{E}_T\tilde{D} = (1 - \theta)D\left[1 - \frac{1}{2n}(1 + f)\right]$$
$$= (1 - \theta_L)D$$

for total expectation.

Determining the variance of average diversity requires knowledge of the covariance of squared sample gene frequencies for pairs of loci. To the indicator variable x_{jk} for gamete k in individual j carrying allele A at locus **A** is added y_{jk} for the gamete carrying allele B at locus **B**. The product of sample frequencies can then be written as

$$\tilde{p}_A\tilde{p}_B = \frac{1}{4n^2}\left(\sum_{j=1}^{n}\sum_{k=1}^{2} x_{jk}\right)\left(\sum_{j=1}^{n}\sum_{k=1}^{2} y_{jk}\right)$$

Within populations

From this last expression, the expected value of the product within populations is

$$\mathscr{E}_W\tilde{p}_A\tilde{p}_B = p_A p_B + \frac{1}{2n}\Delta_{AB}$$

Here Δ_{AB} is the composite linkage disequilibrium (Weir 1979) between the two loci. If P_{AB} and $P_{A/B}$ are the frequencies of alleles A and B on the same or different gametes within individuals, then

$$\Delta_{AB} = P_{AB} + P_{A/B} - 2p_A p_B$$

The expected value of the product of squared sample frequencies within populations is, to order n^{-1},

$$\mathscr{E}_W \tilde{p}_A^2 \tilde{p}_B^2 = p_A^2 p_B^2 + \frac{1}{2n} [p_A p_B (p_A + p_B) + 4p_A p_B (P_{AB} + P_{A/B}) + (p_A^2 P_{BB} + p_B^2 P_{AA}) - 12 p_A^2 p_B^2]$$

and the covariance is

$$\mathrm{Cov}_W(\tilde{p}_A^2, \tilde{p}_B^2) = \frac{2}{n} p_A p_B \Delta_{AB}$$

The within-population variance of average gene diversity requires sums over loci and over alleles within loci. To simplify notation, suppose that A indexes alleles at locus ℓ and B indexes alleles at locus ℓ'. The variance is

$$\mathrm{Var}_W(\tilde{D}) = \frac{2}{m^2 n} \left[\sum_{\ell=1}^{m} (1 + f) [\sum_A p_A^3 - (\sum_A p_A^2)^2] + \sum_{\ell=1}^{m} \sum_{\ell' \neq \ell}^{m} \sum_A \sum_B p_A p_B \Delta_{AB} \right] \tag{8}$$

with simplification in the case of two alleles at each locus to

$$\mathrm{Var}_W(\tilde{D}) = \frac{2}{m^2 n} \left[\sum_{\ell=1}^{m} (1 + f) p_A (1 - p_A)(1 - 2p_A)^2 + \sum_{\ell=1}^{m} \sum_{\ell' \neq \ell}^{m} 2(1 - 2p_A)(1 - 2p_B) \Delta_{AB} \right]$$

Between populations

For the total variance, it is necessary again to work with joint frequencies of gene combinations. The notation that genes in different individuals are separated by a vertical rule is retained, and genes on different gametes within an individual are separated by a slash. With this convention,

$$\mathscr{E}_T \tilde{p}_A \tilde{p}_B = P_{A|B} + \frac{1}{2n} (P_{AB} + P_{A/B} - 2P_{A|B})$$

and

$$\mathscr{E}_T\tilde{p}_A^2\tilde{p}_B^2 = P_{A|A|B|B} + \frac{1}{2n}\Big[P_{A|A|B} + P_{A|B|B} + 4(P_{AB|A|B} + P_{A/B|A|B}) + P_{AA|B|B} + P_{A|A|BB} - 12P_{A|A|B|B}\Big]$$

The covariance of squared sample gene frequencies is quite involved:

$$\begin{aligned}\mathrm{Cov}_T(\tilde{p}_A^2, \tilde{p}_B^2) = {} & (P_{A|A|B|B} - P_{A|A}P_{B|B}) \\ & + \frac{1}{2n}\Big[(P_{A|A|B} - p_B P_{A|A}) + (P_{A|B|B} - p_A P_{B|B}) \\ & + (P_{AA|B|B} - P_{AA}P_{B|B}) + (P_{A|A|BB} - P_{A|A}P_{BB}) \\ & + 4(P_{AB|A|B} + P_{A/B|A|B} + P_{A|A}\,P_{B|B}) - 12P_{A|A|B|B}\Big]\end{aligned}$$

Equation 9 is the complete expression (to order n^{-1}) for the variance of sample gene diversity. Double sums for alleles at the same locus include the cases of the same and different alleles.

$$\begin{aligned}\mathrm{Var}_T(\tilde{D}) = {} & \frac{1}{m^2}\Bigg\{\sum_\ell\Bigg[\sum_A\sum_{A'} P_{A|A|A'|A'} - \Big(\sum_A P_{A|A}\Big)^2\Bigg] \\ & + \sum_\ell\sum_{\ell'\neq\ell}\Bigg[\sum_A\sum_B P_{A|A|B|B} - \Big(\sum_A P_{A|A}\Big)\Big(\sum_B P_{B|B}\Big)\Bigg]\Bigg\} \\ & + \frac{1}{m^2 n}\Bigg\{\sum_\ell\Bigg[2\sum_A P_{A|A|A} - \sum_A P_{AA}\sum_A P_{A|A} + 2\Big(\sum_A P_{A|A}\Big)^2 \\ & + \sum_A\sum_{A'}(2P_{AA'|A|A'} + P_{AA|A'|A'} - 6P_{A|A|A'|A'})\Bigg] \\ & + \sum_\ell\sum_{\ell'\neq\ell}\Bigg[\sum_A\sum_B(2P_{AB|A|B} + 2P_{A|B|A|B} + P_{AA|B|B} - 6P_{A|A|B|B}) \\ & - \sum_A P_{AA}\sum_B P_{B|B} + 2\sum_A P_{A|A}\sum_B P_{B|B}\Bigg]\Bigg\} \qquad (9)\end{aligned}$$

In the mixed mating system, frequencies of genes in different individuals are just the products of corresponding within-individual frequencies, and the total variance of average gene diversity is the same as that within populations, with f replaced by F.

SPECIAL CASES

Random union of gametes

The joint-frequency formulation is convenient for setting up transition equations between generations, although another view is given by the use of two-locus descent measures. Providing random union of gametes is assumed, it is sufficient to work with Θ, Γ, and Δ^* for the joint identity of pairs of genes at two loci, when the four genes are located on two, three or four gametes, respectively. If linkage disequilibrium is ignored, the covariance of squared sample gene frequencies is

$$\mathrm{Cov_T}(\tilde{p}_A^2, \tilde{p}_B^2) = p_A(1 - p_A)p_B(1 - p_B)\left[(\Delta^* - \theta^2) + \frac{1}{2n}(\theta^2 + 2\Gamma - 3\Delta^*)\right]$$

This expression, summed over all pairs of alleles A, B at each pair of loci can be added to Equation 7 to give the total variance of $\tilde{D}$ in descent measure formulation. Allowing for linkage disequilibrium increases the complexity.

In the "infinite-alleles" model, the initial population is supposed to be drawn from an infinite population in which every allele is unique. In effect sums of powers of allelic frequencies are ignored. Still ignoring terms of order n^{-1}, the variance is found to be

$$\begin{aligned}\mathrm{Var_T}(\tilde{D}) = {} & (\Delta^* - \theta^2) + \frac{1}{m}(\Delta - \Delta^*) + \frac{1}{n}(2\Gamma^* - 3\Delta^* + \theta^2) \\ & + \frac{1}{mn}(2\gamma - 3\Delta - 2\Gamma + 3\Delta^*)\end{aligned} \tag{10}$$

Drift-mutation balance

The major effect of mating system on the sampling properties of gene diversity has already been pointed out, but other forces may be acting. Mutation is likely to be a force of importance for large populations, and one mutation model will be considered here.

In a finite random-mating population, if mutation is always to new alleles at a locus, a balance will be set up between the loss of identity by mutation and the gain of identity by drift. At this balance, the descent measures can be expressed in terms of $\phi = 4N\mu$ for populations of size N and loci with mutation rate μ

$$\theta = \frac{1}{1 + \phi}$$

$$\gamma = \frac{2}{(1 + \phi)(2 + \phi)}$$

$$\delta = \frac{6}{(1 + \phi)(2 + \phi)(3 + \phi)}$$

$$\Delta = \frac{6 + \phi}{(1 + \phi)(2 + \phi)(3 + \phi)}$$

If all loci have the same mutation rate, and if loci are independent so that $\Gamma = \Delta^* = \theta^2$

$$\mathrm{Var_T}(\tilde{D}) = \frac{1}{m}\frac{2\phi}{(1 + \phi)^2(2 + \phi)(3 + \phi)}\left(1 + \frac{\phi}{n}\right)$$

The greatest reduction in variance comes from increasing m, the number of loci scored. Increasing n, the number of individuals sampled, has relatively little effect.

Self and random mating

The opposite effect of increasing numbers of individuals as opposed to the number of loci is found for the mixed mating system. It has already been stated that there is no between-population contribution to variances in this system, and Equation 8 suggests that n, which is generally larger than m, will play the dominant role in reducing variance. Note that the m^2 in the denominator of Equation 8 is offset by the m terms in the first sum and the m^2 terms in the second sum.

A concrete example is the infinite alleles case, when the variance (B. S. Weir, unpublished) becomes

$$\mathrm{Var}(\tilde{D}) = \frac{1}{n^3}\left[(\bar{F}_{11} - F^2) + \frac{1}{m}(F - \bar{F}_{11})\right]$$

where $\bar{F}_{11}$ is the two-locus inbreeding coefficient averaged over all pairs of loci. At equilibrium, for loci with linkage parameter λ,

$$F_{11} = \frac{s[2(1 + \lambda^2) + s(1 - 3\lambda^2)]}{(2 - s)\,[4 - s(1 + \lambda^2)]}$$

(Weir and Cockerham 1973). When all m loci are unlinked

$$\bar{F}_{11} = \frac{s(2 + s)}{(2 - s)\,(4 - s)}$$

With selfing tending to hold blocks of genes together, as though they were linked, it is better to increase the number of individuals sampled than the number of loci scored in order to reduce the variance of gene diversity.

ESTIMATING VARIANCE OF DIVERSITY

Estimation of the variance of gene diversity for a single population proceeds by substituting observed values into the equations derived above. A more convenient expression than Equation 8 is

$$\mathrm{Var}_W(\tilde{D}) = \frac{2}{n}\left\{\sum_{\ell}\left[\sum_{A}(p_A^3 + p_A^2 P_{AA} - 2p_A^4) + \sum_{A}\sum_{A' \neq A} p_A p_{A'}(P_{AA'} - 4p_A p_{A'})\right]\right.$$

$$+ \sum_{\ell} \sum_{\ell' \neq \ell} \sum_{A} \sum_{B} p_A p_B \, \Delta_{AB} \Bigg\} \qquad (11)$$

Estimation of the total variance requires data from more than one population. It has often been suggested that replicate loci can play the role of replicate populations, however, so that the variance among diversities at different loci could serve as an estimate of the total variance. This approach has merit in some circumstances, but not others, as is now shown.

The estimate of total variance for average gene diversity $\tilde{D}$ is taken to be $1/m$ times the sample variance of the m single-locus diversities. If the gene diversity at locus ℓ is written as d_ℓ, so that

$$\tilde{D} = \frac{1}{m} \sum_{\ell} d_\ell$$

then the variance estimate is taken to be

$$s_{\tilde{D}}^2 = \frac{1}{m(m-1)} \sum_{\ell=1}^{m} (\tilde{d}_\ell - \tilde{D})^2$$

Taking expectations provides

$$\begin{aligned} \mathscr{E}_T s_{\tilde{D}}^2 &= \mathrm{Var}_T(\tilde{D}) \\ &\quad + \frac{1}{m(m-1)} \sum_{\ell} \left(\mathscr{E}_T \tilde{d}_\ell - \frac{1}{m} \sum_{\ell} \mathscr{E}_T \tilde{d}_\ell \right)^2 \\ &\quad - \frac{1}{m(m-1)} \sum_{\ell} \sum_{\ell' \neq \ell} \mathrm{Cov}_T(\tilde{d}_\ell, \tilde{d}_{\ell'}) \end{aligned} \qquad (12)$$

Evidently s^2 can serve as an unbiased estimator of the total variance of $\tilde{D}$ strictly only when each locus has the same expected gene diversity, and when the diversities at different loci have zero covariance. These conditions are met for the situation discussed earlier for random mating populations and independent loci in the drift-mutation balance model. Probably the main indication when the approach is not valid will be given by evidence of linkage disequilibrium between loci. In the mixed self and random mating situation, the covariance between diversities is directly proportional to these linkage disequilibria:

$$\mathrm{Cov}_T(\tilde{d}_\ell, \tilde{d}_{\ell'}) = \frac{2}{n} \sum_{A} \sum_{B} p_A p_B \, \Delta_{AB}$$

Therefore s^2 should not be used to estimate variance if observed Δ_{AB} values are significantly different from zero. The work of Zhang and Allard (1986) and Swofford and Selander (1981) should therefore be treated with care.

EXAMPLE

Gene diversity is of great use in characterizing variability in selfing species. Table 1 shows some data collected for esterase loci A, B, and C (*Est1*, *Est2*, and *Est4*) in generation 4 of Barley Composite Cross V (Allard et al. 1972). The table shows two-locus counts obtained by collapsing alleles at each locus into the most common versus the rest.

From that table, the following statistics for each of the three loci separately can be calculated—recall that there is no distinction between within-population and total variances in this situation.

ℓ	$\tilde{d}_\ell$	$n\hat{\mathrm{Var}}(\tilde{d}_\ell)$	$\tilde{d}_\ell - \tilde{D}$
A	0.4984	0.0026	0.1359
B	0.1710	0.1982	−0.1915
C	0.4182	0.1216	0.0557

Clearly the level of diversity differs among these loci.

For the three pairs of loci, the disequilibrium and covariance estimates are

ℓ,ℓ'	Δ_{AB}	$n\hat{\mathrm{Cov}}(\tilde{d}_\ell, \tilde{d}_{\ell'})$
A,B	0.0478	−0.0043
A,C	0.0117	−0.0005
B,C	0.0540	0.0354

and the linkage disequilibrium is significantly nonzero for pairs **A,B** and **B,C** as tested by the methodology of Weir and Cockerham (1988).

Using Equation 11 the (total) variance of $\tilde{D}$ is estimated to be $(0.0059)^2$, compared to $(0.0985)^2$ found from the variance among the three single-locus diversities. If these variances were to be used as a basis for testing differences among diversities, it is likely that different conclusions would be reached with the two estimates.

DISCUSSION

In spite of the widespread use of gene diversity, little careful attention has been paid to its sampling properties. In this chapter, means and variances have been determined and several general principles illustrated.

In the first place, even though gene diversity is defined in terms of gene frequencies, it is estimated from genotypic frequencies. Expressions for both

TABLE 1. Two-locus counts for barley esterase loci.

	BB	**Bb**	**bb**	**Total**
AA	518	3	20	543
Aa	72	6	1	79
aa	502	38	72	612
Total	1094	47	93	1234

	CC	**Cc**	**cc**	**Total**
AA	375	25	143	543
Aa	44	27	8	79
aa	391	61	160	612
Total	810	113	311	1234

	CC	**Cc**	**cc**	**Total**
BB	770	83	241	1094
Bb	4	18	25	47
bb	36	12	45	93
Total	810	113	311	1234

means and variances therefore involve expected genotypic frequencies, and hence can be expressed in terms of gene frequencies *and* descent measures.

Second, the sampling properties of any statistic, including sample gene frequency, depend on the mating structure of the population. This is a consequence of the involvement of genotypic frequencies and it means that a given sample size can result in quite different variances for systems as different as finite random-mating and infinite partial-selfing populations.

Third, the sampling variance attached to an estimate needs to take account of the desired scope of inference. Within-population variances refer to variation among repeated samples from the same population. They can be estimated from single-population data by the means discussed here or by numerical resampling methods such as jackknifing or bootstrapping. These variances are appropriate if only the present population is of interest.

When the population sampled is to be used as a basis for inference about a wider collection of populations, the variation between populations caused by genetical sampling is also relevant. Total variances are appropriate. The theoretical values for such total variances need to be used in sample size calculations for future samples. Estimation of total variances is not possible from single-population data unless a set of loci has independent and identical expected gene diversities. In that case the variance of sample single-locus diversities can lead to an estimate of the total variance of the average diversity over loci. Evidence of significant linkage disequilibrium is a clue that such a procedure should not be invoked.

Nei and his colleagues have given much attention to the sampling properties of gene diversity and related statistics. Their work is almost exclusively for random-mating populations, and often invokes drift-mutation balance. Li and Nei (1975), for example, advocate using the variance among loci to estimate the total variance of diversity but, as discussed above, this is correct for their assumptions of identical and independent single-locus diversities.

Finally, the reporting of gene diversities continues to be of interest to population geneticists. The chapter by Hamrick and Godt in this volume bears eloquent testimony to this. The variances reported in that chapter were generally calculated as the sample variance over populations and so refer to total variance.

APPENDIX

Expected value of powers of $\tilde{p}_A$

From the expression for sample gene frequency as an average of indicator variables

$$\tilde{p}_A = \frac{1}{2n} \sum_{j=1}^{n} \sum_{k=1}^{2} x_{jk}$$

the next three powers of the frequency are found to be

$$\tilde{p}_A^2 = \frac{1}{2n}\tilde{p}_A + \frac{1}{4n^2}\left(\sum_{j=1}^{n}\sum_{k=1}^{2}\sum_{k'\neq k} x_{jk}x_{jk'} + \sum_{j=1}^{n}\sum_{j'\neq j}\sum_{k=1}^{2}\sum_{k'=1}^{2} x_{jk}x_{j'k'}\right)$$

$$\tilde{p}_A^3 = \frac{3}{2n}\tilde{p}_A^2 - \frac{1}{2n^2}\tilde{p}_A + \frac{1}{8n^3}\left(3\sum_{j=1}^{n}\sum_{j'\neq j}\sum_{k=1}^{2}\sum_{k'\neq k}\sum_{k''=1}^{2} x_{jk}x_{jk'}x_{j'k''}\right.$$
$$\left. + \sum_{j=1}^{n}\sum_{j'\neq j}\sum_{j''\neq j,j'}\sum_{k=1}^{2}\sum_{k'=1}^{2}\sum_{k''=1}^{2} x_{jk}x_{j'k'}x_{j''k''}\right)$$

$$\tilde{p}_A^4 = \frac{3}{n}\tilde{p}_A^3 - \frac{11}{4n^2}\tilde{p}_A^2 + \frac{3}{4n^3}\tilde{p}_A$$
$$+ \frac{1}{16n^4}\left(3\sum_{j=1}^{n}\sum_{j'\neq j}\sum_{k=1}^{2}\sum_{k'\neq k}\sum_{k''=1}^{2}\sum_{k'''\neq k''} x_{jk}x_{jk'}x_{j'k''}x_{j'k'''}\right.$$

$$+ 6\sum_{j=1}^{n}\sum_{j'\neq j}\sum_{j''\neq j,j'}\sum_{k=1}^{2}\sum_{k'\neq k}\sum_{k''=1}^{2}\sum_{k'''=1}^{2} x_{jk}x_{jk'}x_{j'k''}x_{j''k'''}$$

$$+ \sum_{j=1}^{n}\sum_{j'\neq j}\sum_{j''\neq j,j'}\sum_{j'''\neq j,j',j''}\sum_{k=1}^{2}\sum_{k'=1}^{2}\sum_{k''=1}^{2}\sum_{k'''=1}^{2} x_{jk}x_{j'k'}x_{j''k''}x_{j'''k'''})$$

Taking expectations over samples and replicate populations, and keeping only terms in n^{-1}

$$\mathscr{E}_T\tilde{p}_A = p_A$$

$$\mathscr{E}_T\tilde{p}_A^2 = P_{A|A} + \frac{1}{2n}(p_A + P_{AA} - 2P_{A|A})$$

$$\mathscr{E}_T\tilde{p}_A^3 = P_{A|A|A} + \frac{3}{2n}(P_{A|A} + P_{AA|A} - 2P_{A|A|A})$$

$$\mathscr{E}_T\tilde{p}_A^3 = P_{A|A|A|A} + \frac{3}{n}(P_{A|A|A} + P_{AA|A|A} - 2P_{A|A|A|A})$$

Replacing joint frequencies of genes in different individuals by the appropriate products of gene frequencies:

$$P_{A|A} = p_A^2, \quad P_{A|A|A} = p_A^3, \quad P_{A|A|A|A} = p_A^4,$$
$$P_{AA|A} = p_A P_{AA}, \quad P_{AA|A|A} = p_A^2 P_{AA}$$

leads to the within-population expectations

$$\mathscr{E}_W\tilde{p}_A = p_A$$

$$\mathscr{E}_W\tilde{p}_A^2 = p_A^2 + \frac{1}{2n}(p_A + P_{AA} - 2p_A^2)$$

$$\mathscr{E}_W\tilde{p}_A^3 = p_A^3 + \frac{3}{2n}(p_A^2 + p_A P_{AA} - 2p_A^3)$$

$$\mathscr{E}_W\tilde{p}_A^3 = p_A^4 + \frac{3}{n}(p_A^3 + p_A^2 P_{AA} - 2p_A^4)$$

The expectation of products of powers of frequencies of different alleles is more complicated. For convenience, the indicator variables x_{jk} and y_{jk} for gene k in individual j refer to alleles A and A′, respectively. The product of two squares can be expressed as

$$16n^4\tilde{p}_A^2\tilde{p}_{A'}^2 = \sum_{j=1}^{n}\sum_{k=1}^{2}\sum_{k'\neq k} x_{jk}y_{jk'} + \sum_{j=1}^{n}\sum_{j'\neq j}\left[\sum_{k=1}^{2}\sum_{k'=1}^{2} x_{jk}y_{j'k'}\right.$$

$$+ \sum_{k=1}^{2}\sum_{k'\neq k}\sum_{k''=1}^{2}(2x_{jk}y_{jk'}(x_{j'k''} + y_{j'k''})$$

$$+ x_{jk}x_{jk'}y_{j'k''} + x_{j'k''}y_{jk}y_{jk'})$$

$$+ \sum_{k=1}^{2}\sum_{k'\neq k}\sum_{k''=1}^{2}\sum_{k'''\neq k''} x_{jk}y_{j'k''}(2y_{jk'}x_{j'k'''} + x_{jk'}y_{j'k'''})\Bigg]$$

$$+ \sum_{j=1}^{n}\sum_{j'\neq j}\sum_{j''\neq j,j'}\Bigg[\sum_{k=1}^{2}\sum_{k'=1}^{2}\sum_{k''=1}^{2} x_{jk}y_{j'k'}(x_{j''k''} + y_{j''k''})$$

$$+ \sum_{k=1}^{2}\sum_{k'\neq k}\sum_{k''=1}^{2}\sum_{k'''=1}^{2} (4x_{jk}y_{jk'}x_{j'k''}y_{j''k'''} + x_{jk}x_{jk'}y_{j'k''}y_{j''k'''}$$

$$+ y_{jk}y_{jk'}x_{j'k''}x_{j''k'''})\Bigg]$$

$$+ \sum_{j=1}^{n}\sum_{j'\neq j}\sum_{j''\neq j,j'}\sum_{j''\neq j,j'}\sum_{j'''\neq j,j',j''}\sum_{k=1}^{2}\sum_{k'=1}^{2}\sum_{k''=1}^{2}\sum_{k'''=1}^{2} x_{jk}x_{j'k'}y_{j''k''}y_{j'''k'''}$$

Taking expectations over samples and populations, and retaining only terms in n^{-1}, gives

$$\mathscr{E}_T\tilde{p}_A^2\tilde{p}_{A'}^2 = P_{A|A|A'|A'} + \frac{1}{2n}\Big[(P_{A|A'|A'} + P_{A|A|A'})$$
$$+ 4P_{AA'|A|A'} + (P_{AA|A'|A'} + P_{A'A'|A|A}) - 12P_{A|A|A'|A'}\Big]$$

Replacing joint frequencies by products of single-individual frequencies gives the within-population expectation:

$$\mathscr{E}_W\tilde{p}_A^2\tilde{p}_{A'}^2 = p_A^2p_{A'}^2 + \frac{1}{2n}\Big[p_Ap_{A'}(p_A + p_{A'})$$
$$+ 4p_Ap_{A'}P_{AA'} + p_{A'}^2P_{AA} + p_A^2P_{A'A'} - 12p_A^2p_{A'}^2\Big]$$

ACKNOWLEDGMENTS

This is Paper Number 11944 of the Journal Series of the North Carolina Agricultural Research Service, Raleigh, NC 27695-7601. This investigation was supported in part by NIH Grant GM11546. Helpful comments were received from Drs. A. Hastings and T. Prout.

CHAPTER 3

ALLOZYME DIVERSITY IN PLANT SPECIES

J. L. Hamrick and M. J. W. Godt

ABSTRACT

We review the plant allozyme literature to update previous reviews of variation within and among plant populations, and the patterns of allozyme variation at the species level. On average, 50% of a plant species' loci are polymorphic and mean genetic diversity is 15%. Within an average population, 34% of the loci are polymorphic and the mean diversity is 11%. Genetic variation at polymorphic loci is partitioned such that most (78%) of the diversity is found within populations, while a smaller fraction (22%) accounts for population differentiation. Taxa were classified for eight traits: (1) major phyletic group, (2) life form, (3) geographic range, (4) regional distribution, (5) breeding system, (6) seed dispersal mechanism, (7) mode of reproduction, and (8) successional status. Geographic range accounted for the largest proportion of variation in species level genetic diversity. Species with widespread ranges had significantly higher levels of diversity than more narrowly distributed species. At the population level, the plant breeding system in combination with geographic range accounted for the greatest proportion of variance in genetic diversity. Predominantly outcrossed species had significantly higher levels of genetic diversity than plants that selfed or had a mixed-mating system. Variation in diversity among populations was influenced mainly by plant breeding systems. Selfing species had 51% while outcrossed wind-pollinated species had less than 10% of their total genetic diversity among populations. Mean genetic diversity values did not change greatly from those previously reported, despite a 4-fold increase in the number of studies reviewed. In addition, the major associations previously found between genetic diversity and species' traits have been substantiated by the enlarged data base. However, several new insights concerning genetic diversity resulted from this study. The first, and perhaps most interesting, is that genetic diversity at the population level reflects, fairly accurately, diversity at the species level. Second, species with limited geographic ranges tend to have less genetic diversity, yet they partition this variation in much the same way as more widespread species. Third, the majority of the differences in levels of allozyme diversity are due to the proportion of loci polymorphic rather than differences in diversity at individual polymorphic loci.

Potential for evolutionary change is dependent on the existence of genetic variation. As a result, studies of genetic variation in natural populations are a major focus of population geneticists. Beginning in the late 1960s the application of electrophoresis to the measurement of genetic variation at enzyme loci led to an explosion of studies in natural populations. Since then, allozymes have provided the most abundant source of information regarding genetic diversity in natural populations.

With the advent of isozyme studies came the demonstration of surprising levels of genetic diversity in natural populations. In response, theoretical geneticists developed mathematical models attempting to explain the maintenance of this variation. Evolutionary biologists, on the other hand, have spent considerable time describing levels of allozyme variation in natural populations and determining how this variation is distributed among populations. In addition, ecological geneticists have searched for associations between levels of genetic variation and the ecological and life history traits of species. Such patterns are important because they suggest mechanisms for the maintenance of different levels of genetic variation within and among populations.

Earlier reviews of the allozyme literature found significant correlations between ecological and life history characters of species and the levels and distribution of allozyme diversity (Hamrick et al. 1979; Nevo et al. 1984; Loveless and Hamrick 1984). In this chapter, we provide an updated review of the plant allozyme literature. Since earlier reviews (see above) the number of plant species analyzed for allozymes has increased 4-fold. Because plant allozyme studies have traditionally lagged far behind animal studies in number, the inclusion of a larger number of plant studies in this review permits a more convincing comparison to be made between plant and animal data. In addition, the expanded data base allows a more thorough search for patterns between the levels and distribution of genetic diversity and species characteristics.

We examine variation at three levels: at the species level, at the population level, and among populations within species. Variation within and among populations has been examined previously (Brown 1979; Hamrick et al. 1979; Gottlieb 1981; Loveless and Hamrick 1984), but variation at the species level has not been reviewed. Important questions are how accurately does genetic variation at the population level reflect variation at the species level and what relationships exist between levels of genetic diversity and the characteristics of species.

METHODS

Data were obtained from papers published during the period 1968 to 1988 that reported allozyme variation for gymnosperms and angiosperms. Only those papers with genetic interpretations of electrophoretic banding patterns were included. Genetic parameters could not be calculated at the species, population, and among population levels for every study. Studies that surveyed both monomorphic and polymorphic loci provided data to calculate genetic parameters at each level. Studies that were limited to polymorphic loci provided data for calculations of among population variation but were not used to estimate variation at species or population levels. Surveys of variation in single populations were not used to estimate species level values. Taxa at the subspecies level or above were considered. Species that were the focus of more than one study were represented in the data more than once.

Species characteristics

Each species was classified for eight characteristics: taxonomic status (gymnosperms, dicotyledons, monocotyledons), regional distribution (boreal–temperate, temperate, temperate–tropical, tropical), geographic range (endemic, narrow, regional, widespread), life form (annual, short-lived perennial—either herbaceous or woody, long-lived perennial—herbaceous or woody), mode of reproduction (sexual, sexual and asexual), breeding system (selfed, mixed mating—either animal- or wind-pollinated, outcrossed—either animal- or wind-pollinated), seed dispersal mechanism (gravity, gravity and animal-attached, explosive, wind, animal-ingested, animal-attached), and successional status (early, mid, late). The classification of each species was based on information gleaned from descriptions in the original papers or by consulting pertinent floras.

Genetic parameters

Four genetic parameters were calculated at the species and the within-population levels: percentage polymorphic loci, mean number of alleles per locus, effective number of alleles per locus, and genetic diversity (Weir, this volume).

At the species level the percentage of polymorphic loci (P_s) was calculated by dividing the number of loci polymorphic in at least one population by the number of loci analyzed. The mean number of alleles per locus (A_s) was determined by summing the alleles observed over all loci and dividing by the number of loci. Genetic diversity (H_{es}) was calculated for each locus by $H_{es} = 1 - \Sigma p_i^2$, where p_i is the mean frequency of the ith allele. Mean genetic diversity was obtained by averaging the H_{es} values over all loci. The effective number of alleles was calculated by $A_{es} = 1/(1 - H_{es})$.

At the population level the percentage polymorphic loci (P_p) was the proportion of loci polymorphic in each population averaged over all populations. The number of alleles per locus (A_p) was determined for each population and a mean value was obtained by averaging over all populations. Genetic diversity (H_{ep}) was calculated for each locus and population by $H_{ep} = 1 - \Sigma p_i^2$ where p_i is the frequency of the ith allele in each population. The mean H_{ep} value was obtained for each locus by averaging over all populations and an overall mean (H_{ep}) was obtained by averaging over all loci. The effective number of alleles was calculated by $A_{ep} = 1/(1 - H_{ep})$.

Variation among populations was estimated with Nei's (1973) genetic diversity statistics. Total genetic diversity (H_T) and mean diversity within populations (H_S) were calculated for each polymorphic locus. Genetic diversity due to variation among populations (D_{ST}) was related to the total diversity (H_T) to determine the proportion residing among populations (G_{ST}).

Statistical analyses

Statistical analyses generally followed procedures used by Hamrick et al. (1979) and Loveless and Hamrick (1984) with a few exceptions. Means and standard errors of the means were calculated for each category and genetic parameter. Levels of statistical significance among the categories of each trait were determined by performing separate one-way ANOVAs. In this review we also employed a Duncan's multiple range analysis to test whether categories of the eight traits differed significantly. Multiple regression models were constructed [using GLM procedures of SAS (SAS 1987)] to determine characteristics that best predicted variation in H_{ep}, H_{es}, and G_{ST}. This last analysis was not used in the previous studies but provides additional insights into the influence of the eight traits on the genetic parameters.

Since many of the eight traits were correlated (e.g., gymnosperms are outcrossing, long-lived species), we performed a principal component analysis (SAS 1987) to determine whether particular combinations of traits explained significant proportions of the variation in genetic diversity.

RESULTS

Data were provided by 653 studies which included 449 species representing 165 genera. Since only one-third of the studies had a complete data set, effective sample sizes varied from 400 to 500 for any genetic parameter. This sample size is approximately four times that previously used to review within population variation (Hamrick et al. 1979) and more than twice the sample size used to review the distribution of allozyme variation among populations (Loveless and Hamrick 1984).

The analyses demonstrated that plant species are polymorphic at approximately 50% of their isozyme loci and that the mean genetic diversity at the species level was nearly 15% (Table 1). Less variation occurs within plant populations with 34% of the loci polymorphic and a mean genetic diversity of 11%. Approximately 78% of the total allozyme diversity at polymorphic loci occurs within populations.

TABLE 1 Levels of allozyme variation for plant taxa.[a]

Level	N	Mean no. populations	Mean no. loci	P	A	A_e	H_e
Species	473	12.7 (1.3)[b]	16.5 (0.4)	50.5 (1.4)	1.96 (0.05)	1.21 (0.01)	0.149 (0.006)
Population	468	12.7 (1.3)	16.5 (0.4)	34.2 (1.2)	1.53 (0.02)	1.15 (0.01)	0.113 (0.005)

	N	Mean no. populations	Mean no. loci	H_T	H_S	G_{ST}
Among populations	406	12.7 (1.3)	16.5 (0.4)	0.310 (0.007)	0.230 (0.007)	0.224 (0.012)

[a]N, number of taxa; P, percentage polymorphic loci; A, number of alleles per locus; A_e, effective number of alleles per locus; H_e, genetic diversity index; H_T, total genetic diversity; H_S, genetic diversity within populations; G_{ST}, proportion of the total diversity among populations. See text for more complete definitions.
[b]Standard errors are in parentheses.

Variation at the species level

Differences among the categories of taxonomic status, life form, geographic range and seed dispersal were significant ($p < 0.05$) for the four genetic parameters (Table 2). Breeding system categories were significantly different for P_s, A_s, and H_{es}. Significant differences occurred among the categories of regional distribution and reproductive mode for two genetic parameters (P_s and A_s) while the categories of successional stage were significantly different for P_s.

Gymnosperms had a high proportion of their loci polymorphic (P_s), high numbers of alleles per locus (A_s), high effective numbers of alleles (A_{es}), and high levels of gene diversity (H_{es}) (Table 2). Monocots had a significantly lower P_s value but were equal or somewhat higher than the gymnosperms in terms of A_s, A_{es}, and H_{es}. Dicots had significantly lower values for all four parameters. Using H_{es} as the most comprehensive measure of genetic diversity, the data show that gymnosperms and monocots differ by 5% but that their genetic diversity exceeds that of dicots by approximately 30%. The results also indicate that a larger proportion of gymnosperm loci are polymorphic but that gymnosperms have fewer alleles per polymorphic locus (2.89) than the monocots (3.26). This difference can be attributed, in part, to the large number of alleles maintained at polymorphic loci in *Zea mays* (Goodman and Stuber 1983), *Triticum dicoccoides* (Nevo et al. 1982) and *Hordeum vulgare* (Brown and Munday 1987).

TABLE 2 Levels of allozyme variation at the species level for species with different attributes.[a]

Categories	N	Mean no. populations	Mean no. loci	P_s	A_s	A_{es}	H_{es}
TAXONOMIC STATUS				***	***	*	**
Gymnosperms	55	8.5 (0.9)[c]	16.1 (1.3)	70.9a[b] (3.6)	2.35a (0.12)	1.23ab (0.02)	0.173a (0.011)
Monocots	111	18.7 (3.1)	15.5 (0.9)	59.2b (3.4)	2.38a (0.17)	1.27a (0.03)	0.181a (0.015)
Dicots	329	11.9 (1.7)	16.8 (0.4)	44.8c (1.5)	1.79b (0.04)	1.19b (0.01)	0.136b (0.007)
LIFE FORM				***	***	**	***
Annual	190	18.5 (3.0)	14.9 (0.5)	50.7b (2.2)	2.07a (0.09)	1.24a (0.02)	0.161a (0.009)
Short-lived perennial							
Herbaceous	152	8.8 (1.2)	17.2 (0.7)	41.3c (2.2)	1.70b (0.06)	1.15b (0.01)	0.116b (0.009)
Woody	17	18.1 (6.7)	25.2 (1.4)	41.8-[d] (6.0)	1.54- (0.12)	1.12- (0.03)	0.097- (0.020)
Long-lived perennial							
Herbaceous	4	6.0 (1.7)	13.8 (3.8)	39.6- (16.5)	1.42- (0.13)	1.28- (0.12)	0.205- (0.084)
Woody	110	9.3 (1.4)	17.0 (0.9)	64.7a (2.7)	2.19a (0.09)	1.24a (0.02)	0.177a (0.010)
GEOGRAPHIC RANGE				***	**	***	***
Endemic	81	6.5 (0.9)	17.8 (0.6)	40.0c (3.2)	1.80b (0.08)	1.15b (0.04)	0.096c (0.010)
Narrow	101	8.8 (1.0)	16.9 (0.8)	45.1bc (2.8)	1.83b (0.08)	1.17b (0.02)	0.137b (0.011)
Regional	193	10.4 (1.1)	16.7 (0.7)	52.9ab (2.1)	1.94b (0.06)	1.20b (0.01)	0.150b (0.008)
Widespread	105	25.5 (5.2)	14.6 (0.9)	58.9a (3.1)	2.29a (0.16)	1.31a (0.03)	0.202a (0.015)
REGIONAL DISTRIBUTION				***	***	NS	NS
Boreal-temperate	19	7.8 (1.4)	16.2 (2.0)	79.7a (5.7)	2.64a (0.23)	1.26a (0.03)	0.186a (0.018)
Temperate	348	11.6 (1.3)	15.6 (0.4)	48.5b (1.5)	1.91b (0.05)	1.21a (0.01)	0.146a (0.006)
Temperate-tropical	30	37.3 (12.1)	15.1 (1.3)	58.8b (6.4)	2.53a (0.39)	1.23a (0.04)	0.170a (0.026)
Tropical	76	10.6 (2.8)	21.3 (1.0)	49.2b (3.6)	1.81b (0.10)	1.21a (0.03)	0.148a (0.015)

TABLE 2 (Continued)

Categories	N	Mean no. populations	Mean no. loci	P_s	A_s	A_{es}	H_{es}
BREEDING SYSTEM				***	***	NS	**
Selfing	123	20.3 (3.8)	16.2 (0.7)	41.8b (2.9)	1.69b (0.09)	1.18a (0.02)	0.124b (0.011)
Mixed-animal	64	8.9 (2.1)	14.4 (0.8)	40.0b (3.5)	1.68b (0.08)	1.16a (0.02)	0.120b (0.015)
Mixed-wind	9	10.0 (3.1)	12.5 (3.6)	73.5a (9.3)	2.18ab (0.21)	1.28a (0.06)	0.194a (0.038)
Outcrossing-animal	172	10.7 (2.1)	17.7 (0.7)	50.1b (2.0)	1.99ab (0.07)	1.24a (0.02)	0.167ab (0.010)
Outcrossing-wind	105	10.7 (1.6)	16.7 (0.9)	66.1a (2.7)	2.40a (0.13)	1.21a (0.02)	0.162ab (0.009)
SEED DISPERSAL				***	***	**	***
Gravity	198	10.1 (1.1)	16.9 (0.6)	45.7b (1.9)	1.81cd (0.05)	1.19ab (0.02)	0.136bc (0.008)
Gravity-attached	15	29.2 (15.8)	18.6 (1.8)	69.3a (3.5)	2.42a (0.22)	1.23ab (0.07)	0.166ab (0.037)
Attached	55	20.8 (5.7)	16.5 (0.9)	68.8a (3.9)	2.96a (0.28)	1.30a (0.05)	0.204a (0.019)
Explosive	27	12.4 (2.7)	18.6 (1.4)	30.4c (4.7)	1.48d (0.08)	1.12b (0.02)	0.092c (0.017)
Ingested	67	17.6 (6.5)	13.2 (1.1)	45.7b (3.9)	1.69cd (0.08)	1.25a (0.03)	0.176ab (0.019)
Wind	111	8.7 (0.9)	16.6 (0.9)	55.4b (3.0)	2.10bc (0.09)	1.19ab (0.02)	0.144bc (0.010)
MODE OF REPRODUCTION				*	*	NS	NS
Sexual	407	13.1 (1.4)	16.7 (0.4)	51.6a (1.5)	2.00a (0.05)	1.21a (0.01)	0.151a (0.006)
Sexual and Asexual	66	9.2 (3.0)	15.0 (1.1)	43.8b (3.7)	1.69b (0.08)	1.20a (0.03)	0.138a (0.016)
SUCCESSIONAL STATUS				**	NS	NS	NS
Early	226	18.5 (2.7)	16.6 (0.6)	49.0b (2.0)	1.98a (0.08)	1.21a (0.02)	0.149a (0.008)
Mid	152	7.6 (0.8)	16.4 (0.6)	47.6b (2.3)	1.86a (0.06)	1.20a (0.02)	0.141a (0.010)
Late	95	9.4 (1.5)	16.3 (0.9)	58.9a (3.0)	2.08a (0.09)	1.22a (0.02)	0.161a (0.011)

[a]N, number of taxa; P_s, percentage polymorphic loci; A_s, number of alleles per locus; A_{es}, effective number of alleles per locus; H_{es}, genetic diversity. See text for more complete definitions. Levels of significance: *, $p < 0.05$; **, $p < 0.01$; ***, $p < 0.001$; NS, not significant.
[b]Means followed by the same letter in a column are not significantly different at the 5% probability level.
[c]Standard errors are in parentheses.
[d]A (-) indicates that these data were excluded from statistical tests because of small sample sizes.

The Duncan multiple range analysis for life forms was affected by low sample sizes in the long-lived herbaceous perennial and the short-lived woody perennial classes. When these groups were removed from the analysis significant differences occurred among the three remaining classes for the four genetic parameters (Table 2). Annual species and long-lived species (both woody and herbaceous) have high levels of genetic diversity and differ by 11%. Short-lived perennials (either woody or herbaceous) have lower values and differ from the previous two categories by 41 and 56%, respectively. The data suggested that longevity of perennials may have a more marked effect on genetic diversity than woody versus herbaceous growth habit. Caution must be exercised in interpreting these results, however, as sample sizes are unequal between the herbaceous and woody species within the two longevity classes.

Geographic range had a marked effect on levels of genetic diversity within species (Table 2). Endemic species had less than 50% of the genetic diversity of widespread species and 70 and 64% of the genetic diversity of narrowly and regionally distributed species. Widespread species had 47% more genetic diversity than narrowly distributed species and 35% more than regionally distributed species.

Regional distribution had much less effect on genetic parameters than the preceding traits (Table 2). Cold climate species (boreal–temperate) tended to have higher P_s and A_s values than species from warmer regions. In terms of A_{es} and H_{es} there were no differences among the regions. This suggests that the boreal–temperate species have a greater abundance of low-frequency alleles than species from other regions. These results are affected by the high number of conifers in the boreal-temperate group.

Among breeding system categories, selfing species and animal-pollinated mixed-mating species have the lowest values for all four genetic parameters (Table 2). Mixed-mating, wind-pollinated species have the highest levels of allozyme variation (but note the low number of taxa) whereas outcrossed species that are animal- or wind-pollinated have somewhat less variation. Differences in H_{es} values between selfing and animal-pollinated, mixed-mating species, on the one hand, and outcrossing species, on the other, only approach significance but differ on average by 31–39%.

Considering seed dispersal mechanisms, species whose propagules are dispersed by becoming attached to animals or are moved by humans have the highest levels of variation (Table 2). Species with explosively dispersed seeds have the lowest genetic diversity while gravity-dispersed, animal-ingested, and wind-dispersed species have intermediate values. The H_{es} values for species with attached seeds were 122% higher than those for species with explosively-dispersed seeds.

Mode of reproduction had a weak influence on genetic diversity (Table 2). Sexual species had significantly higher P_s and A_s values but mean H_{es} and A_{es} values were equal between the two classes. Completely asexual species were so rare that they were lumped with the sexual–asexual class.

Successional status had little effect on genetic diversity (Table 2). Late successional species had higher P_s values but differences among the categories for the other three parameters were not significant.

Overall, the eight traits explained 24% of the variation in genetic diversity at the species level. Geographic range accounted for the largest proportion of this variation (32%); life form accounted for an additional 25%, and breeding system and seed dispersal mechanism each added approximately 17%.

Variation within populations

Differences among the categories of taxonomic status, life form, geographic range, and breeding system were significant ($p < 0.05$) for all four genetic parameters (Table 3). Categories for regional distribution and seed dispersal mechanism were significantly different for P_p, A_p, and H_{ep} while the categories of successional stage were significantly different for P_p and A_p. None of the genetic parameters was significantly different among the two mode of reproduction categories.

Gymnosperms had significantly higher values for P_p and A_p than did monocots or dicots (Table 3). Gymnosperms had the highest H_{ep} value but it was not significantly higher than the monocots. These results indicate that gymnosperms tend to have more alleles at lower frequencies in their populations than monocots, perhaps due to their greater potential for long-distance pollen movement. Dicot species had the lowest values for the four genetic parameters. The mean H_{ep} value for the dicots was 67 and 60% of the values for the monocots and gymnosperms, respectively.

The small sample sizes of the long-lived herbaceous plants and the short-lived woody species led to difficulties in the interpretation of differences among the life form classifications (Table 3). When these two classes were removed, significant differences occurred among the three categories. The H_{ep} value of long-lived woody species was 43 and 55% higher than those of annual and short-lived herbaceous perennials. Differences in genetic diversity within populations of annual and short-lived species were not significant.

Of the four geographic range categories, endemics had the lowest levels of variation within their populations (Table 3). Narrow and regionally distributed species had intermediate levels of variation while widespread species had the greatest amount. The percentage of polymorphic loci found within populations of regionally distributed species did not differ significantly from the value for widespread species. The H_{ep} value for widespread species was 152% higher than the value for endemic species and 51 and 35% higher than values for narrowly and regionally distributed species. Widespread species had more alleles per polymorphic locus and allele frequencies probably were less skewed.

Regional distribution had little influence on the levels of variation found within populations (Table 3). Generally, there were two significance classes:

TABLE 3. Levels of allozyme variation at the population level for species with different attributes.[a]

Categories	N	Mean no. populations	Mean no. loci	P_p	A_p	A_{cp}	H_{cp}
TAXONOMIC STATUS				***	***	***	***
Gymnosperms	56	8.5 (0.9)[c]	16.1 (1.3)	57.7a[b] (3.4)	1.93a (0.09)	1.21a (0.02)	0.160a (0.011)
Monocots	80	18.7 (3.1)	15.5 (0.9)	40.3b (3.0)	1.66b (0.08)	1.21a (0.03)	0.144a (0.012)
Dicots	338	11.9 (1.7)	16.8 (0.4)	29.0c (1.3)	1.44c (0.02)	1.13b (0.01)	0.096b (0.005)
LIFE FORM				***	***	**	***
Annual	187	18.5 (3.0)	14.9 (0.5)	30.2b (1.9)	1.48b (0.04)	1.15b (0.02)	0.105b (0.008)
Short-lived perennial							
Herbaceous	159	8.8 (1.2)	17.1 (0.7)	28.0b (1.8)	1.40b (0.03)	1.12b (0.01)	0.096b (0.008)
Woody	11	18.1 (6.7)	25.2 (1.4)	31.3-[d] (6.7)	1.55- (0.12)	1.11- (0.03)	0.094- (0.021)
Long-lived perennial							
Herbaceous	4	6.0 (1.7)	13.8 (3.8)	39.3- (16.2)	1.44- (0.20)	1.14- (0.05)	0.084- (0.028)
Woody	115	9.3 (1.4)	17.0 (0.9)	50.0a (2.5)	1.79a (0.06)	1.21a (0.02)	0.149a (0.009)
GEOGRAPHIC RANGE				***	***	***	***
Endemic	100	6.5 (0.9)	17.8 (0.6)	26.3c (2.1)	1.39c (0.03)	1.09c (0.01)	0.063c (0.006)
Narrow	115	8.8 (1.0)	16.9 (0.8)	30.6bc (2.2)	1.45bc (0.05)	1.13bc (0.01)	0.105b (0.009)
Regional	180	10.4 (1.1)	16.7 (0.7)	36.4ab (2.0)	1.55b (0.04)	1.16b (0.02)	0.118b (0.007)
Widespread	85	25.5 (5.2)	14.6 (0.9)	43.0a (3.3)	1.72a (0.07)	1.23a (0.02)	0.159a (0.013)
REGIONAL DISTRIBUTION				***	***	NS	**
Boreal-temperate	22	7.8 (1.4)	16.2 (2.0)	64.5a (5.1)	2.08a (0.13)	1.25a (0.04)	0.184a (0.019)
Temperate	361	11.6 (1.3)	15.6 (0.4)	32.6b (1.3)	1.51b (0.03)	1.15b (0.01)	0.109b (0.005)
Temperate-tropical	20	37.3 (12.1)	15.1 (1.3)	35.9b (5.9)	1.52b (0.11)	1.16ab (0.04)	0.123b (0.024)
Tropical	66	10.6 (2.8)	21.3 (1.0)	32.7b (3.0)	1.45b (0.05)	1.13b (0.02)	0.109b (0.012)

TABLE 3. (Continued)

Categories	N	Mean no. populations	Mean no. loci	P_p	A_p	A_{ep}	H_{ep}
BREEDING SYSTEM				***	***	***	***
Selfing	113	20.3 (3.8)	16.2 (0.7)	20.0c (2.3)	1.31b (0.05)	1.10b (0.03)	0.074d (0.010)
Mixed-animal	85	8.9 (2.1)	14.4 (0.8)	29.2bc (2.5)	1.43b (0.04)	1.12b (0.02)	.090cd (0.010)
Mixed-wind	10	10.0 (3.1)	12.5 (3.6)	54.4a (8.9)	1.99a (0.19)	1.28a (0.07)	0.198a (0.041)
Outcrossing-animal	164	10.7 (2.1)	17.7 (0.7)	35.9b (1.8)	1.54b (0.03)	1.17b (0.01)	0.124bc (0.008)
Outcrossing-wind	102	10.7 (1.6)	16.7 (0.9)	49.7a (2.6)	1.79a (0.06)	1.19ab (0.02)	0.148b (0.009)
SEED DISPERSAL				***	***	NS	**
Gravity	199	10.1 (1.1)	16.9 (0.6)	29.8bc (1.7)	1.45ab (0.03)	1.14ab (0.01)	0.101ab (0.007)
Gravity-attached	12	29.2 (15.8)	18.6 (1.8)	34.4ab (4.6)	1.64a (0.11)	1.16ab (0.04)	0.127a (0.028)
Attached	68	20.8 (5.7)	16.5 (0.9)	42.1ab (2.9)	1.68a (0.08)	1.20a (0.04)	0.137a (0.012)
Explosive	34	12.4 (2.7)	18.6 (1.4)	21.3c (3.1)	1.25b (0.04)	1.08b (0.02)	0.062b (0.011)
Ingested	54	17.6 (6.5)	13.2 (1.1)	32.4abc (3.3)	1.48ab (0.07)	1.17ab (0.02)	0.129a (0.015)
Wind	105	8.7 (0.9)	16.6 (0.9)	42.9a (3.0)	1.70a (0.06)	1.16ab (0.02)	0.123a (0.010)
MODE OF REPRODUCTION				NS	NS	NS	NS
Sexual	413	13.1 (1.4)	16.7 (0.4)	34.9a (1.3)	1.53a (0.03)	1.16a (0.01)	0.114a (0.005)
Sexual and asexual	56	9.2 (3.0)	15.0 (1.1)	29.4a (3.3)	1.47a (0.06)	1.14a (0.02)	0.103a (0.013)
SUCCESSIONAL STATUS				***	**	NS	NS
Early	198	18.5 (2.7)	16.6 (0.6)	29.6b (1.9)	1.46b (0.04)	1.14b (0.01)	0.107b (0.008)
Mid	182	7.6 (0.8)	16.4 (0.6)	33.8b (1.9)	1.52b (0.04)	1.14b (0.01)	0.106b (0.007)
Late	103	9.4 (1.5)	16.3 (0.9)	43.9a (2.8)	1.67a (0.06)	1.19a (0.02)	0.133a (0.010)

[a]N, number of taxa; P_P, percentage polymorphic loci; A_p, number of alleles per locus; A_{ep}, effective number of alleles per locus; H_{ep}, genetic diversity index. See text for more complete definitions. Levels of significance: *, $p < 0.05$; **, $p < 0.01$; ***, $p < 0.001$; NS, not significant.
[b]Means followed by the same letter in a column are not significantly different at the 5% probability level.
[c]Standard errors are in parentheses.
[d]A (-) indicates that these data were excluded from statistical tests because of small sample sizes.

one representing the boreal–temperate species (mostly conifers) and a second that included species from temperate and tropical regions. Average differences in H_{ep} values among these two groups ranged from 50 to 69%.

Comparing breeding systems, populations of selfing species had the lowest proportion of polymorphic loci, the fewest number of alleles per locus, and the lowest effective number of alleles per locus (Table 3). The H_{ep} value for the selfing species was 82% of the value for the mixed-mating animal-pollinated species, 60% of the value for outcrossed, animal-pollinated species, 50% of the value for outcrossed wind-pollinated species, and 37% of the value for mixed-mating, wind-pollinated species. Animal-pollinated species maintain intermediate levels of variation with the mixed-mating species being somewhat closer to the selfers and outcrossed species being closer to the wind-pollinated species.

Plant species that disperse their seeds by wind, animal ingestion, or animal attachment maintain higher amounts of variation within their populations than species with other seed dispersal mechanisms (Table 3). Species with explosive seeds or those with gravity-dispersed seed had lower than average levels of allozyme variation. Average differences in H_{ep} values between the explosive category and the other four categories ranged from 63 to 121%.

Significant differences were not observed between sexual species and species that reproduce by sexual and asexual means (Table 3). Some differences were seen among species representing different stages of succession. Species of the later stages of succession had somewhat higher values for all four genetic parameters.

Based on the GLM analysis the eight traits explained 28% of the variation in population genetic diversity. Breeding systems (33%) and geographic range (28%) accounted for the largest proportion of this variation. Life form (13%), taxonomic status (12%), and seed dispersal mechanism (11%) accounted for significant but smaller proportions of this variation.

Variation among populations

Relatively little variation occurred among the eight traits for total genetic diversity at polymorphic loci (H_T) (Table 4). Only geographic range ($p < 0.01$), seed dispersal mechanism ($p < 0.001$) and successional status ($p < 0.001$) were significant. The magnitude of H_T was dependent on the number of alleles and the evenness of allele frequencies at polymorphic loci. Thus, since H_{es} is also a function of the proportion of polymorphic loci, the higher levels of significance seen among categories for H_{es} must be a function of differences in the proportion of polymorphic loci. Differences in H_S and G_{ST} indicate that plant species with different combinations of characteristics partition allozyme variation in quite different ways. The largest differences in H_S values occurred between species with different breeding systems. Selfing species had a mean H_S value of 0.149 while wind-pollinated mixed-mating species had a value of 0.342. Species with different seed dispersal mechanisms also had large differences in H_S values. Significant differences in G_{ST} values occurred among the categories of taxonomic status, life form, regional distribution, breeding system, and stage of succession.

For taxonomic status, dicot species had the smallest H_S value and the largest G_{ST} value. Monocots had an intermediate H_S value and their G_{ST} value closely resembled that of the dicots. Gymnosperms, on the other hand, had a G_{ST} value one-fourth of either angiosperm class. This difference can be attributed to the presence of selfing species in angiosperms and the predominance of wind pollination in gymnosperms.

Annual species and short-lived herbaceous species had the lowest H_S values and highest G_{ST} values among the five life form categories. Woody perennials had G_{ST} values that were 3- to 4-fold lower than annual or herbaceous species.

Values of H_T and H_S were significantly different among species with different geographic ranges. In contrast, geographic range was not significantly associated with G_{ST} values. Considering only polymorphic loci, endemic species had the lowest levels of diversity at the population and species levels, while widespread species had the highest levels of diversity. These results suggest that endemic species have fewer alleles per polymorphic locus and/or more skewed allele frequencies than more widespread species. However, endemic species partition their variation in much the same way as more widespread species.

There were no significant differences for H_T or H_S among species with different regional distributions but the G_{ST} value of the boreal–temperate species (mostly conifers) was significantly lower than the other three classes.

Selfing species were characterized by a relatively high H_T value, low H_S value, and high G_{ST} value. Outcrossed, wind-pollinated species were at the other extreme with a somewhat lower H_T value and a G_{ST} value 5-fold lower than the selfing species. Animal-pollinated species had an intermediate G_{ST} value. These results indicate that selfing species maintain as much diversity at their polymorphic loci as outcrossing species, but that a much larger proportion of this variation is found among populations.

The Duncan multiple range analysis on the categories of seed dispersal mechanism was difficult to interpret although the H_T, H_S, and G_{ST} values varied significantly among the six classifications. Ingested species had the highest H_T and H_S values while gravity-attached species had the lowest values for these parameters. Gravity-dispersed species had the highest G_{ST} value while gravity-attached and wind-dispersed species had the lowest. Thus, the more vagile species appear to have relatively more variation within and less among their populations. Species of late successional stages had intermediate H_T values, high H_S values, and low G_{ST} values. Early successional species had the reverse combination of genetic parameters.

TABLE 4 Distribution of allozyme variation among populations of species classified according to their traits.[a]

Categories	N	Mean no. populations	Mean no. loci	H_T	H_S	G_{ST}
TAXONOMIC STATUS				NS	**	***
Gymnosperms	80	8.5 (0.9)[c]	16.1 (1.3)	0.297a[b] (0.013)	0.271a (0.013)	0.068b (0.013)
Monocots	81	18.7 (3.1)	15.5 (0.9)	0.320a (0.019)	0.238ab (0.017)	0.231a (0.023)
Dicots	246	11.9 (1.7)	16.8 (0.4)	0.311a (0.009)	0.214b (0.008)	0.273a (0.017)
LIFE FORM				NS	***	***
Annual	146	18.5 (3.0)	14.9 (0.5)	0.330a (0.012)	0.200b (0.012)	0.357a (0.024)
Short-lived perennial						
Herbaceous	119	8.8 (1.2)	17.2 (0.7)	0.300a (0.013)	0.222b (0.013)	0.233b (0.019)
Woody	8	18.1 (6.7)	25.2 (1.4)	0.292-[d] (0.033)	0.266- (0.032)	0.088- (0.024)
Long-lived perennial						
Herbaceous	2	6.0 (1.7)	13.8 (3.8)	0.346- (0.018)	0.282- (0.024)	0.213- (0.144)
Woody	131	9.3 (1.4)	17.0 (0.9)	0.298a (0.012)	0.269a (0.011)	0.076c (0.010)
GEOGRAPHIC RANGE				**	***	NS
Endemic	52	6.5 (0.9)	17.8 (0.6)	0.263c (0.023)	0.163c (0.016)	0.248a (0.037)
Narrow	82	8.8 (1.0)	16.9 (0.8)	0.300bc (0.015)	0.215b (0.013)	0.242a (0.024)
Regional	186	10.4 (1.1)	16.7 (0.7)	0.308ab (0.010)	0.236ab (0.010)	0.216a (0.019)
Widespread	87	25.5 (5.2)	14.6 (0.9)	0.347a (0.013)	0.267a (0.014)	0.210a (0.025)
REGIONAL DISTRIBUTION				NS	NS	***
Boreal-temperate	28	7.8 (1.4)	16.2 (2.0)	0.272a (0.017)	0.260a (0.017)	0.036b (0.007)
Temperate	322	11.6 (1.3)	15.6 (0.4)	0.318a (0.008)	0.228a (0.008)	0.246a (0.015)
Temperate-tropical	15	37.3 (12.1)	15.1 (1.3)	0.301a (0.045)	0.219a (0.039)	0.233a (0.049)
Tropical	41	10.6 (2.8)	21.3 (1.0)	0.278a (0.023)	0.228a (0.017)	0.173a (0.021)

TABLE 4 (Continued)

Categories	N	Mean no. populations	Mean no. loci	H_T	H_S	G_{ST}
BREEDING SYSTEM				NS	***	***
Selfing	78	20.3 (3.8)	16.2 (0.7)	0.334ab (0.017)	0.149c (0.016)	0.510a (0.035)
Mixed-animal	60	8.9 (2.1)	14.4 (0.8)	0.304b (0.022)	0.221b (0.017)	0.216b (0.024)
Mixed-wind	11	10.0 (3.1)	12.5 (3.6)	0.378a (0.057)	0.342a (0.054)	0.100c (0.022)
Outcrossing-animal	124	10.7 (2.1)	17.7 (0.7)	0.310b (0.010)	0.243b (0.010)	0.197b (0.017)
Outcrossing-wind	134	10.7 (1.6)	16.7 (0.9)	0.293b (0.011)	0.259b (0.011)	0.099c (0.012)
SEED DISPERSAL				***	***	***
Gravity	161	10.1 (1.1)	16.9 (0.6)	0.306b (0.010)	0.207b (0.011)	0.277a (0.021)
Gravity-attached	11	29.2 (15.8)	18.6 (1.8)	0.211c (0.038)	0.171b (0.031)	0.124b (0.031)
Attached	52	20.8 (5.7)	16.5 (0.9)	0.325b (0.024)	0.236ab (0.021)	0.257ab (0.032)
Explosive	23	12.4 (2.7)	18.6 (1.4)	0.302b (0.021)	0.217b (0.023)	0.243ab (0.048)
Ingested	39	17.6 (6.5)	13.2 (1.1)	0.394a (0.020)	0.305a (0.022)	0.223ab (0.033)
Wind	121	8.7 (0.9)	16.6 (0.9)	0.292b (0.012)	0.241ab (0.011)	0.143ab (0.020)
MODE OF REPRODUCTION				NS	NS	NS
Sexual	352	13.1 (1.4)	16.7 (0.4)	0.311a (0.007)	0.229a (0.007)	0.225a (0.013)
Sexual and asexual	54	9.2 (3.0)	15.0 (1.1)	0.305a (0.019)	0.236a (0.018)	0.213a (0.027)
SUCCESSIONAL STATUS				*	***	***
Early	165	18.5 (2.7)	16.6 (0.6)	0.328a (0.011)	0.221b (0.011)	0.289a (0.021)
Mid	121	7.6 (0.8)	16.4 (0.6)	0.287b (0.012)	0.205b (0.011)	0.259a (0.022)
Late	121	9.4 (1.5)	16.3 (0.9)	0.308ab (0.012)	0.270a (0.011)	0.101b (0.013)

[a]N, number of taxa; H_T, total genetic diversity; H_S, genetic diversity within populations; G_{ST}, proportion of the total diversity among populations. See text for more complete definitions. Levels of significance: *, $p < 0.05$; **, $p < 0.01$; ***, $p < 0.001$; NS, not significant.
[b]Means followed by the same letter in a column are not significantly different at the 5% probability level.
[c]Standard errors are in parentheses.
[d]A (-) indicates that these data were excluded from statistical tests because of small sample sizes.

The eight traits explained 47% of the heterogeneity in G_{ST} values. Breeding system and life form were most closely associated with this variation, together accounting for 84% of the variation explained by the model.

Multivariate analyses

Multivariate analyses were performed on the eight species traits and species level genetic diversity (H_{es}), population level diversity (H_{ep}), and G_{ST}. The eight traits were only loosely associated with one another. The highest correlation ($r = 0.72$) was between life form and successional stage. A second cluster of correlations ranged between 0.25 and 0.50 and involved taxonomic stage, life form, breeding system, seed dispersal, and successional stage.

Variation in H_{es} values was most highly correlated with geographic range ($r = 0.27$) and to a lesser extent with taxonomic status ($r = 0.15$) and breeding system ($r = 0.16$). The multivariate analysis indicated that variation in species level diversity was strongly influenced by geographic range and the regional distribution of species.

Genetic diversity within populations was most highly correlated with breeding system ($r = 0.28$), geographic range ($r = 0.28$), taxonomic status ($r = 0.23$), and life form ($r = 0.18$). Variation in H_{ep} values was moderately influenced by the geographic range and the regional distribution of the species.

Variation in G_{ST} values was highly correlated with the breeding system ($r = 0.53$) and life form ($r = 0.46$) and was somewhat less closely associated with taxonomic status ($r = 0.31$), seed dispersal ($r = 0.21$), and successional stage ($r = 0.31$).

Variation in genetic diversity at the species (H_{es}) and population (H_{ep}) levels and the distribution of genetic variation among populations (G_{ST}) were each associated with a suite of species characters, one extreme of which was characterized by late-successional, long-lived, wind-pollinated gymnosperms with high H_{es} and H_{ep} values and low G_{ST} values.

Associations among genetic parameters

Associations among H_{es}, H_{ep}, H_T, H_S, and G_{ST} were examined by calculating a correlation matrix based on individual species values. The H_{es} values were highly correlated with H_{ep} values ($r = 0.89$; $p < 0.0001$), and moderately correlated with H_T and H_S ($r = 0.56$ and $r = 0.58$; $p < 0.0001$) but negatively correlated with G_{ST} ($r = -0.11$; $p < 0.05$). Values of H_{ep} were highly correlated with H_S ($r = 0.77$; $p < 0.0001$) and moderately correlated with H_T and G_{ST} ($r = 0.43$ and $r = -0.45$; $p < 0.0001$). The H_T values were positively correlated with H_S ($r = 0.70$; $p < 0.0001$) and weakly correlated with G_{ST} ($r = 0.15$; $p < 0.01$). Values of H_S were negatively correlated with G_{ST} ($r = -0.55$; $p < 0.0001$).

These results suggest that genetic diversity at the species level is associated with genetic diversity at the population level. However, the association between H_{es} and H_{ep} and genetic diversity at the polymorphic loci (H_T) is much lower, indicating that much of the variation in genetic diversity among species is the result of differences in the proportion of polymorphic loci and not in the number and frequency of alleles at polymorphic loci.

The G_{ST} values were negatively correlated with H_{ep} and H_S, corroborating our previous observation that differences among species in population level genetic diversity are a product of the level of genetic diversity found in the species and how this variation is partitioned among populations.

DISCUSSION

Our results demonstrate that allozyme diversity within species and populations is as much a characteristic of individual plant species as other characteristics such as growth form or flower structure. Species level genetic diversity ranged from approximately 0.40 to 0.00. Species with high levels of genetic variation included *Alseis blackiana* (Hamrick and Loveless in press), *Echium plantagineum* (Burdon and Brown 1986), *Oryza latifolia*, (Sécond 1985), *Plectritis congesta* (Carey and Ganders 1987), and *Solanum pennelli* (Rick and Tanksley 1981). Species with low levels of variation included *Agastache rugosa* (Vogelmann and Gastony 1987), *Citrullus lanatus* (Zamir et al. 1984), *Limmanthus fluccosa* (McNeil and Jain 1983), and *Pinus resinosa* (Simon et al. 1986). Within population diversity ranged from approximately 0.35 [*Alseis blackiana* (Hamrick and Loveless in press), *Echium plantagineum* (Burdon and Brown 1986), *Picea abies* (Lundkvist and Rudin 1977), *Pinus longaeva* (Hiebert and Hamrick 1983)] to 0.0 (see above). The G_{ST} values covered the entire possible range with some species having no variation among their populations [*Cucumis metuliferus* (Staub et al. 1987)] to a few having all their genetic diversity among populations [*Clarkia similis* (Smith-Huerta, 1986); *Emex spinosa* (Marshall and Weiss 1982)].

However, even in the face of such heterogeneity, certain generalizations can be made. First, plant species, on average, maintain higher levels of allozyme variation within their populations (0.113) than either invertebrates (0.100) or vertebrates (0.054) (Nevo et al. 1984). Species level comparisons are not possible, since, to our knowledge, reviews of species level variation are not available in the animal literature.

In contrast to the earlier review (Hamrick et al. 1979), we have excluded all studies based solely on polymorphic loci from calculations of population and species level variation. This decision may have reduced the overall within population genetic diversity from 0.141 (Hamrick et al. 1979) to 0.113. This difference in mean genetic diversity may have also resulted from an increase in the number of endemic species analyzed during the past 10 years. Hamrick et al. (1979) reviewed 17 endemic species in their study of 113 taxa (15%)

whereas we included 100 endemic taxa (28 genera) among the 480 species (21%) that supplied population level data. Concern has also been expressed that the inclusion of studies with few loci would increase estimates of genetic diversity. In response to this concern, we tested the relationship between genetic diversity (H_{cs} and H_{cp}) and the number of loci analyzed; the correlations accounted for only 2 and 1% of the variation in genetic diversity.

The second major generalization concerns the relationship between genetic diversity and the characteristics of species. Earlier reviews (Hamrick et al. 1979; Brown 1979; Gottlieb 1981; Loveless and Hamrick 1984) found that long-lived, outcrossing, wind-pollinated species of the later stages of succession have higher levels of allozyme variation within populations and less among population variation than species with other combinations of traits. Our results support this conclusion and, because of the increased data set, expand the generality of these results. Our results also support the conclusion of Ellstrand and Roose (1987) that predominantly clonal species may maintain as much genetic diversity within populations as sexually reproducing species. One exception to the general agreement between this review and earlier reviews was the observation in the earlier review that widespread species had, on average, somewhat less genetic diversity within their populations than regionally or narrowly distributed species (H_{cp} = 0.120, 0.185, and 0.158, respectively; Hamrick et al. 1979). In this study widespread species had the highest H_{cp} values (Table 3). This inconsistency was due in part to the addition of several studies of allozyme variation in land races of cultivated species. For example, 10 studies of *Zea mays* had a mean H_{cp} value of 0.190 and 4 studies of *Hordeum vulgare* had a mean H_{cp} of 0.236.

At the species level, geographic range is the best predictor of levels of allozyme variation. Endemic species have the lowest genetic diversity whereas regionally distributed and widespread species maintain the most diversity. Species with regional and widespread geographic ranges may have a history of large, continuous populations that are less susceptible to losses of genetic variation due to genetic drift. Endemic species, on the other hand, might be expected to consist of smaller, more ecologically limited populations that have experienced population bottlenecks during their evolutionary history.

At the population level, two traits (geographic range and the breeding system) are the best predictors of the genetic parameters. Again, the more widespread species have, on average, greater genetic diversity within populations than more geographically restricted species. The second factor affecting the level of variation within populations is the vagility (breeding system and seed dispersal) of the species. Wind-pollinated species have the highest within population variation. Species with wind- and animal-dispersed propagules also tend to have high levels of within population diversity.

Geographic range plays almost no role in the distribution of genetic variation among populations—endemic species distribute their variation in much the same way as widespread species. This result is somewhat surprising and may be due to the interaction of several factors. The more local distribution of endemic species should increase gene flow among populations and reduce differentiation. On the other hand, endemic species often occur in small isolated populations—a condition that should produce greater differentiation.

The gene flow potential of species had the predominant influence on the partitioning of allozyme variation among populations. This conclusion is supported by the relationship between G_{ST} and the species characteristics. Analyses of the distribution of private alleles among populations (Slatkin 1985) lent further support for this conclusion. Hamrick (1987) and Hamrick et al. (in press) using a portion of this data base, demonstrated that outcrossed, wind-pollinated species have much higher estimated numbers of migrants per generation (Nm; where N is the population size and m is the rate of migration) than selfing or animal-pollinated species. Furthermore, there is a close relationship between Nm values calculated from the Slatkin procedure and estimates of Nm based on G_{ST} values. These results indicate that widespread, long-lived species with potential for long-range gene movement should have the highest levels of variation within populations. Our results are generally consistent with this prediction.

An important aim of this review was to compare the amount of variation at the species, population, and interpopulation levels. The finding that variation at the species level was positively and significantly associated with variation at the population level has important implications for the development of sampling strategies. It was also clear that species level variation was not closely associated with the distribution of variation among populations. Population-level variation was, on the other hand, strongly influenced by levels of among population diversity. These differences hint that different evolutionary processes influence variation at the species and population levels.

An important question is whether patterns of allozyme variation can be used to predict patterns of genetic variation in other traits. The results of the few studies that compare morphometric and allozyme variation have not demonstrated consistently positive associations among the different types of traits (Hamrick, in press). Price et al. (1984) compared estimates of population differentiation based on allozyme polymorphism and morphometric traits in the self-pollinated species *Avena barbata*, *Hordeum vulgare*, and *H. jubatum*, and an outcrossing species, *Clarkia williamsonii*. Rank correlations between morphometric and allozyme distance measures either approached significance or were highly significant for the selfing species. The rank correlation for *C. williamsonii* was not significant suggesting that isozyme loci may provide more information about the distribution of other genes in selfing plants than in outcrossers. This conclusion has a firm foundation in theory since the breeding system employed by a species should affect how allozyme variation is linked with other traits. Selfing species experience slower decay of linkage disequilibrium, increases in the variance of coancestry and increases in direct gene expression due to increased homozygosity—all factors that should increase associations among isozyme and phenotypic variation (Brown and Burdon 1987).

CONCLUSIONS

Allozyme markers are often used to investigate systematic problems or to measure levels of variation within and among populations. These descriptive studies have collectively made an important contribution to plant systematics, evolutionary ecology, population genetics, plant breeding, and the conservation of genetic resources. However, in certain areas of plant population genetics the results of isozyme analyses have not met original expectations. Soon after isozyme analyses were introduced, it became apparent that it would be difficult to demonstrate that selection acts directly on variation at particular enzyme loci (Lewontin 1974). Also, the inability to find significant associations between the distribution of allozyme variation and quantitative traits has limited the generalization of allozyme analyses to other genetic traits. On the other hand, allozyme analyses continue to play an important role in studies of plant evolution by providing genetic markers with which to study evolutionary processes.

Isozyme loci have demonstrated that genetic variation within natural plant populations is patchily distributed (Allard et al. 1972; Hamrick and Holden 1979; Linhart et al. 1981; Epperson, this volume). In some cases, this heterogeneity is associated with environmental features and may be influenced by selection (Allard et al. 1972; Hamrick and Holden 1979) while in others the patchy genetic structure results from limited pollen or seed dispersal (Linhart et al. 1981). Isozymes have also been used to demonstrate that populations of selfing species often consist of genotypes in gametic phase disequilibrium (Allard et al. 1972; Hamrick and Holden 1979; Brown et al. 1980).

In the last 10 years, quantitative estimates of plant mating systems have increased dramatically due to the availability of isozyme markers (Brown, this volume). Furthermore, multilocus estimation procedures that provide more accurate estimates of outcrossing rates have been developed, at least in part, in response to the ready availability of multilocus allozyme genotypes (Shaw et al. 1981; Ritland and Jain 1981). Studies of gene movement have also been greatly advanced by the availability of isozyme loci (Barrett and Husband, this volume). Isozyme loci have been used to follow pollen and seed movement (e.g., Muller 1977; Smyth and Hamrick 1986; Schaal 1980) with mixed success. The rather recent development of estimation procedures based on the frequency of unique alleles (Slatkin 1981, 1985) to estimate the number of migrants per generation has made use of the abundant allozyme survey data. These analyses have shown that measures of Nm are predictably associated with the pollen and seed dispersal characteristics of the species (Hamrick 1987; Hamrick et al. in press). Finally, studies of gene movement based on the determination of parentage have provided previously unavailable details of the breeding structure of plant populations (Meagher 1986; Neale and Adams 1985; Ellstrand and Marshall 1985).

Several poorly studied questions can be approached by applying isozyme analyses to natural plant populations. We have little information concerning annual fluctuations in the mating system or in rates of gene flow. Nor do we understand how these fluctuations are affected by interactions among genetic and ecological factors. By reconstructing the genealogy of plant populations via multi-locus parentage analysis it may be possible to document differences in the reproductive success of individuals (Ennos, this volume). It may also be possible to demonstrate differences in the survival and reproduction of individuals with different pedigrees (Hamrick and Loveless 1986).

An understanding of the way genetic variation is partitioned among populations is of primary importance for the conservation of genetic diversity (and hence the evolutionary potential) of species. Surveys of allozyme variation provide data that are critical for the establishment of strategies designed to preserve genetic diversity. Ideally, each species of concern would be analyzed extensively. This goal is impossible given the current rate of species extinction and population destruction. Reviews such as this may be used to provide general guidelines for the establishment of effective management practices. It should be stressed, however, that while reviews can provide broad generalizations concerning average levels of genetic variation and the associations between species traits and genetic diversity, there is still a great deal of unexplained heterogeneity among species. Evidence of this is the relatively small proportion of the variation among species (30–50%) that is accounted for by our models.

In conclusion, the availability of isozyme loci has increased profoundly our knowledge of the genetic structure of plant populations. The distribution patterns of genetic variation within and among plant populations are today much clearer than they were 20 years ago. In addition, new insights concerning how various species characteristics interact to shape genetic variation within plant populations are emerging. In the future, allozyme variation will be used in conjunction with other genetic markers (Clegg, this volume) to provide further insights into the influence of plant mating systems, breeding structure, seed dispersal, and differential survival on the genetic structure of plant populations.

ACKNOWLEDGMENTS

We are grateful to Vikas Kumar and Jenna Hamrick for their assistance on this project. Portions of this work were supported by NSF Grants BSR 86–00083 and BSR 87–18803.

CHAPTER 4

GENETIC DIVERSITY OF SEED STORAGE PROTEINS IN PLANTS

Paul Gepts

ABSTRACT

Seed storage proteins have been studied intensively because they are important nutritionally and they provide a biological model system for the temporal and spatial regulation of gene expression. Genetic diversity of seed storage proteins is evident from various types of electrophoresis that detect charge as well as size differences. As genetic markers, these proteins are characterized by a high level of polymorphism, limited environmental influence on their electrophoretic pattern, a simple genetic control, a complex molecular basis for genetic diversity, and homologies between storage proteins that extend across taxa. The simple genetic control of seed storage proteins is, however, a limitation because it restricts their genome coverage. The complex molecular basis for their diversity, on the other hand, suggests the uniqueness of each pattern and their role as an evolutionary marker. Questions that have been addressed using seed storage protein diversity include the organization of genetic diversity within and between populations, the process of plant domestication (the identity of the wild ancestor and the fate of genetic diversity on domestication), and homologies between genomes (identification of the genome donors in amphiploids). Seed storage protein diversity is also studied directly as a breeding objective (e.g., bread-making quality, insect resistance, nutritional qualities) and indirectly as a genetic marker linked to a trait of interest or for the purpose of varietal identification.

The first comprehensive studies on seed proteins were performed by Osborne (1907). Through systematic analyses in many species, he established chemical properties of seed proteins such as their solubility in various solvents and precipitation or denaturation by various factors. He classified seed proteins into four general solubility groups. Proteins soluble in water were called *albumins*, those soluble in a salt solution *globulins*, and those soluble in aqueous alcohol *prolamines*. Proteins remaining after removing these fractions were extracted with dilute acids or alkali and were called *glutelins*. A comparison of seed proteins in various species indicated that most cereal seeds contained primarily prolamine-type proteins whereas leguminous seeds contained globulins. Cereal prolamines contain low levels of lysine and tryptophan and legume globulins contain low levels of the sulfur-containing amino acids, methionine and cysteine. These deficiencies are detrimental to monogastric animals (including humans) that have to rely on a supply of these amino acids in their diet.

Seed proteins can also be classified according to their biological function. Although seeds contain metabolic and structural proteins, the major protein fraction consists of the so-called storage proteins, which can account for 50% or more of total protein in the seed. Higgins (1984) defined seed storage proteins as "any protein accumulated in significant quantities in the developing seed which on germination is rapidly hydrolyzed to provide a source of reduced nitrogen for the early stages of seedling growth." Storage proteins do not exhibit any known enzymatic activity (Shewry and Miflin 1985), with the possible exception of cucurbitin in pumpkin seeds (Hara et al. 1976, cited by Casey et al. 1986). Storage proteins are accumulated in a tissue-specific manner (in the endosperm in cereal and in cotyledons in legumes) and they are sequestered in specialized inclusions called protein bodies (Miège 1982). The major seed storage protein of a given species has been given a vernacular name that often but not always reflects the taxonomic status of the species: e.g., phaseolin (*Phaseolus* sp.) and hordein (*Hordeum* sp.).

Since the seminal studies of Osborne (1907), the information on seed storage proteins has expanded considerably, and for two major reasons. First, seed proteins are of considerable economic importance as a component of human food and animal feed. According to FAO data cited by Swaminathan (1983), plants provide on the average two-thirds of the per capita daily protein supply for humans and some 45% of cereal grains in the world are used for livestock feeding. More recently, bread-making quality has been correlated with certain seed protein components of common wheat (Payne 1987). Conversely, seed proteins also include antinutritional components, mainly phytohemagglutinins (lectins) and enzyme inhibitors, such as trypsin and chymotrypsin inhibitors (e.g., Sgarbieri and Whitaker 1982).

A second reason for the sustained interest in seed proteins is that they provide an excellent biological model for the study of the molecular and cellular biology of gene expression. The expression of seed protein genes is highly regulated both temporally (a specific stage of seed development) and spatially (a specific seed tissue, such as the endosperm in Poaceae and cotyledons in Fabaceae). In addition, seed proteins undergo a series of modifications before their deposition in protein bodies that include removal of a signal peptide, glycosylation, posttranslational cleavages, etc. (see below). Because of the abundance of the protein product (and its mRNA), seed protein genes were among the first cloned plant genes (e.g., Sun et al. 1981; Goldberg et al. 1981).

Information on the molecular biology of seed protein genes can be used to improve their nutritional qualities (Larkins 1983; Slightom and Chee 1987). The nucleotide sequence of genes encoding seed proteins can be modified so that they include a higher level of limiting essential amino acids. This requires a knowledge of the nonconserved regions in the gene sequence. These regions are presumably not subject to functional constraints related to deposition during maturation and mobilization during germination. Alternatively, heterologous seed protein genes can be transferred that encode polypeptides with a higher amount of the limiting amino acids. A third possibility is to identify native proteins that have an adequate amino acid balance but occur in low amounts. Modifications in their regulatory sequences can then be made that would enhance their expression.

The strategies for the nutritional improvement of seed protein outlined above require, besides plant transformation systems, information on the nucleotide and amino acid sequence of seed protein genes (and particularly the location of nonconserved regions), the developmental and tissue-specific regulatory controls, the number of seed protein genes and their distribution in the genome, and the process of protein deposition in protein bodies (including co- and posttranslational modifications). It is this type of information that also allows us to interpret data from surveys on seed storage protein variability and to assess their value in answering questions pertaining to population genetics, germplasm conservation, and crop improvement. Although molecular features of seed storage proteins will be mentioned when necessary, it is beyond the scope of this chapter to detail the extensive studies on the molecular and cellular biology of seed storage protein synthesis and deposition. The reader is referred to a number of excellent volumes or reviews on the topic: Gottschalk and Müller (1983), Shewry and Miflin (1985), Casey et al. (1986), and Slightom and Chee (1987).

ANALYTICAL TOOLS USED TO CHARACTERIZE SEED STORAGE PROTEIN DIVERSITY

Many analytical methods have been used to characterize protein in general and seed storage proteins in particular. These include centrifugation, affinity chromatography, high-performance liquid chromatography (HPLC), amino acid analysis, and various electrophoretic techniques. Electrophoretic separation has contributed more than any other technique to our knowledge of seed storage protein variability, partly because of its rapidity, relatively low cost, and capacity to handle a large number of samples compared to other techniques. Stegemann and Pietsch (1983) outline the major electrophoretic methods, including one-dimensional starch gel electrophoresis, polyacrylamide gel electrophoresis (PAGE; with or without sodium dodecyl sulfate: SDS), isoelectric focusing (IEF), and two-dimensional polyacrylamide gel electrophoresis. Polyacrylamide gel electrophoresis in an SDS-containing medium (SDS–PAGE) offers the advantage that all proteins are solubilized, negatively charged, and randomly coiled. The denatured protein is then separated by size. Separation by SDS-PAGE has, therefore, become one of the more widely used electrophoretic separation techniques used in seed storage protein analysis.

Because the number of samples analyzed by one-dimensional methods can become quite large, attempts have been made at computerizing the data processing. A recent example of such an attempt (Mansur Vergara et al. 1984) involved interfacing a densitometer to a microcomputer, scanning the gels at 570 mm, and digitizing this information before transfer to the microcomputer. Software de-signed for this application read and stored the data, assigned a molecular weight and relative intensity value to each of the bands (independent of the protein amount loaded onto the gel), and calculated frequencies for each band.

Compared to DNA sequence polymorphisms, seed protein diversity will not be as informative. Analyses of DNA sequence polymorphism by, for example, restriction analysis and Southern hybridization can potentially reveal sequence modifications such as silent nucleotide substitutions or insertions/deletions in coding or flanking regions, or sequence divergence in nonexpressed DNA such as repeated sequences. Analyses of DNA sequence polymorphisms by Southern hybridization are, however, much more cumbersome, time-consuming, and expensive than electrophoresis of seed storage proteins or isozymes.

CHARACTERISTICS OF SEED STORAGE PROTEINS AS GENETIC MARKERS

In order to understand the information provided by seed storage protein diversity, it is necessary to appreciate their properties as markers. The following aspects will be dealt with: intraspecific seed storage protein polymorphism, environmental influence, genetic control, molecular complexity, and relationships among seed storage proteins from different taxa.

Intraspecific polymorphism for seed storage proteins

Polymorphisms for seed storage proteins have been identified in several species, all or most of which are cultivated species or wild species related to cultivated species: *Phaseolus* beans (Brown et al. 1981b; Gepts et al. 1986; Schinkel and Gepts 1988), field beans (*Vicia faba*; Utsumi et al. 1980), wheat (*Triticum* sp.; Autran and Bourdet 1975; Vallega and Waines 1987), barley (*Hordeum* sp.; Shewry et al. 1978; Nevo et al. 1983), soybean (*Glycine max*; Mori et al. 1981), pea (*Pisum sativum*; Casey et al. 1986), maize (*Zea mays*; Wilson 1985), and peanut (*Arachis hypogea*; Tombs and Low 1967). Table 1

TABLE 1 Examples of polymorphisms for seed storage proteins.

Species	Protein ("locus")[a]	Sample size examined	Number of variants	Sources[b]
Phaseolus vulgaris cultivars	Phaseolin (*Phs*)	>1000	6	1
	Lectin (*Lec*)	>100	12	2
Phaseolus vulgaris wild	Phaseolin (*Phs*)	>100	23	3
	Arcelin (*Arl*)	>100	4	
Hordeum vulgare	Hordein C (*Hor-1*)	163	8	5
	Hordein B (*Hor-2*)	163	17	
Hordeum spontaneum	Hordein C (*Hor-1*)	3	20	6
	Hordein B (*Hor-2*)	3	17	
Triticum aestivum	HMW glutenins *Glu-A1*	300	3	7
	HMW glutenins *Glu-B1*	300	11	
	HMW glutenins *Glu-D1*	300	6	
Triticum turgidum	HMW glutenins *Glu-A1*	11	11	8
	HMW glutenins *Glu-B1*	11	15	

[a]Locus usually refers to a complex locus or gene cluster consisting of several linked and homologous genes.
[b]1: Brown et al. (1981b), Lioi and Bollini (1984), Gepts et al. (1986), Gepts and Bliss (1986), unpublished results; 2: Brown et al. (1981c); 3: Gepts et al. (1986), Gepts and Bliss (1986); 4: Osborn et al. (1986); 5: Shewry et al. (1977, 1978); 6: Doll and Brown (1979); 7: Payne and Lawrence (1983); 8: Nevo and Payne (1987).

illustrates some specific examples of electrophoretic variation for seed storage proteins.

A comparison of seed protein and isozyme polymorphism in *Phaseolus vulgaris*, *Triticum turgidum* var. *dicoccoides* (Table 2), and *Hordeum spontaneum* reveals that the high levels of seed storage protein diversity contrast with the relatively lower level of isozyme diversity. For example, Doll and Brown (1979) estimated, using the sampling theory of neutral alleles, that hordeins are 10–30 times more variable than isozymes. There are several reasons, however, why the levels of diversity of isozymes and seed storage proteins are not strictly comparable. First, the analytical methods used in these studies are not identical and will detect different causes of polymorphism. Second, seed storage proteins are encoded by small multigene families (see below) that provide additional opportunities for polymorphism. A third possibility is a lower level of functional constraints acting on seed storage proteins compared to isozymes.

Table 2 Comparison of intrapopulation diversity for HMW glutenins and isozymes in 11 populations of *Triticum turgidum* var. *dicoccoides*.

	HMW glutenin[a]	Isozymes[b]
Mean no. of alleles/locus	2.50 (1.0–5.5)	1.33 (1.2–1.46)
Proportion of polymorphic loci per population	0.77 (0.0–1.0)	0.25 (0.16–0.38)
Proportion of heterozygous loci per population	0.0006 (0.0–0.038)	0.070 (0.03–0.12)
Genic diversity (He)	0.36 (0.00–0.70)	0.097 (0.06–0.17)

[a]Nevo and Payne (1987).
[b]Nevo et al. (1982).

Environmental influence on seed storage protein diversity

The environment is known to influence quantitative parameters of seed storage proteins such as total seed protein content. However, qualitative aspects such as the protein banding pattern after electrophoresis are much less subject to environmental influences. A study by Smith and Smith (1986) quantified the effects of the maize genotype (inbred), the source of seeds, and the protein extraction on zein chromatograms. They found that genotypic differences accounted for 60–90% of the variation, the source of seeds (i.e., the environment in which they were grown) for 10–15%, and the extraction procedure for 2–5%. This finding confirms earlier evidence that prolamine electrophoretic profiles exhibit little or no environmental effect and can be extended to other seed storage proteins such as albumins (Adriaanse et al. 1969) and phaseolin (Gepts, unpublished results).

Although in general environmental factors do not influence electrophoretic patterns, sulfur deficiency does affect various storage protein fractions or polypeptides differentially in legume seeds (e.g., Gaylor and Sykes 1985). Despite the relative insignificance of environmental effects, they cannot be ignored. It is desirable, at the outset of a study using seed storage proteins as genetic markers, to determine the extent of environmental influence on the electrophoretogram or chromatogram of the seed proteins.

Genetic analyses of electrophoretic patterns of seed storage proteins

Genetic analyses have shown that the electrophoretic patterns of seed storage proteins are generally inherited codominantly and in a simple manner, that involves a limited number of genes (Table 3).

All of the data presented in Table 3 were obtained by analyses of segregating generations with protein gel electrophoresis, with the exception of the data for soybean legumin, which were derived from a restriction fragment length polymorphism (RFLP) analysis. It remains to be determined whether the two methods of analysis are equivalent. RFLP analysis may, in the end, provide more information if gene-specific probes can be identified, especially in the case of zein whose numerous structural genes appear to be scattered over chromosomes 4 and 7, and, to a lesser extent, on chromosome 10, in *Zea mays*. An additional benefit of RFLP analyses would be to determine that the segregations of the protein banding patterns represent the segregation of the structural loci for the storage protein or the segregation of protein-modifying enzymes such as glycosylating, deglycosylating, or proteolytic enzymes.

Genetic analyses have also established linkage relationships between several loci[1] controlling seed storage protein electrophoretic patterns or between these loci and loci controlling morphological traits, disease resistance, etc. In *Phaseolus vulgaris*, the *Lec* and *Arl* locus are tightly linked ($r < 0.003$; Osborn et al. 1986). In barley, the *Hor-1*, *Hor-2*, and *Hor-3* loci are located on chromosome 5 of barley. The *Hor-1* and *Hor-2* loci are about 10 cM apart and bracket the *Ml-a* locus, a locus controlling resistance to powdery mildew (*Erysiphe graminis*) (see below). In soybean, the *Cgy2* and *Cgy3* loci are linked to each other but are unlinked to *Cgy1* (Davies et al. 1985). In pea, the *r* locus determining the wrinkledness versus nonwrinkledness of the seed is linked to both the *Lg-1α* locus (legumins) and *Vc-1* locus (vicilin) on chromosome 7 (Matta and Gatehouse 1982; Mahmoud and Gatehouse 1984); also, two legumin loci are linked to the *a* locus, controlling pigment synthesis on chromosome 1 (Matta and Gatehouse 1982). The zein genes for the "M_r 19,000" zein polypeptides and one coding for a "M_r 15,000" zein polypeptide are present in a region of 30 crossover units on the short arm of chromosome 7 (Soave et al. 1981; Soave and Salamini 1982) and nine genes for the "M_r 22,000" zein polypeptides and three coding for the "M_r 19,000" zein polypeptides are located on chromosome 4 (Soave et al. 1982). The genes for the "M_r 22,000" polypeptide are scattered on both arms of chromosome 4, whereas those for the "M_r 19,000" are linked in a region encompassing five crossover units onto the short arm.

Molecular basis for genetic variation in seed storage proteins

The molecular analysis of seed storage proteins has provided insights into the sources of the widespread and abundant polymorphisms of seed storage proteins. In this section these causes will be outlined (Table 4) and illustrated

[1]The term locus is used here even though molecular genetic data show that seed storage protein "loci" consist of multigene families and should therefore be called more appropriately complex loci or gene clusters.

TABLE 3 Genetic control of electrophoretic patterns of seed storage proteins.

Species (chromosome number)	Protein	Loci[a] Number	Name	Source[b]
Phaseolus vulgaris ($2n=2x=22$)	Phaseolin	1	*Phs*	1
	Lectin	1	Lec	1
	Arcelin	1	*Arl*	2
Hordeum vulgare, ($2n=2x=14$)	B-Hordein	1	*Hor-2*	3
	C-Hordein	1	*Hor-1*	
	D-Hordein	1	*Hor-3*	
Triticum aestivum ($2n=6x=42$)	γ-, β-, ω-Gliadins	3	*Gli-A1*, *Gli-B1*, and *Gli-D1* (short arm of chromosomes 1A, 1B, and 1D)	4
	γ-, β-, α-Gliadins	3	*Gli-A2*, *Gli-B2*, and *Gli-D2* (short arm of chromosomes 6A, 6B, and 6D)	4
	HMW glutenins	3	*Glu-A1*, *Glu-B1*, and *Glu-D1* (long arm of chromosomes 1A, 1B, and 1D)	5
Triticum tauschii (*Aegilops squarrosa*) ($2n=2x=14$)	Gliadins	2	*Gli-D1* (chrom. 1D) *Gli-D2* (chrom. 6D) *Glu-D1* (chrom. 1D)	6
	HMW glutelins	1		
Zea mays ($2n=2x?=20$)	Zein	?	Chromosomes 4, 7 and 10	7
Glycine max[c] ($2n=2x?=40$)	Glycinin	4	*Gy1*, *Gy3-Gy5*	8
	β-Conglycinin	3	*Cgy1* (α′ sub-unit), *Cgy2* (α), *Cgy3* (β)	9
Pisum sativum ($2n=2x=14$)	Legumin	>4	*Lg-1*α, *Lg-2*α, *Lg-3*α, *Lg-3*β	10
	Vicilin	1	*Vc-1*	11
	Covicilin	1		12

[a]The word locus refers usually to a complex locus of gene cluster consisting of several linked and homologous genes.
[b]1: Brown et al. (1981a); 2: Osborn et al. (1986); 3: Oram et al. (1975), Shewry et al. (1978, 1980), Doll and Brown (1979); 4: Wrigley and Shepherd (1973), Kasarda et al. (1976), Payne et al. (1982); 5: Lawrence and Shepherd (1981a, b), Payne et al. (1980); 6: Lagudah and Halloran (1988a, b); 7: Viotti et al. (1980, 1982), Soave et al. (1981, 1982), Valentini et al. (1979); 8: Davies et al. (1987); 9: Davies et al. (1985); 10: Matta and Gatehouse (1982); 11: Mahmoud and Gatehouse (1984).
[c]Determined by restriction fragment lengths polymorphism analysis instead of gel electrophoresis.

with two examples—phaseolin of *Phaseolus vulgaris* and zein of *Zea mays*. The reader is referred to exhaustive reviews for specific details and literature references (cited in Tables 3 and 4).

Phaseolin polypeptides of the type identified as "T" phaseolin have similar molecular weights, isoelectric points, amino acid compositions, and proteolytic and chemical cleavage profiles. These similarities suggest that they are encoded by a small and conserved multigene family. These genes are tightly linked as a gene cluster at a complex locus (*Phs*).

Nucleotide sequencing of phaseolin cDNA has shown that the phaseolin polypeptides are encoded by two gene families (α and β), which differ in their coding regions by the presence or absence of two direct repeats of different size (15 and 27 bp) and 16 nucleotide substitutions; overall there is 98% homology between the α- and β-type genes.

After translation, phaseolin polypeptides are subjected to a series of processing steps including the cleavage of a signal peptide, and cotranslational glycosylations at one or two canonical *N*-glycosyl sites. Transport of these glycosylated polypeptides through the Golgi complex and their ultimate deposition in the protein bodies are accompanied by further processing steps. It is the glycosylating reactions that are largely responsible for the molecular weight heterogeneity characteristic of the various phaseolin banding patterns.

Zein polypeptides separated by SDS–PAGE have M_r ranging from 19,000 to 26,000, with two major bands called Z19 and Z22. Cloned zein cDNA sequences can be divided into three to six subfamilies on the basis of their cross-hybridization behavior and their restriction endonuclease maps. Each subfamily encodes predominantly a single size class of polypeptides (Z19 or Z22); the degree of homology of nucleotide sequences is 60–90% between subfamilies and over 90% within them. *In situ* hybridization of cloned zein cDNA sequences to chromosome preparations indicates that subfamilies may be associated with specific structural loci. The organization of zein genes into four structural areas—the signal peptide, the N-terminal sequence, the central region, and the C-terminal sequence—is similar to the organization of

Table 4 Molecular causes of seed storage protein diversity.[a]

Multigene families resulting from duplication and divergence of ancestral genes:

Nucleotide substitutions, insertions-deletions, variable number of intragenic repeats

Modifications to the protein products during and after translation:

Cotranslational removal of transit peptide, co- and posttranslational glycosylations, posttranslational proteolytic cleavage

[a]Casey et al. (1986), Shewry and Miflin (1985).

prolamine genes that has been observed in other cereals such as barley (hordeins B and C) and wheat (gliadins and HMW glutenins). The repeated units in the different groups do not appear to be related, however, suggesting independent origins of these multiple repeat sequences.

The phaseolin and zein examples illustrate the wide variety of causes that can lead to the observed seed storage protein heterogeneity. This high level of heterogeneity is somewhat paradoxical in light of the numerous constraints that presumably operate on seed storage proteins. These constraints include subunit interactions for proper holoprotein assembly, packaging of holoproteins into protein bodies, membrane transit during synthesis, general stability and solubility constraints during dissociation, and proteolytic cleavage during germination (Doyle et al. 1986). The data seem to indicate that while these constraints may lead to a certain degree of sequence conservation in particular regions of the polypeptides, sequence divergence can occur in other regions of the storage polypeptides. In at least two *Phaseolus* species, the major seed storage proteins are not essential constituents of the seed. A phaseolin-deficient mutant was identified in *P. coccineus* (Gepts and Bliss 1984); this trait was transferred to *P. vulgaris* where it was combined to the lectin-deficient trait (Osborn and Bliss 1985). Together, the two proteins would account for 60% of total seed nitrogen. The genetic removal of phaseolin, or phaseolin and lectin, does not affect total seed protein content and germination appreciably (Gepts and Bliss 1984; Osborn and Bliss 1985; F. A. Bliss, personal communication).

The high degree of sequence homology (85–95%) within gene families of a number of different genera such as *Phaseolus*, *Pisum*, *Glycine*, and *Zea* (Messing et al. 1983; Doyle et al. 1986) suggests either the presence of strong species-specific constraints or the homogenization of sequences through concerted evolution (Slightom et al. 1980). The organization of seed storage protein genes in complex loci composed of repeated copies of the genes together with the presence of internal repeated sequences such as in the cereal prolamin genes (Miflin et al. 1984) and to a lesser extent in pea legumin genes (Lycett et al. 1984) and bean phaseolin genes (Slightom et al. 1985) would increase the probability of unequal crossing over. The latter phenomenon can in turn lead to expansion or contraction of gene families and, with or without gene conversion, lead to homogenization of sequences within a genotype.

Finally, the complexity of the biosynthesis and deposition of seed storage proteins has an important consequence when seed storage protein data are used as genetic markers (Gepts 1988a). Because the banding or spotting patterns observed after electrophoresis result from a complex series of steps at the molecular level, it is unlikely that any of them would have appeared more than once. Each of these patterns should therefore have a single origin and genotypes exhibiting the same pattern should have a common evolutionary origin. This argument resembles that for haplotypes of restriction endonuclease recognition sites at the DNA sequence level. The probability that a complex haplotype is unique is very high. One can therefore use specific haplotypes to identify the geographic origin of a particular trait. Because seed storage proteins are encoded by a limited number of complex loci, similar electrophoretic patterns will, strictly speaking, reflect only the common ancestry of the complex loci and the nearby chromosome region. This limitation is mitigated in self-pollinating species, which tend to display extensive multilocus associations extending beyond the linkage groups (Allard 1975).

Homologies among seed storage proteins belonging to different taxa

The molecular biology and chemistry of seed storage proteins have also revealed homologies among protein components belonging to widely different taxa (e.g., Borroto and Dure 1987). Prolamines are found in the endosperm of most Poaceae (Table 5) while globulins have been identified in almost all higher plants including gymnosperms (Blagrove et al. 1984) (Table 6), suggesting that the origin of globulins predates the divergence between gymnosperms and angiosperms. Furthermore, globulins generally consist of two major classes, the "11 S" (or "legumin-like") and "7 S" (or "vicilin-like") proteins, both of which are present in the Fabaceae and Poaceae (Danielsson 1949). Homology among various seed storage proteins within each class (11 S or 7 S) has been deduced from a number of arguments, including N-terminal and C-terminal amino acid sequences, cDNA nucleotide sequences, peptide mapping, structural composition, and immunological reactions.

Based on nucleotide and amino acid sequence data and hydropathy analyses, Borroto and Dure (1987) found that all of the sequenced globulin storage proteins of angiosperms available had vestigial sequence homology with either the 11 S or the 7 S globulins of legumes. Thus they emanated from two genes that existed at the beginning of angiosperm evolution. Kreis et al. (1985) reported that regions in the nonrepeated domains of the S-rich prolamines (e.g., B-hordein; α-, β-, γ-gliadins; γ-secalins) and the N- and C-termini of HMW glutenins are homologous with each other, with trypsin and α-amylase

Table 5 Homologies among prolamin seed storage proteins of barley, rye, and wheat.[a,b]

Species	HMW prolamins	S-poor prolamins	S-rich prolamins
Hordeum vulgare	D-Hordein	C-Hordein	B-Hordein
Secale cereale	HMW secalin	ω-Secalin	γ-Secalin (40K, 75K)
Triticum aestivum	HMW subunits	ω-Gliadins	α-, β-, γ-Gliadins

[a]Modified from Shewry and Miflin (1985).
[b]Proteins in each column exhibit homology.

Table 6 Homologies between globulin seed storage protein fractions.[a,b]

Family	Species	11 S globulin, legumin-like	7 S globulin, vicilin-like
Fabaceae	*Pisum sativum*	Legumin	Vicilin
	Vicia faba	Legumin	Vicilin
	Glycine max	Glycinin	β-Conglycinin
	Phaseolus vulgaris		Phaseolin
	Lupinus angustifolius	α-Conglutin	β-Conglutin
Brassicaceae	*Brassica napus*	Cruciferin	
	Arabidopsis thaliana	Globulin	
	Raphanus sativus	Cruciferin	
Asteraceae	*Helianthus annuus*	Helianthinin	
Malvaceae	*Gossypium hirsutum*	β-Globulin	α-Globulin
Ranunculaceae	*Nigella damascena*	Legumin-like	Vicilin-like
Magnoliaceae	*Magnolia grandiflora*	Legumin-like	Vicilin-like
Betulaceae	*Corylus avellana*	Legumin-like	Vicilin-like
Oleaceae	*Fraxinus excelsior*	Legumin-like	Vicilin-like
Scrophulariaceae	*Digitalis purpurea*	Legumin-like	Vicilin-like
Poaceae	*Avena sativa*	Globulin	
	Oryza sativa	Glutelin	

[a]Modified from Borroto and Dure (1987). Additional sources: Jensen (1984), Gayler et al. (1984), Johnson et al. (1985), Laroche-Raynal and Delseny (1986), Schwenke et al. (1979).
[b]Proteins included within each column are homologous.

inhibitors from cereals, and with 2 S storage proteins from oil seed rape (*Brassica napus* L.) and castor bean (*Ricinus communis* L). The authors suggested that the structural genes from all these proteins evolved from a single ancestral gene. According to Kreis et al. (1985), this gene may have been a protein inhibitor with a single domain structure. Gene duplication gave two identical domains that diverged to give the double-headed Bowman-Birk inhibitors. Further duplication of one domain within this gene or a separate duplication of an ancestral gene gave rise to a gene encoding three homologous domains. This gene then diverged to give rise to the genes for the other proteins including the cereal prolamines. The major evolutionary event for the prolamines was the insertion of additional sequences giving rise to the large repetitive domain. Because prolamines are found only in grasses that are thought to have appeared about 100 million years ago, these inserts must be relatively recent. The addition of the repeats, which are rich in proline and glutamine, appear to have resulted in the unusual physical properties of the proteins, among which is their insolubility in water.

Is the genetic variation for seed storage protein adaptive?

The question of the adaptiveness of the polymorphisms for seed storage proteins has been raised by a number of authors. For example, Nevo et al. (1983) and Levy and Feldman (1988) have claimed that some of the variation they observed for seed proteins (as well as isozymes and total protein content) in *Hordeum* sp. and *Triticum* sp. conferred adaptedness to specific environments. Their statistical analyses indicated correlations between certain seed storage protein phenotypes and environmental attributes such as soil type (Nevo et al. 1982). Multiple regression analyses showed correlations between glutenin diversity and an a posteriori combination of environmental parameters. These results, however, provide no conclusive proof of the adaptiveness of seed storage protein polymorphisms. Such correlations can be attributed to a variety of causes, in addition to a possible adaptive role. They can appear as a consequence of stochastic processes such as migration or genetic drift or linkage between the seed storage protein loci and actual loci confirming adaptedness (hitchhiking effect). An additional mechanism, especially in self-fertilizing species such as *Hordeum* sp. and *Triticum* sp., is the presence of extensive multilocus associations beyond the physical linkage groups. These correlations, however, provide working hypotheses that can be verified experimentally, for example by synthesizing artificial populations containing alternate alleles, establishing these populations in sites with contrasting environments, and monitoring gene frequencies over a number of years.

To summarize this section, seed storage proteins present the following characteristic as genetic markers: (1) they are highly polymorphic, even within individual populations in certain cases; (2) their polymorphism is in large part genetically determined; (3) the genetic control of qualitative genetic variation is simple and involves a limited number of loci of the nuclear genome; (4) the molecular sources of seed storage protein polymorphisms are known; they show that most protein variants are unique and can therefore be used as evolutionary markers; and (5) homologies have been established between seed storage proteins of different taxa. With these characteristics in mind, the different uses of seed storage proteins as markers will be reviewed in the next section.

QUESTIONS ADDRESSED WITH SEED STORAGE PROTEINS AS GENETIC MARKERS

Seed storage proteins have been used as genetic markers in four main areas: (1) analyses of genetic diversity within and among populations, (2) plant domestication in relation to genetic resources conservation and breeding, (3) genome relationships, especially in polyploid series, and (4) as a tool in plant breeding.

Organization of genetic diversity within and among populations

In *Hordeum spontaneum*, B- and C-hordeins are polymorphic within populations (Doll and Brown 1979; Nevo et al. 1983) (see Table 1). Most of the hordein variants identified by Doll and Brown (1979) occurred in a single locality, suggesting strong population differentiation. Genetic diversity for HMW glutenins in wild emmer (*Triticum turgidum* var. *dicoccoides*) was distributed equally within and between populations (Nevo and Payne 1987). Most glutenin alleles occurred locally or sporadically. The selfing nature of *Hordeum spontaneum* and *Triticum turgidum* var. *dicoccoides* accounts at least in part for these strong population differences.

An additional aspect of this protein diversity is the presence of multilocus associations. Doll and Brown (1979) observed an association between the two hordein loci (*Hor-1* and *Hor-2*) due either to epistatic selection or to linkage, self-fertilization, and the historical effects of population bottlenecks. In the cultivated CC XXI population, these multilocus associations appeared to have been attenuated.

In common bean, multilocus associations occur between the unlinked loci for phaseolin and lectin. The "T" lectin type was associated almost exclusively with "T" and "C" phaseolin and the other lectin types with "S" phaseolin (Brown et al. 1982). Strong correlations (r = 0.7–0.8) occur among certain isozyme loci and between certain isozyme loci and the phaseolin locus both in wild and cultivated populations (R. Koenig and P. Gepts unpublished results; S. P. Singh, R. Nodari, and P. Gepts unpublished results). The simplest explanation for these multilocus associations is a lack of recombination due to geographic isolation (Mesoamerica versus Andes) and strong self-fertilization.

Study of plant domestication in relation to germplasm conservation and breeding

Identification of the wild ancestor of cultigens The wild ancestor and the precise area of domestication of a crop is of interest because it may indicate which wild populations did—and perhaps more importantly, which populations did not—contribute their genetic diversity to the cultivated gene pools. For some crops, the identification of the wild ancestral form represents perhaps the single most important question addressed with seed storage proteins (Table 7). The wild ancestor is usually classified in a different species or genus, because of the strong morphological differences separating wild ancestor and cultivated descendant. However, hybrids between the two are usually viable and fertile and they belong therefore to the same biological species.

Phaseolin electrophoretic patterns in cultivated *Phaseolus vulgaris* and its wild conspecific ancestor reveal that this species has an unusual domestica-

Table 7 Crops whose wild ancestor has been identified wholly or in part by seed protein markers.

Crop	Wild ancestor	Source[a]
Lentil (*Lens culinaris*)	*Lens nigricans*	1
Chickpea (*Cicer arietinum*)	*Cicer reticulatum*	2
Pigeon pea (*Cajanus cajan*)	*Atylosia* sp.	3
Maize (*Zea mays* subsp. *mays*)	*Zea mays* subsp. *mexicanum*	4
Watermelon (*Citrullus lanatus*)	*Citrullus lanatus* var. *citroides*	5
Common bean (*Phaseolus vulgaris*)	*Phaseolus vulgaris* var. *mexicanus* and var. *aborigineus*	6
Tepary bean (*Phaseolus acutifolius*) var. *acutifolius*	*P. acutifolius* var. *acutifolius*	7
Peanut (*Arachis hypogea*)	*Arachis monticola*	8

[a]1: Ladizinsky (1979); 2: Ladizinsky and Adler (1975); 3: Ladizinsky and Hamel (1980); 4: Paulis and Wall (1977); 5: Navot and Zamir (1987); 6: Gepts et al. (1986); 7: Schinkel and Gepts (1988); 8: Cherry (1975).

tion pattern. The cultivars result from multiple domestications along the large distribution range of the wild ancestor, which extends from northern Mexico to northwestern Argentina. There are two major domestications—one in Mesoamerica leading to "S" phaseolin cultivars and the other in the southern Andes, leading principally to "T" and "C" phaseolin cultivars (Gepts et al. 1986). A minor domestication in Colombia or Central America gave rise to "B" phaseolin cultivars that are related to the Mesoamerican "S" phaseolin cultivars (Gepts and Bliss 1986). The cultivated gene pool of the common bean consists of two entities—Mesoamerican versus Andean—which appear to exhibit contrasting agronomic traits such as seed type, growth habit, disease resistance genes, and adaptation (reviewed in Gepts 1988b).

Is it possible to define the precise domestication areas of the common bean? Based on the complexity principle presented in the previous section, Mesoamerican common bean cultivars may have been domesticated in a specific region in Mexico. More than 15 different phaseolin patterns occur in wild common beans in Mexico (Gepts et al. 1986). Of these, only one—the "S" pattern—is found among cultivars. Since the series of molecular events leading to each pattern is complex, we infer that each pattern appeared only once. Of about 100 wild common bean lines from Mexico that have been evaluated, only five exhibited the "S" phaseolin pattern (without any morphological signs of hybridization with cultivars). These five accessions all originated in an area of west-central Mexico including the current states of Jalisco and Guanajuato (Gepts 1988a). Thus the phaseolin patterns provide a working hypothesis for analyses with additional genetic markers such as isozymes and RFLPs or for future archaeological investigations.

Effect of domestication on genetic diversity In barley and common bean, it has been possible to examine the fate of genetic diversity upon domestication based on data for the diversity of seed storage proteins. Doll and Brown (1979) observed high levels of polymorphisms for the B- and C-hordeins in wild and cultivated barley populations. Cultivars, represented by composite cross XXI, which resulted from pollination of male-sterile "Atlas" by some 6000 fertile cultivars, were less variable than three wild barley populations from Israel.

In common bean, the overall diversity of phaseolin is lower than that for hordeins. Nevertheless, the loss of genetic diversity in phaseolin on domestication is readily apparent in Mexico. As mentioned earlier, of the 15 to 20 phaseolin patterns observed in Mexican wild bean populations, only one pattern is found among Mexican cultivars (Gepts et al. 1986). In the tepary bean (*Phaseolus acutifolius* A. Gray), which is distributed in Mexico and the southwestern United States, a similarly strong reduction in phaseolin diversity occurred during domestication (Schinkel and Gepts 1988).

These reductions of seed protein diversity on domestication raise several issues. A first issue is the representivity of the seed protein results. Do other traits in the same or other crops also indicate this type reduction in diversity? Similar data for isozymes and rDNA (Nevo et al. 1979; Allard 1988) indicate a reduction in diversity upon domestication of barley. In common bean, wild populations did not exhibit high levels of isozyme diversity, hence little reduction upon domestication was evident (R. Koenig, S. P. Singh, and P. Gepts, unpublished results; S. P. Singh, R. Nodari, and P. Gepts, unpublished results). However, in the tepary bean, isozymes showed a reduction in diversity (C. Schinkel and P. Gepts, unpublished results). Several additional examples of reduction in genetic diversity in rice, tomato, and watermelon are provided by Ladizinsky (1985).

In several crops, biochemical markers (such as isozymes and seed proteins) show little correlation with morphological traits. In particular, morphological diversity seems to increase on domestication: e.g., spike morphology in barley (Nevo et al. 1979) and seed type (color, size, shape) and growth habit in common bean (Gepts and Debouck, In press). Morphological traits are under direct human selection whereas biochemical markers presumably are not. Morphological traits distinguishing wild and cultivated types are usually highly heritable and controlled by a limited number of genes with a major phenotypic effect; these traits are also highly heritable (e.g. common bean, Gepts and Debouck, in press). These characteristics probably reflect the domestication process. A farmer would more easily notice and select a trait with a marked phenotypic effect; in the progeny he or she would more likely recover this trait if it had a high heritability. Yet, such genes with a strong phenotypic effect and high heritability are hardly representative of all the genes included in a plant genome. Therefore, the current results indicating a reduction in diversity on domestication need testing with additional markers, such as RFLPs.

A second issue is the cause of the reduction in genetic diversity. The reduction could be the consequence of a stochastic process such as genetic drift. A limited number of domestications could lead to such a reduction. The alleles found in the cultivated gene pool were included in this gene pool merely because they occurred in the source wild populations in which the last hunters–gatherers (or the earliest farmers) started the process of domestication. Alternatively, this reduction is a consequence of selection either on the seed proteins (or isozymes) or on products of genes linked to those of seed proteins (or isozymes). It is intriguing that the Mesoamerican cultivars of common and tepary bean each displays a single phaseolin pattern despite being sympatric and cross-compatible with their wild ancestors. Thus, information on the linkage relationships of the diversity markers with genes for traits affecting adaptedness to certain environments is needed.

Genome relationships of different species and identity of genome contributors to amphiploids

Seed storage proteins can be used to determine relationships between genomes belonging to different species. One example is provided by a comparison of the chromosomal location of the *Gli-1* and *Glu-1* loci in *Triticum aestivum* (chromosome 1A, 1B, and 1D) with that of the *Hor-1*, *Hor-2*, and *Hor-3* in *Hordeum vulgare*, which code for related prolamine (see Tables 2 and 4) and are located in a similar position on chromosomes 1A, 1B, and 1D and chromosome 5, respectively. The similar position for two series of homologous proteins confirms the common ancestry of the wheat and barley genomes, both species belonging to Triticeae tribe (Kreis et al. 1985).

Electrophoretic profiles of seed storage proteins have identified the donor genomes of *Aegilops cylindrica* (Johnson 1967), *Avena* sp. (Murray et al. 1970), and *Solanum* sp. (Edmonds and Glidewell 1977).

Seed storage protein markers as breeding tools

Linkage to genes for agronomic traits Jensen et al. (1980) determined that the *Hor-1* and *Hor-2* loci bracketed the *Ml-a* locus, coding for powdery mildew resistance. Because the recombination fraction on either side of the *Ml-a* locus is low ($r < 0.05$), hordein phenotypes could be used to backcross new resistance genes into cultivars without the need for time-consuming testing of reactions to mildew. Yunushkanov and Ibragimov (1984) established a linkage between a gene for a seed protein distinguishing *Gossypium hirsutum* and *Gossypium barbadense*, and a gene determining fiber development on cotton seeds.

Technological qualities of seed transformation products Specific alleles at the *Glu-1* locus are associated with bread-making quality (Burnouf and Bourriquet 1980). This association holds the promise of accelerating the breeding of cultivars with improved dough strength. A simple screening during segregating generations could replace cumbersome quality tests. This is true not only for the production of bread, but also of cookies, pastas, etc.

Nutritional qualities of seed storage proteins Regulatory genes that modify the relative proportions of the various protein fractions of the seed can affect the amino acid balance (Mertz et al. 1964; Gepts and Bliss 1984). Progress in molecular genetics and gene transfer techniques may enable the transfer and expression of native or modified genes in homologous or heterologous species (Larkins 1983; Slightom and Chee 1987).

Insect resistance Certain wild common beans from Mexico possess a novel seed protein called arcelin, which confers resistance to a seed weevil (*Zabrotes subfasciatus*), a major seed storage pest. The locus encoding this protein is closely linked to the lectin locus ($r < 0.003$) (Osborn et al. 1986). Arcelin shows 80% sequence homology with lectins and reacts immunologically with them. The two proteins have similar subunit molecular weights (Osborn 1988; Osborn et al. 1988). The similarities between arcelin and lectins and their close linkage suggest that they are evolutionarily related, possibly through duplication and divergence. This gene is now being introduced into improved cultivars (Bliss, this volume).

Varietal identification Because of their high levels of polymorphism and their environmental stability, seed storage proteins can identify cultivars either alone or in conjunction with other markers. Varietal identification through seed proteins is possible in the following crops: wheat (Autran and Bourdet 1975), faba beans (Barratt 1980), barley (Shewry et al. 1977), and maize (Wilson 1985).

CONCLUSIONS

Seed storage proteins offer five advantages as genetic markers: (1) they are easily (and cheaply) analyzed by electrophoresis; (2) electrophoresis usually reveals high levels of diversity; (3) this diversity appears to be mostly genetically controlled; (4) a complexity of molecular events underlies each specific electrophoretic pattern, which is likely therefore to have a unique origin; and (5) they can be used to evaluate phylogenetic relationships and genome homologies because of the wide taxonomic distribution of certain storage proteins. Their main disadvantage is the limited coverage of the genome.

Questions that have been addressed using seed storage proteins as markers include (1) the organization of genetic diversity within and between populations, (2) the process of plant domestication (identification of wild ancestors, fate of genetic diversity and multilocus associations on domestication, and (3) homologies between genomes or chromosomes (e.g., identification of genome donors to amphiploids). Furthermore, seed storage proteins are also used in breeding programs: (1) directly, to improve their nutritional and technological properties or, in the case of common bean, as a source of insect resistance, and (2) indirectly, for varietal identification.

Seed storage proteins provide excellent models for the study of the evolution of multigene families. So far, this evolution has been studied mostly within individual genotypes. Yet substantial variability exists between genotypes and the precise steps at the molecular level involved in this intergenotypic differentiation are not well known. The genomic context of the seed storage protein loci (i.e., their linkage relationships with genes coding for survival traits) is not well known, as for many other marker genes.

ACKNOWLEDGMENTS

I am most grateful to E. Martinelli for typing the manuscript and to J. Kami for helpful suggestions. Our work on the genetic diversity of *Phaseolus* beans is partially funded by the C. A. Lindbergh Fund and the International Board for Plant Genetic Resources.

CHAPTER 5

EVOLUTION OF MULTIGENE FAMILIES: THE RIBOSOMAL RNA LOCI OF WHEAT AND RELATED SPECIES

Jan Dvořák

ABSTRACT

Loci coding for rRNA in wheat and related species are favorable model systems for investigation of the evolution of localized multigene families. The genes comprise two families: the *Nor* family, coding for the 18 S–5.8 S–26 S rRNA and the *5SDna* family, coding for the 5 S rRNA. The genes of the two families are tandemly arranged in separate loci. The examination of molecular organization of the *Nor* loci suggested that several mechanisms—saltatory replication, transposition, nucleotide sequence conversion, and, to a lesser extent, unequal crossing over—are responsible for concerted evolution of genes and spacers both within and between loci. Within the *5SDna* loci concerted evolution of spacers occurs principally by unequal crossing over or conversion leading to hypervariation of the loci. No evidence for concerted evolution of spacers between *5SDna* loci was obtained. This showed that different multigene families may utilize different molecular mechanisms for their evolution. Superimposed on the concerted evolution of each family is joint evolution of the families. Joint evolution is apparent in genomes of allopolyploid species of *Triticum* in which the *Nor* and *5SDna* loci are either jointly retained or jointly eliminated. It is suggested that the fate of the loci in polyploid evolution is related to their expression in primitive allopolyploids. Deletion is a dominant stochastic tendency in tandemly arranged multigene families leading to elimination of redundant loci with dispensable functions.

INTRODUCTION

Genetic redundancy is a strategy with which eukaryotic cells cope with demands for high levels of transcription of specific genes. Redundant genes with identical function and identical or similar nucleotide sequence form a multigene family. The genes of a multigene family can be either interspersed or localized in chromosomal loci where they are separated from each other by spacers. The localized families can be further subdivided according to the characteristics of their spacers, which are either homogeneous, i.e., have similar nucleotide sequences, or heterogeneous, i.e., have largely unrelated nucleotide sequences.

In a single-copy gene a change in a single nucleotide may have a profound phenotypic effect. On the contrary, in a multigene family a single nucleotide change is likely to have little or no effect. The change would become phenotypically apparent only if the mutation spreads into a significant number of the repeated genes in the family. The spread of a mutation through a multigene family is made possible by the process of homogenization, the primary function of which is to preserve homogeneity in the family (Birky and Skavaril 1976; Dover 1982). Perhaps the most obvious mechanism by which repeated sequences can maintain homogeneity is by repeated cycles of saltatory replication of one or several repeats (Britten and Kohne 1968). A gradual turnover of genes in the locus by unequal crossing over (Tartof 1975; Smith 1976) would also maintain some degree of homogeneity within the locus. Since many localized repeated gene families are in two or more loci their homogenization by unequal crossing over would result in structural rearrangements of chromosomes that would reduce fitness of the progeny. A strong candidate for homogenization of gene families localized in two or more loci that does not lead to chromosome rearrangements is gene conversion (Jinks-Robertson and Petes 1985), which involves a temporary formation of a heteroduplex between two related nucleotide sequences. The DNA heteroduplex creates a potential for a transfer of sequences between the parental sequences by repair of mismatched nucleotides. A gene conversion can also increase or decrease the number of repeats in a repeated array (Fogel et al. 1984). Computer simulations revealed that homogenization by gene conversions is a conservative process in that most unique mutations are eliminated from a family (Birky and Skavaril 1976; Dvořák et al. 1987). This conservative tendency of homogenization by gene conversion is reduced if a conversion shows a disparity in mismatch correction toward one parental sequence (Birky and Skavaril 1976). Dover (1982) emphasized the directional tendency of disparity in gene conversion and named this stochastic evolutionary mode the molecular drive.

The distribution of mutations in localized arrays of repeated nucleotide sequences suggests that gene conversions may not be random in the arrays of repeated nucleotide sequences. The likelihood that two sequences in the array will be involved in a heteroduplex may be inversely proportional to their distance from each other in the array (Lassner and Dvořák 1986; Dvořák et al. 1987). The model predicts that as new variants spread through a family homogenized by conversions the family acquires a clustered appearance. This model also predicts that homogenization of flanking repeats with the internal repeats is reduced. Consequently, flanking repeats appear to accumulate mutations more rapidly than the internal repeats. Clustering of variants in a localized gene family is also expected if the genes in a locus are homogenized by unequal crossings over (Kimura and Ohta 1979).

The hierarchy of homogenization within and among loci in multigene families will be illustrated here for two multigene families coding for rRNA in wheat and related species. Finally, joint evolution of the families and factors affecting the processes will be discussed.

Wheats are classical allopolyploids. Tetraploid macaroni wheat, *Triticum turgidum* ($2n = 4x = 28$), contains one pair of genomes (AA) of einkorn wheat *T. urartu*, and the second pair (B^cB^c) from an unknown species closely related to *T. speltoides*. The sibling species, *Triticum timopheevii* ($2n = 4x = 28$), also contains pair of *T. urartu* genomes but has genome pair B^tB^t (often designated GG) derived principally from *T. speltoides*. Hexaploid bread wheat, *T. aestivum* ($2n = 6x = 42$), evolved from the hybridization of *T. turgidum* with *T. tauschii* (genomes DD) (Kihara 1944; McFadden and Sears 1946; Sarkar and Stebbins 1956; Nishikawa 1983; Dvořák et al. 1988).

Genes coding for rRNA comprise two multigene families in wheat and its relatives: the *Nor* family, coding for the 18 S–5.8 S–26 S rRNA (further rDNA), and the *5SDna* family, coding for the 5 S rRNA (further 5 S DNA). The genes of the two families are localized at separate loci on short arms of chromosomes 1, 5, and 6. In bread wheat, thousands of *Nor* genes are at loci *Nor-B1* and *Nor-B2* on chromosome 1B and 6B, respectively, and several hundred genes are at locus *Nor-D3* on chromosome 5D (Flavell and O'Dell 1979; Appels et al. 1980; Lassner et al. 1987). An additional locus appears to be on chromosome 1A in some cultivars (Miller et al. 1980). In barley rDNA loci are on chromosomes 6 (*Rn1*) and 7 (*Rrn2*) (Saghai-Maroof et al. 1984). Within each locus the genes are arranged in tandem arrays in which the gene units are separated by homogeneous complex external spacers (henceforth spacers). The most conspicuous feature of each spacer is an array of tandemly arranged repeats 133 bp long in the wheat B genome and 120 bp long in the wheat D genome and barley (Appels and Dvořák 1982a; Saghai-Maroof et al. 1984; Lassner et al. 1987). These repeats will be further designated "130 bp" repeats.

Wheat *5SDna* gene family is localized at the *5SDna-1* and *5SDna-2* loci on the short arms of chromosomes 1 and 5, respectively (Appels et al. 1980; Lassner and Dvořák 1985). The 120-bp-long genes are tandemly arranged and are separated by homogeneous simple spacers about 380 bp long on chromosomes 5A, 5B, and 5D, and about 290 bp long on chromosomes 1B and 1D. The former will be referred to as long-spacer 5 S DNA units while the latter will be referred to as short-spacer 5 S DNA units.

CONCERTED EVOLUTION

Intralocus homogenization

Examination of populations of wild macaroni wheat, *T. turgidum* spp. *dicoccoides*, related *Triticum* species and barley showed that the *Nor* loci are highly polymorphic in the length of the rDNA units, which include the 18 S–5.8 5–26 S genes plus the spacers. This polymorphism is chiefly caused by variation in the number of the "130 bp" repeats in the spacers (Appels and Dvořák 1982a; Dvořák and Appels 1982; Saghai-Maroof et al. 1984; Lassner and Dvořák, 1986; Lassner et al. 1987). Haplotypes that show the same overall restriction pattern are considered to be the same multigene allele. Because a single homogenization event is likely to alter the characteristics of a haplotype, an allele may comprise a single, several, or many closely related haplotypes in plant populations. Three alleles have been recorded at the *Nor-B1* locus and five at the *Nor-B2* locus in a survey of 25 bread wheat cultivars (May and Appels 1987). A similar survey of the polymorphism at the *Nor-D3* locus in 208 individuals of diploid *T. tauschii* yielded a total of 21 alleles (J. Dvořák and H.-B. Zhang, in preparation). In barley 12 alleles were observed at the *Nor* locus *Rrn1* and eight alleles at the *Nor* locus *Rrn2* (Saghai-Maroof et al. 1984; Allard 1988). The high polymorphism among individuals contrasts with the relative paucity of variation within individual genomes where the several thousand gene units are usually represented by single or a few spacer-length variants. For example, when DNA of cultivar Chinese Spring is digested with a restriction enzyme recognizing restriction sites flanking the "130 bp" repeats in the spacer, such as *Tag*I, and probed with a DNA fragment from the spacer, only five spacer variants are apparent. One of them maps to the *Nor-D3* locus on chromosome 5D (*Nor-D3a* allele) and two each map to the *Nor-B1a* and *Nor-B2a* alleles on chromosomes 1B and 6B, respectively (Appels and Dvořák 1982a). The nucleotide sequences of two spacers, one from each of the two variants that are present at the *Nor-B2a* allele, have been sequenced (Lassner and Dvořák 1986; Barker et al. 1988). Except for the array of "130 bp" repeats, the nucleotide sequences of the rest of the spacer, including remnants of the "130 bp" repeats that flank the array, were very similar. This indicated that the spacers within a *Nor* allele tend to be similar in both the length and the nucleotide sequence. While uniformity of the spacer lengths within a locus is the prevalent pattern in *Triticum* and *Hordeum*, in several plant species, e.g., *Vicia faba*, the *Nor* loci may show spacer length heterogeneity indicating a high degree of variation in the numbers of the repeats in the spacers within a single haplotype (Rogers et al. 1986).

The examination of the *5SDna* loci on chromosomes 1A, 1B, 1D, 5A, 5B, and 5D in bread wheat and its diploid ancestors showed a remarkably different picture. Digestion of genomic DNAs with restriction enzymes cleaving the spacers and not being sensitive to plant DNA methylation, such as

TaqI, resulted in ladders of monomers, dimers, trimers and higher order multimers (Figure 1), indicating that the spacers separating genes within each haplotype are relatively variable. The *5SDna* loci are extremely variable at the population level. A sample of 208 plants from a germplasm collection of *T. tauschii* yielded a minimum of 41 patterns of *TaqI* fragments (some are shown in Figure 1) (J. Dvořák and H.-B. Zhang, unpublished). Ladders of bands shown in Figure 1 differed extensively in the intensities of individual bands, which reflects differences in the abundance of specific oligomers cleaved by *TaqI* from the array of the 5 S DNA units. There must be many similar types of units at each *5SDna* haplotype. Their relative abundance must be subjected to extensive fluctuations. For example, a tetramer of 500 bp units (long-spacer units) is abundant in *T. tauschii* accessions C, F, H, I, and K but rare in accessions A, B, E, and G. In these accessions a long-spacer pentamer is abundant. The abundance of higher oligomers does not decrease or increase gradually, which would be expected if the distribution of *TaqI* sites were random in each haplotype, but there are apparent discontinuities in most of the ladders (Figure 1). This is expected if the various types of units were distributed nonhomogeneously in a haplotype. Clustering of like units in an array of

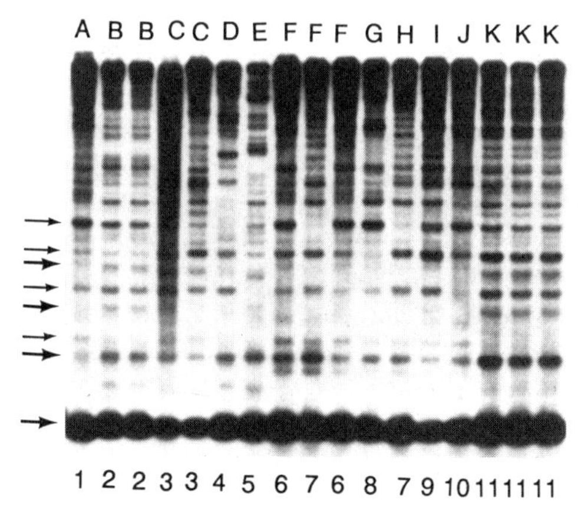

FIGURE 1. DNAs were isolated from single plants from 11 accessions (designated by letters) of *Triticum tauschii*, digested with *TaqI*, and probed with a 5 S DNA unit cloned in pTa794 (Gerlach and Dyer 1980). Monomers, dimers, trimers, and higher order multimers (specified by large arrows) of 410 bp units (short-spacer units) presumably originated from chromosome 1D and the dimers, trimers, and higher order multimers (specified by small arrows) of the 500 bp units (long-spacer units) from chromosome 5D. Eleven different profiles (designated by numbers) were observed among the accessions. Individuals within an accession showed identical profiles except for accession F from Turkmenia that was polymorphic; and one plant shared a similar profile with accession H, also from Turkmenia.

repeated units is an expected outcome if homogenization in the array occurs by repeated rounds of unequal crossing overs or gene conversions (Kimura and Ohta 1979; Dvořák et al. 1987). Hypervariation of tandemly repeated arrays, such as the *5SDna* loci, could be a valuable tool in varietal identification and in characterization of accessions in germplasm collections, provided that it is relatively stably inherited.

The contrast between the wheat *Nor* and *5SDna* loci in the intraallelic homogeneity is striking and suggests that the rates or mechanisms of homogenization of the families are not the same. Variation within haplotypes of the *5SDna* family is compatible with the assumption that repeated rounds of unequal crossing over or gene conversion is the principal mechanism of homogenization and concerted evolution within the loci. On the contrary, the relative paucity of variation within the *Nor* haplotypes indicates that unequal crossings over are rare at these loci. This was also suggested in an attempt to find homologous recombination in wheat rDNA loci. No homologous crossing over and one putative sister chromatid exchange at the *Nor-B2* locus was detected among 446 chromosomes (Dvořák and Appels 1986). Similar results were obtained for the *Nor-B1* locus (Snape et al. 1985) and *Rrn1* and *Rrn2* loci of barley (Saghai-Maroof et al. 1984). This difference between *Nor* and *5SDna* loci is not absolute since unequal crossings over must be frequent at the *Nor* loci of species, such as *Vicia faba*, which show a multitude of spacer length variants within individual *Nor* haplotypes. The factors causing the different behavior of the *Nor* loci in Triticeae on one hand and *V. faba* on the other hand are currently unknown.

A plausible reason for the homogeneity within *Nor* alleles could be saltatory replication of a single or several repeated units. If recombination occurs between neighboring or nearby repeated units (Dvořák et al. 1987) in a *Nor* allele, a circular molecule is excised that can replicate as a rolling circle (Hourcade et al. 1973) and an array of identical copies can be produced. Recombination of the new units with those at the existing *Nor* haplotype would result in an insertion of a cluster of homogeneous units in the same orientation into the same or different haplotypes.

Although this model is currently speculative, it is compatible with data obtained by sequencing of the spacers of two *Nor-D3* units, one cloned from the *T. aestivum Nor-D3a* allele and the other from the *Nor-D3a* allele from *T. tauschii* (Lassner et al. 1987; I. Vinitsky, R. Appels, and J. Dvořák, unpublished). The spacers of the two rDNA units, although apparently selected randomly from large numbers of units within each haplotype, had the same number of repeats and were almost identical in nucleotide sequence. If the origin of new spacer variants within loci were solely dependent on the spread of mutations throughout the hundreds or thousands of rDNA units within *Nor* loci by constantly occurring unequal crossing over, then it would be extremely unlikely to find this remarkable similarity between the spacers of the two independent clones. This is particularly true since the two rDNA units have been isolated from each other by a minimum of 8000 years, the estimated time since the origin of *T. aestivum* (Helbaeck 1966).

Interlocus homogenization

Three lines of evidence show that concerted evolution in wheat occurs between *Nor* loci on different chromosomes within a genome. Additionally, these lines of evidence show that the mechanism is actual transfer of nucleotide sequences.

The first line of evidence comes from comparing the divergence rates of spacers between *Nor* loci in related species. The spacers evolve rapidly and show abundant variation at the interspecific level (Appels and Dvořák 1982a; Dvořák and Appels 1982). The *Nor* loci of the wheat B genome have diverged from those of the diploid *Triticum* species of the B-genome group by 4–9% of nucleotides in the array of the "130 bp" repeats in the spacer and by 3–5% of nucleotides in the promoter and the transcribed portion of the spacer. Yet, the spacers of the *Nor-B1* and *Nor-B2* loci in the wheat B genome differ from each other by only 1.2% of nucleotides. Thus, the two *Nor* loci in the wheat B genome must be evolving in concert although both are rapidly diverging from those of closely related species. In *T. monococcum*, rDNA is at loci *Nor-A1* and *Nor-A3* on chromosomes 1A and 5A, respectively (Miller et al. 1983). Examination of the spacer nucleotide sequences at the two loci with a thermal dissociation technique failed to distinguish them from each other, indicating a close similarity of the spacer sequences (Gill and Appels 1988). Thus, the *Nor* loci in *T. monococcum* are also evolving in concert. However, in *T. timopheevii* the same technique readily distinguished the nucleotide sequences of the spacers at the *Nor-B1* and *Nor-B2* loci and in *T. speltoides* the nucleotide sequences of spacers in different spacer length variants (Gill and Appels 1988). From these data Gill and Appels (1988) suggested that the high degree of homology between nonhomologous rDNA loci in wheat and *T. monococcum* reflects recent translocations between *Nor*-bearing chromosomes.

The second line of evidence comes from the observation that the same spacer variants that are present as major variants at one *Nor* locus may also be detectable at the other *Nor* locus or elsewhere in the genome. For example, the *Nor-B2a* allele in cv. Chinese Spring contains two equally abundant spacer variants. A minor number of copies of the same two variants are also on another Chinese Spring chromosome (see Figure 6 in Dvořák and Appels 1986).

The third and most direct line of evidence comes from sequencing of spacers from paralogous *Nor-B2* and *Nor-D3* loci. Two repeats at the 5′ end of the array of the "130 bp" repeats in the spacer are degraded relative to the rest of the array by numerous point mutations and deletions. Four spacers have been sequenced, two from a haplotype of the *Nor-B2a* allele of wheat, one from a haplotype of the *Nor-D3a* allele in wheat, and one from a haplotype of the *Nor-D3a* allele in diploid *T. tauschii*. All four sequences have the same

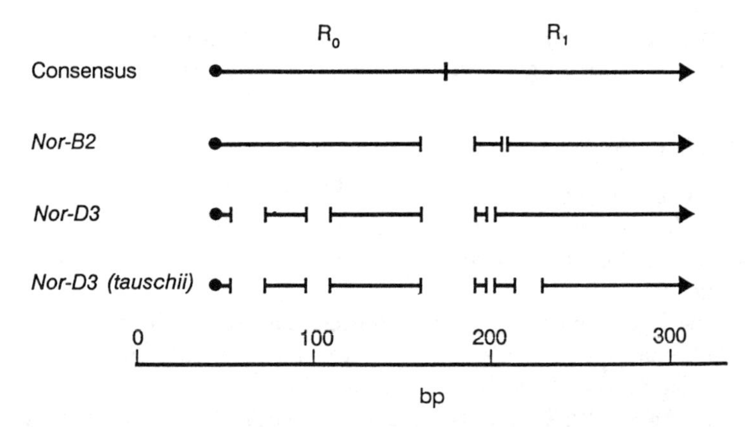

FIGURE 2. Comparison of two most 5′ repeats R_0 and R_1 in the array of "130 bp" repeats in the spacers of two clones from the *T. aestivum Nor-B2* locus (Lassner and Dvořák 1986), one clone from the *T. aestivum Nor-D3* locus on chromosome 5D (Lassner et al. 1987), and one clone from the *Nor-D3* locus on chromosome 5D of *T. tauschii* (I. Vinitzky, R. Appels, and J. Dvořák, unpublished). The gaps in the solid lines represent deletions relative to the consensus nucleotide sequence of the "130 bp" repeats in the B genome.

deletion at the junction of the degraded repeat R_0 and R_1 (Figure 2). Since *Nor-1*, *Nor-2*, and *Nor-3* loci occur throughout the tribe Triticeae, the triplication of the *Nor* locus must precede the divergence of the tribe. The existence of the same deletion in the spacers of paralogous *Nor-2* and *Nor-3* loci in *Triticum* genomes showed that the concerted evolution of nonhomologous loci occurs by actual transfer of sequences between the loci.

Recombination or translocations between nonhomologous *Nor* loci was suggested to be responsible for spreading a variant from its original locus into other *Nor* loci (Arnheim 1983; Coen and Dover 1983; Gill and Appels 1988). Reciprocal translocations of distally located loci are expected to follow such a process. However, the loci distal to *Nor-B1* and *Nor-B2* in wheat show no sign that they were involved in recurrent translocations. Nor is there evidence that loci distal to *Nor-2* and *Nor-3* on chromosomes 5 and 6 have been translocated during evolutionary radiation of the genus *Triticum*. The occasional saltatory replication of one or several rDNA units by the rolling circle and their insertion into an rDNA haplotype, the mechanism suggested above as a possible reason for intralocus homogeneity, could also lead to transposition of rDNA units between loci if the amplified copies were inserted into a nonhomologous *Nor* locus. This seems to be a more satisfactory model of the apparent transfers of rDNA nucleotide sequences between loci in *Triticum* than reciprocal recombination or translocations between chromosomes. This model accounts for the absence of translocations of distal loci and provides a unifying explanation for both the intralocus homogeneity and the rDNA transpositions between loci. If the amplification events were rare the originally high spacer homology between the ancestral and from it amplified new rDNA variant would have time to decay. Perhaps the various degrees of spacer sequence homology between specific rDNA variants in wheat, *T. monococcum*, *T. timopheevii*, and *T. speltoides* reported by Appels and Dvořák (1982a) and Gill and Appels (1988) reflect the ages of individual amplification events.

The pattern of the evolution of the nonhomologous *Nor* loci contrasts with that in the *5SDna* family. In wheat and related diploid species, the *5SDna* family contains two types of repeated units differing by the length of the spacers. Analysis of restriction site differences among wheat aneuploid stocks and interspecific chromosome substitution lines showed that the short-spacer repeated gene units are in most genomes exclusively at the *5SDna-1* locus on chromosome 1, whereas the long-spacer repeated gene units are at the *5SDna-2* locus on chromosome 5. The only apparent exception is *T. umbellulatum* in which one of two short-spacer 5 S DNA variants was detected on chromosome 5 (J. Dvořák, unpublished). Sequencing the short-spacer and long-spacer variants from a number of species of the tribe Triticeae showed low similarity between the two classes of spacers (Gerlach and Dyer 1980; Scoles et al. 1988). This suggests that the spacers of the two *5SDna* loci are not evolving in concert. Since the coding sequences of the short- and long-spacer units are relatively homogeneous the coding sequences of the *5SDna* repeated units appear to evolve in concert. The homogenization of interspersed 5 S rRNA genes in *Neurospora* also appears to be limited to the coding regions (Morzycka-Wroblewska et al. 1985). The reasons why spacers are involved in the interlocus homogenization in the *Nor* family but not in the *5SDna* family is unknown.

The *Nor* and *5SDna* loci appear to differ in both the intralocus and the interlocus homogenization. The nature of rules determining which mechanism of homogenization prevails in a specific family is enigmatic. Additionally the coding sequences are more conserved than spacers during evolution. This could be caused by natural selection, which eliminates mutations that occurred in the coding sequences. Higher rates of homogenization in the transcribed sequences (gene regions) than in the less frequently or untranscribed sequences (spacers) could be another reason for the slower evolutionary rates of the coding sequences (Appels and Dvořák 1982b; Morzycka-Wroblevska et at. 1985; Lassner and Dvořák 1986). These possibilities are not mutually exclusive. The observation that the yeast *Nor* promoter and enhancer sequences stimulate recombination (Roeder et al. 1986) indicates that transcribed portions of repeated gene families may homogenize more efficiently than nontranscribed portions, such as the spacers of the *5SDna* family.

JOINT EVOLUTION BETWEEN MULTIGENE FAMILIES

The *Nor* and *5SDna* multigene families are located at different loci and are transcribed by different RNA polymerases. The major factor that the families have in common is their function in the formation of ribosomes. This common function may place constraints on their evolution because of the need for common regulation. In diploid plants the evolution of each family is a gradual process and, consequently, their joint evolution may be difficult to demonstrate. Allopolyploidy suddenly brings together multigene families that have previously diverged from each other during differentiation of parental species, and thus provides a unique opportunity to examine the joint evolutionary process.

In the allotetraploid macaroni wheats, *T. turgidum* and *T. timopheevii*, most copies of the *Nor* family are in the B-genome loci (Appels and Dvořák 1982a; Frankel et al. 1987; Gill and Appels 1988). The same is true for the *5SDna* family (Lassner and Dvořák 1985 and Figure 3). In the allohexaploid bread wheat the B genome again retains the copies of both the *Nor* (Appels and Dvořák 1982a) and the *5SDna* families (Appels et al. 1980 and Figure 3). The 5 S DNA in the *T. aestivum* D genome shows reduced numbers of gene copies (Lassner and Dvořák 1985).

A similar situation was observed in two other allotetraploid *Triticum* species in which the *Nor* and *5SDna* loci could be unequivocally allocated to respective genomes. The allotetraploid *Triticum cylindricum* (genomes CCDD) originated from hybridization of *T. dichasians* (genomes CC) with *T. tauschii* (genomes DD) (Kihara and Matsumura 1941; McFadden and Sears 1946). In *T. cylindricum* most of the copies of the *Nor* and *5SDna* gene families are in the C genome and the numbers of gene copies in the D genome are greatly reduced (Figure 4, other not shown). In *T. triunciale* (genomes UUCC), which originated from hybridization of *T. umbellulatum* (genomes UU) with *T. dichasians* (genomes CC) (Kihara and Kondo 1943), four of five accessions investigated had more *Nor* copies in the C genome than in the U genome (not shown). An identical pattern was found in the *5SDna* family (Figure 4, Table 1).

Thus, in the genomes of these five polyploid species the *Nor* and *5SDna* families have a common fate, either both being retained or both being eliminated. It is apparently advantageous to a polyploid organism to retain those *Nor* and *5SDna* families that are coadapted from their previous joint evolution. This would suggest that the *Nor* and *5SDna* loci within a genome may have some common characteristics that *Nor* and *5SDna* loci from different genomes may not have. The existence of such common characteristics would suggest that multigene families have been evolving jointly during the speciation of diploid *Triticum* species.

A characteristic feature of the *Nor* loci is nucleolar dominance in interspecific hybrids (Navashin 1934). Nucleolar dominance is most easily detected as the absence of secondary constrictions in the nucleolus organizing chromosomes. It is tempting to speculate that loci that are retained in a polyploid are those that were dominant in the primitive hybrid. In an attempt to explain the loss of the A-genome *Nor* loci in allotetraploid wheat Frankel et

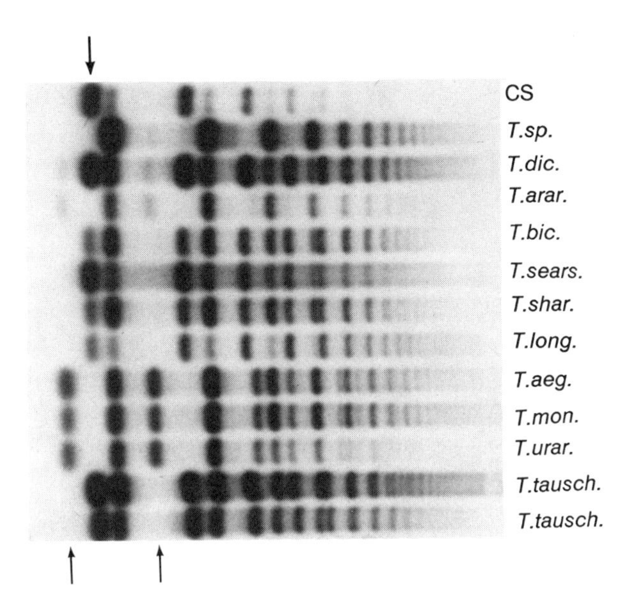

FIGURE 3. DNAs of *T. aestivum* cv. Chinese Spring (genomes AAB^cB^cDD), *T. speltoides* (B^sB^s), *T. turgidum* ssp. *dicoccoides* (AAB^cB^c), *T. timopheevii* ssp. *araraticum* (AAB^tB^t), *T. bicorne* (B^bB^b), *T. searsii* ($B^{se}B^{se}$), *T. sharonense* ($B^{sh}B^{sh}$), *T. longissimum*-(B^lB^l), *T. monococcum* ssp. *aegilopoides* (AA), *T. m.* ssp. *monococcum* (AA), *T. urartu* (AA), and *T. tauschii* (DD) were digested with *Sau*3A and probed with a 5 S DNA unit from pTa794. The repeated units at the *5SDna-1* locus on chromosome 1A of *T. monococcum* and *T. urartu* are detectable by the presence of the 360 and 600 bp fragments in their multimers (designated by small arrows). These fragments are greatly underrepresented in tetraploid *T. turg.* ssp. *diccocoides*, *T. tim.* ssp. *araraticum*, and absent in the hexaploid *T. aestivum*, indicating reduction of the *5SDna* copies in the A genomes of all polyploid species. The profiles of *T. turg.* ssp. *dicoccoides*, *T. aestivum*, and those of the B-genome diploid species, except for *T. speltoides*, are similar. The profile of *T. tim.* ssp. *araraticum* is similar to that of *T. speltoides* in that they both entirely lack the short-spacer *5SDna* copies (specified by a large arrow) on chromosome 1B.

al. (1987) investigated the expression of the *Nor* loci in artificial allotetraploids synthesized from diploid species related to those that contributed the wheat A and B genomes. More than four nucleoli appeared in some cells, indicating that both the A- and B-genome *Nor* loci were expressed. The authors, therefore, suggested that the loss of the A-genome *Nor* loci in the allotetraploid wheat may be due to initial hybridization event involving an A-genome genotype with few *Nor* gene copies. In synthetic allooctoploids constructed by hybridizing bread wheat (genomes AABBDD) with four different *T. tauschii* genotypes (genomes DD) the secondary constriction on the 5D chromosome, the site of the *Nor-D3* locus, is apparent. However, the con-

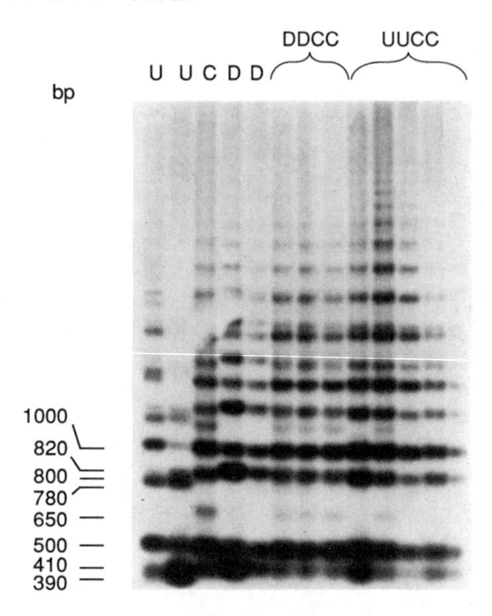

FIGURE 4. DNAs of *Triticum umbellulatum* (genomes UU), *T. dichasians* (CC), *T. tauschii* (DD), *T. cylindricum* (DDCC), and *T. triunciale* (UUCC) were digested with *Bam*H1 and probed with a 5 S DNA unit cloned in pTa794. Note that the restriction fragment patterns in three accessions of *T. cylindricum* are like those of *T. dichasians* rather than like those of *T. tauschii*. Note also that the profiles of the *T. triunciale* accessions are more similar to that of *T. dichasians* than to those of *T. umbellulatum*.

striction is smaller and discernible in fewer cells than are the secondary constrictions at the B-genome *Nor* loci (J. Dvořák, unpublished). This suggests that the nuclear dominance among the A-, B-, and D-genome *Nor* loci may be partial rather than complete.

TABLE 1 Expression of *Nor* loci in artificial hybrids or polyploids and retention of rRNA genes in polyploid *Triticum* species[a]

			Gene retention	
Polyploid	Genomes	*Nor* expression	Nor	*5SDna*
T. turgidum	AAB^cB^c	B>A?	B>A	B>A
T. timopheevii	AAB^tB^t	B>A?	B>A	B>A
T. aestivum	AAB^cB^cDD	B>A? B>D	B>A B>D	B>A B>D
T. cylindricum	CCDD	C>D	C>D	C>D
T. triunciale	UUCC	C>U	C≥U	C>U

[a]J. Dvořák (unpublished).

Nucleolar dominance is also evident in the tetraploid *T. cylindricum* (genomes CCDD) and *T. triunciale* (UUCC). In hybrids between *T. dichasians* and *T. tauschii* (genomes C and D) and between *T. umbellulatum* and *T. dichasians* (genomes U and C), the C-genome chromosomes expressed secondary constrictions in root-tip cells almost exclusively (J. Dvořák, unpublished). These observations suggest that the *Nor* loci that show a tendency to be lost during evolution are those that were poorly expressed in the primitive allopolyploids (Table 1).

DELETION OF MULTIGENE LOCI

An intriguing question is whether the elimination of *Nor* and *5SDna* gene copies from some of the genomes is caused by the same mechanisms that homogenize the gene families or by some other mechanisms. Changes in the gene copy numbers per locus are usually attributed to unequal crossing overs (Tartof 1975; Smith 1976; Williams and Strobeck 1985). Since unequal crossing over is reciprocal, a loss in one haplotype is complementary to a gain in the other haplotype. It is hard to see a reason why one product would have a priori an enhanced likelihood of survival than the other. It is, however, possible that stochastic variation in the copy number by an unequal crossing over could result in chance fixation of a loss. Since the gain cannot be fixed, redundant loci could show a stochastic tendency to be lost.

Another possibility is that deletions may naturally outnumber duplications in repeated DNA, leading to ultimate loss of dispensable repeated DNA. This tendency is apparent in degraded repeats flanking the array of the "130 bp" repeats in the spacers of the wheat B- and D-genome *Nor* loci (Figure 2). These repeats are homologous with the internal repeats of the array and their sequences originally were similar to the consensus sequence of the "130 bp" repeats. Because homogenization of the flanking repeats is less effective, mutations that are removed from the internal repeats by homogenization remain in the flanking repeats (Lassner and Dvořák 1986; Dvořák et al. 1987; Barker et al. 1988). Numerous deletions were detected in the R_0 and R_1 repeats (Figure 2; Table 2). All deletions occurred in the region of the "130 bp" repeats that evolved by elongation of a simple dodecamer. This region still shows a great deal of redundancy (Lassner et al. 1987; Barker et al. 1988). A deletion

TABLE 2 Percentages of the degraded "130 bp" repeats R_0 and R_1 at the 5′ end of the repeated array in the *Nor* spacers lost by deletions.

Origin of clone	Species	R_0 (%)	R_1 (%)
Nor-B2	*T. aestivum*	11	9
Nor-D3	*T. aestivum*	36	16
Nor-D3	*T. tauschii*	36	30

found in a cloned rDNA fragment from the *Nor-D3a* allele in *T. tauschii* did not occur in this region, but the deleted DNA was, nevertheless, found to involve a short direct repeat (I. Vinitzky, R. Appels, and J. Dvořák, unpublished). Short repeats in the direct orientation have been shown to be potent sources of deletions due to their recombination or slippage during DNA replication or conversion (Farabaugh et al. 1978; Albertini et al. 1982; Fogel et al. 1984; Scoles et al. 1988).

A possible mechanism leading to potentially massive interstitial deletions is intrachromosomal recombination between repeats with the same orientation in the multigene locus (Albertini et al. 1982; Pont et al. 1988). Such recombination leads to excision of a circular molecule involving DNA between recombined repeats (Dvořák et al. 1987; Pont et al. 1988). If the likelihood of reinsertion of the circle would be low, intrachromosomal recombination would tend to delete multigene loci. The propensity for massive deletions in multigene loci is apparent not only on the evolutionary scale but also in the experimental studies (Table 3). Of 10 cases of new rDNA or

TABLE 3 Nature of new *Nor* and *5SDna* alleles originating from a known existing allele.

Species	Change	Agent	Source
Maize	New restriction sites? (*Nor* locus)	Hybridization	Lin et al. (1985)
Teosinte	Restructuring of spacer (*Nor* locus)	Hybridization	Zimmer et al. (1988)
Teosinte	Reductions in copy number (*5SDna* locus)	Hybridization	Zimmer et al. (1988)
Potato	70% reduction (*Nor* locus)	Tissue culture	Landsmann and Uhrig (1985)
Wheat	30% reduction (*Nor* locus)	Recombination	Dvořák and Appels (1986)
Wheat	30% reduction (*Nor* locus)	Spontaneous	R. S. Kota (unpublished)
Wheat	80% reduction (*5SDna-B1*) and elimination (*5SDna-D1*)	Absence of *Phl* gene	J. Dvořák and H.-B. Zhang (unpublished)
Wheat	Length of the spacer	Tissue culture	Rode et al. (1987)
Triticale	80% reduction	Tissue culture	Brettell et al. (1986)

5 S DNA alleles that originated under experimental conditions, 7 were a reduction or elimination of gene units (Table 3). No cases of an increase have been reported. It is possible that some of the massive losses of gene units from *Nor* and *5SDna* loci originally attributed to other causes may have arisen from intrachromosomal recombination.

Whether a deletion can be tolerated or not must ultimately be determined by natural selection. Because of the polysomic inheritance, polyploids are expected to be more tolerant of deletions than diploids. There is no doubt, however, that the same process occurs in diploids. For example, diploid *T. speltoides* has lost all of the short-spacer 5 S DNA that is at the *5SDna-1* locus on chromosome 1B of species closely related to it such as *T longissimum*, *T. searsii*, *T. sharonense*, and *T. bicorne* (Figure 3).

CONCLUSIONS

The investigation of concerted evolution of the *Nor* and *5SDna* multigene families in wheat and related species has revealed several levels of homogenization within each family. The major strategies of homogenization of the two families differ. The genetic and molecular bases determining the dominant strategy require further investigation. The two families appear to evolve jointly but, again, the specific molecular factors responsible for joint evolution of these, otherwise very different, gene families are unknown. Finally, the loci in both multigene families show a general tendency to be eliminated if rendered dispensable by polyploid evolution. Whether this tendency is unique to these two multigene families or general remains to be explored.

CHAPTER 6

MOLECULAR DIVERSITY IN PLANT POPULATIONS

Michael T. Clegg

ABSTRACT

Levels of genetic diversity within populations and within species govern rates of adaptive evolution and limit rates of advance in conventional crop improvement. The science of molecular biology has provided new tools to study variation, with greater power to resolve different classes of mutational change than all previous experimental methods. Despite these powerful new tools, current knowledge of molecular diversity within plant populations is limited.

The objectives of this chapter are to review our understanding of molecular diversity within plant species and to raise questions about future plant population genetic research based on the results of contemporary research in both plant and *Drosophila* population genetics. Single-copy gene diversity is considered first, followed by a discussion of variation associated with multigene families; the chapter concludes with a brief review of intraspecific variation in organelle genomes. The mutational processes that generate DNA sequence variation and the forces that determine the fate of different classes of mutational variants in populations receive particular attention.

INTRODUCTION

The analysis and characterization of genetic diversity have always been a primary concern of population geneticists. Genetic diversity plays a fundamental role in evolutionary theory because natural selection chooses among the variants that occur within populations, based on their adaptation to the immediate environment. Without genetic variation there can be no adaptive evolution. The ultimate result is that variation within populations is converted into variation between populations and, finally, as evolution proceeds, into variation between species. The situation is analogous for crop improvement. Here too the goal is to fix agronomically useful genetic variants within cultivars by selective breeding. As a consequence, both evolutionists and agriculturalists are fundamentally concerned with the quality and extent of genetic diversity.

Over the years the methods for detecting and analyzing genetic diversity have expanded from Mendelian analyses of discrete morphological and cytological variants, to statistical analyses of quantitative variation, to biochemical assays, and, finally, to molecular assays. The molecular study of genetic variation has revealed a number of previously unsuspected genetic phenomena and it has raised a host of questions and applications for population genetics (reviewed by Clegg and Epperson, 1985). This chapter reviews the current state of knowledge about molecular genetic diversity in plant populations and points to open questions, unsolved problems, and new research opportunities. The primary focus will be on molecular diversity within populations or at the intraspecific level. The term molecular diversity will be taken to mean DNA sequence variation detected by the methods of molecular biology. (For discussions of protein diversity see chapters by Gepts and Hamrick and Godt, this volume.) The discussion will begin with studies of variation in "single-copy" genes, will then take up population analyses of diversity in multigene families, and close with population studies of diversity in organelle genomes.

SINGLE-COPY GENES

The survey of restriction site differences is a major method for the characterization of genetic variation at single-copy genes. To assess its utility, it is necessary to consider two questions: first, what can be learned about population diversity from molecular studies that was not already apparent from isozyme surveys? And second, what new insights into evolutionary processes can be obtained at the molecular level that were not revealed at other levels of investigation? Very few detailed studies of restriction site polymorphism for plant single-copy genes have been published to date. As a consequence, we turn to portions of the animal literature to seek answers to these questions.

Contrasts between isozyme and molecular methods for the study of genetic diversity

To address our first question, consider the contrast between isozyme and molecular methods (Table 1). Lewontin and Hubby (1966), in their classical paper on isozyme variation in *Drosophila pseudoobscura*, discussed the disadvantages and advantages of the isozyme method for studies of variation. Their discussion is as relevant today as it was over 20 years ago. Two major disadvantages of isozymes are (1) variation can be detected only for protein-coding genes; and (2) only a fraction of all mutational events (i.e., those changing protein mobility in a gel) can be resolved. Molecular methods overcome both

TABLE 1 Contrasts between isozyme and molecular methods for the detection and study of genetic variation.

Isozyme analyses	Southern analyses
Limited to protein-coding genes	Can assay any component of genome
Detects variants causing mobility differences in gel assay	Can detect any mutational change
Relatively rapid and inexpensive assay	Relatively slow and expensive assay
Can accommodate large samples	Limited to modest samples with consequent loss of statistical power for testing population genetic hypotheses
Limited number of genetic loci can be detected	Unlimited number of genetic loci can be detected

of these disadvantages because the detection of variation is not limited to coding regions and all categories of mutational events can, in principle, be detected.

A major advantage of the isozyme method is that the assay is relatively simple and rapid, and, as a consequence, it permits the screening of large samples at modest cost. In contrast, molecular methods are time-consuming and relatively expensive, so that sample sizes are usually small and the power to test statistical hypotheses is limited. While the sample of individuals assayed in molecular studies is limited, the number of loci that can be detected is often one or two orders of magnitude greater than that for isozymes. For example, more then 800 RFLP (restriction fragment length polymorphism) markers have been mapped on the maize genome (Stuber, this volume), while the maximum number of isozyme loci available for study is much fewer (Hamrick and Godt, this volume). Both isozyme and molecular methods allow the provisional assignment of gene products (proteins in the isozyme case) or restriction sites (in the case of Southern analyses) to genetic loci, independent of preexisting variation. This "unbiased" assignment permits a direct assessment of the level of genetic variation in a population or species.

Levels of variation inferred from restriction site surveys

Let us now consider the second question posed above, concerning new insights into evolutionary processes from molecular studies of genetic diversity. We shall begin with a discussion of the results of some population surveys of *Drosophila*. Table 2 summarizes the results of six recent surveys of restriction site variation in two species of *Drosophila*. The most intensive surveys, with respect to both size of gene region investigated and sample size, have been conducted with *D. melanogaster* (Table 2). These sample sizes are still small by the standards of most population genetic work, but the studies attempted to obtain a representative sample of chromosomes, to permit inferences about population or species diversity from the sample statistics. In addition, fine scale restriction maps of the gene regions studied allowed the unambiguous classification of allelic differences into site changes (assumed to result from nucleotide substitutions) and insertion/deletion polymorphisms.

The size of the gene region mapped in each study, which is substantial, ranges from 13 to 60 kb (kilobase pairs). Owing to this substantial size, estimates of allelic diversity are based on a reasonable number of mutational changes. This provides some statistical power (albeit weak) to address several important questions of particular interest: (1) Are estimates of restriction site diversity heterogeneous over gene regions? (2) Are estimates of diversity heterogeneous over species and what does such heterogeneity suggest about the recent evolutionary history of the species? (3) What is the magnitude of linkage disequilibrium (D) among polymorphic sites within a gene region and are estimates of D heterogeneous over gene regions? (4) What is the frequency of individual insertion and deletion polymorphisms in the sample and do these frequencies indicate that insertion/deletion polymorphisms are on average disfavored by selection?

An estimate of the expected heterozygosity, H (the probability that two randomly drawn chromosomes will differ at a given restriction site, averaged

TABLE 2 Studies of *Drosophila* restriction site variation associated with single-copy genes.

Species	Gene region	H^a	Number of large inversions	D^b	kb[c]	N[d]	Source[e]
melanogaster	*Amy*	0.006	9	≈0	15	85	1
	Adh	0.006	11	≠0	13	49	2
	White	0.014	14	≈0	45	38	3
	Notch	0.007	5	≈0	60	37	4
	Rosy	0.003	4	—	40	60	6
pseudoobscura	*Adh*	0.021	2	≈0	32	20	5
simulans	*Rosy*	0.019	0	—	40	30	6

[a]H is average heterozygosity of restriction sites averaged over all restriction sites.
[b]D is the conventional measure of pairwise linkage disequilibrium = $f(A^+, B^+) - f(A^+)f(B^+)$, where A and B are two polymorphic restriction sites and + denotes presence of the site versus absence (−). The joint frequency of presence at both sites is $f(A^+, B^+)$ and $f(A^+)$, $f(B^+)$ are marginal frequencies.
[c]kb is the size of the region in kilobase pairs.
[d]N is the number of chromosomes sampled.
[e]1. Langley et al. (1988); 2. Aquadro et al. (1986); 3. Langley and Aquadro (1987); 4. Schaeffer et al. (1988); 5. Schaeffer et al. (1987); 6. Aquadro et al. (1988).

over sites), for each of the six surveys is given in Table 2. The statistic, H, is a useful measure of genetic diversity, because it reflects the frequency of site polymorphisms taken over the sample of chromosomes. The estimates of H for the *D. melanogaster* gene regions are fairly consistent, although H may be somewhat higher for the white locus region. Langley et al. (1988) have concluded that the average value of H for *D. melanogaster* seems to converge to a value of about 0.008. In addition, site polymorphisms appear to be randomly distributed within the particular gene regions studied.

Comparisons among different *Drosophila* species with respect to genetic diversity suggest that *D. melanogaster* has low levels of restriction site diversity, but high levels of insertional polymorphism (Table 2). Levels of isozyme polymorphism also appear to be poorly correlated with restriction site diversity. For example, *D. pseudoobscura* has a significantly higher estimate of H (=0.021 for restriction site diversity); this value represents nearly a 3-fold increase over *D. melanogaster*, despite the fact that both species exhibit essentially equivalent levels of isozyme polymorphism (Schaeffer et al. 1987). Aquadro et al. (1988) have reported a similar discrepancy between levels of isozyme polymorphism and restriction site diversity based on a study of the *rosy* locus region from *D. melanogaster* and *D. simulans*. The per locus heterozyosity for allozymes is 0.102 and 0.096 for *D. melanogaster* and *D. simulans*, respectively, while restriction site diversity for the 40-kb *rosy* locus region is 0.003 and 0.019 for these two respective comparisons (Table 2). These findings raise the question of whether protein polymorphism will be a good predictor of overall levels of genetic diversity.

For neutral polymorphisms the statistic $H = 4N_e\mu/(4N_e\mu + 1)$. There is no compelling reason to suppose that the mutation rate (μ) differs between the two species, so assuming that the site polymorphisms are selectively neutral, we may conclude that differences in effective population size (N_e) account for the reduced diversity in *D. melanogaster*. The reasons for the lower diversity and, by inference, a smaller value of N_e are unknown. Perhaps the reduced diversity reflects past genetic bottlenecks that occurred prior to the association of *D. melanogaster* with human habitation.

Estimates of linkage disequilibrium from small samples have large standard errors (e.g., Weir and Brooks 1986; Hill and Weir 1988), and, consequently, the analysis of patterns of association of the *Drosophila* data must be regarded as provisional at best. Nevertheless, most of the gene regions give little evidence for high values of D. All four possible gametic types (++, +−, −+, −−) are frequently observed for pairwise sets of restriction sites, suggesting that recombination has occurred between sites. The one notable exception is the *Adh* region of *D. melanogaster*, where significant values of D for pairwise site comparisons are more common (Aquadro et al. 1986). A considerable body of evidence indicates that the *Adh*-F allele is subject to selection in ethanol environments (Anderson and McDonald 1983). It is possible that the higher values of D associated with the *Adh* region are the result of recent hitchhiking selection. Theoretical analyses show that hitchhiking selection can generate substantial linkage disequilibrium (Asmussen and Clegg 1981) and the rate of decay of transiently generated disequilibria should be slow for very tightly linked sites.

The *Drosophila* data on large insertions (>300 base pairs) are particularly interesting. In the vast majority of cases insertions are found to be unique, that is a particular insertion is observed just once in a sample. This stands in marked contrast to restriction site polymorphisms, where individual site variants are often in moderate to high frequencies in population samples. Evidently, these two classes of polymorphism obey different evolutionary dynamics.

A large body of theory exists on the expected behavior of polymorphisms that are selectively neutral. A comparison of observed frequency distributions to theoretical expectations indicates that the observed frequency distribution of insertions is inconsistent with simple neutral dynamics (Aquadro et al. 1986). More complex models can allow excision independent of copy number as might be the case for some families of transposable elements. Studies of such models lead to the conclusion that the observed frequency distribution of insertions is not at insertion/excision equilibrium (Golding et al. 1986). The problem is that all insertion polymorphisms are quite rare in populations, suggesting their relatively rapid removal. There are two possibilities: (1) insertion polymorphisms are mildly deleterious and are removed by natural selection; or (2) some other genetic mechanism is removing insertions. Langley et al. (1988) suggest that meiotic recombination may play a role in the removal of insertions from populations, although no direct evidence is available to support this suggestion.

The selection hypothesis can be tested by investigating the frequency distribution of X-linked insertions. If insertions are deleterious, they should be removed more rapidly from the hemizygous X chromosome. Studies of molecular variation in the region of the X-linked *Amy* locus indicate that the frequency of insertions is about the same as that seen for autosomal gene regions (Table 2). However, the data are few and it is probably premature to discard the selection hypothesis.

We have considered the results of *Drosophila* population surveys in detail, because there is little comparable work in plant population genetics. The most extensive survey of molecular variation for a plant single-copy gene so far reported is that of Gepts and Clegg (1989). A sample of 53 cultivated and 25 wild lines of pearl millet (*Pennisetum glaucum*) was surveyed for restriction fragment polymorphism associated with the *Adh1* locus, using a maize *Adh1* cDNA clone as a probe. The sample of wild lines was chosen to span the geographic range of the Sahel region of Subsaharan Africa where *P. glaucum* is native and where the crop was domesticated. In addition, cultivated lines were chosen from all regions of the world where pearl millet is cultivated.

Restriction maps spanning approximately 10 kb of DNA associated with the *Adh1* region were inferred for four restriction enzymes from several wild and cultivated lines. Three distinct haplotypes could be identified in this limited sample and several of the map differences appeared to be the result of insertion/deletion polymorphism; however, the map resolution did not permit a precise localization of insertions. The entire sample of 78 lines was then screened with a single enzyme (*Eco*RV) that did not cut within the *Adh1* region. This permitted the detection of polymorphism as fragment mobility differences, but it did not allow the separate classification of insertion/deletion versus site changes. The results showed high levels of restriction fragment polymorphism in both the wild and cultivated materials. The calculation of statistics on site diversity was not possible and the frequency distribution of insertion events could not be determined.

The *Drosophila* work illustrates some of the empirical issues that can be investigated in plant populations, but it would be unwise, at this rudimentary state of knowledge, to generalize from *Drosophila* to plants. Why are there no comparable studies of plant molecular diversity? There are several reasons for the lack of information about molecular diversity associated with plant single-copy genes. First, much is known about the molecular genetics of *Drosophila* and this knowledge provides a context for the population work. Second, the level of research activity in *Drosophila* population genetics exceeds that of all plant population genetics. And third, work on plants has focused strongly on the practical problems of developing RFLP maps of important crop species (e.g., Bernatzky and Tanksley 1986; Helentjaris et al. 1986; Landry et al. 1988).

RFLP diversity within plant species

RFLP mapping programs have the objective of uncovering polymorphism for a wide spectrum of probes and exploiting this preexisting polymorphism in conventional genetic mapping. For this purpose, the construction of physical restriction maps is unnecessary and far too time consuming. Moreover, the most efficient way to uncover polymorphism is to screen several genetically diverse lines. However, inferences about the genetic characteristics of the population (or species) of interest are compromised by the deliberate selection of widely divergent lines, when such inferences assume a random sample. Despite these limitations, some useful information can be obtained from RFLP analyses.

Several groups have screened large numbers of anonymous clones against a small sample of cultivars or wild lines to ask what proportion of single-copy clones is associated with fragment polymorphism. It can be seen from Table 3, which summarizes the results of several of these studies, that polymorphism is common, although the species, source of clone (cDNA versus genomic), and materials surveyed all influence the level of polymorphism detected. In one

TABLE 3 Fraction of anonymous clones that are polymorphic.

Species	Genomic	cDNA	Number of lines surveyed	Source[a]
Lettuce	10/82 = 12%	25/103 = 24%	3 cultivars and 1 landrace	1
Soybean	60/300 = 20%	—	2 cultivars	2
Brassica	253/361 = 70%	—	Subspecies	3

[a]1. Landry et al. (1987); 2. Apuya et al. (1988); 3. Figdore et al. (1988).

such study, Landry et al. (1987) compared genetic distance rankings for isozymes with those determined for RFLPs and found that the two methods did not always give the same distance rankings. This finding is similar to the results for *D. pseudoobscura* and *D. simulans* cited above where the level of isozyme diversity did not predict the level of restriction site diversity.

In a survey of RFLP variation in soybean, Apuya et al. (1988) report that rearrangements of genomic DNA appear to be the cause of polymorphism for more than 50% of polymorphic probes. The detection of fragment variation associated with two or more enzyme digests for a given probe was the criterion used to test for rearrangements. This is a reasonable and simple test to apply, but it provides no information on whether the putative rearrangements are insertions or deletions versus inversions.

RFLPs are likely to play a significant role in plant improvement programs (see Stuber, this volume). One argument in favor of RFLP markers is that, unlike morphological mutants, RFLPs have no adverse fitness effects, and, consequently, the transfer of particular RFLP markers into adapted cultivars is not expected to be disruptive. While this argument appears reasonable, it is important to note that a substantial proportion of RFLP variation may be the result of insertional events (e.g., Johns et al. 1983; Apuya et al. 1988), and, as we have just seen, population surveys from *Drosophila* raise the possibility that insertional mutations are either mildly deleterious or unstable. If insertional mutations turn out to be mildly deleterious, then their utility in breeding programs could be compromised. If, on the other hand, insertional mutations are unstable, a loss of markers may be experienced. In either case, it is desirable to conduct careful population surveys of insertional variation as a first step in addressing the potential fitness effects associated with random RFLPs.

Complete DNA sequences of genes from population samples

What can be learned about population diversity from complete DNA sequence data that is not already apparent from surveys of restriction site polymorphism? To address this question, we again turn to *Drosophila* work for a paradigmatic example. Krietman (1983) reported complete DNA sequence data for a 2659-bp region containing the *Adh* locus from 11 *D. melanogaster* chromosomes. The sample of chromosomes assayed was chosen from a worldwide distribution and was divided between the two major electrophoretic alleles (*Adh*-F and *Adh*-S).

The resulting data showed that 8 of the 11 genes differed by one or more nucleotide substitutions and were thus distinct alleles. This very high allelic diversity is consistent with predictions of the infinite alleles model for an organism with a value of $N_e \gg \mu^{-1}$ (Crow and Kimura 1970). Interestingly, the data also revealed much more nucleotide diversity within the *Adh*-S allelic class than within the *Adh*-F class, leading to the inference that the *Adh*-S allele predated the *Adh*-F allele. This inference is further supported by the fact that the sibling species of *D. melanogaster, D. simulans*, and *D. mauritana* are monomorphic for the S allele (Bodmer and Ashburner 1984).

The distribution of nucleotide substitutions was classified by functional region (e.g., intron, exon, 5′ flanking sequences, 3′ flanking sequences). This classification revealed a surprisingly low level of nucleotide diversity associated with the 3′ flanking regions of *Adh*; however, no marked differences in nucleotide diversity were observed between introns and exons. It has recently been shown that the low level of nucleotide diversity associated with the 3′ flanking regions can be accounted for by coding exons for an adjacent gene that was previously unrecognized (Schaeffer and Aquadro 1987). Indeed the low diversity value observed for this region stimulated the search for an adjacent coding function.

Perhaps the most interesting result noted by Kreitman (1983) relates to intragenic recombination. The distribution of nucleotide substitutions within the *Adh*-F and *Adh*-S classes is distinct in most sequences. However, in several sequences the substitutions characteristic of the 5′ region of *Adh*-S are associated with *Adh*-F, and conversely, the substitutions associated with the 3′ region of *Adh*-F are associated with *Adh*-S. Evidently, as many as three distinct alleles in the sample had an origin through intragenic recombination rather than mutation (Stephens and Nei 1985). Previously, intragenic recombination had not been regarded as a major force in the production of allelic novelties, so this result is of major significance for population genetic theory. Other examples of intragenic recombination, or gene conversion, producing allelic novelties are to be found in the human genetics literature (e.g., the γ-globin locus, Stephens 1985).

How do plant studies compare with those of *Drosophila* with regard to several sequences of a gene? Is intragenic recombination an important source of allelic novelties in plants? So far there are no published reports of complete DNA sequence data for single-copy genes from population samples for a plant species. In at least three cases, two or more alleles at a locus have been sequenced from plant nuclear single-copy genes, but these alleles are not random samples from a population, and so do not permit the kinds of inferences discussed above. Three alleles of the *bronze* locus of maize that codes for the enzyme UFGT (UDPglucose:flavonal O^3-D-glucosyltransferase) have been sequenced (Ralston et al. 1988). Two of these alleles (*Bz-McC* and *Bz*-W22) are wild type and produce a purple phenotype; the third allele does not produce enzyme protein and results in a bronze color (*Bz-R*). Both *Bz-R* and *Bz-McC* have different large insertions in their 5′ regions that are presumed to be transposable elements. In addition, *Bz-R* has a large 340-bp deletion that includes 285 bp of coding DNA that is assumed to account for the absence of enzyme. The proportion of silent nucleotide substitutions observed among alleles is comparable to that observed for the *Adh* alleles of maize (discussed below). However, a higher proportion of missense substitutions among *bronze* alleles than among *Adh* alleles suggests that the UFGT enzyme is more tolerant of amino acid replacements. Moreover, thermal stability differences are observed among UFGT enzymes and some of these appear to be associated with amino acid replacements, suggesting that subtle phenotypic differences are conferred by some substitutions.

A genomic clone of the *shrunken* (*Sh*) locus of maize (encoding the enzyme sucrose synthase) has been sequenced by Werr et al. (1985), together with a nearly full length cDNA clone obtained from a different maize line. The gene is large (≈5.4 kb) and has 15 introns, most of which are small. A comparison of the genomic and cDNA clones allows an analysis of substitutions in 2100 bp of coding DNA. A total of 16 silent and zero missense substitutions were detected, indicating that, unlike UFGT, there is strong conservation of amino acid sequence. Hudspeth and Grula (1989) have sequenced a cDNA clone of phospho*enol*pyruvate carboxylase (PEPCase) from maize and compared this to a second PEPCase cDNA sequence. They find that 3.4% of synonymous sites have experienced a nucleotide substitution, a value that is quite similar to that reported for *bronze*, *shrunken*, and *Adh* in maize.

Sachs et al. (1986) have sequenced both major electrophoretic alleles of the alcohol dehydrogenase-1 locus of maize (*Adh*-1F and *Adh*-1S). The data provide a number of intriguing insights into molecular diversity in the *Adh*-1 region. First, the *Adh*-1 gene is contained within a strongly conserved region of about 4 kb. Outside of this region and 3′ to the gene there appear to be rearrangements between the two sequences, tandem duplications, one transposable element-like insertion, and one deletion. The distribution of nucleotide substitutions by gene function (introns, exons, 5′ and 3′ flanking regions) within the conserved region indicates a rather uniform level of diversity, independent of function. However, virtually all sequence similarity ends between the two *Adh* alleles at a point downstream from the numerous translation stop signals, where multiple sequence rearrangements occur. Moreover, rearrangements are also suspected to occur upstream from the conserved region, which may account for the fact that the two alleles differ in developmental expression (Sachs et al. 1986).

These results present a very complicated picture of the sequence differences between two allozymes of *Adh*, because they suggest that the DNA outside the conserved region has undergone many complex mutational changes. It would be of great interest to know how representative this picture is of DNA sequence diversity in maize or in other plant species. The *Adh* situation may not be atypical because it has been shown in *Antirrhinum majus* that the transposable element Tam3 can generate deletions of genomic DNA (presumably owing to imprecise excision) and Tam3 also causes complex rearrangements, perhaps owing to recombination between nearby Tam3 elements (Martin et al. 1988). Plant species with active transposable element systems may be subject to frequent DNA rearrangements in noncoding regions. It would be very interesting to know whether these rearrangements have fitness consequences.

POPULATION SURVEYS OF MULTIGENE FAMILY DIVERSITY

Surveys of nuclear-encoded ribosomal gene variation

The majority of population surveys of plant molecular diversity have focussed on the nuclear encoded ribosomal RNA (rDNA) gene family. These genes are tandemly repeated in large arrays (blocks) that occur at one or more chromosomal locations (Dvořák, this volume). The typical repeating unit consists of five regions: internal and external transcribed spacers, an intergenic (nontranscribed) spacer, the 26 S ribosomal RNA sequence, the 18 S ribosomal RNA sequence, and the 5.8 S ribosomal RNA sequence. The intergenic spacer region (which may in fact be transcribed) is itself composed, in part, of a number of subrepeats that range in size from about 100 bp to about 400 bp for different plant species. The subrepeats contain sequences implicated in polymerase binding and, as a consequence, may play a role in gene expression. Much of the variation associated with the rDNA blocks is the result of variations in number of copies of subrepeats and is therefore resolved as length variants of the basic repeating unit. For example, several species of rice and carrot exhibit two length classes (Oono and Sugiura 1980; Kato et al. 1982), soybean has a single length variant (Doyle and Beachy 1985), and barley has 20 length variants (Allard 1988; Saghai-Maroof et al. 1984).

An individual plant can be heterogeneous with respect to rDNA for at least three reasons: (1) two or more length variants occur within a single block, (2) different length variants are associated with different blocks, and (3) heterozygosity—that is, different length variants on homologous chromosomes. This complexity makes the precise analysis of rDNA variation difficult. Perhaps the best documented study of rDNA variation is that of Saghai-Maroof et al. (1984) where a careful analysis of the inheritance of the rDNA variants was initially undertaken. Two separate *Rrn* loci were identified (*Rrn1* and *Rrn2*) associated with chromosomes 6 and 7. Twelve different length variants were shown to be alleles of *Rrn1* and eight length alleles map to *Rrn2* (Allard 1988). Frequency changes in length variants were then followed in a barley composite cross population (CCII) that had been maintained at Davis, CA for many years (see Allard 1988 for a description of the composite cross populations). Marked changes in the frequencies of the various length variants were documented from seed samples spanning a period of approximately 40 years. In addition, the frequencies of length classes have been surveyed in a worldwide sample of wild barleys. The wild materials are characterized by greater levels of rDNA diversity than are the cultivated barleys, possibly owing to selection and reductions in population size associated with domestication (Allard 1988).

A number of population surveys of rDNA variation in crop plants and their wild relatives have appeared in the past several years (wild wheat, *Triticum dicoccoides*, Flavell et al. 1986; barley, *Hordeum vulgare* and *H. vulgare* ssp. *spontaneum*, Allard 1988; pearl millet, *Pennisetum glaucum*, Gepts and Clegg, 1989). All of these studies agree in showing that variation in spacer length is common (Table 4). Pearl millet appears to have the lowest levels of length variation, perhaps owing to reductions in species' N_e associated with the expansion of the Sahara. Some studies of natural populations of wild plant species also indicate abundant spacer length variation (Schaal et al. 1987; Learn and Schaal 1987).

The mechanism generating spacer length variation is believed to be unequal recombination, and possibly gene conversion between subrepeats within rDNA gene copies (Dvořák et al. 1987). Both of these mechanisms can cause the spread of mutant sites between adjacent gene copies. This type of spread leads to a pattern of "concerted" change in the gene family (Arnheim et al. 1980). Dover (1986) has argued that these asymmetric exchanges cause long-term evolutionary change. This change, termed "molecular drive," is the result of the genetic system rather than natural selection. Whatever the long-

TABLE 4 rDNA length diversity in surveys of plant populations.

Species	Number of length variants	N	Source[a]
Barley	20	>250	1
Wild wheat	>10	112	2
Pearl millet	4	88	3
Clematis fremontii	7	217	4
Phlox divaricata	12	65	5

[a]1. Allard (1988); 2. Flavell et al. (1986); 3. Gepts and Clegg (1989); 4. Learn and Schaal (1987); 5. Schaal et al. (1987).

term implications, when genetic change occurs through concerted evolution, we expect to see cases where two or more variants are associated with a single rDNA block and, as a consequence, individual genomes may display several different rDNA variants. This expectation is confirmed by virtually all studies when two or more length variants occur within populations. For example, the average individual of *Phlox divaricata* ssp. *divaricata* has 2.47 rDNA length variants per diploid genome (Schaal et al. 1987). In tetraploid wheat (a self-fertilizing species that is expected to be highly homozygous), some populations have as many as 3.3 rDNA length variants per average individual (Flavell et al. 1986). Thus we may conclude that rDNA variation is common in plant populations and that there is a continuous flux of rDNA types in plant genomes.

Hypervariable sequences in plant populations

Jeffreys et al. (1985) found that the genome of man contains classes of hypervariable minisatellite sequences that permit the unique identification of individuals in a population ("genetic fingerprints"). The minisatellite sequences consist of dispersed arrays of short tandem repeats that are believed to be "hot spots" for recombination. It is thought that the hypervariability is caused by unequal recombination within the dispersed arrays and that the frequency of recombination per kilobase is more than an order of magnitude greater than observed for typical regions of the genome. Hypervariability is seen as changes in fragment mobility, owing to changes in fragment length generated by unequal recombination. Since this discovery, the "genetic fingerprinting" of individuals has found a number of applications in human genetics, including forensic medicine and in paternity analysis.

Do hypervariable sequences also occur in plant populations? The tentative answer seems to be yes. Remarkably, a human minisatellite probe has been shown to hybridize to rice genomic DNA and to yield a complex fragment pattern that seems to be highly variable (Dallas 1988). In addition, a portion of the M13 bacteriophage genome that had previously been shown to hybridize to human hypervariable sequences has recently been shown to reveal highly variable patterns in DNAs from several plant species (Rogstad et al. 1989). Finally, RFLP studies frequently turn up repetitive sequences and some of these may be candidates for hypervariable sequences (Apuya et al. 1988). It is obvious that hypervariable probes will be useful in plant population genetics for studies of plant mating systems, studies of plant neighborhood size, pollen and seed dispersal characteristics, migration, etc. It is equally obvious that such probes will also be of great utility in agriculture. For example, the pollen parent involved in crosses leading to important cultivars is frequently unknown in avocado and in some other tree crops where the breeding system is difficult to manipulate. Hypervariable probes provide a potential means of retrospectively identifying successful pollen parents, so that they may be incorporated into future breeding efforts.

Rapid change of plant genome composition

One of the more remarkable discoveries of plant genetic research is that components of plant nuclear genomes can undergo large, heritable changes within single generations. (Progeny plants that exhibit such differences are refered to as genotrophs.) The most extensive evidence for rapid genomic change initially came from research with flax where specific stress environments were shown to elicit major phenotypic changes that were heritable (Durrant 1962). Cullis (1987) has shown that flax genotrophs are characterized by large changes in DNA content (up to 15% of total DNA) and these changes involve highly repetitive and intermediately repetitive fractions of the genome (Table 5). Rapid genomic changes have also been documented in other plant species (maize, Walbot and Cullis 1985; *Microseris* species, Price et al. 1983; *Helianthus*, Cavallini et al. 1986).

The mechanism(s) that cause rapid copy number changes in some multigene families is (are) unknown and the role of environmental stresses in evoking changes is also obscure. What is well established (at least in the case of flax) is that substantial quantitative changes in phenotype, that are subsequently inherited in a stable fashion, can be induced. The fact that large phenotypic changes in traits like plant height and growth rate occur argues strongly that the changes can have fitness consequences. The general significance of these observations for plant population genetics has yet to be investigated.

ORGANELLE GENOMES

Chloroplast genome variation in populations

It is now well established that the chloroplast genome (cpDNA) evolves at a conservative rate (Curtis and Clegg 1984; Palmer 1987). Analyses of DNA sequence data show that the evolutionary rates of chloroplast-encoded genes are considerably below those observed for animal or plant nuclear genes (Zurawski and Clegg 1987; Wolfe et al. 1987). In view of the slow rates of gene evolution, it is not surprising that population diversity is frequently low (reviewed by Birky 1988). However, population surveys are difficult and time consuming for the reasons already discussed concerning surveys of nuclear gene diversity, and, as a consequence, only a few population surveys of cpDNA variation have appeared.

Intraspecific surveys of cpDNA variation reveal, not surprisingly, that species differ with regard to cpDNA diversity. For example, surveys of cpDNA revealed no variation in pearl millet (Clegg et al. 1984a; Gepts and Clegg,

TABLE 5 Variations in DNA content among flax genotrophs.[a]

Plant source	Probe			
	rDNA (2,400)[b]	5 S DNA (120,000)[b]	Satellite-1 (62,000)[b]	Satellite-2 (131,000)[b]
	% of nuclear DNA			
Parent	1.5	3.0	5.6	2.2
Large genotroph	1.6	3.2	5.8	1.7[c]
Small genotroph	0.8[d]	2.2[d]	5.6	1.2[c]

[a]After Cullis (1987).
[b]Copy number per diploid genome.
[c]Differs significantly from the parental value ($p < 0.01$).

1989). In contrast, several studies have uncovered cpDNA polymorphism within wild and cultivated barley, with somewhat greater levels of variation observed in wild barley (Clegg et al. 1984b; Holwerda et al. 1986; Neale et al., 1989). In a detailed study of 100 *Lupinis texensis* plants, Banks and Birky (1985) found that three restriction sites and one small insertion were polymorphic.

Insertion or deletion polymorphism seems to account for a substantial fraction of cpDNA fragment variation in the few population studies where intraspecific sample sizes are greater than ten genomes and where mapping of site variation allows the identification of moderate sized insertions/deletions (Table 6). Complete DNA sequence data for the tobacco chloroplast genome (Shinozaki et al. 1986) show that most of the genome has a coding function, but there are numerous intergenic regions, often of several hundred base

TABLE 6 Intraspecific cpDNA variation.

Species	Site changes	Insertion	N[a]	Source[b]
Pennisetum glaucum	0	0	78	1
Lupinus texinesis	3	1	100	2
Hordeum vulgare	3	1	64	3
Zea mays	2	1	13	4

[a]N = sample size.
[b]1. Gepts and Clegg (1989); 2. Banks and Birky (1985); 3. Neale et al. (1988); 4. Doebley et al. (1986).

pairs, that do not code for protein products. Presumably the insertion/deletion changes observed in population studies are confined to this relatively minor fraction of the genome. (This assumption is based on the fact that protein function would most likely be disrupted by the relatively large insertions/deletions resolved in Southern analyses.) If this is the case, there may be a relatively high frequency of insertion/deletion mutation, because the target regions where insertions or deletions can be accepted constitute a small fraction of the genome, compared to site changes that are assumed to result from nucleotide substitutions and may occur throughout the genome. Despite this small target area, insertions/deletions appear to account for a third to a half of all population variation in fragment mobility. Additional studies of intraspecific variation may help to resolve questions about the relative rates of these different classes of mutations. Intraspecific studies are also important because the use of cpDNA for biosystematic and phylogenetic studies must properly account for intraspecific variation to avoid biases in phylogenetic inference, caused by unrepresentative samples.

A very interesting and potentially useful recent discovery is paternal inheritance of cpDNA in several conifer species (Neale et al. 1986; Wagner et al. 1987). This stands in contrast to those studies of angiosperm cpDNA transmission that indicate maternal or biparental transmission (Sears 1983). Different asymmetric modes of organelle transmission could have great utility for the study of various population genetic questions. For example, paternally transmitted markers could, in combination with maternally transmitted markers (e.g., mitochondrial DNA polymorphisms discussed below), be used to dissect pollen and seed components of gene flow. In addition, questions about population structure and hybridization could be addressed by resolving different paternal and maternal lineages within and between populations.

Population variation in mitochondrial DNA

The plant mitochondrial genome (mtDNA) constitutes an evolutionary enigma. It has a large genome size relative to animal mitochondria and large size variations are observed among closely related taxa; it is capable of integrating foreign DNA sequences (e.g., from the choloroplast genome); it is characterized by frequent rearrangements, but has very slow rates of nucleotide substitution (at least in protein-coding genes); and it is sometimes associated with linear supernumerary DNA molecules that appear to be transposon-like (reviewed by Sederhoff 1987). While several studies of mtDNA variation have been published (e.g., Timothy et al. 1979; Holwerda et al. 1986), these complexities have tended to preclude inferences about population processes.

The most detailed investigation of plant mtDNA diversity published to date is that of Palmer (1988), who carried out careful restriction site mapping of mtDNA molecules in several lines from each of eight *Brassica* species. In this work, no intraspecific site variation was detected, although two length polymorphisms were uncovered, as was a single large (62 kb) inversion polymorphism. This work provides little evidence for high rates of mtDNA variation within plant species. It does show, however, that most of the variation detected results from insertions/deletions or rearrangements, as opposed to nucleotide substitutions. This means that mtDNA markers, useful for addressing questions of population structure, can probably be uncovered if sufficient effort is invested.

CONCLUSIONS

With respect to resolving power, it is not an exaggeration to compare molecular techniques to the invention of the microscope, because we can now detect and quantify genetic diversity at the most elemental level—that of DNA sequences. This greatly enhanced resolving power allows the precise characterization of mutational differences among individuals and it thereby yields a much more detailed view of the mutational processes that generate genetic variation. Despite its great power, the molecular study of plant genetic diversity is in its infancy and detailed investigations of single-copy gene variation in plants are extremely limited. It is necessary to turn to examples from population surveys of *Drosophila* to highlight some of the questions that need to be investigated in plant population studies. One especially pressing problem concerns the average fitness effects of insertional polymorphism. Studies from *Drosophila* suggest that insertional polymorphisms may be weakly deleterious or unstable. Because much RFLP variation detected in plant species appears to be insertional, it is important to address the question of fitness effects to determine whether RFLPs will have some negative effects when used as markers in plant breeding programs. Population surveys of complete DNA sequence data from *Drosophila Adh* reveal that intragenic recombination plays a major role in the generation of allelic novelties. No comparable data are presently available from plants, thus precluding an assessment of the significance of intragenic recombination as a force in the generation of allelic novelties in plant species.

The class of plant genes most studied at the intraspecific level are the genes encoding the ribosomal RNAs (rDNA). Most rDNA variation is associated with length changes mapping to the nontranscribed spacer region. Many length variants are observed in some plant species and it is common to find that individual plant genomes harbor several length classes. Hypervariable minisatellite sequences appear to be present in plant genomes (as in the human genome) and these variants may prove very useful in addressing numerous population genetic questions. Perhaps most intriguing are studies of rapid changes in copy number in a variety of multigene families that are apparently associated with specific stress environments. These changes can cause marked phenotypic differences and they appear to be inherited in a stable fashion. The implications of rapid genomic change for plant evolution are profound, but at this stage little is known about the general significance of the phenomenon.

Studies of organelle variation have been largely conducted at the interspecific level; however, several population surveys have been reported of cpDNA diversity. These studies reveal that up to 50% of restriction fragment polymorphisms are the result of moderate-sized insertion/deletion changes. Because the relatively small noncoding portions of the cpDNA molecule are the presumed target sites for these changes, it is possible that insertion/deletion changes occur at a relatively rapid rate, compared to nucleotide substitutions, in noncoding portions of the genome. Current very limited data on the mitochondrial genome (mtDNA) indicates that intraspecific variation is largely the result of insertion/deletion and rearrangement changes. Different patterns of maternal (mtDNA) and paternal (cpDNA) transmission of organelle genomes in conifers may prove especially useful in addressing a variety of population genetic problems in these species.

ACKNOWLEDGMENTS

I thank Drs. Gerald Learn, Edward Golenberg, Glenn Furnier, and Paul Gepts for comments on an earlier draft of this chapter. Supported in part by NSF Grants BSR-8500206 and BSR-8614608.

CHAPTER 7

QUANTITATIVE CHANGES IN MAIZE LOCI INDUCED BY TRANSPOSABLE ELEMENTS

Oliver E. Nelson

ABSTRACT

Transposable elements have long been been recognized as effective inducers of genetic change, but it has been only recently that it has been possible to interpret the effects in molecular terms and correlate the molecular changes with the biochemical effects. Insertion and subsequent excision of *Ac,Ds* and *Spm,dSpm* elements can create a variety of wild-type isoalleles in addition to null alleles. In the absence of a transactive *Spm,* certain insertions of a *dSpm* into coding sequences also constitute wild-type isoalleles. Specific instances at such well-studied loci as *bronze-1*, *shrunken-1*, and *waxy* are discussed.

INTRODUCTION

The study of transposable elements in plants has been transformed over the past few years by molecular techniques, and many of the baffling questions raised by the genetic investigations initiated by McClintock (1950) and joined subsequently by others have been answered wholly or in part. The emerging answers have also often supported hypotheses concerning the structure and behavior of transposable elements, based on the results of genetic experiments. At the same time, transposable elements can enable via transposon tagging the cloning of genes that are otherwise difficult to access (Fedoroff et al. 1984; O'Reilly et al. 1985; Schmidt et al. 1987; Cone et al. 1988). The structural changes that are also attributable to the action of transposable elements have been discussed by McClintock (1978) and Schwarz-Sommer et al.

(1985). Both the insertions and excisions of transposable elements are capable of creating *de novo* wild-type isoalleles that vary quantitatively or both qualitatively and quantitatively from the progenitor allele in which the insertion originally occurred. MacKay (1987) has reported the large increase in variance for a quantitative trait (abdominal bristles) in *Drosophila* resulting from P element-induced mutagenesis over that resulting from X-ray-induced mutagenesis. In this chapter, I intend to focus on transposable element insertions and excisions with an emphasis on the quantitative changes that have been shown to occur to functional alleles at a few well-studied loci.

As a preamble to this discussion, a brief summary of the genetic and molecular knowledge concerning maize transposable elements is essential. There are a number of families of different elements. The best known are the *Activator, Dissociation* (*Ac,Ds*) family first identified by McClintock (1950) and the *Suppressor-Mutator* (*Spm*) family identified independently by McClintock (1954) and by Peterson (1953) who designated it as the *Enhancer, Inhibitor* (*En,I*) system. The *Mutator* (*Mu*) family detected by Robertson has also been intensively studied, but salient points concerning the system remain to be elucidated (Lillis and Freeling 1986). This system will not be further considered here. The *Ac,Ds* and *Spm* families have two components. One is an autonomous transposable element that is capable of catalyzing its own excision from one location in the genome and reinsertion at another location. For the *Ac,Ds* family, the autonomous element is *Ac*, and for the *Spm* family, it is *Spm*. The second member of these transposable element families [*Ds* for the *Ac,Ds* family and a defective *Spm* (*dSpm*) for the *Spm* family] is a nonautonomous element that is usually derived from the autonomous by an internal deletion. In addition, there exist anomalous *Ds* elements (*Dsl's*) that bear little resemblance to *Ac* except for a limited number of base pairs at either end of the element and are unlikely to have arisen from *Ac* by an internal deletion. These nonautonomous elements can transpose only when an autonomous element is also present in the genome. Thus the insertion of a nonautonomous element in the coding region of a gene creates a mutant allele that is stable in the absence of an autonomous element of the cognate family. In the presence of the relevant autonomous element, one observes genetic instability revealed somatically by patches of nonmutant tissue superimposed on a mutant background and germinally by the production of nonmutant gametes. The nonmutant gametes and the somatic sectors both result from excisions of the nonautonomous element that either restore the nucleotide sequence existing before the element insertion or create a new sequence in the area of the insertion that still codes for sufficient gene activity to condition a nonmutant phenotype. One observes the same spectrum of effects when an autonomous element is present in the coding region of a gene. A detailed review of the genetic aspects of maize transposable elements has been prepared by Fedoroff (1983).

Details of the molecular genetics of transposable elements have been steadily emerging since the first report of the sequence of a transposable element, which was the *Dsl* element present in the *adhl-Fm335* allele (Sutton et al. 1984). To summarize these advances briefly, we know that the transposable elements are flanked by inverted repeats of a characteristic size and sequence for a family (11 bp for the *Ac,Ds* family and 13 bp for the *Spm* family). On insertion, a duplication of host sequences is created. For *Ac,Ds*, this is 8 bp and for *Spm* the duplication is 3 bp. On excision of a transposable element, all, part, or none of the host duplication may remain, and one or several nucleotides at the junction between the original sequence of the gene and the duplication may be changed. A model by Saedler and Nevers (1985) accounts for these changes in terms of a staggered cut such as occurred on insertion followed by exonuclease action on the single-stranded ends and DNA polymerase acting to fill the gaps and possibly switching templates. The various events that may ensue on excision account for the fact that null, partially functional, and fully functional stable alleles are isolated as germinal derivatives of transposable element-controlled alleles.

Doering and Starlinger (1986) have reviewed molecular aspects of maize transposable elements. Since that review, considerable data have accumulated concerning the ability of the maize splicing system to remove most of the sequence of a nonautonomous transposable element from the pre-mRNA in which it is present. Wessler et al. (1987) have shown that *Ds* insertions in *waxy* alleles are transcribed and subsequently removed by splicing events in which one of two donor sequences present in the first 27 bases at the 5′ side of the *Ds* element is utilized as a donor site. Since there is no acceptor site in the 3′ side of the *Ds*, the first acceptor site in the *waxy* sequence 3′ to the *Ds* insertion is used. A similar splicing event removes the *Dsl* sequence present in the 5′ transcribed but untranslated leader of the *adhl-335* primary transcript (Dennis et al. 1988).

A different type of splicing event has been reported for a *dSpm* insertion in a *Bz* allele that has created the *bz-m13(CS9)* allele (Kim et al. 1987). The functional alleles at the *bz* locus catalyze a late step in anthocyanin biosynthesis. Null alleles at the locus result in the presence of characteristic bronze pigmentation in tissues that would have anthocyanin pigmentation if a *Bz* allele were present. One copy of the CS9 allele (*bz/bz/CS9*) conditions a nonmutant phenotype in the triploid endosperm of the maize seed. *CS9* results in the production of 40–50% as much UDPglucose:flavonol O_3-D-glucosyltransferase (UFGT, EC 2.4.1.91), the enzyme for which the *Bz* alleles code (Dooner and Nelson 1977a; Larson and Coe 1977) as does the progenitor allele. The 990-bp *dSpm* in *CS9* is inserted in the second exon of the *Bz* allele 38 bp 3′ to the 3′ side of the single intron present in the *bz* locus. In the processing of the primary transcript, which includes the *dSpm* sequence, the donor site of the intron is utilized, but the acceptor site of the intron is skipped. The spliceosome uses an acceptor site two bases from the 3′ end of the *dSpm* so that the intron, the 38 bases of *Bz* sequence, and all of the *dSpm* except the two terminal bases on the 3′ side are removed. While 38 bases of *Bz* sequence are lost in processing, the three-base duplication of *Bz* sequence created upon the *dSpm* insertion remains, as do the terminal two bases of the *dSpm*, so there is a net loss of 33 bases. The reading frame is preserved, and a functional protein is produced in spite of the loss of 11 amino acids and change of one, from alanine to valine at the junction (Figure 1). In effect, the insertion of the *dSpm* has created a new intron.

THE GENERATION OF SEQUENCE DIVERSITY

It is clear that transposable elements constitute a mechanism for creating complex rearrangements as they move about the genome (McClintock 1977), and several have been analyzed in detail (Courage-Tebbe et al. 1983; Klein et

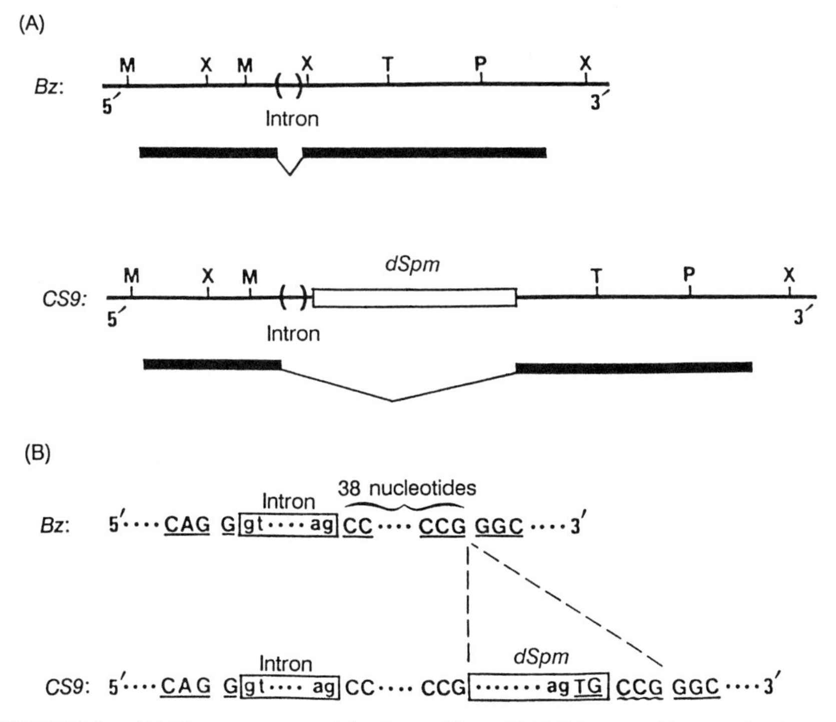

FIGURE 1. (A) The structures of the *Bz* and *bz-m13(CS9)* pre-mRNAs and mRNAs. (B) The nucleotide sequences surrounding the intron and the site of the *dSpm* insertion in *Bz′-3* and *CS9*. The wavy line indicates the duplication of *Bz* sequence, and the underlined trinucleotides indicate codons in the *Bz* reading frame. From Kim et al. (1987).

al. 1988). Deletions of various sizes are also a possible consequence of transposable element excision (Dooner et al. 1988). However, in an analysis of *bz'* derivatives of *bz-m13*, Schiefelbein (1987) found that only one of 13 derivatives involved a deletion of the *Bz* sequence. We have adopted the convention of referring to the stable alleles derived from a transposable element-controlled *bz-mutable* allele as *bz'* or *Bz'* alleles. It is apparent that the majority of changes to a coding sequence that occur upon excision of a transposable element are a result of the seemingly random events that ensue. The transposable element is excised precisely, but the duplication of host sequences may be left intact, partially eliminated, or wholly eliminated. The net result is obviously a function of the characteristic size of the duplication created by a transposable element family and the number of bases in that duplication that are eliminated in the repair process following excision as related to the reading frame. For the *Ac,Ds* family, which produces an 8-bp duplication on insertion, losses of 2, 5, or 8 bp of the duplication (or the possible gain of one) preserve the reading frame. While the elimination of 8 bp would restore the original reading frame, the amino acid sequence may be altered owing to the change in one or two bases at the insertion point. When one or two amino acids are added to the polypeptide as a consequence of the extra bases remaining after transposable element excision, the results are unpredictable.

The data are scanty concerning the results of excision events that restore sufficient gene function to the locus to condition a nonmutant phenotype. Dooner and Nelson (1979) reported that of 15 *Bz'* derivatives from *bz-m2 (DI)*, which has a *Ds* insertion in a *Bz* allele, only 5 derivatives were indistinguishable from the progenitor *Bz* allele. The properties tested were specific UFGT activity in mature seeds, thermal stability at 55°C, stability of enzymatic activity during electrophoresis, and amount of protein that was cross-reactive (CRM) with antiserum against the *Bz* protein. Of the other 10 derivatives, 9 differed markedly from the progenitor *Bz* allele in all these parameters. Enzymatic activity ranged from 0 to 5%, half-lives at 55°C about 1/5, CRM from 0 to 12%. There was no activity detectable after polyacrylamide gel electrophoresis. One derivative was intermediate between these two classes in that it had 20% as much enzymatic activity and 43% CRM as the *Bz* progenitor but also the reduced thermal stability and loss of enzymatic activity during electrophoresis observed in those derivatives with the very reduced enzymatic activity. Of the seven *Bz'* derivatives isolated from *bz-m13*, only one produced enzymatic activity equal to the *Bz* progenitor (Nelson and Klein 1984). This derivative is one that is known to have had the preexisting *Bz* sequence restored upon the *dSpm* excision (Schiefelbein et al. 1985). It should be apparent from the examples cited that excisions of a transposable element from a coding sequence in events creating derivative alleles that condition a nonmutant phenotype restore the sequence existing previous to the insertion only in a minority of instances. The majority of these functional derivative alleles are wild-type isoalleles differing both quantitatively and qualitatively from their progenitor alleles.

There is evidence concerning the generation of quantitative changes by transposable elements at other loci in maize, notably the *waxy* locus. Wessler et al. (1986) have reported that two germinal derivatives of *wx-ml* (which has a *Dsl* insertion in an exon of a *Wx* allele), *wx-S5* and *wx-S9*, have 9 and 6 bp, respectively, left at the site of the *Ds* insertion. The *S5* allele conditions the production of starch with 5% amylose (as compared to 25% in *Wx* starch) and has 32% of *Wx*'s enzymatic activity. The *S9* allele conditions the production of starch with 9% amylose and has 53% of *Wx*'s enzymatic activity. It is clear that the addition of two or three amino acids to the *Wx* protein at this point is detrimental to the functioning of the protein. Schwarz-Sommer et al. (1985) have recorded 16 excision events from *wx-m8* (a *dSpm* inserted into an exon of a *Wx* allele) in somatic tissues in the presence of a transactive *Spm*. Two excisions restored the preexisting sequence; six created frameshift mutations; seven events would result in the addition of a leucine to the amino acid sequence; and one would result in the deletion of lysine and phenylalanine from the polypeptide. Since these derivative alleles encoding proteins with altered sequences were isolated only in somatic tissues, it was not possible to ascertain whether the proteins would be functional.

A special class of wild-type isoalleles resulting from transposable element insertions should be discussed here. These differ from those discussed above that result from the insertion and subsequent excision of a transposable element. The evidence that transposable element insertions can be transcribed as the gene is transcribed and then removed from the pre-mRNA has been discussed briefly. In the case of *Ds* insertions in a coding sequence, no instance is known in which the allele with the insertion can condition a nonmutant phenotype in the absence of *Ac*.

The situation is different with respect to the *Spm* (*I,En*) family of transposable elements. McClintock (1954, 1965) noted early in her investigations of the system that some associations of the nonautonomous member (which we now know to be a defective *Spm*) with a gene create new alleles functional to some extent in the absence of the transactive *Spm*. Some of these alleles can condition a nonmutant phenotype. In the presence of a transactive *Spm*, such alleles are suppressed in their activity, leading to a fully mutant phenotype on which background are superimposed sectors of nonmutant tissue resulting from excisions of the *dSpm* that restore function to the locus. It is on the basis of such observations that McClintock discerned two functions of the transactive element, those of suppressing any activity of an allele with a *dSpm* insertion and then catalyzing the excision of the *dSpm* from the allele. Hence the designation of the autonomous element as *Suppressor-mutator (Spm)*. The *bz-m13* allele and almost all changes of state derived from it are examples of *dSpm* insertions in a functional allele creating new alleles that in the absence of a transactive *Spm* still condition nonmutant phenotypes. This capability for CS9 is enabled by the splicing reaction explained earlier. CS9 (45–50% *Bz* activity) was derived from *bz-m13*, which conditions much lower enzymatic activity (5–10% *Bz* activity), by a large deletion in the *dSpm* that removed an internal splicing acceptor site present in the *dSpm* in *bz-m13*. This internal site is used at least as often as the terminal acceptor site in processing *bz-m13* pre-mRNAs, and its use produces a larger message with a stop codon in the *dSpm* sequence still remaining in the mRNA (Kim 1987). The only other derivative of *bz-m13* analyzed to date that has enhanced enzymatic activity (20–25% *Bz* activity) is CS3 (Figures 2 and 3). CS3 also has a large deletion in the 5′ half of its *dSpm*, but the deletion has not removed the internal acceptor site. However, this site is not used in processing the pre-mRNA for reasons that are not now apparent, although we have not yet shown that the

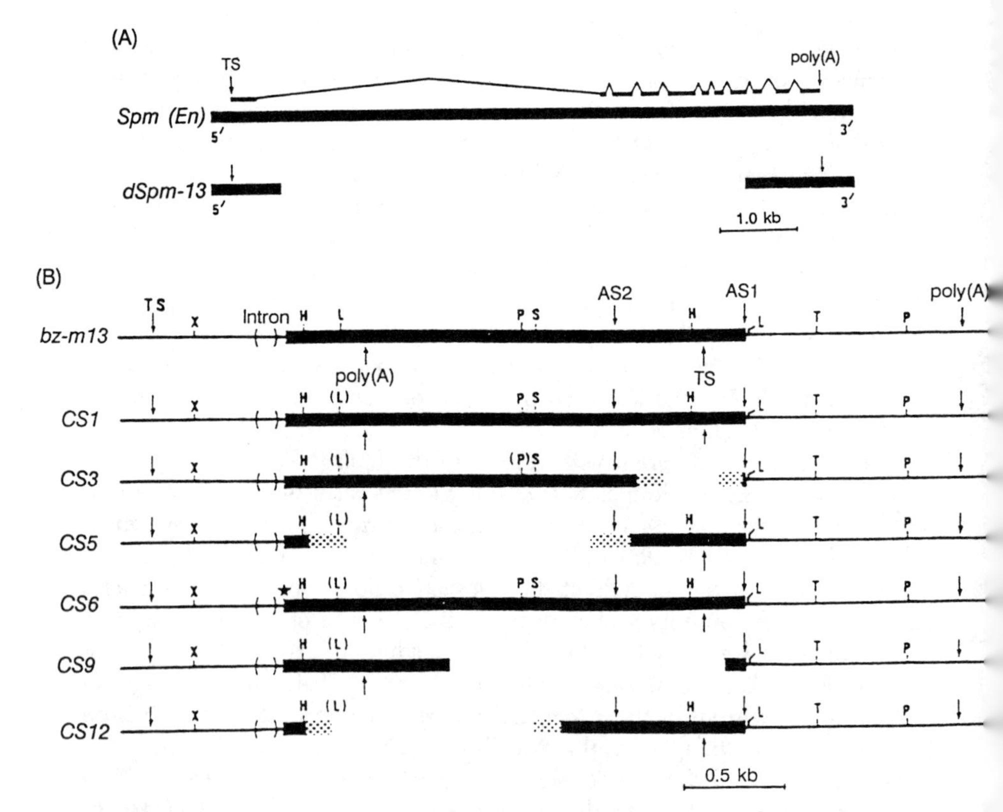

FIGURE 2. Structure of *bz-m13* and change of state derivatives. The restriction endonuclease maps show the portions of the 2.2-kbp *dSpm* present (solid bars) and absent (open areas) in a derivative. The lightly shaded areas represent areas of uncertainty as to whether that portion is present or not. From Schiefelbein et al. (1985).

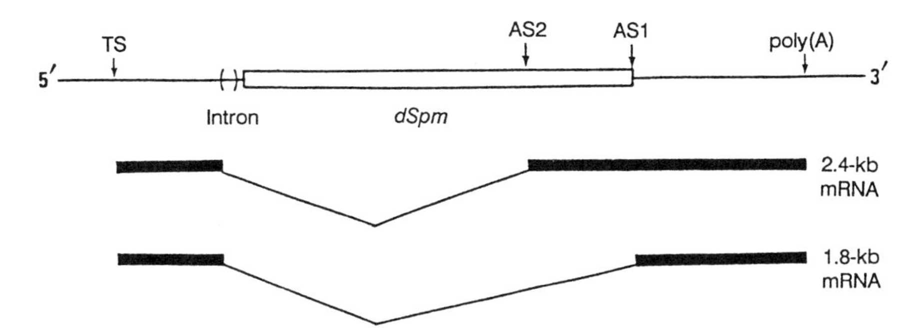

FIGURE 3. The structure of the two poly(A) mRNAs produced by *bz-m13* indicating the internal acceptor splice site, the use of which produces the 2.4-kb message. TS marks the start of transcription. From Kim (1987).

acceptor site has not been altered. It seems clear that we do not understand all the rules governing the splicing of pre-mRNAs containing a transposable element. The *bz-m13* allele and its change of state derivatives are the only *dSpm*-controlled alleles with partial function that have been analyzed biochemically and molecularly. It is not certain that the explanation for their activity is pertinent for other *dSpm*-controlled alleles that retain some activity in the absence of a transactive *Spm*. My conjecture is that splicing reactions of the type observed for the *bz-m13* alleles underlie all such active alleles resulting from the insertion of a *dSpm* in the coding region of a gene.

The excisions of transposable elements inserted in coding sequences result in stable null, partially active, or fully active alleles. When the resultant allele has activity and the excision event has not restored the sequence existing before the insertion (by eliminating completely the duplication of host gene sequences created by the insertion), that allele almost certainly differs from its progenitor wild-type allele both quantitatively and qualitatively as in the examples already discussed. There are, however, reported instances in which a mutable allele (functional allele with a nonautonomous insertion) in the absence of its transactive element produces a reduced amount of a gene product that is indistinguishable from that produced by its progenitor. The cases reported to date involve insertions in the 5′ noncoding regions of genes. Sutton et al. (1984) noted that the *Adhl* mRNA produced by the *Adh1-Fm335* allele was the approximate size of the mRNA produced by the wild-type progenitor in spite of the presence of a 403-bp *Dsl* insertion between the start of transcription and translation start site. They suggested that either the insertion was processed out after transcription or a new start site for transcription within the *Ds* sequence was used that resulted in a message that was wild type in size. Subsequent work by the same laboratory (Dennis et al. 1988) has shown that the insertion is indeed transcribed and then processed out using a donor splice within the first 15 bases of the *Ds* and an acceptor site at the 5′ junction of the *Ds* and *Adh1*. The *Adh1* mRNA in *Adh1-Fm335* is only 1/100 of that present in nonmutant genotypes in spite of an equal transcription rate so the pre-mRNA may be inefficiently processed or the mRNA may be markedly unstable.

Another informative case is that of *Bz-wm*, which was isolated some years ago by McClintock (1962) and which is a *Ds*-controlled mutable allele. In the absence of *Ac*, a single copy of *Bz-wm* (*bz*/*bz*/*Bz-wm*) conditions the production of 60% as much anthocyanin pigment as a *Bz* allele and a noticeably paler aleurone. In the presence of *Ac*, the pigment is still produced, and one observes deeper spots of color marking the clones of cells resulting from excision events that restore a higher level of gene function. Dooner and Nelson (1977b) reported that this allele produced markedly lower levels of enzymatic activity than the presumed *Bz* progenitor and that the enzyme was decidedly more thermal-labile. The molecular analysis of *Bz-wm* has shown that these two changes have different bases (Schiefelbein et al. 1988). The *Bz-wm* allele descended from *bz-m2* (an *Ac* insertion in the second exon of a *Bz* allele) in two steps. The first step was the derivation of a *Bz′* allele conditioning a nonmutant (fully pigmented) phenotype as a result of the *Ac* excision. From this allele, there was derived in a subsequent generation *Bz-wm*, which was shown to have come under *Ds* control. The *Ds* insertion is 63 base pairs 5′ to the transcription start site (Figure 4) and results in much reduced message production and hence lower enzyme production but is apparently not the cause of the lower thermal stability. Our hypothesis is that this change in thermal stability is ascribable to the "footprint" left on excision at the site of the *Ac* insertion in *bz-m2*. Sequencing across this site has shown that 3 bp of the 8-bp duplication created on insertion of *Ac* were left following excision (Figure 5). This would result in an additional glycine being inserted in the

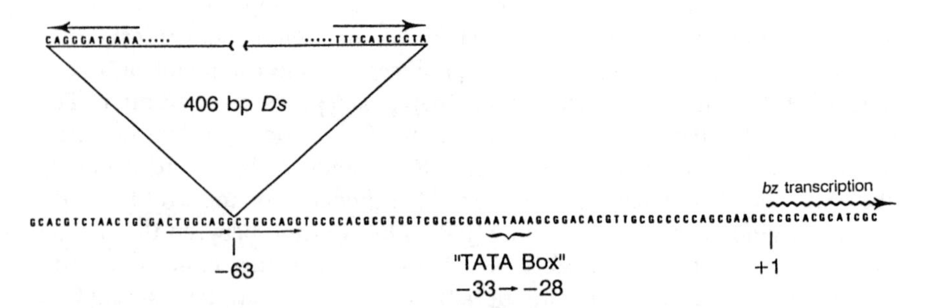

FIGURE 4. The DNA sequence near the insertion site of the *Ds1* in *Bz-wm*. The 8-bp duplication of *Bz* sequence and the 11-bp inverted repeats of the *Ds1* are marked by arrows. From Schiefelbein et al. (1988).

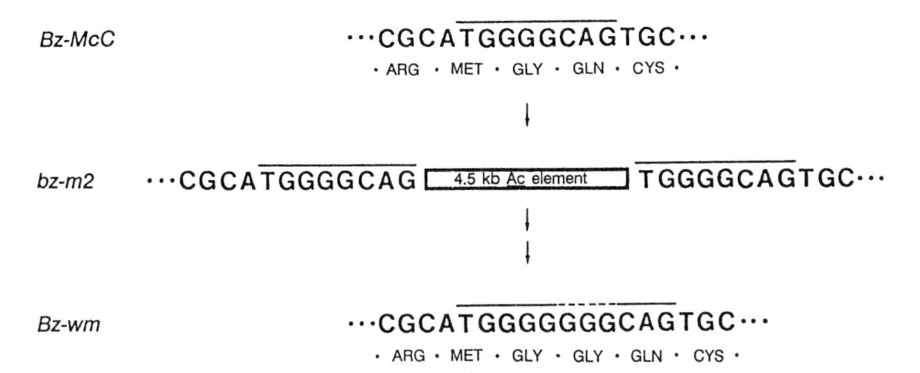

FIGURE 5. The nucleotide sequence of *Bz-McC* and *Bz-wm* surrounding the site of the *Ac* insertion in *bz-m2*. The dashed line over the *Bz-wm* sequence marks the three nucleotides left following *Ac* excision. From Schiefelbein et al. (1988).

UFGT polypeptide, and this is apparently sufficient to destabilize the protein to some extent. Corroboration for this conclusion comes from a study of three *Bz′* derivatives of *Bz-wm* in which the *Ds* has been excised from the 5′ non-transcribed leader in which it is located. All three derivatives retain the lowered thermal stability characteristic of *Bz-wm*.

A reasonable question concerning the changes in sequence of a locus brought about by the insertion and subsequent excision of a transposable element is whether these perturbations always result in a protein product that is reduced quantitatively or affected negatively in some parameter. On this point, the data are scanty. Most studies of the derivatives of transposable element-controlled genes have simply noted the relative frequency of stable null versus stable nonmutant derivatives without inquiring further into the manner in which these nonmutant derivatives relate to a wild-type progenitor allele. The wild-type alleles at most loci have a large margin of excess activity over and above the activity required to condition a nonmutant phenotype, and, as we have seen, almost all derivatives in which there has been a sequence change relative to the progenitor are less effective in some way. However, Tuschall and Hannah (1982) have reported the examination of five *Sh2′* derivatives from *sh2-ml* (a *Ds*-controlled mutable allele of a structural gene for a subunit of the endosperm-specific ADPglucose pyrophosphorylase in maize). They found one derivative allele that produces 140% of the enzymatic activity of the progenitor *Sh2* allele. A second derivative coded for an enzyme that is markedly less inhibited by inorganic phosphate than is the enzyme encoded by the progenitor allele. These interesting derivatives are only now being investigated at the molecular level to ascertain the basis for these changes.

DISCUSSION

It is apparent that the insertion in and subsequent excision from a location within the coding sequence of a gene is very likely to effect sequence changes at that location. If the derivative allele is one that conditions a nonmutant phenotype it is very likely to be altered in a manner that results in a gene product that is qualitatively and quantitatively different from that of the progenitor allele. When the insertion is located in the 5′ nontranslated region, the excision could produce a change that has a quantitative but not a qualitative effect on the gene product. This can also be true of an insertion in the 5′ nontranscribed region as shown by *Bz-wm*. A second path to the generation of alleles altered qualitatively and quantitatively is by the insertion of a nonautonomous member (a *dSpm*) of the *Spm* family in a coding sequence. Such insertion mutants, in the absence of a transactive *Spm*, sometimes retain sufficient function to lead to a nonmutant phenotype as pointed out by McClintock (1965). In the case of *bz-m13*, we understand how this is possible. The point to be made is that without an autonomous *Spm* present, such insertion mutants behave as wild-type isoalleles.

In maize, where transposable elements were first detected, it is apparent that within the transcription units examined numerous polymorphisms exist that seem to have been created by previous visitations by transposable elements. Zack et al. (1986) compared the sequence of a *Shrunken-1* allele (the functional alleles at this locus encode sucrose synthase) isolated from Black Mexican sweet corn with that of another *Sh1* allele sequenced by Werr et al. (1985). There are seven differences that seem to have arisen from transposable element visitation since at four sites one allele has a direct duplication of a sequence present in only one copy in the other. In addition, the two alleles differed by three insertions that were flanked by direct repeats (6–7 bp) of a sequence found in only one copy in the other allele. Similar polymorphisms have been found in the *Bz* alleles analyzed by Ralston et al. (1988) and Furtek et al. (1988). What is true of maize, where a number of different transposable element families are present (Peterson 1988), may not be true of other plants in which transposable elements have yet to be detected. Okagaki and Wessler (1988) have in restriction mapping of *waxy* mutants of rice found none that resulted from insertions. This contrasts markedly with the *waxy* mutants of maize, a number of which were found to result from either insertions or deletions (Wessler and Varagona 1985). The majority of these *waxy* mutants occurred spontaneously in nature, in corn breeding nurseries, or in increase plots in progenies in which transposable elements were not known to be present. A few were generated in mutagenesis experiments (Nelson 1968). Those species, such as maize and snapdragon, in which the effects of transposable elements are readily discernible, are likely to differ in the extent of polymorphisms in coding sequences and in the occurrence of wild-type isoalleles from species in which there is not yet convincing evidence of transposable elements.

ACKNOWLEDGMENTS

This is Paper No. 3007 from the Laboratory of Genetics, University of Wisconsin–Madison. The research has been supported by the College of Agriculture and Life Sciences and by Grant DCB-8507895 from the National Science Foundation. It is a pleasure to acknowledge the contributions of Douglas Furtek, Hwa Yeong Kim, John Schiefelbein, and Victor Raboy, who garnered most of the data presented here, and Russell Huseth for his invaluable assistance with the field work.

CHAPTER 8

DNA POLYMORPHISM AND ADAPTIVE EVOLUTION

Masatoshi Nei

ABSTRACT

Recent advances in the study of DNA polymorphism and adaptive evolution are discussed. The extent of DNA polymorphism as measured by nucleotide diversity usually ranges from 0.001 to 0.01 in eukaryotic genes. This indicates that if flanking regions and introns are included, virtually every eukaryotic gene is polymorphic for one or more nucleotide sites. However, most of these nucleotide differences occur in noncoding regions or silent nucleotide positions, and nonsynonymous (amino acid altering) nucleotide differences are a minority. This is in accordance with predictions from Kimura's neutral theory, since in this theory nonsynonymous substitutions are subject to purifying selection more often than synonymous substitutions. Purifying selection is known to be particularly strong in the active center of a protein. There are, however, exceptions. One of the most interesting examples is the major histocompatibility complex (MHC) loci in mammals, which are known to be extremely polymorphic. Hughes and Nei (1988, 1989) recently showed that in the antigen-recognition site (ARS) of MHC molecules the rate of nonsynonymous nucleotide substitution is significantly higher than that of synonymous substitution. This finding and other considerations led them to conclude that the high polymorphism of MHC loci is mainly caused by overdominant selection that occurs at the ARS region. A relatively high rate of nonsynonymous substitution is also observed at the active center of serine protease inhibitors. There are now many examples of adaptive evolution established at the amino acid or nucleotide level. These examples show that adaptive change of a gene generally occurs by a small proportion of nucleotide substitutions in the gene, other substitutions being more or less neutral.

The decade between 1966 and 1975 was an exciting period for the study of population genetics. In this period, the electrophoretic study of protein polymorphism was introduced and led to the discovery of unprecedented levels of genetic polymorphism in many plant and animal species. The discovery of high levels of protein polymorphism generated a great controversy over the mechanism of maintenance of genetic polymorphism, particularly over Kimura's (1968) neutral theory. This controversy has not yet been completely resolved, but we now know that the neutral theory can explain a wide range of data on protein polymorphism and molecular evolution (Kimura 1983).

In the history of biology, the year 1977 will be remembered as the one in which a technique of rapid DNA sequencing, together with the gene cloning technique, revolutionized molecular biology, leading to the discovery of many new features of genes such as introns, exons, flanking regions, pseudogenes, and transposons. These techniques have also proved to be useful for studying evolution at the most fundamental level of genetic organization, i.e., at the DNA level. An equally useful technique that was introduced in the late 1970s for the study of evolution is restriction enzyme analysis. This technique identifies only a limited portion of nucleotide sequences, but the simplicity of the technique makes it appropriate for a large-scale study of genetic variation at the DNA level.

During the last 10 years, these two techniques have been applied to various groups of organisms, and new insights into the nature of genetic variability and evolution of genes have been obtained. Particularly important are the finding of an even higher level of genetic variability at the DNA level than at the protein level and the inference about the evolutionary forces that generate genetic variability. We now have a better understanding of adaptive evolution at the molecular level. In the following, I would like to discuss some of these new findings with emphasis on our own work.

EXTENT OF DNA POLYMORPHISM

One of the important tasks for population geneticists is to evaluate the extent of genetic variability in populations. At the DNA level, the extent can be measured most conveniently by nucleotide diversity (π), which is defined as the average number of nucleotide differences per site between two randomly chosen alleles (Nei 1987). This is estimated by

$$\hat{\pi} = \sum \pi_{ij}/n_c,$$

where π_{ij} is the proportion of different nucleotides between the ith and jth allelic sequences in the populations and n_c is the total number of comparisons available. This quantity has been estimated for many genes from various organisms, and some examples are presented in Table 1. In most eukaryotic genes, $\hat{\pi}$ is less than 0.011. In the major histocompatibility complex (MHC) genes in humans and mice, however, the nucleotide difference between alleles is very high, and $\hat{\pi}$ goes up to 0.08. These high values suggest that MHC loci are subject to an evolutionary force that is different from that of other loci. Indeed, as will be discussed later, the antigen-recognition site of MHC molecules seems to be subject to overdominant selection.

TABLE 1 Estimates of nucleotide diversity ($\hat{\pi}$).[a]

DNA or gene region	Organism	Method	n	bp	$\hat{\pi}$
mtDNA	Man	R	100	16,500	0.004
mtDNA	*D. melanogaster*	R	10	11,000	0.008
β-Globin	Man	R	50	35,000	0.002
Growth hormone	Man	R	52	50,000	0.002
Notch gene region	*D. melanogaster*[b]	R	37	60,000	0.005
White locus region	*D. melanogaster*[c]	R	38	45,000	0.011
Adh locus (C)	*D. melanogaster*	S	11	765	0.006
Prochymosin (C)	Bovine[d]	S	8	1,146	0.004
Growth hormone (C)	Pig[d]	S	6	651	0.007
Class I MHC (*HLA-A*)	Man[e]	S	5	274	0.043
Class I MHC (*H2-K*)	Mouse[e]	S	4	273	0.077
Hemagglutinin	Influenza virus	S	12	320	0.510

[a]n, sample size; bp, base pairs; R, restriction enzyme technique; S, DNA sequencing; C, coding region.
[b]Schaeffer et al. (1988).
[c] Langley and Aquadro (1987).
[d]D. J. McConnell and P. M. Sharp (unpublished).
[e]Hughes and Nei (unpublished); see Nei (1987) for others.

Table 1 shows that the hemagglutinin gene in the influenza virus A also has an extremely high $\hat{\pi}$ value. However, this high value is caused by a high mutation rate rather than by balancing selection. Influenza viruses are RNA viruses, and the mutation rate in RNA viruses is known to be about one million times higher than that in DNA organisms (Holland et al. 1982). Indeed, Saitou and Nei (1986) have estimated that the mutation rate of the hemagglutinin gene in the influenza virus A is about 1–2% per nucleotide site per year, compared with a value of about 5×10^{-9} of nuclear genes in higher organisms.

SILENT POLYMORPHISM

From the standpoint of the study of natural selection, it is interesting to examine the extent of DNA polymorphism that is not expressed at the amino acid level. In the neutral theory, the level of silent polymorphism is expected to be high compared with the polymorphism revealed at the amino acid level, because silent mutations are subject to purifying selection less often than nonsilent mutations. On the other hand, if polymorphism is actively maintained by natural selection and the effect of genetic drift is unimportant, one would expect silent polymorphism to be lower than nonsilent polymorphism. One way of testing this hypothesis is to examine the polymorphism at the first, second, and third positions of codons. All nucleotide changes at the second position of codons lead to amino acid replacement, whereas at the third position only about 28% of changes are expected to affect amino acids because of the degeneracy of the genetic code. At the first position, about 95% of nucleotide substitutions result in amino acid changes. Therefore, if the neutral theory is valid, the extent of DNA polymorphism is expected to be highest at the third position and lowest at the second position.

The available data indicate that this is indeed the case for most genes. For example, the F and S alleles at the alcohol dehydrogenase locus in *Drosophila melanogaster* are electrophoretically distinguishable because of the amino acid difference (threonine versus lysine) at the 192nd residue (codon change from ACG to AAG). The nucleotide sequences of about a dozen alleles at this locus indicate that there is no other amino acid substitution between the F and S electromorphs but that there are many third position substitutions that are silent (Kreitman 1983). Similar examples of a high proportion of silent substitutions are found in the tryptophan operon genes in *Escherichia coli*, the β-globin gene complex in man, and the conserved region (domain 3) of class I MHC loci in humans and mice (see Nei 1987; Hughes and Nei 1988). The view that amino acid substitutions are usually subject to purifying selection rather than to balancing selection is also supported by the observation that nucleotide substitutions between closely related species are usually synonymous rather than nonsynonymous (Kimura 1983; Nei 1987; Zurawski and Clegg 1987).

OVERDOMINANT SELECTION AT MHC LOCI

There are, however, exceptional genes, in which the rate of nonsynonymous (amino acid altering) nucleotide substitution is higher than that of synonymous substitution. The most conspicuous are the MHC loci in humans and mice. The MHC genes encode glycoproteins that are expressed on cell surfaces and function as agents for recognizing foreign antigens produced by viruses and other parasites (Klein 1986). Thus, they are essential for protection from infectious diseases in vertebrates. The MHC genes can be classified into two groups, i.e., class I and II. Class I loci are composed of classical class I and nonclassical class I loci. The former are extremely polymorphic in humans and mice, the heterozygosity being 80–90%, whereas the latter are virtually monomorphic. Class II loci are also composed of a few highly polymorphic loci and a few monomorphic loci.

The maintenance of high polymorphism at MHC loci has been studied for more than two decades, but no consensus has been reached. There are four major hypotheses proposed for explaining the polymorphism (see Klein 1986): (1) a high mutation rate, (2) gene conversion or interlocus genetic exchange,

(3) overdominant selection, and (4) frequency-dependent selection. In an attempt to establish which of these hypotheses is correct, we initiated a study of the pattern of nucleotide substitution among different alleles.

The class I MHC molecule consists of the extracellular portion, transmembrane portion, and cytoplasmic tail (Klein 1986). The extracellular portion includes two variables domains (α_1 and α_2) and a third conserved domain (α_3) that associates noncovalently with β_2-microglobulin. The α_1 and α_2 domains are known to include 57 amino acid residues that constitute the antigen-recognition site (ARS) of the MHC molecule (Bjorkman et al. 1987). Therefore, if positive Darwinian selection operates at MHC loci, it should occur at these amino acid residues. With this reasoning in mind, we studied the rates of synonymous and nonsynonymous substitution in codons involved in the ARS and other gene regions separately, using published DNA sequences (Hughes and Nei 1988). For this purpose, we compared sequences for 12 alleles from 3 human MHC loci (*HLA-A*, *B*, and *C*) and 8 alleles from 2 mouse MHC loci (*H2-K* and *L*).

Figure 1 shows the nucleotide sequences of the α_1 and α_2 domains for the five alleles from the *HLA-A* locus. The pattern of nucleotide substitution between these alleles is quite different from that of the genes discussed earlier. Here we see many substitutions in the first and second codon positions that are obviously nonsynonymous. Particularly at the codons in ARS (marked with + in Figure 1), the rate of nonsynonymous substitution is very high. To quantify the differences in the pattern of nucleotide substitution among ARS, the remaining region of the α_1 and α_2 domains, and the conserved α_3 domain, we computed the numbers of nucleotide substitutions per *synonymous* site and per *nonsynonymous* site for each of the above three regions of the gene by using Nei and Gojobori's (1986) method I. Since the number of sequence comparisons is very large, I present only the mean numbers of synonymous (d_S) and nonsynonymous (d_N) substitutions (Table 2).

As expected from the sequence data in Figure 1, d_N is much higher than d_S in ARS of the *HLA-A* locus. In the other regions of the gene, however, d_S is higher than d_N. The same pattern of nucleotide substitution is observed also for the *HLA-B* and C loci. It is interesting to note that d_S is more or less the same for all three regions, whereas d_N in ARS is significantly higher than d_S in any region. This clearly indicates that amino acid substitutions in ARS

FIGURE 1. Nucleotide sequences for domains α_1 and α_2 of the five alleles from the human *HLA-A* locus. Domains α_1 and α_2 of class I MHC molecules (encoded by the *HLA-A* locus in humans) are known to contain the antigen-recognition site (ARS). The amino acid residues involved in ARS are marked with +. Note that the codon positions with + marks show many nucleotide substitutions that result in amino acid substitutions. The coding region of domain α_3, which is not shown here, is known to show more synonymous nucleotide substitutions than nonsynonymous substitutions. "-" indicates that the nucleotide is identical with that of allele A2.

```
      α1
                     +       +       +                                                       20
A2    GGC TCT CAC TCC ATC AGG TAT TTC TTC ACA TCC GTG TCC CGG CCC GGC CGC GGG GAG CCC
A3    --- --C --- --- --G --- --- --- --- --- --- --- --- --- --- --- --- --- --- ---
A11   --- --C --- --- --G --- --- --- -A- --C --- --- --- --- --- --- --- --- --- ---
Aw24  --- --C --- --- --G --- --- --- -C- --- --- --- --- --- --- --- --- --- --- ---
Aw68  --- --C --- --- --G --- --- --- -A- --C --- --- --- --C --- --- --- --- --- ---

                                                                                             40
          +       +       +
A2    CGC TTC ATC GCA GTG GGC TAC GTG GAC GAC ACG CAG TTC GTG CGG TTC GAC AGC GAC GCC
A3    --- --- --- --C --- --- --- --- --- --- --- --- --- --- --- --- --- --- --- ---
A11   --- --- --- --G --- --- --- --- --- --- --- --- --- --- --- --- --- --- --- ---
Aw24  --- --- --- --C --- --- --- --- --- --G -GC --- --- --- --- --- --- --- --- ---
Aw68  --- --- --- --C --- --- --- --- --- --- --- --- --- --- --- --- --- --- --- ---

                                                                                             60
                                                                      +   +   +
A2    GCG AGC CAG AGG ATG GAG CCG CGG GCG CCG TGG ATA GAG CAG GAG GGT CCG GAG TAT TGG
A3    --- --- --- --- --- --- --- --- --- --- --- --- --- --- --- --G --- --- --- ---
A11   --- --- --- --- --- --- --- --- --- --- --- --- --- --- --- --G --- --- --- ---
Aw24  --- --- --- --- --- --- --- --- --- --- --- --- --- --- --- --G --- --- --- ---
Aw68  --- --- --- --- --- --- --- --- --- --- --- --- --- --- --- --G --C --- --- ---

                                                                                             80
       +   +   +   +   +   +   +   +   +   +   +   +   +   +   +   +   +               +
A2    GAC GGG GAG ACA CGG AAA GTG AAG GCC CAC TCA CAG ACT CAC CGA GTG GAC CTG GGG ACC
A3    --- CA- --- --- --- --T --- --- --- --G --- --- --- G-- --- --- --- --- --- ---
A11   --- CA- --- --- --- --T --- --- --- --G --- --- --- G-- --- --C --- --- --- ---
Aw24  --- -A- --- --- G-- --- --- --- --- --- --- --- --- G-- --- -A- A-G --- C-- -T-
Aw68  --- C-- A-C --- --- --T --- --- --- --G --- --- --- G-- --- --- --- --- --- ---

                                                    α2                                      100
       +   +       +                                                  +       +       +
A2    CTG CGC GGC TAC TAC AAC CAG AGC GAG GCC | GGT TCT CAC ACC GTC CAG AGG ATG TAT GGC
A3    --- --- --- --- --- --- --- --- --- --- | --- --- --- --- A-- --- -TA --- --- ---
A11   --- --- --- --- --- --- --- --- --- -A- | --- --- --- --- A-- --- -TA --- --- ---
Aw24  GC- -T- C-- --- --- --- --- --- --- --- | --- --- --- --- C-- --- -T- --- -T- ---
Aw68  --- --- --- --- --- --- --- --- --- --- | --- --- --- --- A-- --- -T- --- --- ---

                                                                                            120
                                                          +       +
A2    TGC GAC GTG GGG TCG GAC TGG CGC TTC CTC CGC GGG TAC CAC CAG TAC GCC TAC GAC GGC
A3    --- --- --- --- --- --- G-- --- --- --- --- --- --- -GG --- G-- --- --- --- ---
A11   --- --- --- --- C-- --- G-- --- --- --- --- --- --- -GG --- G-- --- --- --- ---
Aw24  --- --- --- --- --- --- G-- --- --- --- --- --- --- --- --- --- --- --- --- ---
Aw68  --- --- --- --- --- --- G-- --- --- --- --- --- --- -GG --- G-G --- --- --- ---

                                                                                            140
A2    AAG GAT TAC ATC GCC CTG AAA GAG GAC CTG CGC TCT TGG ACC GCG GCG GAC ATG GCA GCT
A3    --- --- --- --- --- --- --C --- --- --- --- --- --- --- --- --- --- --- --G ---
A11   --- --- --- --- --- --- --C --- --- --- --- --- --- --- --- --- --- --- --- ---
Aw24  --- --- --- --- --- --- --- --- --- --- --- --- --- --- --- --- --- --- --G ---
Aw68  --- --- --- --- --- --- --- --- --- --- --- --- --- --- --- --- --- --- --- ---

                                                                                            160
               +       +   +   +       +   +   +   +       +   +   +   +   +   +
A2    CAG ACC ACC AAG CAC AAG TGG GAG GCG GCC CAT GTG GCG GAG CAG TTG AGA GCC TAC CTG
A3    --- -T- --- --- -G- --- --- --- --- --- --- -A- --- --- --- --- --- --- --- ---
A11   --- -T- --- --- -G- --- --- --- --- --- --- -C- --- --- --- CA- --- --- --- ---
Aw24  --- -T- --- --- -G- --- --- --- --- --- --- --- --- --- --- CA- --- --- --- ---
Aw68  --- --- --- --- --- --- --- --- --- --- --- --- --- --- --- -G- --- --- --- ---

                                                                                            180
       +   +   +       +   +   +       +       +
A2    GAG GGC ACG TGC GTG GAG TGG CTC CGC AGA TAC CTG GAG AAC GGG AAG GAG ACG CTG CAG CGC ACG
A3    --T --- --- --- --- --- --- --- --- --- --- --- --- --- --- --- --- --- --- --- --- ---
A1    --- --- CG- --- --- --- --- --- --- --- --- --- --- --- --- --- --- --- --- --- --- ---
Aw24  --- --- --- --- --- --C G-- --- --- --- --- --- --- --- --- --- --- --- --- --- --- G--
Aw68  --- --- --- --- --- --- --- --- --- -C- --- --- --- --- --- --- --- --- --- --- --- ---
```

TABLE 2 Mean numbers of percentage nucleotide substitutions per synonymous site (d_S) and per nonsynonymous site (d_N), with their standard errors, between alleles from the same and different class I MHC polymorphic loci in humans (*HLA-A, B,* and *C*) and mice (*H2-K* and *D*).[a]

Locus (number of sequences)	Antigen-recognition site (ARS) ($N = 57$)		Remaining codons in α_1 and α_2 ($N = 124, 125$)[b]		Domain α_3 ($N = 92$)	
	d_S	d_N	d_S	d_N	d_S	d_N
HUMAN						
A (5)	3.5 ± 2.0	13.3 ± 2.2***	2.5 ± 1.2	1.6 ± 0.5	9.5 ± 3.0	1.6 ± 0.7**
B (4)	7.1 ± 3.1	18.1 ± 2.8**	6.9 ± 2.0	2.4 ± 0.7*	1.5 ± 1.1	0.5 ± 0.4
C (3)	3.8 ± 2.5	8.8 ± 2.2	10.4 ± 2.8	4.8 ± 1.1	2.1 ± 1.5	1.0 ± 0.6
Overall mean	4.7 ± 2.6	14.1 ± 2.4***	5.1 ± 2.1	2.4 ± 0.8	5.8 ± 2.0	1.1 ± 0.6**
MOUSE						
K (4)	15.0 ± 4.5	22.9 ± 3.3	8.7 ± 2.2	5.8 ± 1.0	2.3 ± 1.3	4.0 ± 1.0
L (4)	11.4 ± 4.0	19.5 ± 3.1	8.8 ± 2.3	6.8 ± 1.2	0.0 ± 0.0	2.5 ± 0.8**
Overall mean	13.2 ± 4.3	21.2 ± 3.2*	8.8 ± 2.3	6.3 ± 1.1	1.2 ± 0.9	3.6 ± 0.9**

[a]The difference between d_S and d_N is significant at the 5% level (*), 1% level (**), or 0.1% level (***). *N*, number of codons compared.
[b]*N* is 124 for the mouse and 125 for the human.

are enhanced by positive Darwinian selection. By contrast, d_N is much smaller than d_S in the non-ARS regions, suggesting that amino acid substitutions are generally subject to purifying selection in these regions.

Table 2 also includes the results of allelic comparisons at the mouse *H2-K* and *L* loci. The d_N in ARS is again substantially higher than the d_S in any region. By contrast, d_N is quite low in the other two regions. Therefore, the pattern of nucleotide substitution for the mouse MHC loci is essentially the same as that for the human MHC loci. (The unusually low value of d_S in α_3 may have been caused by interlocus genetic exchange; Hughes and Nei 1988.)

We conducted a similar statistical analysis for class II MHC loci (Hughes and Nei 1989). The class II MHC molecule is composed of an α-chain and a β-chain, which are noncovalently associated. Both α- and β-chains include domain 1 (D1), domain 2 (D2), the transmembrane portion, and the cytoplasmic tail (Klein 1986). In the case of class II molecules, the antigen-recognition site is known to be located in the first domains of the α- and β-chains. Recently, Brown et al. (1988) proposed a putative antigen-recognition site that involves 19–20 amino acid residues in D1 of the α-chain and 15–16 residues in D1 of the β-chain. We therefore examined d_S and d_N in D1, the putative ARS, and D2 of both α- and β-chains.

In the mouse, there are eight class II loci in the $H\text{-}2^b$ haplotype (perhaps seven in some other haplotypes) (Klein and Figueroa 1986). However, only three of them (Aα, Aβ, and *E*β) are polymorphic, others being either monomorphic or unexpressed. The number of class II loci in the human is considerably larger than that in the mouse, but only three loci (or groups of loci) are polymorphic (*DQ*α, *DQ*β, and *DR*β). Others are again either monomorphic or unexpressed. For some reasons, the β-chain loci are more polymorphic than the α-chain loci. In our study, we considered only polymorphic loci.

The d_S and d_N values obtained are presented in Table 3. In both mouse and human polymorphic loci, d_N is again much higher than d_S in ARS, though the difference is not necessarily significant because of the small number of codons involved. In the present case, even the entire domain 1 including both ARS and non-ARS regions generally show a higher d_N value than d_S. Some amino acid residues outside the putative ARS may be involved in antigen recognition. In the conserved D2 region, however, d_N is always smaller than d_S. Therefore, class II polymorphic loci show the same pattern of nucleotide substitution as that of class I loci.

TABLE 3 Mean numbers of nucleotide substitutions per synonymous site (d_S) and per nonsynonymous site (d_N) expressed as percentages, with their standard errors, between alleles at mouse and human class II MHC loci.[a]

Locus (number of sequences)	Putative recognition site (ARS) (N = 15–20)		Domain 1 excluding ARS (N = 64–78)		Domain 2 (N = 94)	
	d_S	d_N	d_S	d_N	d_S	d_N
			β-Chain loci			
MOUSE						
Aβ (7)	0.0 ± 0.0	30.0 ± 6.7***	4.0 ± 1.6	6.7 ± 1.2	7.3 ± 2.3	1.3 ± 0.5*
*E*β (4)	11.6 ± 9.1	41.5 ± 9.5*	1.8 ± 1.3	5.2 ± 1.3	0.9 ± 1.1	0.6 ± 0.4
HUMAN						
*DP*β (3)	3.9 ± 5.5	19.0 ± 6.4	2.4 ± 1.7	2.8 ± 1.1	5.3 ± 2.4	0.6 ± 0.4
*DQ*β (8)	13.7 ± 7.7	26.5 ± 5.4	8.5 ± 2.3	6.7 ± 1.3	5.6 ± 2.0	1.6 ± 0.5
*DR*β (14)	15.0 ± 8.5	45.7 ± 6.2**	8.0 ± 1.9	4.5 ± 0.9	8.3 ± 1.8	3.3 ± 0.6**
			α-Chain loci			
MOUSE						
Aα (6)	3.2 ± 3.0	23.7 ± 4.9***	2.8 ± 1.7	2.6 ± 0.8	7.2 ± 2.2	0.7 ± 0.4**
HUMAN						
*DQ*α (4)	21.7 ± 11.8	27.0 ± 6.7	8.0 ± 3.2	4.3 ± 1.3	4.0 ± 1.9	2.4 ± 0.8

[a]The difference between mean d_S and d_N is significant at the 5% level (*), 1% level (**), or 0.1% level (***). N, number of codons compared.

As mentioned earlier, some authors have advocated the hypothesis of a high mutation rate or that of interlocus genetic exchange. The present data are, however, incompatible with either of these hypotheses, because these hypotheses are not capable of explaining the relationship of $d_N > d_S$ in ARS and $d_N < d_S$ in the non-ARS regions. The first hypothesis can also be rejected by the observation that the number of nucleotide substitutions between the human and mouse MHC genes is not particularly high compared with other genes (Klein and Figueroa 1986). In the case of multigene families like the MHC genes, however, one cannot exclude the possibility of interlocus genetic exchange. Indeed, there is indirect evidence that genetic exchange has occurred at the MHC loci (Hughes and Nei 1988). Nevertheless, the contribution of this factor to MHC polymorphism seems to be small for the following reasons. First, if interlocus genetic exchange is important, the extent of polymorphism is expected to increase with the number of loci involved (Ohta 1983; Nagylaki 1984). Yet, the classical class I gene cluster composed of only three loci is much more polymorphic than the nonclassical class I cluster including a large number of loci (Klein and Figueroa 1986). Second, the class II cluster genes consist of a few highly polymorphic loci and many monomorphic loci mixed, and it is hard to explain how interlocus genetic exchange operates only on polymorphic loci. Third, this hypothesis is not able to explain why polymorphism is concentrated on the antigen-recognition site (Figure 1). Fourth, it cannot explain the higher value of d_N than d_S in ARS, as mentioned above.

Polymorphism at the MHC loci is extraordinary in several respects, and any theory of MHC polymorphism must be able to explain the following four observations: (1) extremely high heterozygosity, (2) a high degree of nucleotide differences between alleles (Klein and Figueroa 1986), (3) persistence of polymorphic alleles (allelic lineages) for a long time (at least 10 million years; Figueroa et al. 1988; McConnell et al. 1988), and (4) $d_N > d_S$ in ARS. This points to overdominant selection as the major factor of MHC polymorphism. Indeed, as long as mutation occurs at a normal rate, overdominant selection increases heterozygosity and the number of alleles tremendously compared with the neutral level (Kimura and Crow 1964; Wright 1966). Overdominant selection is also known to maintain polymorphic alleles for a long time. Takahata and Nei (unpublished) recently showed that when population size is large (say 10^4–10^5) and selective advantage is a few percent, some polymorphic allelic lineages may persist in the population for over 10 million years so that Figueroa et al.'s (1988) observation can easily be explained. Furthermore, Maruyama and Nei (1981) have shown that under overdominant selection the rate of codon substitution in a DNA sequence is accelerated substantially. This explains both observations (2) and (3) above, since overdominant selection would work only for amino acid substitution.

What is the biological basis of overdominant selection in MHC loci? This question can be answered if we know the function of MHC molecules. As mentioned earlier, these molecules act as agents for recognizing foreign antigens. In the case of class I molecules, MHC molecules are known to bind to a foreign antigen (peptide), and this combined entity is then recognized by the cytotoxic T cell. This cytotoxic T cell will then kill the infected cells. Since a particular class I molecule may provide enhanced recognition of a particular pathogen, individuals expressing various types of class I antigen will have a selective advantage in a population exposed to a diverse array of pathogens; such individuals will be those which are heterozygous at several loci. Essentially the same explanation is provided for class II MHC polymorphism (Hughes and Nei 1989).

It should be noted that the acceleration of amino acid substitution can also be explained by some kind of frequency-dependent selection. A popular model of frequency-dependent selection is to assume that host individuals carrying a recently arisen mutant allele have a selective advantage because pathogens will not have had time to evolve the ability to infect host cells carrying a new mutant antigen (Damian 1964; Snell 1968; Bodmer 1972). This will result in a constant turnover of alleles in the population because old alleles lose resistance to pathogens. While this model generates a higher value of d_N than d_S, it can explain neither the high degree of polymorphism nor the long-persisting polymorphism at the MHC loci (Hughes and Nei 1988; N. Takahata and M. Nei, unpublished). There are some other models of frequency-dependent selection, but they do not seem to be appropriate for biological reasons (N. Takahata and M. Nei, unpublished).

The above discussion suggests that the major cause of MHC polymorphism is overdominant selection. If this conclusion is correct, this will be the first case of overdominant selection established at the molecular level since the discovery of the sickle-cell anemia mutation.

POSITIVE SELECTION AT THE SERINE PROTEASE INHIBITOR LOCI

Some readers might wonder what the general significance of the study of MHC polymorphism is. Although the MHC genes are important for self–nonself discrimination and play an important role in the vertebrate immune system, the genes involved are of a special type and occupy only a small portion of the genome. In particular, no genes homologous to the MHC seem to exist in plants. However, there are several other gene families that are involved in defense mechanisms against pathogens, and it is possible that many of these gene families undergo similar adaptive evolution. Indeed, we have obtained evidence that the variable region genes of the immunoglobulin heavy chain and the light κ-chain are subject to overdominant or diversity-enhancing selection (Gojobori and Nei 1984; Tanaka and Nei 1988). Here, however, let me discuss the evolutionary change of serine protease inhibitor genes.

Both animals and plants contain various kinds of protease inhibitors. They are encoded by a gene family consisting of a large number of member genes. However, the target enzyme for each inhibitor is not known except for a very few inhibitors. As early as 1972, Green and Ryan (1972) suggested that the role of many plant protease inhibitors is defense against invading insects, inhibiting the function of proteases released by the insects. Hilder et al. (1987) recently transferred a cowpea protease inhibitor gene into tobacco by genetic engineer- ing and showed that the tobacco plants with this gene are much more resistant to infestation by larvae of the lepidopteran species *Haliothis virescens*.

Interestingly, the active center of serine protease inhibitors is known to show an unusually high rate of amino acid substitution (Laskowski et al. 1987). Hill and Hastie (1987) computed the rates of synonymous and nonsynonymous substitution for the active center and other regions of the gene comparing the four genes from the mouse, rat, and human. Unfortunately, their results included many cases where estimates were not obtainable. I therefore reexamined the rates by using a different statistical method, i.e., Nei and Gojobori's (1986) method I. In this case I computed the proportions of nucleotide differences per synonymous (p_S) and nonsynonymous (p_N) sites. The results obtained are presented in Table 4. In the active center (region 2) p_N is very high, yet its magnitude is similar to that of p_S. Therefore, one need not invoke an accelerated rate of nonsynonymous substitution. As noted by Graur and Li (1989), amino acid substitution might have occurred just by

TABLE 4 Percentage nucleotide differences per synonymous site (p_S) and per nonsynonymous site (p_N) between serine protease inhibitor genes from the mouse (*con*), rat (s-2.1 and s-2.2), and human (*achy*).

Comparisons	*Region 1 (172–174 codons)*[b]		*Region 2*[a] *(11–14 codons)*[b]		*Region 3 (27 codons)*	
	p_S	p_N	p_S	p_N	p_S	p_N
Con vs s-2.1	26 ± 4	13 ± 2	58 ± 16	57 ± 10	13 ± 8	32 ± 6
Con vs s-2.2	26 ± 4	14 ± 2	68 ± 15	66 ± 9	44 ± 12	32 ± 6
Con vs *achy*	46 ± 5	19 ± 2	56 ± 16	55 ± 10	67 ± 12	22 ± 5
s-2.1 vs s-2.2	31 ± 4	12 ± 2	56 ± 16	83 ± 7	41 ± 12	35 ± 6
s-2.1 vs *achy*	45 ± 5	21 ± 2	52 ± 16	81 ± 8	80 ± 10	31 ± 6
s-2.2 vs *achy*	44 ± 4	17 ± 2	59 ± 14	43 ± 9	90 ± 7	31 ± 6

[a]Active center.
[b]The number of codons varied with comparison because there were some deletions/insertions and the codons involving stop codons in the evolutionary pathways were excluded (see Nei and Gojobori 1986). Data from Hill and Hastie (1987).

mutation and drift without any positive selection. However, if we note that there are some highly conserved amino acids even in the active center, it is possible that at other amino acid residues amino acid substitution has been enhanced by positive Darwinian selection. In the present case, the DNA sequences compared are quite divergent, so that it is not easy to detect positive selection. If more closely related sequences are available, one might be able to make a more definitive conclusion.

At any rate, if protease inhibitors are involved in the defense mechanism against insect infestation, it would be important to know the amino acid sequence of the active center of the protein in many different plant species. Some species may have a gene for resistance to a particular species of insect. One can then transfer the gene into an agricultural crop that lacks the gene, as in the case of tobacco. Available data suggest that interspecific variation in the active center of protease inhibitors is high.

OTHER EXAMPLES OF ADAPTIVE EVOLUTION AT THE DNA LEVEL

Until recently, it was not easy to produce a clear-cut case of adaptive evolution at the molecular level except in microorganisms (Mortlock 1984). This led Wilson (1975) to propose that most adaptive or morphological evolution occurs by mutations at regulatory loci rather than at structural loci. In recent years, however, the number of examples of adaptation by structural gene mutations is increasing. Interestingly, these examples show that adaptive change is usually caused by one or a few amino acid changes at the active center. Table 5 shows some of the examples.

Microorganisms still lead in this field. One of the most interesting examples from microorganisms is the development of resistance to antibiotics. β-Lactam antibiotics kill bacteria by inactivating a set of penicillin-binding proteins (PBPs) that are essential for cell division. Some mutants of *E. coli* are resistant to these antibiotics because of the reduction in affinity between antibiotics and PBPs. Hedge and Spratt (1985) have shown that this reduction in affinity is caused by one to four amino acid substitutions in the active center of a PBP and that the majority of other amino acid substitutions do not affect the susceptibility to the antibiotics.

One of the most interesting examples of adaptive change of enzymes or proteins in eukaryotes is that of crocodilian hemoglobin. This hemoglobin maintains high activity under the conditions of increased blood acidity that occurs when crocodiles stay under water for a prolonged period of time. This adaptation can be explained by five amino acid substitutions. This number is a small fraction of the total number of amino acid differences (123) that exist between the crocodilian and human hemoglobins (Perutz et al. 1981). The convergent evolution of stomach lysozyme of ruminants and the langur

TABLE 5 Examples of molecular changes that affect adaptive characters

Character	Gene (substitutions)	Source
1. Resistance to antibiotics	β-Lactamase (PBP) (1–4 amino acid substitutions)	Hedge and Spratt (1985)
2. Crocodilian Hb	Hemoglobin (5 amino acid substitutions)	Perutz et al. (1981)
3. Stomach lysozyme in ruminants and langur monkey	Lysozyme c (5 amino acid substitutions)	Stewart et al. (1987)
4. Courtship song rhythm in *Drosophila*	Period (*per* gene) (1 nucleotide substitution)	Yu et al. (1987)
5. Heterochrony in *C. elegans*	Cell division (single mutation)	Ambros and Horvitz (1984)
6. Swimming speed of fish	*Ldh* locus (1 amino acid substitution?)	DiMichele and Powers (1982)
7. Resistance to herbicides	Chloroplast *psbA* gene (1 nucleotide substitution)	Hirschberg and McIntosh (1983)

monkey having a foregut can also be explained by five amino acid changes (Stewart et al. 1987).

Another interesting example is herbicide resistance in plants. Many commercially important herbicides kill plants by inhibiting photosynthesis. This inhibition of photosynthesis occurs because herbicides bind to a thylakoid-membrane protein encoded by a chloroplast gene, *psbA*. Several plant species have mutant strains that are resistant to herbicides. Hirschberg and McIntosh (1983) sequenced the *psbA* genes from a normal and a mutant strain of the pigweed *Amaranthus hybridus* and found that there is only one codon difference between the two genes; the normal gene has a serine codon (AGT) at the 264th codon, whereas the mutant gene has a glycine codon (GGT). Clearly the A → G mutation has occurred. Interestingly, the serine codon AGT is evolutionarily conserved and is shared by many plants, and the same A → G mutation has occurred in other plants (Table 6). However, some lower plants or cyanobacteria have different codons, though they code for the same amino acid (serine), and the mutant codons are for alanine. These findings suggest that the serine at the 264th residue is responsible for herbicide binding and that the replacement of this serine by some other amino acid prevents herbicides from binding to the chloroplast thylakoid membranes.

CONCLUDING REMARKS

The progress of the study of molecular evolution during the last two decades has been spectacular. We now know what kinds of genetic change occur at the DNA level. We also know general patterns of nucleotide substitution and ge-

TABLE 6 The codons and amino acids at the 264th residue of the chloroplast gene *psbA* in herbicide-sensitive (wild-type) and herbicide-resistant plants.

Species	Wild type (sensitive)	Mutant (resistant)	Source
DICOTYLEDONS			
Pea	AGT (Ser)		Oishi et al. (1984)
Alfalfa	AGT (Ser)		Aldrich et al. (1986a)
Soybean	AGT (Ser)		Spielmann and Stutz (1983)
Tobacco	AGT (Ser)	AAT (Asn)	Pay et al. (1988)
Petunia	AGT (Ser)		Aldrich et al. (1986b)
Solanum nigrum	AGT (Ser)	GGT (Gly)	Goloubinoff et al. (1984)
Mustard	AGT (Ser)		Link and Langdridge (1984)
Brassica napus	AGT (Ser)	GGT (Gly)	Reith and Straus (1987)
Spinach	AGT (Ser)		Zurawski et al. (1982)
Pigweed	AGT (Ser)	GGT (Gly)	Hirschberg and McIntosh (1983)
Senecio vulgaris	AGT (Ser)	GGT (Gly)	Blyden and Gray (1986)
MONOCOTYLEDONS			
Rice	AGT (Ser)		Wu et al. (1987)
Barley	AGT (Ser)		Efimov et al. (1988)
Poa annua	AGT (Ser)	GGT (Gly)	Barros and Dyer (1988)
Phalaris paradoxa	AGT (Ser)	GGT (Gly)	Schoenfeld et al. (1987)
Liverwort	AGC (Ser)		Ohyama et al. (1986)
Euglena	TCG (Ser)	GCG (Ala)	Johanningmeier and Hallick (1987)
Chlamydomonas	TCT (Ser)	GCT (Ala)	Erickson et. al. (1984)
CYANOBACTERIA			
Anabaena	TCC (Ser)		Curtis and Haselkorn (1984)
Fremyella	TCC (Ser)		Mulligan et al. (1984)
Synechococcus	TCG (Ser)	GCG (Ala)	Golden and Haselkorn (1985)

nome change. Yet we are profoundly ignorant about the adaptive change of morphological and physiological characters at the molecular level. What we know now is only a glimpse of the entire picture. In the near future, however, I expect that much progress will be made in this direction as the molecular biology of gene regulation and morphogenesis advances.

This volume is partly concerned with plant breeding or the utilization of plant resources for human welfare. Recently plant breeding has become an exciting area of human endeavor. At least theoretically one can introduce desirable genes into any commercial crop. There are still some technical problems, but I am sure they will be solved rather rapidly. However, it is important to keep in mind that useful genes may exist not only in commercial plants but also in wild plants or even in other kingdoms. In the past, plant breeding has depended almost exclusively on hybridization within species or between closely related species. Recent biotechnology is about to remove this barrier, and one can now use desirable genes from distantly related organisms. It is therefore important to search for desirable genes from a wide range of organisms. In this search, the study of evolution will give some guidance, because it will tell us how genes and proteins have changed in the evolutionary process. In the future, therefore, the relationship between plant breeding and the study of molecular evolution is expected to be closer than ever before. After all, plant or animal breeding is nothing but artificially controlled evolution.

ACKNOWLEDGMENTS

I thank Tatsuya Ota and Ken Wolfe for their assistance in preparing Tables 4 and 6 and Dan Graur, Manolo Gouy, Austin Hughes, and Paul Sharp for their comments on an earlier draft of this paper. This study is supported by research grants from the National Institutes of Health and the National Science Foundation.

SECTION 2

EVOLUTIONARY PROCESSES

This section considers in detail the structure of genetic variation among loci, among individuals, and among populations, together with the evolutionary processes of mating system, selection, recombination, genetic drift, and migration that shape these patterns of genetic diversity. Experimental, methodological, and theoretical approaches are developed. A theme that runs through this section is that population genetic studies of plants need to encompass ecological and demographic variables. Thus population genetics is seen as an essential component of modern population biology.

The diversity of plant mating systems is perhaps the most distinctive feature of the population genetics of plants. Brown reviews both the range of plant mating systems and methods for the estimation of their parameters. The basic procedure depends on the use of single gene markers such as allozymes and the model of mixed selfing and random mating. Progress is evident in the development of more refined and elaborate models, a wider array of genetic markers, and a more comprehensive analysis of mating events.

Frequencies of pairs of genes in populations may differ from the products of the two single frequencies, with the difference being known as linkage disequilibrium. Hastings reviews the joint effects of selection on pairs of loci and linkage between them in causing such disequilibrium. The importance of mating system to plant population studies is evident in that linkage disequilibrium is more common in selfing than outcrossing species. Hastings considers both the question of how much disequilibrium may be maintained under specified regimes of selection and linkage, and how much selection is necessary to lead to the observed disequilibrium levels for loci with certain recombination values.

The correlation between discrete marker genes and the loci controlling quantitative characters can be approached with a very different purpose in mind. The underlying idea is to infer breeding structure and thus estimate the genetic parameters for quantitative traits. Ritland develops such techniques to answer important issues in plant population biology such as the extent of inbreeding depression. The approach has been used in studies of human disease risk and in plant and animal breeding, and it offers much to the study of plant demography. Estimates are possible for plants in natural populations without disturbing the environment. Indeed the heritability of quantitative traits is estimated by using the relationship between markers for individuals close enough physically to presume common ancestry.

The challenge to chart the course and the causes of selection in natural populations remains a fundamental one for experimental population genetics. Ennos contrasts the approaches based either on discrete genes or on continuously varying characters. More ecologically based studies of selection are afforded by interacting systems of coevolving species. The interactions between plants as hosts and the pathogens that attack them are good examples, and Burdon and Jarosz consider host-pathogen interactions in both natural populations and farmers' fields. In such systems genetic diversity within the host species can reduce the incidence of disease and have substantial effects on the genetic structure of the pathogen population. Mixed cultivars are a possible means of disease control, although some studies have shown that increases in yield have not accompanied any lowering of disease. Burdon and Jarosz argue that such studies should use experimental conditions that more closely reflect those found in agricultural stands.

The spatial arrangement of resistant components in a mixed cultivar can affect disease progress. Indeed the sessile habit of plants invites general study of the spatial patterns of genetic variation within a population. Epperson reviews the statistical procedures for measuring such spatial patterns in plants. Some of these procedures, such as those using spatial autocorrelations, are relatively new and should see increased use in studies of plant populations.

Turning to patterns among populations of a species, Barrett and Husband consider the processes leading to geographic diversity, with emphasis on different modes of colonization. They review studies that have measured genetic divergence resulting from continent-island, between-island, and central-marginal migration patterns. These studies highlight the key roles of small population size and migration rates. Yet surprisingly few formal estimates of these parameters are available, so that reviews of geographic patterns (e.g., Hamrick and Godt's review in Section 1) must rely on broad species attributes. Also, temporal studies in both colonizing and colonized populations are lacking. Such studies, together with those called for in the other chapters of this section, would provide a much richer genetic demography of plant populations.

CHAPTER 9

GENETIC CHARACTERIZATION OF PLANT MATING SYSTEMS

A. H. D. Brown

ABSTRACT

Mating systems determine the mode of transmission of genes from one generation to the next, and plant species exhibit a great variety of such systems. Their genetic impact is measurable in terms of models (e.g., mixed self-fertilization and random mating), using the segregation of marker genes (e.g., isozymes) in progeny arrays. For experimental purposes, plant species conveniently fall into five classes of mating: predominantly self-fertilizing, predominantly outcrossing, mixed selfing and outcrossing, partially apomictic, and partially selfing of gametophytes. Recent research on such species has extended the mixed mating model and produced new methods based on multilocus approaches, the effective selfing model, the analysis of correlated matings, estimates of male fertility, and the paternity of progeny. These methods form a diverse and impressive set for the precise study of plant mating systems and their integration into plant population biology.

INTRODUCTION

A crucial step in the dynamics of plant populations is the generation of new individuals for tomorrow from those of today. To characterize this step in genetic terms has long been important in experimental population genetics and plant breeding. Mating systems largely determine the number and genetic make-up of offspring that form the raw material for future selection. From the perspective of the population, "regulation of the amount of inbreeding plays a major role in adjusting the manner in which genetic variability is organized in populations" (Allard 1975). Furthermore, mating systems are under genetic control, and themselves subject to selection. Consequently, it is important to have an accurate account of mating processes—both their natural history and their genetic realization.

The patterns of transmission of marker genes through one mating cycle can reflect the matings actually taking place in the population. We use such markers to estimate the parameters of models that summarize the mating behavior of plant populations or of their individual members. The models can then specify the mode of transmission of any gene to the next generation at the population level. Furthermore, differences among species, among populations, or among individuals for their estimated values in the models lead to a study of the selective agents responsible.

There has been much recent research in this area, as seen in recent reviews and summaries (Brown et al. 1985 and in press; Clegg and Epperson 1985). In this chapter, a brief historical note precedes a review of the current major research questions and the techniques available to answer them.

HISTORICAL PHASES

Research on the genetic analysis of plant mating systems falls into three distinct periods. The first period, lasting until about 1960, was largely one of survey, the second of exact model analysis, particularly of experimental populations, and the third, from 1970, began with the use of isozyme markers in mating system studies.

Survey period

This preliminary phase was essentially summed up by Fryxell's (1957) comprehensive review, which catalogued the mating system of over 1200 plant taxa. The evidence was drawn from morphology, tests for autofertility, and occasionally the behavior of marker genes in progenies from experimental planting. Most of the quantitative estimates, as opposed to categoric assessments, were for cultivated agricultural or horticultural species.

Development of model systems

The 1960s saw a new school of plant population genetics emerge at Davis, emphasizing predominantly inbreeding species. The work, reviewed by Allard et al. (1968), developed precise models of microevolutionary dynamics to account for the polymorphism unexpectedly found in inbreeding populations. Selection theory described the transition in marker frequencies over generations, and the existence of stable equilibria in experimental populations of barley and lima bean. This theory showed that very low levels of outcrossing were critical in predicting the future evolution of populations. Hence it was essential to measure the precise level of outcrossing in selfing populations, especially to obtain unbiased estimates of selection intensities.

Outcrossing was estimated using refinements of the open-pollinated progeny testing method, or from estimating the fixation index, both given by Fyfe and Bailey (1951). The procedures employed morphological markers (in lima bean by Allard and Workman 1963; Harding and Tucker 1964). Such markers have the problem of dominant expression, which lowers statistical precision, and the possibility that selection acts on the marker itself. Yet these experiments gave rise to concepts about variation in mating systems in plant populations, which were to emerge more clearly with better markers. In addition, studies began on genes that affect the mating system, e.g., male sterility in barley (Jain and Suneson 1966) and flower color mutants in wild lupins (Horovitz and Harding 1972).

The electrophoretic breakthrough

The development of isozyme markers began a new era in genetic studies of plant mating systems (Brown and Allard 1970; Marshall and Allard 1970; Clegg 1980). Most natural populations could now be studied, because isozyme polymorphism is common and readily detectable. Zymograms of homologous isozyme loci and their variation in different taxa resemble one another. Thus genetic hypotheses as to the inheritance of allozyme variants are easier to frame than are hypotheses for morphological markers, which can have cryptic duplicate genes, background effects, etc. Finally, codominant expression of isozymes removes the need for progeny testing to uncover heterozygosity, greatly increasing the statistical power of these genetic markers. Thus both predominant outbreeders and inbreeders in natural, cultivated, or experimental populations of an increasing range of plant species have since been studied.

MAJOR MODES OF MATING IN PLANTS

As rightly stressed by some authors (e.g., Jain 1984), plant mating systems should be thought of as a continuum of possibilities rather than as strict types. Yet it is helpful to speak of major modes of mating, because the more important issues and approaches differ between them. Table 1 summarizes the modes together with the research questions and relevant procedures within each mode.

Predominant selfing (with outcrossing rate, $t < 0.10$)

About 20% of higher plant species reproduce mainly through self-fertilization, with only occasional outcrossing (Fryxell 1957; Brown et al. 1988). In populations of such species, the great bulk of the seeds are selfed progeny, and the major questions are (1) whether any outcrossing at all is evident, and (2) what are the patterns of variation in outcrossing rate within and among populations.

Except for mutation, the finding of variation for polymorphic marker loci within a progeny array proves that outcrossing can occur. The variation can

TABLE 1 Modes of plant mating systems, their major research questions, and relevant procedures.

Mating system	Issues	Procedures
Predominant self-fertilization	Whether any outcrossing is detectable Outcrossing rate and its variation	Mixed mating model
Predominant outcrossing	Whether selfing occurs Subpopulation structure Biparental inbreeding Male fertility variance	Multilocus t Effective s Paternity Transition matrix
Mixed selfing and outcrossing	Variation in outcrossing Progeny disposition in space and time	Correlated mating Paired fruit
Apomixis	Variation in outcrossing	Mixed model
Intragametophytic selfing	Variation in outcrossing Structure of fruiting bodies	Mixed model

stem from new outcrosses, or from segregation in the progeny of heterozygous maternal plants. (Sometimes the level of polymorphism in selfing populations can be inadequate for detecting outcrossing, especially in populations of colonizing species.) Such direct genetic proof that outcrossing occurs at all is most important.

The long-term significance of rare outcrossed individuals far exceeds their current trivial frequency. Essentially they give coherence to the population, which otherwise would steadily fragment into innumerable lines each of unit effective size. Outcrossed progeny give rise, in subsequent generations, to a stream of new segregants (Stebbins 1957).

The direct genetic observation of an outcrossing event (heterozygous progeny from a homozygous maternal plant) requires no statistical inference. Conversely a statistical statement is needed for the power of the experimental test of the question: how high might the true, yet unknown rate of outcrossing (t) be, and yet remain undetected at a given level of probability because of limited polymorphism? For this question, and for estimating rates of outcrossing in predominant selfers, the mixed mating model is particularly suited, as discussed below.

The pattern of variation in outcrossing rate in predominantly inbreeding species is the second major focus for research. Only limited information is available on spatial and temporal variation and on the frequency distribution of outcrossing. Even less is known on the genetics of this variation or environmental factors that affect it. Yet such variation can bias estimates of fixation index, and its patterns can vary markedly among species (Brown and Albrecht 1980).

In some fungi (e.g., *Ustilago bullata*) selfing may occur through the union of gametes derived from the same meiotic event. Such *intratetrad* selfing retards the loss of heterozygosity compared with intertetrad selfing, particularly for loci linked to a diallelic mating-type locus (Kirby 1984).

Predominant outcrossing (self-fertilization rate, $s < 0.05$)

At the other end of the spectrum are the majority of plant species, which rarely, if ever, self-fertilize. For these species, the mixed mating model has a more limited role. It is a useful first step in establishing whether selfing or other forms of inbreeding are occurring. However, independent direct proof that self-incompatibility is incomplete should be sought, by testing whether an isolated plant sets seed, with or without facilitated self-pollination. Returning to the population, assay of progeny arrays usually cannot prove that actual selfing is occurring. Only a statistical statement is possible: the genotype frequencies in the arrays suggest that a certain level of inbreeding is present. Other factors, such as population subdivision and selection, are subsumed under selfing.

The question of whether any selfing is taking place is a critical one. Genes that promote selfing but without restricting the transfer of pollen to other plants as well (i.e., without pollen discounting) enjoy an selective advantage. The existence of such genes would raise the issue of how they are held in check.

Biparental inbreeding, or mating within distinct subpopulations, constitutes a second deviation from panmixia. Such inbreeding is *relative* to panmixia in the entire local population. This mating pattern can be adaptive, as a mode of evolving reproductive isolation (Antonovics 1968).

Yet the two inbred portions (selfing and biparental inbreeding) comprise only a small fraction of the offspring in the population, even if it is a fraction that has intriguing latent evolutionary potential. Unlike the predominantly selfing case, in which selection occurs largely as differential survival and fecundity of lines, the outbreeding bulk of the population is the raw material for diverse modes of selection, including sexual selection on mating systems. Therefore most interest surrounds the source of the outcrossing pollen. Patterns of gender allocation, of differential male and female fertility, and a solution to the paternity quest are the important issues.

We have so far mentioned two components of the mating system of an individual plant, namely, the kinds of zygotes formed (proportion of selfed, crossed, apomictic seed) and the patterns of male and female fertility (Brown et al. 1988). A third component of mating systems is the spatial and temporal disposition of progeny within and among fruit, for example, whether the paternity of fruit is single versus multiple (Ellstrand 1984). The extent to which single fruit contain progeny that are half-sibs or full-sibs is relevant to the outcome of competition among seedlings from each fruit and selection among fruit. Perhaps the full potential of this type of study will be met in species with intermediate mating systems, where starker contrasts in genotypes of progeny occur.

Mixed selfing and outcrossing

Species with a genuine mixed mating system can achieve an overall intermediate and often variable amount of outcrossing through a mixture of self- and cross-pollination at various levels:

Different separated populations—after long-distance colonization (Brown and Marshall 1981; Glover and Barrett 1986).
Different adjacent populations, or temporal shifts within a population (Jain and Martins 1979).
Different plants within populations, e.g., with the breakdown of heterostyly (Piper et al. 1984).
Different fruit on the same plant, e.g., cleistogamous species.
Different seed within the fruit.

The challenge is to discern the patterns of variation in outcrossing rates and determine the scope and the evidence for various modes of selection. This task includes discerning the levels and patterns of "correlated matings," a topic that came to the fore when experimenters studied animal-pollinated species having appreciable levels of both outcrossing and selfing.

Such data are also needed to construct more realistic models for the evolution of selfing, in particular, models that predict stable equilibria at points other than exclusive selfing or outcrossing. For example, the dual flowering strategy of both cleistogamous and outcrossing chasmogamous flowers shown by all perennial species of the genus *Glycine* has clearly been stable while the group has speciated and diverged into most Australian environments.

Facultative or obligate apomixis

Apomictic species that form seed without meiosis and fertilization (i.e., species capable of agamospermy) are yet to receive the attention they deserve. Like autogamy, apomixis is a uniparental mode of reproduction, but in contrast, apomixis does not forfeit the benefits of heterosis. Thus, an understanding of the patterns of variation in levels of apomixis in facultative apomicts, and the genes that control them, is sought by plant breeders wanting to transform crop plants with such genes to exploit heterosis (Asker 1980).

Rare outcrossing in predominantly apomictic populations may be even more important than rare outcrossing in selfing populations. In the former, the more common apomictic seeds preserve the heterozygosity generated by the occasional outcross, whereas selfing erodes it (Marshall and Weir 1979). In *Panicum maximum*, Useberti and Jain (1978) found that some predominantly asexual lines were more variable than predominantly sexual populations.

Intragametophytic or haploid selfing

At the other extreme to apomixis is the mating system that achieves complete homozygosity in a single step. Dihaploid breeding achieves this artificially by generating haploid plantlets and then doubling their chromosome complement using colchicine. A natural analog of this process is intragametophytic selfing, where a gametophyte develops from a single haploid spore and bears both antheridia, which produce sperm, and archegonia, which produce eggs. In homosporous ferns and allied lower plants, self-fertilization of a single gametophyte can occur, resulting in a fully homozygous diploid. Here "self" refers to the same gametophyte whereas in seed plants it refers to the same sporophyte. The main issue in the homosporous ferns is whether mating is fully outcrossing, or fully haploid selfing, or mixed (Holsinger 1987). If mixed mating is occurring, the question is whether the mixed system includes some haploid selfing or simulates the mixed expectations by employing milder forms of inbreeding.

Having outlined the major modes of mating and the questions of interest we turn now to the tools available to address these questions.

MODELS AND PROCEDURES

Mixed mating model

Table 2 summarizes the familiar mixed mating model for the segregation of a single codominant diallelic locus in the progeny arrays of angiosperms. The model, its assumptions, the estimation of its parameters, and their extension to multiple loci and gymnosperms have been discussed elsewhere (Brown et al 1985, in press and citations). Early formal analyses of the model are those of Haldane (1924) in determining the progress of inbreeding and Fyfe and Bailey (1951) in the estimation of its parameters. Estimates were originally made in inbreeding crops. Subsequently, the model has been widely used in both plantations and natural populations of herbaceous and woody species, with a diverse array of mating systems and with wind or animal pollination. Each of these extensions (from planted to natural populations, from inbreeding to outbreeding species, from annuals to perennials, etc.) has had its own difficulties with the simplifying assumptions of the model. Yet in practice it appears that departures from the assumptions generally result in relatively minor effects.

TABLE 2 Basic maternal/offspring matrix for estimation of the parameters of the mating system of mixed self-fertilization(s) and random outcrossing ($t = 1 - s$) at a diallelic locus with p denoting the A_1 allele frequency in pollen.

	Genotypes		Number of progeny	
	Maternal	Progeny	Observed	Expected
1	A_1A_1	A_1A_1	O_1	$N_{11}(1 - X)$
2		A_1A_2	O_2	$N_{11} X$
3	A_1A_2	A_1A_1	O_3	$N_{12}(1 - X + Y)/2$
4		A_2A_2	O_4	$N_{12}(1 + X - Y)/2$
5	A_2A_2	A_1A_2	O_5	$N_{22} Y$
6		A_2A_2	O_6	$N_{22}(1 - Y)$
	Where	$X = tq$	$N_{11} = O_1 + O_2$	
		$Y = tp$	$N_{12} = O_3 + O_4$	
			$N_{12} = O_5 + O_6$	

Multilocus selfing and "effective" selfing

Several authors have pointed out that subpopulation structure in natural populations will, with mating among neighboring plants, usually lead to increased inbreeding. The use of the simple mixed-mating model, which assumes that all outcrossing is at random, will include some of this inbreeding in the estimate of selfing that is thereby biased up from its true value. A second problem is that this biased estimate includes both true selfing and biparental inbreeding in a combined measure that may not be strictly interpretable.

These effects can be illustrated by considering an example population made up of two equally sized subpopulations. Assume that the frequency of the A_1 allele in each is p_1 and p_2, that the outcrossing is at random *within* each subpopulation, and that each is in separate inbreeding equilibrium. In cases 1 and 2, the selfing rate is 0.5 in both subpopulations, whereas in cases 3 and 4, $s = 0.5$ in subpopulation 1 (SP1) and $s = 0.8$ in subpopulation 2 (SP2). Cases 1 and 2 include allele frequency differentiation, case 3 has variation for the level of inbreeding, and case 4 has both types of variation. Table 3 summarizes the mating system estimates in three cases. The maximum likelihood estimate of the selfing rate using the mixed mating model exceeds its true average value (0.5 or 0.65) in all cases. Also the estimate of allele frequency in the outcrossing pollen is biased by a small amount. In sufficiently large experiments, a test of lack-of-fit of the observed progeny frequencies to those expected under the mixed mating model (Table 2) would give a significant χ^2 in cases 2, 3 and 4. Departure from the model is readily apparent when the proportions in classes 1 (= 1 − X) and 5 (= Y) of Table 2 do not average to the proportion in class 3 [= (1 − X + Y)/2].

To distinguish the true selfing rate from biparental inbreeding requires sufficient polymorphic loci to detect as many as possible of the allogamous matings genetically. The higher the polymorphism, the smaller will be the fraction of undetected outcrosses. The multilocus procedures attempt to delete this undetected fraction from the overall estimate of selfing. The multilocus estimate of outcrossing can then be compared with the mean of single locus estimates and the difference attributed to biparental inbreeding.

Table 4 is a list of several such comparisons from the recent literature. Generally, the estimates show only limited depression of single locus estimates (about 10% of the minimum of *s* or *t*). So this has left the "heterozygosity paradox" of apparent excess heterozygosity in inbreeders and deficiency in outbreeders (Brown 1979) still unresolved. The most marked differences tend to be for intermediate values of outcrossing, and in animal-pollinated

TABLE 3 Estimates of mating system parameters at a codominant, diallelic locus in four model populations consisting of two cryptic, reproductively isolated subpopulations (SP1, SP2) each in inbreeding equilibrium.

	Cases			
	1	2	3	4
A_1 frequency In SP1 (p_1)	0.25	0.25	0.25	0.25
In SP2 (p_2)	0.75	0.50	0.25	0.50
F_{IS} in SP1	0.33	0.33	0.33	0.33
F_{IS} in SP2	0.33	0.33	0.67	0.67
F_{IT} − total fixation	0.50	0.378	0.50	0.556
F_{ST} − differentiation	0.25	0.067	0.0	0.111
Mean selfing	0.50	0.50	0.65	0.65
MIXED MATING MODEL				
Selfing rate (s)	0.67	0.55	0.67	0.72
Pollen allele frequency (p)	0.50	0.373	0.258	0.402
"EFFECTIVE" SELFING MODEL				
Effective selfing rate (E)	0.67	0.55	0.66	0.71
Inbred selfing rate (s_i)	0.67	0.55	0.68	0.73
Outbred selfing rate (s_o)	0.67	0.55	0.64	0.68
Pollen allele frequency (p)	0.50	0.375	0.252	0.389

TABLE 4 Comparison of estimates of outcrossing based on single loci with the multilocus estimate from the same data in a range of species (species are listed in order of the multilocus estimate).

Species	Population	Loci	Average single	Multilocus	Source[a]
Lupinus alba	1	3	0.10	0.09	1
Limnanthes bakeri	1	4	0.21	0.21	2
Eichhornia paniculata—Jamaica	3	3	0.36	0.48	3
Glycine argyrea	1	10	0.38	0.48	4
Bidens menziesii	4	4	0.48	0.55	5
Larix laricina	5	3	0.64	0.73	6
Eucalyptus delegatensis	3	3	0.75	0.77	7
Limnanthes alba	2	3	0.76	0.80	8
Limnanthes douglasii	6	9	0.74	0.82	2
Eichhornia paniculata—Brazil	7	5	0.80	0.85	3
Pseudotsuga menziesii	1	10	0.74	0.89	9
Pseudotsuga menziesii	8	8	0.91	0.90	10
Abies balsamea	4	3	0.91	0.89	11
Pinus jeffreyi	5	10	0.89	0.94	12
Acacia auriculiformis	2	9	0.88	0.93	13
Pithecellobium pedicellare	1	4	0.93	0.95	14
Pseudotsuga menziesii	4	11	0.92	0.96	15
Acacia crassicarpa	2	7	0.98	0.96	13
Echium plantagineum	3	4	0.98	0.96	16
Cynosurus cristatus	4	3	0.98	0.97	17
Pinus monticola	1	6	0.95	0.98	18
Pinus contorta	2	7	1.0	1.0	19

[a]1. Green et al. (1980); 2. Kesseli and Jain (1985); 3. Glover and Barrett (1986); 4. Brown et al. (1986); 5. Ritland and Ganders (1985); 6. Knowles et al. (1987); 7. Moran and Brown (1980); 8. Ritland and Jain (1981); 9. Yeh and Morgan (1985); 10. Shaw and Allard (1982); 11. Neale and Adams (1985b); 12. Furnier and Adams (1986); 13. Moran et al. (in press); 14. O'Malley and Bawa (1987); 15. Neale and Adams (1985a); 16. Burdon et al. (1988); 17. Ennos (1985); 18. El-Kassaby et al. (1987); 19. Epperson and Allard (1984).

species. In *Bidens* species, the average outcrossing rate for bisexual plants was 0.65 in contrast to that for female plants of 0.85 (Sun and Ganders 1988). Nevertheless, the researcher should be wary of samples including too few maternal plants or excessively widespread samples.

The second problem of a more meaningful, single-locus combined measure of inbreeding (lumping selfing with biparental inbreeding) and its unbiased estimation is addressed by Ritland's "effective selfing" model (Ritland 1984,

1986). This model derives two measures of combined inbreeding, one as the selfing rate on "inbred" parents (s_i) and the other on "outbred" parents (s_o), and combines them in an overall measure E (where F is the inbreeding coefficient),

$$E = Fs_i + (1 - F)s_o$$

Estimates of these parameters require information from open-pollinated arrays, as well as the genotypic frequencies of the maternal plants. The model assumes that the allele frequencies are the same in the maternal plants and pollen, and that the maternal plants are a random sample from the population producing the pollen. A large number of families is needed to estimate the maternal fixation index (F) accurately.

Table 3 shows the estimates for effective selfing (E) in four cases. The estimates of selfing (E, s_i, and s_o) differed in the two cases (3,4) with variation of inbreeding in the population. The covariation of selfing rate with fixation index leads to differences between s_i and s_o and between E and s (Ritland and Ganders 1987). In these cases, the effective selfing model gives a less biased estimate of pollen allele frequency. It seems that this procedure is relevant when the small differences between single and multilocus estimates are to be measured, provided that its somewhat restricting assumptions of maternal sampling and constant allele frequencies are met.

Waller and Knight (in press) have estimated the total genetic correlation (M_t) between pollen and ovules in chasmogamous (CH) fruit of *Impatiens capensis*, by comparing the fixation index in progeny of CH fruit (F_{CH}), with the parental fixation (F_p).

$$M_t = 2F_{CH}/(1 + F_p)$$

Assuming no selection is occurring, the genetic correlation has two components—that due to true (geitonogamous) selfing (s, estimated by multilocus procedures) and that due to biparental inbreeding (M_b), computed as

$$M_b = (M_t - s)/(1 - s)$$

The averages of estimates from five populations were $s = 0.45$ and $M_b = 0.24$, indicating significant biparental inbreeding. These are useful summary statistics for describing the differences between single and multilocus estimates despite their sensitivity to problems of multilocus estimation. In *Glycine argyrea*, the estimates are similar to the *Impatiens* study, with $F_p = 0.175$, $F_{ch} = 0.378$, and $s = 0.524$ (data of Brown et al. 1986) so that the correlation due to biparental inbreeding (M_b) is 0.25.

Paternity

The values in Table 4 for populations of predominantly outcrossing species show broadly that single locus estimates do not reveal major departures from panmixia. But can we assume that local panmixia is operative at the level of individual plants? Studies of mating at this level present a different perspective to those at the level of the population (Brown et al. in press). Since outcrossing is the commonest mating event in these species, the task is to unravel the sources of outbreeding pollen. Do individual plants differ in their male mating success? Do they differ in gender allocation? Is there assortative mating for traits like flower color? Answers to such questions require the assignment of a sample of successful male gametes to particular source plants. In small, highly polymorphic natural or contrived populations or plantations, this can be done and migrational gene flow, fruit paternity, intermate distances, male fertility, and gender allocation have been measured.

In *complete* paternity studies, the aim is to use genetic markers to identify the source plant of all pollen grains that successfully achieve fertilization. *Partial* paternity studies are also possible, where the paternity of fruit is investigated without identifying the actual source plant(s) of the pollen (e.g., in *Glycine*, Brown et al. 1986; and *Acacia*, Muona, this volume). The object of such studies is to determine the minimum number of genetically distinct male sources for fruits.

The following discussion assumes that female parentage of each progeny is known directly from harvest, although it is possible to use similar approaches for biparental assignment of established seedlings (Meagher and Thompson 1987). The ideal situation for paternal assignment has sufficient marker polymorphism in a small population so that the deterministic technique of parental exclusion can be used (Ellstrand 1984). For each progeny, the exclusion algorithm takes each locus in turn and excludes any impossible parent diagnosed by that locus, ideally leaving a unique source for that male gamete.

However, this ideal situation rarely occurs. There may be a dilemma about the requirement for variation at one time to validate a single source by a unique allele or combination. Later, the very same allele or combination might confuse the paternity of the next generation between father and grandfather or between two siblings, especially when variation in male fertility is marked. Thus most students of paternity in natural populations will have to deal with the problem of ambiguous assignments. What techniques are available to meet this difficulty?

Discard progeny with ambiguous paternity This is a straightforward approach, which may be permissible if the ambiguous progeny make up only a small fraction. The danger with this remedy is bias based on genotype may be introduced. The male fertility of plants heterozygous for a rare allele may be inflated. In addition, the ambiguous fraction is rarely small.

Employ hypervariable molecular markers New molecular approaches may well provide extremely variable markers. These include hypervariable minisatellites (Jeffreys et al. 1985), which consist of a collection of fragments dispersed through the genome, or polyallelic single loci segregating for a variable number of tandem repeats (Nakamura et al. 1987). Technology is needed to facilitate large samples of progeny and putative male sources. Until then, these techniques might be used as a way to resolve paternity in ambiguous cases, after isozyme exclusion has reduced the number of potential males.

Assign paternity to the highest segregational likelihood Meagher (1986) used the segregational likelihood or conditional probability that a specific parental genotype would produce a male gamete with a specific genotype, and assigned parentage to the plant with maximum likelihood. However, this procedure is biased toward homozygous parental origin (Devlin et al. 1988), because such a plant would have maximum chance of producing the required male gamete. Further, there are no constraints in space or time for the "most likely" parent. The search for the most homozygous source based on segregation likelihood is limited only and arbitrarily by the total sample and not by the proximity of mates or overlap of their flowering. Thus estimate of mean distances to mates will be biased upward.

Assign paternity fractionally in proportion to the segregational likelihood Devlin et al. (1988) examined the effect of assigning paternity in each ambiguous case, not singly to the most likely source, but fractionally to sources in proportion to their segregational likelihoods. Under the assumption (null hypothesis) of equal male fertility, this method removes the genotypic biases of the above methods. The method still requires average exclusion fractions of 90%, and is biased to underestimates of fertility variation when it is present.

Weight the segregational likelihood with the probability of mating and choose the most likely parent based on this combined measure W. T. Adams, A. R. Griffin, and G. F. Moran (unpublished) have modified Meagher's procedure by including an additional weighting factor in the likelihood index. An approximate relationship between probability of mating and distance between mates, an effective pollen dispersal, was computed from the subsample of unique paternal assignments. This distance function was then used with the segregational likelihood to infer paternity in ambiguous cases. The method restrains the homozygosity bias in a trade-off with separation distance, although in a complex, data-generated fashion. In principle, the weighting factor could include other ecological variables such as pollen production and phenological overlap. However, the appropriate relative weighting of the genetic and the ecologicallikelihoods is not clear, and inferences back to genetic or ecological parameters in the mating scheme could be cyclic in reasoning. It seems that likelihood methods based on partial data have many potential pitfalls (Brown et al. in press).

Study paternity ex situ in experimental populations Several studies of genetic differences in male reproductive performances have been made in experimental populations that avoid problems in genetic discrimination of events. Ennos and Dodson (1987) provide a recent example in estimating pollen success in a model planting of six distinct clones of *Cynosurus cristatus*. While such a study shows that highly nonrandom patterns of mating *can* occur in garden experiments, it leaves open the question of whether such pattern *do* occur *in situ*. This is a challenging problem.

Male fertility components

An alternative approach to detecting variation in components of fertility in the population is tracing the transition in genotype frequencies between two stages of the life cycle. Estimates can be ascribed to single- (Horovitz and Harding 1972) or multilocus (Clegg et al. 1978) genotypes. Maximum likelihood estimators transform shifts in genotype frequency into estimates of fertility components of fitness. Two problems of these methods are that intense selection can take place without being detectable as a shift in allele frequencies, and that estimates may relate negatively to other components of fitness in ways that suggest they are sensitive to sampling error.

Schoen and Stewart (1987) have developed a procedure to estimate male fertilities of clones or genotype classes based on linear models. The procedure is particularly suited to gymnosperms where the pollen genotype can be inferred directly from the megagametophyte-embryo combination. They found great variation in male fertility among 30 *Picea glauca* clones in a seed orchard. Fertility estimates also fluctuated between years. The extent to which such variation is genetically determined is unknown, although clones from disparate origin and phenology might be inclined to differ in reproductive performance when planted together. Further, the estimated fertility of a clone that is the sole source of a particular gamete may be exceptionally high or low (Brown et al., in press). The question is then whether exceptional genotypes really do not have exceptional male fertility, or is this a sampling effect?

In general, the spatial distribution of ovules may affect the estimates of male fertility. If there is variation for female fecundity in the population, plants adjacent to a fecund female could have higher estimates of male fertility. In estimating male fertility, should one sample a fixed number of seed from each individual maternal plant or a fixed proportion of the total seed matured by each plant? Hence population structure and limited pollen dispersal may link together male and female reproductive performance. It appears that variation in male fertility is difficult to measure. Yet this component is important to describe for a complete understanding of the mating system of a population.

Correlated mating

The genetic relationship between two members of a sibship depends on four aspects of the mating system: (1) biparental inbreeding in the population, (2) self-fertilization, (3) correlation of selfing (r_s), and (4) correlation of outcrossed paternity (r_p). The correlation of selfing is the extent to which both sibs are more or less likely than random to originate from the same type of mating. This is an aspect of variation in outcrossing rate. Thus differences between individual plants in the population in their outcrossing rate generate such a correlation. The correlation of paternity among outcrossed sibs is the proportion of full-sib pairs among outcrossed progeny. These two modes of correlation were discussed by Brown et al. (1985) as the "*t*-effects" and the "*p*-effects". Ritland (in press) has developed a "sibling pair" model that explicitly distinguishes correlation of selfing from correlation of paternity. To estimate the two parameters, progeny are sampled for their single-locus genotype from the same maternal plant of known or inferred genotype, either from the same fruit or different fruits. Sibling pairs of seed can arise in four main ways (Table 5).

Ritland (in press) gives a single locus model and algorithm to estimate these kinds of pairs. This is a more general model and approach to documenting correlated matings than the single male parent model of Schoen and Clegg (Schoen 1988) that assumes $r_s = 0$ and $r_p = 1$. However, Ritland found that diallelic data lacked statistical power, as the estimates had large standard errors and the likelihood surface was shallow. Estimates of r_s and r_p were negative correlated. He showed that increasing the number of alleles helped to improve the sensitivity, and suggested that highly polymorphic molecular markers may be a solution.

Once again the problem is genetic discernment of mating events, in this case between two different sources of outcross pollen. This is the problem of single versus multiple outcross paternity of fruits. Perhaps an approach based on multilocus data, rather than data from a single polyallelic locus, would be better. Generalizing the formal approach to multiple loci and many progeny would represent a formidable task. However in some cases one can proceed by arguments based on likelihood of paternity as we did in the "paired fruit" study of *Glycine argyrea* (Brown et al. 1986). Figure 1 shows the sampling pattern in this study.

Seeds in legumes of *G. argyrea* were classified as to whether they were presumed self, or outcross to one or more distinct male sources. Testing k ($k = 6$) seed per fruit, estimates of Ritland's classes of mating can be made directly. Pairs of seed within each fruit were compared in $k(k - 1)/2 = 15$ combinations, of which $(k - 1)$ are statistically independent. The frequencies of types of pairs between fruit follow similarly with k^2 combinations ($2k - 1$ independent). In this study, the decision on class of mating was biased in that

TABLE 5 Correlated mating model (Ritland, in press) and estimates of parameters in *Glycine argyrea.*[a]

Type of sibling pair	Proportions: Expected	Proportions: Observed
Both self-fertilized	$s^2(1 - r_s) + sr_s$	A
One self + one outcrossed	$2st\ (1 - r_s)$	B
Both outcrossed		
One male parent	$[t^2(1 - r_s) + tr_s]r_p$	C
Two male parents	$[t^2(1 - r_s) + tr_s](1 - r_p)$	D

Estimates: $s = A + B/2 = 1 - t$
$r_s = A - B/2st$
$r_p = C/(C + D)$

Example: *Glycine argyrea* (data of Brown et al. 1986)

	Within fruit	Between fruit
s	0.62 ± 0.03	0.62 ± 0.03
r_s	0.39 ± 0.07	0.08 ± 0.07
r_p	0.85 ± 0.05	0.58 ± 0.08

[a]Where A, B, C, and D are estimated by pair-wise comparison of possible pairs of seed within and between fruits. If k seed per fruit from $2n$ fruits are assayed and their mating classified, the total number of independent events in the estimats is $2n(k - 1)$ within fruit and $n(2k - 1)$ between fruit. Hence the variance of the maximum likelihood estimates of s and r_s follow by analogy those for codominant diallelic frequencies and fixation index, and for r_p the binomial, e.g., for between fruit:

$$\text{var}(s) = st/2n\ (k - 1) \qquad \text{var}(r_p) = r_p(1 - r_p)\ /\ (C + D)\ 2n\ (k - 1)$$

the number of males involved in a sibship was taken as the minimum possible, and increased only at a level of probability 0.05.

Other special cases where Ritland's model could be tellingly applied would be where biological data allow us to assume (1) that selfing is essentially zero, such as for self-incompatible species, or (2) that $r_s = 0$ or 1. In the former case, r_s is undefined and progeny pairs can then be examined for estimating r_p.

The importance of developing and applying models of correlated matings is the relevance of such models to whenever members of sibships interact or compete.

Mixed models for partial apomixis and haploid selfing

The procedure for estimating partial autogamy is flexible and can be adapted

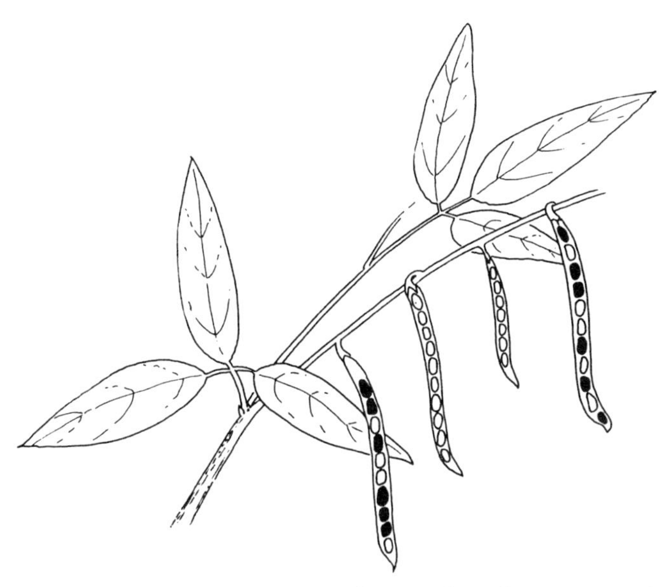

FIGURE 1. Raceme of *Glycine argyrea* showing four chasmogamous legumes. The seeds assayed for the paired fruit analysis of correlated matings are in black.

to estimating partial apomixis (Marshall and Brown 1974). More recently, mixed mating estimation procedures for homosporous ferns (Hedrick 1987; Holsinger 1987) and for fungi (Ennos and Swales 1987) have been devised. In ferns, progeny arrays are generally unavailable, so that estimation of the amount of haploid selfing (s_I), as opposed to random outcrossing, uses the observed sporophyte genotypic frequencies. The expected frequencies are the same as those in terms of Wright's fixation index (F), with F substituted for s_I. Just as in partial autogamy in seed plants, less intense forms of inbreeding (intergametophytic selfing and biparental inbreeding) will increase the fixation index and thus inflate the estimate of s_I. Examples of the extreme range in estimates of s_I by this method include *Bommeria hispida*, $s_I = 0$ (Haufler and Soltis 1984), *Botrychium dissectum*, $s_I = 0.95$ (McCauley et al. 1985), and *Dryopteris expansa*, $s_I = 0.34$ (Soltis and Soltis 1987).

Haploid selfing or less intense forms of inbreeding are also possible and measurable in fungal pathogens. In the discomycete, *Crumenulopsis sororia*, Ennos and Swales (1987) exploited the link between the known haploid maternal genotype and haploid progeny asci scored in an apothecial array. Using a haploid mixed mating model, they estimated outcrossing as complete, probably enforced by a sexual incompatibility system. Further, a single apothecium commonly contains ascogenous hyphae from more than one fertilization event. These conclusions could not be reached from *in vitro* studies as the fungus fails to fruit in culture.

CONCLUSIONS

Quantitative estimates of the mating system parameters in plant populations are needed to understand the present genetic structure of such populations and to predict their future evolution. The mating system also affects how we are to collect optimally, to conserve effectively, and to use efficiently the genetic resources of domesticated plants and their wild relatives. This survey of recent efforts to analyze quantitatively the mating system of plant populations shows this field of research to be rich, intriguing, and important. The mixed mating model, on which much of the early work rested, is still a robust, basic approach, despite nonconformity with several of its simplifying assumptions. It has been adapted to many special cases, sometimes with more restrictive assumptions, and less statistical power. Yet there is a need to refine these models and take greater account of the details of life history and the ecological context of mating systems (Brown et al. in press).

Several modes of inbreeding can operate in natural populations of plants, each at a different level. While it is helpful to speak of inbreeding at one level—such as "effective" selfing rate—it is also important to discern the various contributing levels. Selfing is a botanical absolute whereas biparental inbreeding is a relative concept, because it is relative to the local population. The selective forces that act on different modes of inbreeding will differ. It is important to disentangle these modes if we are to understand fully the evolution of inbreeding in plants.

In outcrossing species, the analysis of mating patterns requires a dissection of the outcrossing component. Marker loci afford the chance to chart outcrossing events. This would provide data fundamental to defining the extent of the local population and thus establishing the theater of action of selection. For example, in studying host-pathogen dynamics, the question may be asked: from what area can a resistance gene come to rescue a population from a new race of pathogen? Yet this new genetic awareness of the population has led to a confusing use of the term "gene flow." Animal population geneticists have used the term to describe this migration *into* the population of organisms with different allele frequencies. Our task is clearly to distinguish between pollen dispersal "within a population" that is part of biparental reproduction and the immigration of new genes. This task will not be helped by using the same term for both processes.

ACKNOWLEDGMENTS

I am very grateful to Drs. W. T. Adams and K. Ritland for reading the manuscript, and to K. Ritland for the use of his Effective Selfing Estimation program.

CHAPTER 10

THE INTERACTION BETWEEN SELECTION AND LINKAGE IN PLANT POPULATIONS

Alan Hastings

ABSTRACT

At equilibrium, epistatic selection causes alleles at different loci to be statistically associated, or in disequilibrium. This chapter reviews this interaction between selection and linkage, and considers other possible causes of disequilibrium. Experimental studies of disequilibrium in plants are reviewed, emphasizing the prevalence of disequilibrium in highly inbred plants, as opposed to the lack of disequilibrium in outbreeding species. None of the experimental evidence is inconsistent with strong epistatic selection, although other explanations for the observed levels of disequilibrium are possible.

INTRODUCTION

Early work on multilocus systems (e.g., Franklin and Lewontin 1970) indicated the potential for a large role of linkage disequilibrium in dynamics and equilibrium behavior. Since this time there have been many attempts to assess the role of disequilibrium in natural and experimental populations. Among several major questions, an overriding one is how prevalent is linkage disequilibrium in natural and experimental populations? A natural question to ask next is whether the action of selection can be deduced from the presence of disequilibrium? An important related question is whether, even with reasonable selection, significant levels of disequilibrium would be expected? A final question is what are the consequences of linkage for the impact of selection on plant populations? Here, one may reasonably ask whether linkage has an important influence even if disequilibrium is not detected.

The interaction between selection and linkage can occur at three different levels of genetic organization: within the gene, at the level of DNA sequences (Clegg, this volume); between loci, at the level of two or a small number of loci; or in terms of quantitative variation (e.g., Jinks et al. 1985), at the level of large numbers of loci. Here, I concentrate on the level of a few loci, although some of the most tantalizing theoretical and experimental questions lie in finding connections among these three different levels of genetic organization.

At all three levels, there are a number of possible causes of disequilibrium. One possible cause is epistatic selection. Other causes of disequilibrium include finite population effects and the role of migration. Transient effects may be especially important in generating observed levels of disequilibrium, particularly when combined with finite population effects (Avery and Hill 1979). The first part of this chapter reviews causes of disequilibrium from a theoretical point of view, emphasizing the consequences of epistatic selection and considering both stationary, or equilibrium, populations and populations that are changing. Then, questions of data interpretation are considered. Finally, data on disequilibrium from both natural and artificial populations of plant species are reviewed.

Plant species that are predominantly self-pollinated exhibit high levels of linkage disequilibrium in experimental and natural populations (e.g., Clegg et al. 1972; Brown et al. 1977; Weir et al. 1974). Since selfing acts in the same manner as does restricting recombination, the case of predominant selfing deserves special emphasis in any study of the interaction between linkage and selection in plant populations. In particular, predominant selfing is a frequent and successful mode of reproduction among plants.

THEORY

Earlier theoretical approaches to understanding the role of linkage disequilibrium in both outcrossing and partially selfing populations were reviewed in Hedrick et al. (1978). Unfortunately, much of this theory studied the dynamics, and more often the statics, only of two-locus models (exceptions began with Franklin and Lewontin 1970).

Outcrossing

The stationary behavior of two loci in outcrossing populations was reviewed by Karlin (1975). One approach is to study particular selection models and examine the consequences of disequilibrium, or association, between two or more loci. An alternate approach consists of specifying information about a set of gametic frequencies and considering the population genetic processes that may account for this set, particularly in a stationary population.

Some of the earliest theoretical investigations of disequilibrium concerned the static behavior of two-locus two-allele models in random mating populations with specified fitnesses. In the two-locus model (e.g., Karlin 1975), let there be two loci with two alleles each: A and a at the A locus and B and b at the B locus. Denote the frequency of the gametes AB, Ab, aB, ab by x_1, x_2, x_3, x_4, respectively. Define the linkage disequilibrium D to be $x_1x_4 - x_2x_3$ and let r be the recombination rate. Let ϵ_i take the values $-1,1,1,-1$ for $i = 1$ to 4, respectively. Denote by w_{ij} the fitness of the individual with the gametes whose frequencies are x_i and x_j. Note that $w_{ij} = w_{ji}$ and if there are no cis-trans effects, we take $w_{14} = w_{23} = 1$. Under this assumption, the fitnesses can be displayed conveniently in the 3×3 matrix listed in Table 1.

Denote the marginal mean fitness of the gamete i by

$$w_i = \sum_{j=1}^{4} x_j w_{ij}$$

and the mean fitness of the population by

$$\bar{w} = \sum_{i=1}^{4} x_i w_i$$

Then the dynamics of this system are given by

$$x_i' = (x_i w_i + \epsilon_i r D)/\bar{w} \tag{1}$$

for a discrete time system with nonoverlapping generations.

Two particular schemes were investigated in great detail, the symmetric viability model and the multiplicative model. One special fitness matrix, the symmetric fitness form, which can have large epistatic effects, was studied in great detail by Karlin and Feldman (1970). However, a special case of the symmetric model introduced by Lewontin and Kojima (1960) illustrates the interaction between linkage and selection (reviewed by Karlin 1975 and Ewens 1979). In the Lewontin-Kojima version the fitnesses of all four homozygotes are the same:

$$w_{11} = w_{22} = w_{33} = w_{44} = 1 - \alpha$$

TABLE 1 Fitnesses in a two-locus two-allele model.

	BB	***Bb***	***bb***
AA	w_{11}	w_{12}	w_{22}
Aa	w_{13}	$w_{23} = w_{14}$	w_{24}
aa	w_{33}	w_{34}	w_{44}

Additionally, it is assumed that

$$w_{12} = w_{34} = 1 - \beta \qquad \text{and} \qquad w_{13} = w_{24} = 1 - \gamma$$

where α, β, γ are positive.

There is always a stationary state with no linkage disequilibrium of the form

$$x_1 = x_2 = x_3 = x_4 = 1/4$$

which is always unstable if

$$|\beta - \gamma| > \alpha$$

and possibly stable otherwise as noted below. Additionally, if recombination is small enough

$$r < (\beta + \gamma - \alpha)/4$$

there are two "symmetric" equilibria of the form that exhibit linkage disequilibrium:

$$x_1 = x_4 = 1/4 + D, \qquad x_2 = x_3 = 1/4 - D$$

and

$$x_1 = x_4 = 1/4 - D, \qquad x_2 = x_3 = 1/4 + D$$

where

$$D = \{[1 - 4r/(\beta + \gamma - \alpha)]^{1/2}\}/4$$

Note that as r approaches zero, the disequilibrium, D, approaches ¼, its maximum value possible. Ewens (1969) showed that these "symmetric" equilibria are stable if and only if

$$4r^2(\beta + \gamma - \alpha) + 2r[2\alpha^2 - \beta^2 - \gamma^2 - \alpha(\beta + \gamma)] \\ + \alpha(\beta + \gamma - \alpha)^2 > 0$$

Moreover, Karlin and Feldman (1970) demonstrated the existence and stability of other equilibria, and Hastings (1985a) demonstrated the possibility of up to four simultaneously stable polymorphic equilibria. The two major biological implications are (1) with epistatic selection and tight linkage (relative to the strength of selection) disequilibrium can be important; (2) for tight linkage, the stationarity behavior of the model depends in a complex way on the recombination, initial conditions, and strength of selection. A contrast is provided by the nonepistatic additive model that has the fitness pattern illustrated in Table 2. Here, in all stationary states, there is no linkage disequilibrium, for any value of the recombination. This equilibrium is also globally stable (see Karlin 1975).

TABLE 2 Fitnesses in an additive two-locus two-allele model.

	BB	*Bb*	*bb*
AA	$w_{AA} + w_{BB}$	$w_{AA} + w_{Bb}$	$w_{AA} + w_{bb}$
A*a*	$w_{Aa} + w_{BB}$	$w_{Aa} + w_{Bb}$	$w_{Aa} + w_{bb}$
aa	$w_{aa} + w_{BB}$	$w_{aa} + w_{Bb}$	$w_{aa} + w_{bb}$

Studies of special fitness patterns do not answer the question of how strong selection must be relative to recombination, nor how strong epistasis must be, to generate linkage disequilibrium. A different approach to studying the stationary behavior of multilocus population genetic systems (Hastings 1981, 1986) answers such questions by specifying information about the stationary population and deducing information about the possible values of the population genetic parameters that lead to this outcome. This approach has two advantages. First, it greatly simplifies approaches for determining information about the behavior of systems with general fitnesses, not just the "special" models discussed above. Second, under this approach experimental observations of gametic frequencies can be used as input, and the results from the analysis are values of parameters that are more difficult to measure, such as fitnesses.

If the stationary population is specified, which is reasonable since it is much easier to measure allele or gametic frequencies than fitnesses, then solving for the fitnesses at stationarity becomes a linear problem. To see this write the equation for stationarity in the two-locus model (1) as

$$\bar{w}x_i = x_i w_i + \epsilon_i r D$$

The fitnesses w_{ij} enter only linearly. The only difficulty with this approach is that the general solution is usually so complex as to be uninformative. One method for obtaining useful information this way is to determine limits to the relationship between fitnesses and equilibria using an optimization technique. Given the underlying linearity of the problem posed this way, by specifying the problem appropriately, linear programming methods can be used.

The application of linear programming methods to this problem can answer the following question (Hastings 1981): for specified stationary gametic frequencies and recombination rate, what is the minimum strength of selection necessary to maintain the population at this state? For given allele frequencies and sign of D, the minimum strength of selection that is the answer to this problem is a monotonic increasing function of the absolute value of the disequilibrium, D. Thus, this analysis also answers the question: given a particular strength of selection, recombination rate, and allele frequencies, what is the maximum value of D possible? If this information is combined

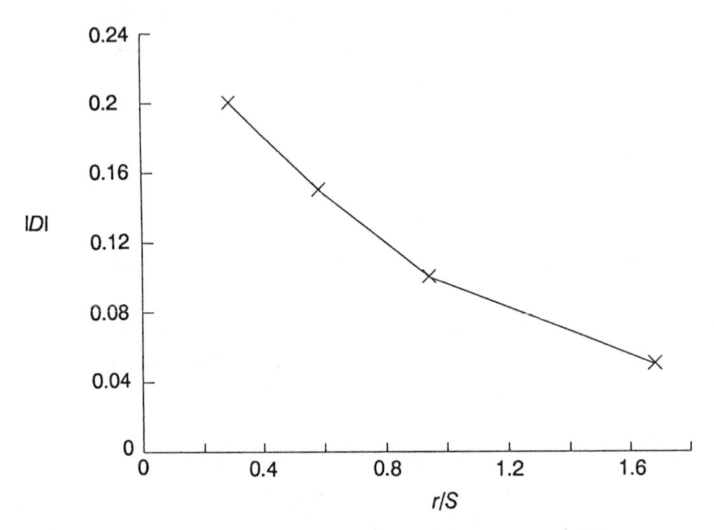

FIGURE 1. Maximum levels of linkage disequilibrium at equilibrium, as a function of the recombination rate divided by the strength of selection, for an outcrossing species wth $p_A = 0.5$, $p_B = 0.5$.

with information about the statistical power of tests used to detect disequilibrium (Brown 1975), one can determine if disequilibrium is likely to be found.

As an illustration of the results possible (Hastings 1981) I have graphed some of the limits to the relationship among D and r/S, where S is defined as the maximum strength of selection—the maximum value of $|w_{ij} - 1|$, where the fitness of the double heterozygote, w_{14} is normalized to be 1. This information is presented in Figures 1 and 2. The results show that only for very closely linked loci is disequilibrium at a stationary state large enough to be easily detected—typically only if r/S is of order one or less.

A related question is whether the levels of epistasis required to generate disequilibrium are so high that only very strongly interacting loci would be likely to lead to high levels of disequilibrium. Most attempts to detect linkage disequilibrium in natural populations have used randomly chosen loci that have shown little evidence for disequilibrium. This problem can again be solved via a linear programming approach. Express the fitnesses in a two-locus model as deviations from an additive model as in Table 3. Define the total epistasis as the maximum of the absolute value of the four quantities $e_1 p_A q_A$, $e_2 p_A q_B$, $e_3 p_B q_A$, $e_4 p_B q_B$. This definition gives a total epistasis where weights are assigned to gametes according to their frequency when there is no disequilibrium. An alternative possibility without weights is to define the total epistasis as the maximum of the absolute value of the four quantities e_i.

Using either definition of epistasis just given, specifying the stationary frequencies and minimizing the total epistasis is a linear programming problem.

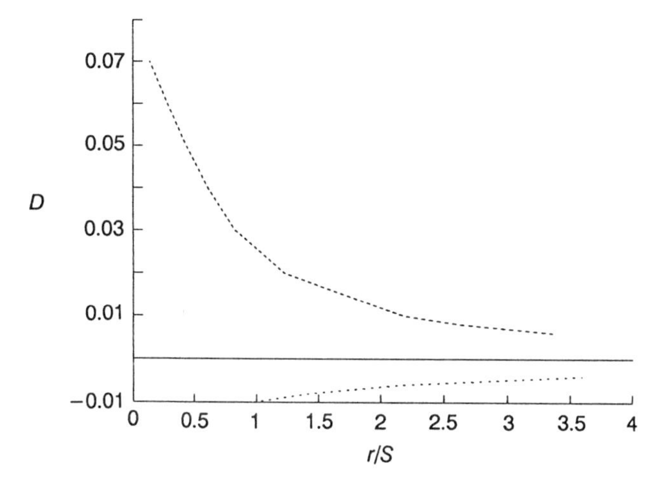

FIGURE 2. Maximum levels of linkage disequilibrium at equilibrium, as a function of the recombination rate divided by the strength of selection, for an outcrossing species with $p_A = 0.8$, $p_B = 0.9$.

The results from this investigation show that strong epistasis (deviations from additivity) is necessary for the maintenance of high values of disequilibrium at equilibrium. Thus only strongly interacting loci are likely to show substantial levels of disequilibrium. In fact, the symmetric model discussed above, with allele frequencies of one-half, generated the highest level of disequilibrium for a given strength of epistasis.

Partial selfing

High levels of disequilibrium have often been found in both natural and experimental populations of inbred plants. Are these large levels of linkage dis-

TABLE 3 Fitnesses for an epistatic two-locus two allele model.[a]

	BB	*Bb*	*bb*
AA	$1 - b_1 - a_1 + e_1$	$1 - a_1$	$1 - b_3 - a_1 - e_2$
A*a*	$1 - b_1$	1	$1 - b_3$
aa	$1 - b_1 - a_3 - e_3$	$1 - a_3$	$1 - b_3 - a_3 + e_4$

[a]The four parameters e_i, with the i corresponding to the labeling of the variable x_i denoting the frequencies, are measures of the epistatic interaction between alleles at loci A and B. The parameters a_i and b_i are measures of the additive effect of selection.

equilibrium a likely outcome at genetic stationarity in selfing species? Theory so far has used the traditional deterministic model describing two loci with two alleles with viability selection and partial selfing. The difference between this model and the outcrossing model is that we need to denote by g_{ij} the frequency of the genotype that in the outcrossing model has frequency $x_i x_j$. (To avoid counting the same genotype twice, let $i < j$.) The relative viability of the genotype whose frequency is given by g_{ij} will be denoted w_{ij}. Also, we do not assume that $w_{14} = w_{23} = 1$. The form of the dynamic model is given in detail in Weir et al. (1972) and Hastings (1985b). Here, it is sufficient to note that in the dynamic model the genotypic frequencies after mating, but before selection (G_{ij}) are given by equations of the form

$$G_{ij} = s\{\text{function of } r \text{ and } g_{ij}\} + \delta_{ij}(1 - s)x_i x_j$$

where δ_{ij} is one if $i = j$ and two otherwise. Then the frequencies in the next generation, after selection, before mating, are given by

$$g'_{ij} = w_{ij}\, G_{ij} / \bar{w}$$

where

$$\bar{w} = \sum_i \sum_{j<i} w_{ij}\, G_{ij}$$

is the mean viability.

This model is much more complex than the model for outcrossers. Early numerical studies by Jain and Allard (1966) pointed out that linkage disequilibrium might be maintained at stationarity for the particular (and strong) selection coefficients they chose. Eshel (1978) and more recently Christiansen and Feldman (1983) have studied the stability of monomorphic equilibria of two-locus multiplicative models with partial selfing. Holden (1979) studied some of the interior equilibria of the completely symmetric two-locus multiplicative model. Models with hitchhiking were studied by Strobeck (1979), Hedrick and Holden (1979), and Hedrick (1980). However, to contrast with the studies of outcrossing species discussed above, we will consider the approach of specifying the equilibrium configuration and determining the minimum strength of selection necessary to maintain this equilibrium (Hastings 1984).

As in the outcrossing model, specifying the equilibrium and finding the fitnesses is a linear problem. The following optimization problem determines the relationship among selfing rate, s, recombination, r, disequilibrium, D, and strength of selection. Define the strength of selection, S, as the maximum of the absolute value of $1-w$, where w is one of the nine genotypic fitnesses. Specify the gene frequencies, the disequilibrium, D, the recombination rate, r, and the selfing rate, s. The goal here is to find the minimum strength of selection, S, subject to the constraint that there be a stable stationary state with all genotype frequencies positive.

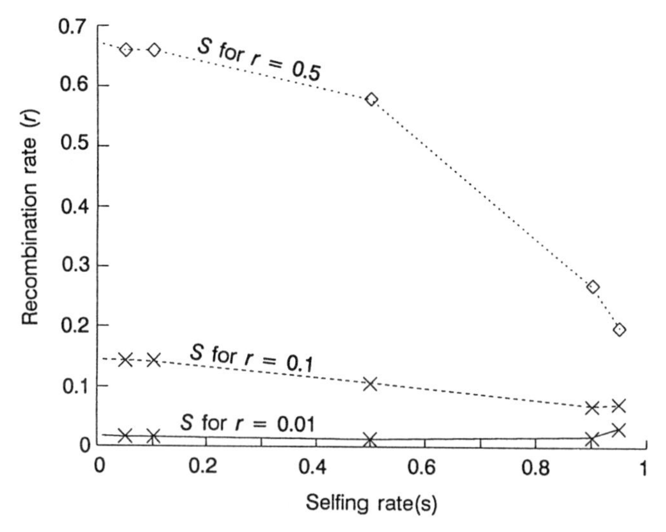

FIGURE 3. Minimum strengths of selection at a stable equilibrium with $p_A = 0.5$, $p_B = 0.5$, $D = 0.1$, as a function of the selfing rate.

The results of this investigation are illustrated in Figures 3 and 4. The outcome is that, at stationarity, the strengths of selection necessary to maintain high levels of disequilibrium are not markedly less than those in outcrossing species. However, with partial selfing, the dependence on recombination rate is much less strong than for the outcrossing case.

The optimization approach is not effective for very small outcrossing rates (less than 0.05). However, perturbation approaches are very effective in studying the partial selfing model for small outcrossing rates, yielding detailed information about stable equilibria with general fitnesses for sufficiently small outcrossing rates (Hastings 1985b). The results are in accord with the results from the optimization approach and can be summarized as follows. Let $t = 1 - s$ be the outcrossing rate. Unless the fitnesses of at least two homozygotes are exactly equal (and assuming that the fitness of heterozygotes is not twice that of homozygotes), no polymorphism is possible [the rare allele at each locus has a frequency of $O(t^2)$ at most] if the selfing rate is high enough. For slightly lower selfing rates, a polymorphism, with all alleles at significant frequencies, may be possible. Given that selfing rates may vary, this effect may have important implications concerning the importance of the mating system.

Another question is what proportion of polymorphic loci exhibits significant linkage disequilibrium? Unless there is very strong selection (a 2-fold advantage for heterozygotes), polymorphism with a high degree of selfing requires that the fitnesses of homozygotes be very similar. With a high degree of selfing, in a two-locus two-allele model, all four alleles will be present at

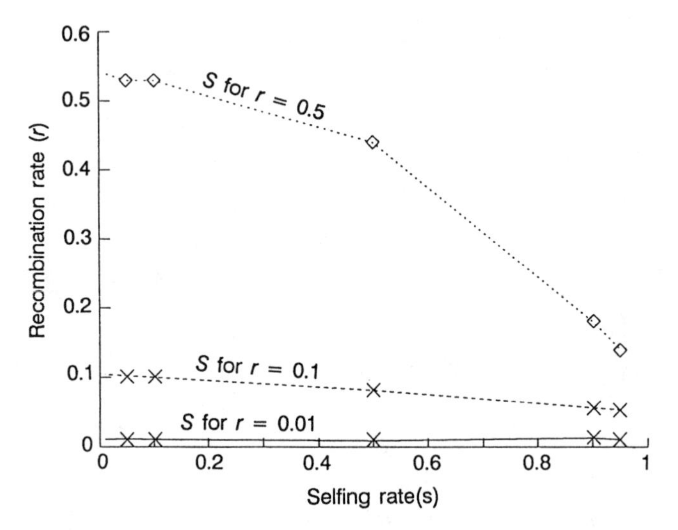

FIGURE 4. Minimum strengths of selection at a stable equilibrium with $p_A = 0.8$, $p_B = 0.9$, $D = 0.05$, as a function of the selfing rate.

equilibrium in only three circumstances. (1) If there is a 2-fold selective advantage for heterozygotes, they will remain in the population. (2) All four homozygotes have fitnesses that are nearly equal, differing by approximately the outcrossing rate or less. (3) Two homozygotes have nearly equal fitnesses and the heterozygotes have a fairly strong selective advantage (roughly 30%). In this final circumstance, very large values of linkage disequilibrium, D, are possible.

Thus, it may be the case that polymorphism is relatively uncommon, but polymorphic loci are typically strongly associated (in linkage disequilibrium) with other polymorphic loci. Perhaps selection is strong enough to maintain the linkage disequilibrium observed as a stationary state. A more complete understanding of this case would require theory including more than two loci. Other possibilities, in particular the role of dynamics in understanding linkage disequilibrium observed, cannot be ruled out and are discussed below.

Dynamics

Knowledge of the dynamic behavior of two-locus models with selection is much more limited than the knowledge of the equilibrium behavior. Insight into dynamic behavior is important for answering the questions raised in the Introduction about the role of epistatic selection in the maintenance of any observed disequilibrium.

Avery and Hill (1979) addressed the question of the dynamics of disequilibrium, D, following a perturbation away from its stationary value. They found that following such a perturbation, D can take hundreds of generations to return to its stationary value in large populations. They also showed that the correlation of D in successive generations is not strongly affected by selection. These results illustrate the limitations of using stationary theory for infinite populations to ascertain the causes of an observed significant value of D.

More recent work has considered the dynamics of D in populations large enough so that deterministic models are appropriate. Nagylaki (1976, 1977) showed that, with weak selection, linkage disequilibrium disappears fairly rapidly in a two-locus model, and is at a low level thereafter. Asmussen and Clegg (1981) have examined the specific dynamics of several models incorporating selection. One question that has received extensive attention is the dynamics of hitchhiking—the role of directional selection at one locus generating disequilibrium at other loci (see Thomson 1977). This occurs because of an initial disequilibrium, which then is maintained by the action of selection. This can be an important force in generating disequilibrium among tightly linked loci.

The dynamics of neutral alleles at two loci in partially selfing species were examined by Weir and Cockerham (1973). They found that recombination and the outcrossing rate played analogous roles. The role of hitchhiking in partially selfing species has been examined using simulation techniques by Strobeck (1979), Hedrick (1980), and Hedrick and Holden (1979). They demonstrate that hitchhiking can be very important in generating disequilibrium in selfing species.

In particular, as might be expected, the major difference between partial selfing and outcrossing species is that hitchhiking can generate disequilibrium in partial selfing systems when the loci are unlinked. Hitchhiking can generate disequilibrium between two neutral loci that are linked to a selected locus as well, if there is initial three-way disequilibrium. One important theoretical challenge is to examine further differences between the dynamics of hitchhiking and epistatic selection.

The dynamics of epistatic selection in partial selfing species have not been studied in great detail, but in principle some information should be quite simple to obtain using the following perturbation approach. The dynamics of a completely selfing system, which are well known, provide the lowest order estimate for the dynamics. If selection is much stronger than the outcrossing rate, this would be included here, otherwise the neutral behavior would be used. The next term in the analysis would include outcrossing, and possibly selection, depending on the relative strength of selection and outcrossing. Even from this verbal description, one can see that the time scale over which an initial disequilibrium is maintained will be very long, as illustrated by the analysis of the dynamics of neutral loci by Weir and Cockerham (1973) (see also Weir et al. 1972).

Other causes of disequilibrium

Two other causes of disequilibrium are important to mention. One is the role of migration, which has been discussed by Prout (1973). Another possible cause of disequilibrium at stationarity is finite population effects. These have been well studied in models of outcrossing species (e.g., Hill and Robertson 1968). The effect here is important only for very closely linked loci. A similar study has been undertaken for partially selfing species by Golding and Strobeck (1980). The major conclusion is that random drift is unlikely to be a major force in the generation of linkage disequilibrium between loosely linked loci.

INTERPRETATION OF DATA

There are two important questions here—the measurement or detection of the presence of associations and the determination of the cause of any association found. These are particularly difficult questions from both an experimental and theoretical point of view.

The detection of linkage disequilibrium by looking for statistically significant departures of D from zero is fairly straightforward (e.g., Brown 1975; Weir 1979). However, very large sample sizes are required for a single pair of loci, as demonstrated by Brown (1975) and Thompson et al. (1988). Brown's results, coupled with the results I have discussed earlier on the maximum value of D possible, shows that only disequilibrium between closely linked loci is likely to be observed, given the sample sizes usually used in experimental studies. In an outcrossing species, if $r = 0.01$, $p_A = 0.9$, $p_B = 0.5$, $S = 0.01$ (a selection strength of about 2%) and D is at its maximum value possible, more than 500 random zygotes are required to give a 50% chance of rejecting the null hypothesis that $D = 0$ (Hastings 1981).

In inbreeding species, where associations are not limited to pairs of closely linked loci, another approach to the measurement of associations that deals with multiple loci has been developed in Brown et al. (1980) and extended to multiple populations by Brown and Feldman (1981). Assume that the genotype at m loci is known. The analysis is based on the variance σ_K^2, of the random variable K, which is the number of loci that is different (heterozygous) when two random gametes from the sample are compared. The variance is

$$\sigma_K^2 = \sum_j^m h_j - \sum_j^m h_j^2 + 2\sum_j^m \sum_{l>j}^m \sum_i \sum_k [2p_{ji}p_{lk}D_{ik}^{jl} + (D_{ik}^{jl})^2]$$

Here p_{ji} is the frequency of the ith allele at the jth locus, and $h_j = 1 - \Sigma_i\, p_{ji}^2$ is the single-locus gene diversity at the jth locus. Denote the frequency of the two-locus gamete with allele i at locus j and allele k at locus l by g_{ik}^{jl}, so

$$D_{ik}^{jl} = g_{ik}^{jl} - p_{ji}p_{lk}$$

is the disequilibrium between allele i at locus j and allele k at locus l. Note that σ_K^2 depends only on single-locus and pairwise effects. The sampling distribution of σ_K^2 can be determined both for a single population and several populations (Brown et al. 1980; Brown and Feldman 1981) under the null hypothesis that allelic distributions among loci are independent. Thus statistical tests for dependence can be derived.

The determination of the cause of any associations among loci that are found is particularly difficult. Among the causes most difficult to separate are hitchhiking and selection. Here, a detailed examination of the pattern of disequilibrium among different pairs of loci, or among different pairs of alleles at the same pair of loci, is the only tool available. The use of information about disequilibrium among different pairs of alleles at the same loci as an indicator of whether hitchhiking or epistatic selection is the cause of observed disequilibrium has been discussed both theoretically and in the context of human HLA data by Klitz and Thomson (1987) and Thomson and Klitz (1987). The development of this approach for partially selfing species is an important development for the future.

In any event, data from several generations can be very useful in distinguishing among several different causes of disequilibrium, as has been obtained for cultivated barley as discussed below.

EXPERIMENTAL OBSERVATIONS

Outcrossing species

There have been several measurements of disequilibrium in populations of outcrossing plant species (Brown and Allard 1971; Baker et al. 1975; Brown et al. 1975; Phillips and Brown 1977; Mitton et al. 1980; Ennos 1982, 1985; Epperson and Allard 1987). As might be expected from the discussion above, most of these studies failed to detect extensive linkage disequilibrium (see Table 4).

One classic case of substantial linkage disequilibrium in plant populations is the gene for heterostyly in primrose, *Primula vulgaris*, as reviewed in Ford (1971). This work showed almost complete disequilibrium among the very closely linked alleles determining style length and anther position.

However, three of the studies of outcrossing plants mentioned above did find extensive evidence for linkage disequilibrium. One was of ponderosa pine (Mitton et al. 1980). Given that the linkage relationships were unknown this result is quite surprising, and not easily explained on the basis of selection. A much more detailed study of *Pinus contorta* (Epperson and Allard 1987) showed significant associations among pairs of loci, each with $r = 0.007$. Es-

TABLE 4 Studies of disequilibrium in outcrossing plant species.

Species	Loci examined	Results	Sources
Cynosurus cristatus L.	4 allozyme loci with one pair loosely linked ($r = 0.36$)	No disequilibrium	Ennos (1985)
Eucalyptus obliqua	3 pairs of isozyme loci from 4 localities	3 out of 12 tests show significant linkage disequilibrium, almost certainly due to sampling procedure	Brown et al., (1975)
Eucalyptus pauciflora	6 pairs of isozyme loci from 3 localities	No tests out of 16 show significant linkage disequilibrium	Phillips and Brown (1977)
Pinus contorta spp. *latifolia*	14 isozyme loci at two localities, sample sizes of 201 and 195 trees	Among 238 tests for disequilibrium between unlinked loci, 9 significant; among linked loci a pair with $r = 0.08$ shows no disequilibrium; significant disequilibrium occurs primarily between two pairs of loci, each with $r = 0.007$	Epperson and Allard (1987)
Pinus ponderosa Laws.	7 polymorphic isozyme loci from 3 localities, linkage relationships unknown	5 out of 30 pairs show significant linkage disequilibrium	Mitton et al. (1980)
Primula vulgaris	Loci for style length and anther position within a supergene	Large, significant disequilibrium	Ford (1971)
Silene maritima	4 isozyme loci, no evidence for linkage	No significant linkage disequilibrium	Baker et al. (1975)
Trifolium repens	2 cyanogenic loci	Significant linkage disequilibrium between unlinked loci	Ennos (1982)
Zea mays	Isozyme loci, one pair closely linked ($r = 0.06$)	Disequilibrium only between closely linked pair, founder effect	Brown and Allard (1971)

timates of population sizes appear to be incompatible with an explanation based on finite population size effects. The primary evidence for disequilibrium was in one population between glutamate oxalacetic transaminase I

(GOT-I) and peroxidase I (PER-I) and between GOT-I and PER-II. However, there is no statistically significant disequilibrium between PER-I and PER-II, for which r is less than 0.002. The size of the observed values of disequilibrium appears to be greater than that which can be generated by hitchhiking (Epperson and Allard 1987). The data are, however, consistent with epistatic selection with reasonably small selection strengths, as suggested by the analysis of Hastings (1981). The fact that not all closely linked loci exhibit disequilibrium is consistent with the discussion of the effects of epistasis given above, where I showed that strong epistasis as well as close linkage is necessary to maintain high levels of disequilibrium.

Another case of strong associations in an outcrossing species is that between cyanogenic loci in white clover studied by Ennos (1982). These loci, which are unlinked, have a strong epistatic interaction because plants possessing the dominant allele at each locus release hydrogen cyanide when their leaves are damaged. The observations of strong associations in natural populations are most easily explained by the action of strong selection.

Inbreeding species

It is in inbreeding plant species that the most dramatic evidence for multilocus associations and the interaction between selection and linkage in both natural and experimental populations have been found (reviewed earlier in Brown 1979) as given in Table 5.

The natural mating system of barley has a selfing rate greater than 0.99, so the theory reviewed above indicates that strong associations among different loci, even unlinked loci, are likely to be found. Investigations of populations of wild barley have confirmed this. In particular the strength of the associations among unlinked loci has led Golding and Strobeck (1980) to conclude that their model of the generation of linkage disequilibrium by finite population effects is not a sufficient explanation for the observed associations. As noted by Brown and Feldman (1981), the associations in a large number of populations in Israel are consistent with an explanation based on selection, but that founder effects may play a role. A quantification of the discussion above concerning dynamics would suggest that founder effects upon linkage disequilibrium might persist for hundreds of generations.

The advantage of studies of cultivated barley is that they have been followed through a number of generations (see Table 5) in a number of studies. Both tests based on examination of pairwise disequilibria and the σ_K^2 measure introduced by Brown et al. (1980) and Brown and Feldman (1981) show significant multilocus associations in a number of studies (Table 5). Moreover, associations appear to be increasing through time (Weir et al. 1974; Brown et al. 1980). The changes appear to be greater than those for neutral loci (Weir et al. 1974) in composite cross CC V in California.

TABLE 5 Studies of disequilibrium in inbreeding plant species.

Species	Loci examined	Results	Sources
Avena barbata	6 isozyme loci, including 3 esterase loci; 3 loci linked, 3 assort independently; several sites	Significant pairwise and higher order interactions among linked and unlinked loci	Allard et al. (1972)
Hordeum spontaneum (wild barley)	4 esterase loci in a sample from 10 locations in Israel; 3 loci are tightly linked,a fourth assorts independently	Many cases of large significant measures of association, even for unlinked loci	Brown et al. (1977)
	20 loci in samples from 28 natural populations in Israel(1179 samples total)	Indication of significant multilocus association using σ_K^2, evidence of regionally specific selection	Brown et al. (1980) Brown and Feldman (1981)
	16 loci in samples from four eastern Mediterranean countries	Significant multilocus associations	Jana and Pietrzak (1988)
Hordeum vulgare	4 esterase loci in generations 5, 17, 26 of one experimental population and generations 7, 8, 41 of a second experimental population	Linkage disequilibrium shows significant increases through time for both linked and unlinked loci	Clegg et al., (1972)
	4 esterase loci in generations 4–6, 14–17, 24–26, of CC V	Linkage disequilibrium changes through time, changes larger than expected for neutral loci	Weir et al., (1972, 1974)
	5 isozyme loci including 4 esterases in generations 8, 13, 23, 45 of CC II	Many significant associations, but some are significant in only some generations	Muona (1982)
	4 esterase loci in 8 generations of selection in Cambridge of populations derived from CC V	Strong multilocus associations, rapid change over 8 generations	Luckett and Edwards (1986)
	16 loci in samples from four eastern Mediterranean countries	Significant multilocus associations	Jana and Pietrzak (1988)
(Ethiopian barley)	5 isozyme loci including 4 esterases	Significant associations at many localitites using the σ_K^2 measure of Brown et al. (1980) (see text)	Bekele (1983)
Phaseolus lunatus	7 loci in generations 5 through 9 of an experimental population	Significant disequilibrium in 3 of 14 cases	Harding and Allard (1969)

Thus, several explanations are possible for the changes in disequilibrium through time of these cultivated barley populations, including hitchhiking (Hedrick and Holden 1979) and strong selection. One way to begin to distinguish between these explanations is to attempt to measure fitnesses directly. The role played by strong selection has been examined directly by Clegg et al. (1978) who estimate that there is strong selection at all phases of the life cycle. As they note, however, it is difficult to assign selection to particular loci if there is strong disequilibrium among many loci. Further theoretical developments along the lines of those in Thomson and Klitz (1987) may permit a more careful assessment of the different dynamic behavior of epistatic selection and hitchhiking. However, assigning the effects of selection to particular loci will be difficult, although the linkage relationship of selected sites to the loci observed may be obtainable.

CONCLUSIONS

The theory that I have reviewed shows that the interaction between selection and linkage can be very important in understanding the dynamic and stationary behavior of both outcrossing and partially selfing plant species. Experimental evidence in both types of species is consistent with the presence of strong epistatic selection, although much work needs to be done to distinguish explanations based on direct selection and hitchhiking.

I would like to emphasize the fact that in the one case where predictions of selection would indicate the possibility of disequilibrium in a study of an outcrossing species, it was found (Epperson and Allard 1987). More studies of closely linked, potentially interacting loci are needed in outcrossing plant populations. In particular, it appears premature to reject the hypothesis that disequilibrium is unimportant in outcrossers. It may certainly be important for closely linked loci.

For inbreeding species, there has been a large body of experimental evidence that demonstrated the pervasiveness of multilocus associations. Here, the theory needs to be developed to understand better dynamic behavior, and selection other than viability selection in partially selfing species. The insights gained from studying these systems, where the interaction of selection and linkage can be so dramatic, may continue to be vital in understanding selection in both natural and artificial populations.

Another area that I have not emphasized here is the potential importance of linkage for understanding the dynamics of quantitative characters. Theoretical results, (e.g., Bulmer 1971) have demonstrated the importance of disequilibrium in the response to selection on a quantitative character for a large enough number of loci. Linkage disequilibrium also can be important in the maintenance of correlations between characters. The experimental work of Jinks and his colleagues (e.g., Jinks et al. 1985) has begun to demonstrate the importance of these effects. Most of the theoretical explorations in this area have concentrated on outcrossing species. Given the vastly different behavior of nearly complete selfers, especially the presence of strong associations among isozyme loci, this is another area deserving of more work in the future.

ACKNOWLEDGMENTS

I thank many colleagues who have helped in the development of these ideas including John Gillespie and especially R. W. Allard. I thank C. Hom and an anonymous referee for comments on the manuscript. Research was supported by PHS Grant 1R01 GM32130

CHAPTER 11

GENE IDENTITY AND THE GENETIC DEMOGRAPHY OF PLANT POPULATIONS

Kermit Ritland

ABSTRACT

The use of genetic markers to study quantitative characters in the field is an unexplored area of plant genetic demography. Estimation formulas that use genetic markers are investigated for (1) the viability of selfed versus outcrossed progeny and (2) the heritability of a quantitative character. Using a formula that assumes inbreeding equilibrium, the viability of selfed progeny relative to outcrossed progeny was found to average 30% across 10 populations in three herbaceous genera. The estimation of heritability involves comparing the covariance of a quantitative character between physically adjacent plants sharing the same marker allele versus different marker alleles. Heritability of capsule number was found to be marginally significant in *Mimulus guttatus*, but its magnitude was uncertain because the correlation of relatedness between marker loci and quantitative loci was difficult to estimate. Populations with moderate levels of inbreeding or relatedness can be used for these inferences, which essentially rely upon gene-identity disequilibrium caused by variation of inbreeding or relatedness.

INTRODUCTION

Quantitative characters are by nature dependent on the environment for their expression. Plant demography is largely concerned with the study of these characters, which include seed dormancy, growth rates, mortality, and fecundity. The best examples of plant genetic demography either involve studies of discrete, single-gene characters largely devoid of environmental effects, such as the cyanogenic polymorphism in *Trifolium repens* and *Lotus corniculatus*

(Jones 1962; Bishop and Korn 1969; Crawford-Sidebotham 1972), or studies of survival and mortality of electrophoretically marked blocks of genes in experimental populations, such as *Avena fatua* and *Hordeum vulgare* (Clegg and Allard 1973; Clegg et al. 1978a). However, the combined study of quantitative characters with populations in natural environments is rare because of the formidable problems with disentangling genes from the environment.

Because of their clear Mendelian expression, isozymes and other types of genetic markers have been used successfully in genetic studies in natural populations, primarily in the context of determining genetic relationships, such as in the estimation of mating systems and gene flow. However, the potential to use the information provided by markers about relationship in the study of quantitative characters has not been realized. This chapter examines how marker genes may be used to study the genetic demography of quantitative traits in unmanipulated natural plant populations. Although their use in this manner may seem questionable to some, at least the problems involved in developing such methods are illustrated and a variety of interesting questions are raised that can be addressed by studies in plant population genetics.

Marker genes have been used in at least three established ways to study quantitative traits. The first of these is largely nonquantitative in nature, and involves their use to distinguish hybrid zones, sibling species, and other evolutionary events, and whose references are too numerous and diffuse to cite. The second, addressed by Stuber in this volume, is to use their linkage with quantitative trait loci to study quantitative inheritance, and even to map the location of quantitative trait loci (Tanksley 1983; Lander and Botstein 1989). The third is to use molecular phylogenies constructed from marker gene variation to infer macroevolutionary trends of morphology (Felsenstein 1985; Sytsma and Gottlieb 1986).

This chapter focuses on a fourth area: the estimation of quantitative genetic parameters in natural plant populations using associations of homozygosity at marker loci. Two features of plant populations make this fourth approach feasible. First, their tolerance for inbreeding (Brown 1979) allows patterns of homozygosity to develop. Second, the sedentary habit of plants allows patterns of relationship to develop, and, furthermore, makes them ideal for demographic studies as they are easy to mark and follow through their life span; they "sit and wait to be counted" (Harper 1977).

One use for genetic markers, developed in this chapter, is the estimation of the genetic component of quantitative variation. The most important quantitative traits in demography are those that determine survival and reproduction, i.e., the "life history strategy" (Stearns 1976). Much of life history theory is based on optimal strategy reasoning, which assumes there is sufficient genetic variability to evolve an optimum strategy. Quantitative genetic models for life history characters have been proposed (Lande 1982) and used for estimation of phenotypic selection intensities in the field (Lande and

Arnold 1983; Arnold and Wade 1984). However, the genetic response to selection (r) cannot be predicted from measurements of phenotypic selection intensity (I) until heritability in the field (h^2) is determined, as the classical selection response formula $r = h^2I$ (Falconer 1982) illustrates.

A large number of studies have estimated heritability of life history traits in wild populations, particularly in animal species (Mousseau and Roff 1987). However, such estimates are usually based on either the growth of individuals in artificial environments or on the results of artificial transplanting of individuals into natural environments. The high level of genotype–environment interaction of traits that determine survival and reproduction (Mitchell-Olds and Rutledge 1986) suggests that alternative methods, involving as little disturbance to the population as possible, are needed to measure the genetic component of life history traits.

A second use of genetic markers developed here is the estimation of inbreeding depression, which is a quantitative character very dependent on the environment. In self-fertilizing plants, inbreeding depression is measured by the relative fitness of selfed progeny compared with outcrossed progeny. Models for the evolution of selfing are critically dependent on its level in natural populations (Lande and Schemske 1985; Schemske and Lande 1985). Although several detailed studies have demonstrated inbreeding depression in the field (Schoen 1983; Mitchell-Olds and Waller 1985; see Charlesworth and Charlesworth 1987), its estimation can involve the transplantation of seedlings into the field, which should be undertaken with care to ensure realism. Such care can involve investments of time that limit the number of populations surveyed. Alternative methods that both reduce the time investment and ensure realism are again warranted.

Genetic demography includes the study of the genetic component of life history traits. It is fundamentally concerned with the age-specific survivorship and fecundity of genotypes (Clegg et al. 1978b) and thus intimately tied to the estimation of fitness. A fundamental component of this field should be the estimation of genetic parameters. Allard and co-workers established a paradigm for the estimation of genetic parameters for discrete characters in their studies of selection (Allard and Workman 1963; Clegg et al. 1978a) and inbreeding (Brown and Allard 1970; Clegg et al. 1978a; Brown 1979). For quantitative characters, Cockerham (1956, 1983) and Cockerham and Weir (1984) developed an extensive theory for inbreeders, where a key feature was the use of gene-identity coefficients (Cockerham 1971; Cockerham and Weir 1973). Coefficients of gene identity serve as common ground between the fields of qualitative genetics and quantitative genetics, and, in our case, are the basis for linking observations at qualitative loci with the properties of quantitative gene loci. Such coefficients will be critical in linking population genetics with plant demography, a union that can produce powerful demographic inferences.

TWO USES OF GENETIC MARKERS

Conceptually, the use of variation at marker loci to study quantitative characters is straightforward. In the first case discussed below (estimation of inbreeding depression), individuals derived from self-fertilization are more homozygous than outcrossed individuals. Hence a correlation between homozygosity and viability provides information about inbreeding depression. In the second case below (estimation of heritability), adjacent pairs of individuals sharing the same marker gene are more likely to be relatives. Hence a correlation between marker-gene similarity and quantitative trait similarity across these pairs provides information about the genetic component of quantitative trait variation, after accounting for the common environment shared by each pair of individuals.

Estimating the viability of selfed versus outcrossed progeny

Since selfed individuals tend to be more homozygous, temporal changes in the level of homozygosity should provide information about the survival of selfed individuals. There are two different ways we can measure this homozygosity: via gene "identity-by-state" as reflected by the proportions of homozygotes, and via measures of gene "identity-by-descent."

Corresponding to these two measures of homozygosity, we will consider two experimental designs to estimate inbreeding depression in the simplified case of a population with discrete generations. These two designs are portrayed in Figure 1. In both designs, we will require knowledge of the "genetic-mating structure" of parents in the previous generation. The genetic-mating structure describes how genotypes give rise to zygotes (Ritland 1985), and if subpopulation structure is not present, is described by Wright's inbreeding co-

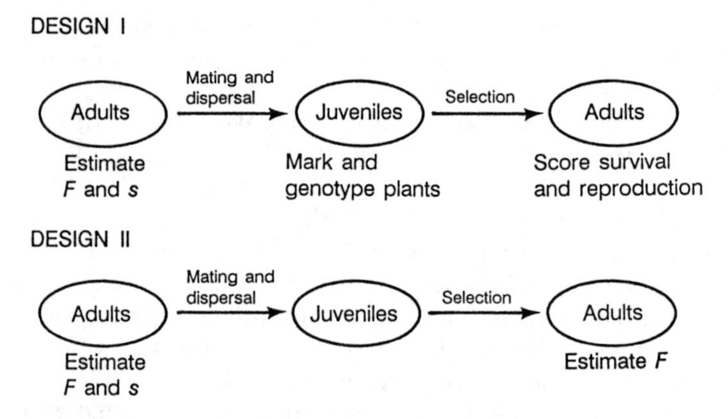

FIGURE 1. Two experimental designs for the estimation of the relative viabilities of selfed versus outcrossed progeny using natural levels of inbreeding.

efficient F (Wright 1969) and the selfing rate s ($= 1 - t$). For our purposes, F is defined as the probability that an individual's alleles are identical-by-descent.

The following will be used in the estimation formulae. A parameter closely related to F is the inbreeding coefficient of offspring zygotes (before selection), which is denoted as F' and depends entirely on the parental parameters s and F through the well-known relation

$$F' = s(1 + F)/2 \tag{1}$$

Second, the proportions of homozygous (P_{hom}) and heterozygous (P_{het}) individuals in the parental population are functions of F and the "expected homozygosity" at the locus, termed J,

$$P_{hom} = \sum_i \left[p_i F + p_i^2(1 - F)\right] = F + (1 - F)J = (1 - J)F + J$$

$$P_{het} = \sum_{i,j} p_i p_j (1 - F) = (1 - F)(1 - J) \tag{2a}$$

where $J = \Sigma p_i^2$ and p_i is the frequency of allele i. Third, the proportions of homozygous versus heterozygous individuals in the progeny generation are

$$P_{hom'} = (1 - J)F' + J$$

$$P_{het'} = (1 - J)(1 - F') \tag{2b}$$

Note that primes are used to denote the offspring generation, which is the generation in which inbreeding depression will be measured.

Individual probabilities of being selfed Before describing the two methods of estimation, we need to derive "individual" probabilities of selfing, conditioned on the observation of homozygosity or heterozygosity at the marker locus. The first step to obtain these probabilities is to specify joint probabilities that individuals are homozygous or heterozygous and selfed or outcrossed, given their *parent's* genotype (homozygous or heterozygous). These are given in Table 1, and were obtained from the mixed-mating probabilities of observing progeny genotypes given maternal genotypes (e.g., Appendix A, Clegg et al. 1978a). No selection is assumed to occur between these parents and progeny.

Table 2 gives the probabilities we seek, namely, the probablilities that homozygotes or heterozygotes were derived by selfing or by outcrossing. These are obtained from Table 1 and Equations 1 and 2 by the rules of conditional probability. For example, the probability a homozygous individual was derived from outcrossing is ("|" denotes "given")

$$P_{t|hom'} = \frac{P_{hom',t|hom}\, P_{hom} + P_{hom',t|het}\, P_{het}}{P_{hom'}}$$

TABLE 1 Probabilities that an individual is homozygous (*hom′*) or heterozygous (*het′*) and derived by selfing (s) or outcrossing (t), given its parent genotype (*hom* or *het*; "|" denotes "given").

Parent homozygous		Parent heterozygous	
Event	**Probability (P)**	**Event**	**Probability (P)**
$hom', s \mid hom$	s	$hom', s \mid het$	$s/2$
$het', s \mid hom$	0	$het', s \mid het$	$s/2$
$hom', t \mid hom$	tJ	$hom', t \mid het$	tJ
$het', t \mid hom$	$t(1 - J)$	$het', t \mid het$	$t(1 - J)$

which, from Table 1 and Equations 1 and 2, equals the probability for the event $t \mid hom'$ in Table 2. Note these individual selfing probabilities are functions of the *parent* parameters F and s (F' is a function of these, Equation 1). Also, since to derive these probabilities selection was assumed absent, fertility selection is not directly incorporated into the following approach.

Using the viabilities of homozygous versus heterozygous individuals to estimate viabilities of selfed versus outcrossed individuals This is the first method, which uses "identity-by-state" of genes. If v_s is the viability of selfed progeny and v_t the viability of outcrossed progeny, the viabilities of heterozygotes and homozygotes are weighted means of selfed and outcrossed viabilities:

$$\begin{aligned} v_{het} &= P_{s \mid het'} v_s + P_{t \mid het'} v_t \\ v_{hom} &= P_{s \mid hom'} v_s + P_{t \mid hom'} v_t \end{aligned} \qquad (3)$$

TABLE 2 Probabilities that an individual was derived from selfing (s) or outcrossing (t), given its genotype (*hom′* or *het′*).

Homozygous individuals		Heterozygous individuals	
Event	**Probability (P)**	**Event**	**Probability (P)**
$s \mid hom'$	$\left(\frac{(1 - J)F' + Js}{(1 - J)F' + J}\right)$	$s \mid het'$	$\left(\frac{s - F'}{1 - F'}\right)$
$t \mid hom'$	$\left(\frac{Jt}{(1 - J)F' + J}\right)$	$t \mid het'$	$\left(\frac{t}{1 - F'}\right)$

After substitution of the values from Table 2 into this pair of equations, one can solve for v_s and v_t and obtain the following estimators for the viabilities of selfed and outcrossed individuals:

$$v_s = v_{hom} + J\left(\frac{1 - F'}{F'}\right)(v_{hom} - v_{het})$$

$$v_t = v_{het} + \left(\frac{s - F'}{t}\right)\left[1 + \left(\frac{1 - F'}{F'}\right)J\right](v_{het} - v_{hom}) \tag{4}$$

With several loci, v_{het} and v_{hom} are computed as averages over loci. It should be noted that F' is entirely determined by the genetic-mating structure of the parental generation (Equation 1), but is included here for simplification. Also note that with infinitely many alleles, expected homozygosity (J) is zero so that v_s equals the mean of homozygotes while v_t equals the mean of heterozygotes plus a correction factor.

The estimators given by Equation 4 assume, of course, that selection does not occur directly at allozyme loci. Fitness differences at allozyme loci arise from a correlation of homozygosity between allozyme loci and fitness loci, together with selection for heterozygotes (outcrosses) at fitness loci. This correlation of homozygosity occurs in partially selfing populations because inbreeding coefficients for individual loci differ *between* individuals (outcrossed individuals have $F = 0$ while selfs have $F > 0$) but not *within* individuals.

The inference of selection intensities from putatively neutral allozyme loci is relevant to studies that attempt to correlate enzyme heterozygosity with fitness (primarily in marine molluscs and conifers, see Mitton and Grant 1984; Zouros 1987). Even among advocates of allozyme neutrality, there appears to be some confusion over whether correlations of heterozygosity with fitness are due to linkage disequilibrium or to associations of homozygosity (identity disequilibrium) between the allozyme and fitness loci. Such studies should determine if correlations of homozygosity between pairs of allozyme loci exist in these species. If they do, the possibility is raised that selection against inbreeding occurs, and can be inferred via neutral marker loci, in these species.

Using changes of inbreeding coefficients to infer the relative viability of selfed progeny This is the second method, which uses "identity-by-descent" of genes. It basically uses changes of the inbreeding coefficient to infer selection against selfing and the sampling strategy is portrayed in "design II" (Figure 1). To develop this method, first note that following viability selection, the proportions of heterozygotes and homozygotes, denoted with double primes, are

$$P_{hom''} = v_{hom}\, P_{hom'}/\bar{v}$$

$$P_{het''} = v_{het}\, P_{het'}/\bar{v} \tag{5}$$

where $\bar{v}$ is the mean viability,

$$\bar{v} = v_{hom} P_{hom'} + v_{het} P_{het'} = sv_s + tv_t$$

One can solve for the relative viability of selfed progeny, v_s/v_t, from the above recursion for $P_{het''}$ after one notes that $P_{het''} = (1 - F'')(1 - J)$. The terms involving J cancel and the estimator for v_s/v_t if found to be

$$\frac{v_s}{v_t} = \frac{tF''}{(F' - F'') + tF''} \tag{6}$$

where F'' is the inbreeding coefficient in the progeny generation following selection. Again it should be emphasized that F' is determined by the genetic-mating structure of the parental generation (Equation 1). Equation 6 can be derived more directly solely by considering changes of F, but the derivation here illustrates use of the "individual-level" probabilities of Tables 1 and 2.

If parental F is constant over generations ($F'' = F$) but F changes *within* generations due to selection against selfed progeny, Equation 6 becomes

$$\frac{v_s}{v_t} = \frac{tF}{\Delta F + tF} \tag{7}$$

where $\Delta F = (F' - F)$ is the change of F from parent to offspring. In species that practice moderate selfing, ΔF is almost always positive (Brown 1979; but note his ΔF differs slightly from the above defined ΔF in a trivial way). Solving for ΔF in Equation 8, one obtains

$$\Delta F = \left(\frac{v_t - v_s}{v_s}\right) tF$$

which shows that observation of positive ΔF is consistent with greater viability of outcrossed progeny relative to selfed progeny. However, we have ignored numerous other factors that may affect levels of F (see Brown 1979).

If one assumes this equilibrium of adult (parental) F, then joint estimates of F and s are sufficient to estimate the relative viability of selfed versed outcrossed progeny. The estimator to use is

$$\frac{v_s}{v_t} = 2\frac{F(1 - s)}{s(1 - F)} \tag{8}$$

which is obtained from Equation 7 by substitution of s and F for F' using Equation 1. Note that if $v_s = v_t$, then $F = s/(2 - s)$, which is the classical relationship for a population at inbreeding equilibrium (Nei and Syakudo 1958). The greatest power to estimate v_s/v_t occurs when F is between 0.1 and 0.4, which are values, in fact, observed in many predominantly outcrossing species.

Estimation of genetic parameters For all three viability-estimation formulas (Equations 4, 6, 8), joint estimates of selfing and inbreeding coefficients for the parental generation are required. By regarding the progeny array as the unit of observation, Ritland (1986) obtained formulas for joint maximum likelihood estimation of s and F using progeny array data (one should note that Wright's F_{IS} is the inbreeding coefficient referred to in this chapter). These arrays may be as small as two offspring, but the optimal array size is about six for outbreeders.

Wright's inbreeding coefficient F was defined as the probability of identity-by-descent of homologous alleles, but the expected estimate of F only *approximates* this probability because this gene identity is relative to an outbred, ancestral population. Fortunately, in a large, partially selfing population, this ancestral population is very recent, at most a few generations previous. Little genetic drift occurs and expected estimates of F would be very close to the true proportion of gene identity. Weir and Cockerham (1984) discuss the estimation of F in more rigorous contexts.

Fertility selection Fertility differences, which can be a major component of selection (Prout 1965), have been assumed absent. These differences include early acting inbreeding depression (Charlesworth and Charlesworth 1987), such as differential seed maturation, which completely escapes detection because it occurs before the seed is electrophoretically assayed and the rate of selfing is computed. Incorporation of other types of fertility selection is possible by including fertility differences into v_{het} and v_{hom} terms in Equation 4, or by weighting estimates of parental F in Equations 6 and 8 by parent fertility (seed output). It may, in fact, be possible to estimate separately both components of selection with inbreeding coefficients.

Estimates of inbreeding depression assuming equilibrium Table 3 gives estimates of selfing rates, parental inbreeding coefficients F, and relative selfing viability v_s/v_t (using Equation 8) for 10 populations across three genera of herbaceous plants. The estimates of v_s/v_t range from 0.0 to 0.84, with a mean of 0.30. A model due to Fisher states that if v_s/v_t is greater than 0.5, genes favoring selfing should increase in frequency, assuming that selfing genes are represented in the outcrossing pollen pool. Thus, outcrossing seems to be favored in most *Mimulus* populations, although there appear to be exceptions.

The ease with which estimates of inbreeding depression were assembled in Table 3 illustrates the economy and power of the population genetic approach to demography. However, it may also have a capability to delude us, as the assumptions made to obtain these estimates reduce their validity. More robust studies, where F is estimated for each generation, are needed.

Estimating heritability in the field

Genetic markers may also potentially be used to estimate the genetic component of quantitative variation in the field. Classical methods to estimate heritability are based on computing phenotypic covariances between relatives of a

TABLE 3 Inferred viabilities of selfed progeny relative to outcrossed progeny (v_s/v_t) across three genera, assuming equilibrium of F among adult generations (Equation 8).

Species	Number of families	Number of progeny	No. of loci	F	Selfing rate s	$\left(\frac{v_s}{v_t}\right)$
Bidens mauiensis[a]	48	463	2	0.03	0.34	0.13
B. amplectens[a]	34	340	2	0.12	0.44	0.35
B. torta[a]	32	609	3	−0.02	0.33	0.00
Limnanthes alba[b]	36	398	3	0.17	0.43	0.54
L. alba[c]	37	355	3	−0.15	0.05	0.00
L. gracilis[d]	42	409	3	0.15	0.49	0.37
Mimulus guttatus[e]	93	558	6	0.13	0.40	0.45
M. guttatus[f]	96	556	6	0.16	0.58	0.28
M. guttatus[g]	71	426	6	0.28	0.48	0.84
M. guttatus[h]	71	426	6	0.01	0.18	0.09

[a]Sun and Ganders (1988); [b]Population, 308; [c]Population 325 (Ritland and Jain 1981); [d]Kesseli and Jain (1985); [e]White Rapids; [f]Nanoose Hill; [g]Horseshoe Bay; [h]Anacortes (Ritland and Ganders 1987a).

certain degree, then relating these covariances to a linear combination of components of genetic variances expected under a particular model (Falconer 1982). Fundamental to this process is the classification of pairs of individuals into categories of relationship. The identification of relationship in natural populations using genetic markers is a topic of current interest, especially for relationships involving paternal parents (Ellstrand 1984; Meagher 1986; Quinn et al. 1987; Devlin et al. 1988).

In natural populations, there are three situations where genetic markers can be used to estimate relationship. First, in small populations, data from a relatively few loci can be sufficient to assign significant probabilities to putative relatives (Ellstrand 1984; Meagher 1986; Devlin et al. 1988). Second, direct DNA assay can vastly increase the number of loci and hence allow relatives to be identified from a large pool of possibilities (Jeffreys et al. 1985; Quinn et al. 1987). The third case, which we examine here, occurs in populations where dispersal is limited and relatives tend to be physically clustered. In this situation, adjacent individuals are more likely to be relatives, and just one or few loci provide information about relationship.

The principles involved in using marker-locus similarity for estimating the genetic component of quantitative variation are illustrated for an idealized case in Figure 2. In this figure, each individual has a marker locus genotype and a quantitative trait value (denoted by the size of the "*"). Each adjacent pair of individuals also shares a local environment (denoted by the

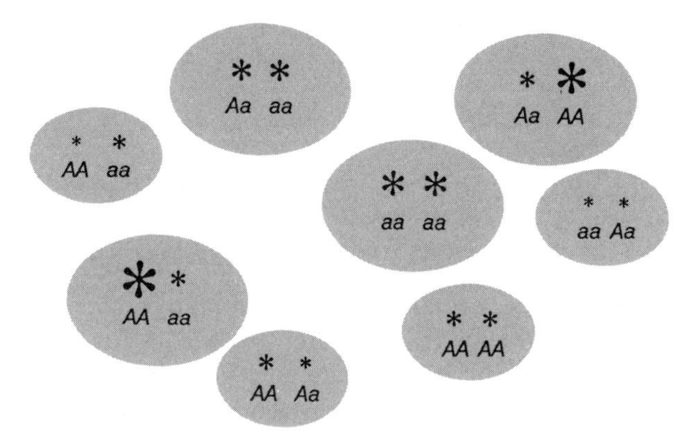

FIGURE 2. An idealized population illustrating the estimation of heritability in the field. Adjacent plants show a correlation of relatedness at the isozyme locus (with alleles "A" and "*a*") with quantitative similarity (size of "*"), but also share common environments, whose effect is proportional to the ellipses.

size of the circle about the pair). The key observation is that adjacent individuals who have similar marker-locus genotypes also have similar trait values. This correlation of similarity across pairs is evidence for a genetic component in the variability of the quantitative trait.

However, to obtain a heritability estimate from this correlation, two factors need to be quantified: (1) the environmental covariance due to sharing of local environments and (2) the correlation of relationship between the marker loci and the loci determining the quantitative trait. What follows is a suggested way to extract the heritability estimate that has several assumptions and raises additional questions, and which is not the only solution.

Components of phenotypic variance We consider an extremely simplified model that incorporates only additive genetic variance. Dominance, epistasis, and the effects of inbreeding are not included.

Consider one locus, the "QT" locus, which contributes to a quantitative trait. Allowing for common environmental effects, the variance of the phenotype at this locus consists of the genetic variance plus two components of environmental variance,

$$V_P = V_A + V_{Eb} + V_{Ew}$$

where V_A is the additive genetic variance, V_{Eb} is the environmental variance among adjacent pairs of individuals, and V_{Ew} is the environmental variance within these pairs ($V_E = V_{Eb} + V_{Ew}$).

Suppose we have two individuals of unknown relationship, and we randomly sample an allele at this locus from each of these two individuals and determine whether these alleles are identical-by-descent. The expected phenotypic covariance between these individuals, denoted "c," conditioned on gene identity between the two sampled alleles, is either

$$c_i = 2V_A + V_{Eb}$$

or

$$c_o = V_{Eb} \tag{9}$$

corresponding to these two alleles being identical ("i") or not identical ("o"). These are obtained from the equation for the covariance between diploid relatives ($2\phi V_A$, Jacquard 1974, p. 135, Equation 23, where ϕ is the probability of identity-by-descent between homologous alleles of different individuals), with the additional consideration that V_{Eb} is added on these covariances.

Conditional probabilities of gene identity Since the c's in Equation 9 depend on gene identity between alleles at QT loci, we need to predict such identity from observation of gene *state* at the marker locus. To do this, we need to first specify the correlation of gene identity (relationship) between marker and QT loci. Then we need to relate gene identity to gene state at the marker locus.

Among two individuals, consider four alleles. Two are the homologous alleles randomly sampled from a marker locus from each diploid individual. The other two are homologous alleles likewise sampled from a QT locus. A "two-gene" measure of relationship between these individuals is the probability that such pairs of homologous alleles are identical-by-descent. This probability, denoted ϕ, is assumed the same for both marker and QT loci. Note that $\phi \leq 0.5$ unless individuals are inbred, when ϕ can be as great as unity.

A correlation of relationship (gene identity) between these loci will exist if some pairs of individuals show higher relationship than other pairs. The concept of "identity disequilibrium" (η) was originally introduced by Weir and Cockerham (1969) to describe the covariance of homozygosity-by-descent between loci of the same individual. For our purposes, it is convenient to base formula on a "four-gene" parameter, the *correlation* of *relationship* between loci, denoted as H ($H = \eta/[\phi(1 - \phi)]$). The frequencies of gene identity between the four sampled genes (two marker alleles, two QT alleles) are then

$$\begin{aligned} P_{ii} &= \phi^2(1 - H) + \phi H \\ P_{oi} &= P_{io} = \phi(1 - \phi)(1 - H) \\ P_{oo} &= (1 - \phi)^2(1 - H) + (1 - \phi)H \end{aligned} \tag{10}$$

where P_{ii} denotes gene identity between alleles at both loci, P_{io} identity at one locus but not the other, and P_{oo} no identity at both loci. Note that H ranges from zero to unity regardless of inbreeding. Equation 10 is an extreme simplification of the four-gene "parameter space." Since we assume individuals are outbred and loci are in linkage equilibrium, many terms that appear in fuller descriptions, such as those based on the descent measures of Cockerham and Weir (1973), are absent.

Table 4 gives the probabilities of gene identity at the QT locus, conditioned upon gene identity at the marker locus. These are derived from Equation 10 using the rules of conditional probability. Table 5 gives the probabilities of gene identity at the marker locus, conditioned on observation of gene state at the marker locus. They are derived from Equation 2b (with $F' = \phi$) in the same way that Table 2 was derived from Table 1.

Using conditional covariances to estimate heritability in the field Let the expected phenotypic covariance between individuals who share the same marker allele be c_{same}, and let the expected phenotypic covariance between individuals with different marker alleles be c_{diff}. These expectations are related to c_i and c_0 through the following conditional probabilities:

$$c_{diff} = P_{i\,|\,diff}\,(P_{i\,|\,i}c_i + P_{o\,|\,i}c_o) + P_{o\,|\,diff}\,(P_{i\,|\,o}c_i + P_{o\,|\,o}c_o)$$

$$c_{same} = P_{i\,|\,same}(P_{i\,|\,i}c_i + P_{o\,|\,i}c_o) + P_{o\,|\,same}(P_{i\,|\,o}c_i + P_{o\,|\,o}c_o)$$

where the P's are given in Tables 4 and 5. Substitution of the P values and the values for c_i and c_o (Equation 9) into these gives

$$c_{diff} = 2\phi(1 - H)V_A + V_{Eb}$$

$$c_{same} = 2\phi\left(\frac{H + (1 - H)J_o}{J_o}\right) V_A + V_{Eb} \qquad (11)$$

where $J_o = \phi + (1 - \phi)J$, and is used for tractability. Note the difference between c_{same} and c_{diff} is $2V_A\phi H/J_o$. If this difference is positive, then both V_A and H are positive, or, in other words, the character has genetic variability *and* a correlation of relationship between QT loci and marker loci exists.

Solving Equation 11 gives the variance components, which can be used as method-of-moments estimators, as

$$V_A = (c_{same} - c_{diff})\left(\frac{J_o}{2\phi H}\right)$$

$$V_{Eb} = c_{diff} - (c_{same} - c_{diff})J_o\left(\frac{(1 - H)}{H}\right) \qquad (12)$$

$$V_{Ew} = V_P - V_A - V_{Eb}$$

Note that as the number of alleles at the locus becomes large, J_o approaches ϕ. With an "infinite" allele locus ($J_o = \phi$) and complete correlation of relationship ($H = 1$), then $V_A = (c_{same} - c_{diff})/2$, $V_{Eb} = c_{diff}$, and $V_{Ew} = V_P - (c_{same} + c_{diff})/2$. Equation 12 was derived for one locus, but it holds for polygenic traits when effects are additive and linkage disequilibrium is absent.

TABLE 4 Probabilities that a quantitative trait locus has alleles identical-by-descent (i) or alleles not-identical-by-descent (o), given gene identity at the marker locus (i or o).

Marker locus inbred		Marker locus outbred	
Event	Probability (P)	Event	Probability (P)
$i\mid i$	$H + (1 - H)\phi$	$i\mid o$	$(1 - H)\phi$
$o\mid i$	$(1 - H)(1 - \phi)$	$o\mid o$	$H + (1 - H)(1 - \phi)$

Estimation of genetic parameters We must assume that genetic parameters at QT loci can be estimated from marker locus data, and in particular, must assume the correlation of relatedness (H) between a marker locus and a QT locus can be estimated as the H between two marker loci. Estimation theory for two-gene coefficients is well developed (Reynolds et al 1983; Weir and Cockerham 1984). However, estimators for four-gene parameters such as H are not well studied. Simple estimators for both ϕ and H can be derived from the equations for the probabilities of drawing two different homologous alleles from each member of a pair (analogous to one-locus heterozygosity) and the probability D of drawing two different homologous alleles at both of two loci (analogous to two-locus heterozygosity):

$$S_{\text{obs}} = S_{\text{exp}}(1 - \phi)$$

$$D_{\text{obs}} = D_{\text{exp}}\,[(1 - \phi)^2(1 - H) + (1 - \phi)H]$$

where $S_{\text{exp}} = (1 - J)$ and $D_{\text{exp}} = (1 - J_a)(1 - J_b)$ are expected frequencies under no relationship (the two loci are denoted "a" and "b" and $J = (J_a + J_b)/2$). Solving for ϕ and for H in these gives their "method-of-moments" estimators

TABLE 5 Probabilities that two homologous alleles are identical-by-descent (i) or not-identical-by-descent (o), given their gene state [same (*same*) or different (*diff*) allele].

Same gene state		Different gene state	
Event	Probability (P)	Event	Probability (P)
$i\mid$*same*	$\left(\dfrac{\phi}{\phi + (1 - \phi)J}\right)$	$i\mid$*diff*	0
$o\mid$*same*	$\left(\dfrac{(1 - \phi)J}{\phi + (1 - \phi)J}\right)$	$o\mid$*diff*	1

as

$$\hat{\phi} = 1 - (S_{obs}/S_{exp})$$

$$\hat{H} = \left[\left(\frac{D_{obs}}{D_{exp}}\right) - (1 - \hat{\phi})^2\right] / [\hat{\phi}(1 - \hat{\phi})] \qquad (13)$$

Note at least two marker loci are needed; the S's can be found as unweighted averages over loci, and the D's as averages over all possible pairings of loci. Monte Carlo simulations indicate the H estimator has an upward bias of about 10%. Further work on unbiased and efficient estimators of H is needed.

The covariances c_{same} and c_{diff} can be estimated by weighting the cross-products of the quantitative trait by the proportion of shared versus unshared marker alleles of each pair, which is computed by considering all four combinations of alleles at a locus, averaged over all loci.

Estimates from a natural population of the common monkeyflower A population of *Mimulus guttatus* in Washington Park, near Anacortes, Washington, was chosen. Compared to other *Mimulus* populations, inbreeding is low in this population (Table 3), so additional complications introduced by inbreeding can be neglected. The sample consisted of 110 pairs of physically proximate plants along a 400 m transect through the population (Figure 3).

Entire, senescent plants were collected and 8–16 offspring per plant were assayed for three electrophoretic loci [aspartic amino transferase (2 loci) and glucosephosphateisomerase; procedures in Ritland and Ganders 1987a]. Two of three loci were triallelic, which is important for estimation of relatedness (Ritland 1987). Maternal genotype was inferred using the Brown and Allard

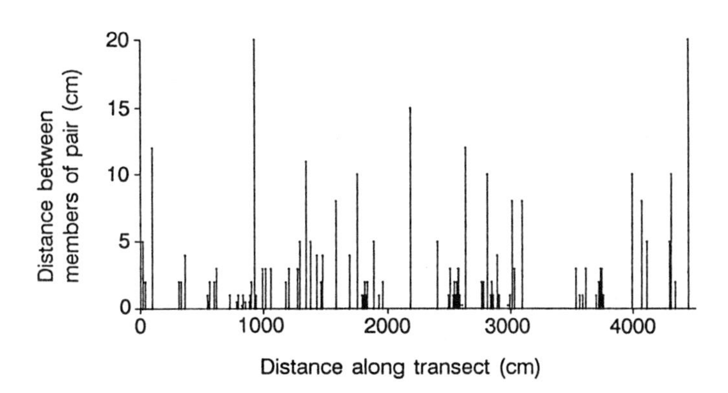

FIGURE 3. Locations of plants collected along a transect through the *Mimulus guttatus* population in Washington Park, near Anacortes, Washington.

(1970) method. Progeny from both members of a pair were assayed in only 89 of the 110 cases. Capsule number on the parent plants was log transformed for analysis. Standard errors of estimates were found using the bootstrap method (Efron 1982), with adjacent pairs as the unit of resampling.

Table 6 gives the actual estimates and Figure 4 gives the distributions of 1000 bootstrap estimates of ϕ, H, $c_{same} - c_{diff}$, and heritability. A moderate amount of polymorphism was present, as single-locus average heterozygosity was 0.399 and two-locus average heterozygosity was 0.145. The two-gene coefficient of relatedness was significantly positive ($p \leq 0.01$ as shown by the proportion of bootstraps falling at or below zero, Figure 4A). The covariance c_{same} was greater than c_{diff}, this difference being marginally significant ($p \approx 0.10$, Figure 4C). Thus, Equation 11 shows that marginally significant evidence exists for (1) additive genetic variance of capsule number and (2) a correlation of relationship H between isozyme loci and loci controlling capsule number.

However, the estimate of H showed large fluctuations among bootstraps (Figure 4B). More efficient estimators of H should be devised. Because of this variance of H, estimates of heritability likewise showed large variance (Figure 4D), so that the rather high estimate for heritability found (0.76, Table 6) should be disregarded. At least, the marginally significant difference between c_{diff} and c_{hom} indicates that both H and V_A are probably positive in reality. The environmental variance within pairs, V_{Ew}, was estimated to be zero, but this estimate would be biased downward by inbreeding.

The limitations of these estimates should be realized in light of the numerous assumptions made in deriving the estimators. Further development of estimators is needed, and inbreeding needs consideration as relatedness tends to correlate with homozygosity in substructured populations (Ritland and Ganders 1987a). Estimates obtained via Equation 12 could also be compared to estimates obtained through artificial plantings of known relatives.

TABLE 6 Estimates of genetic parameters for determination of the heritability of capsule number in *Mimulus guttatus*.

Relationship (ϕ)	0.112
Correlation of relationship (H)	0.378
Same marker covariance (c_{same})	0.231
Different marker covariance (c_{diff})	0.185
Genetic variance (V_A)	0.346
Environmental variance within pairs (V_{Ew})	−0.025
Environmental variance between pairs (V_{Eb})	0.137
Heritability (V_A/V_P)	0.761

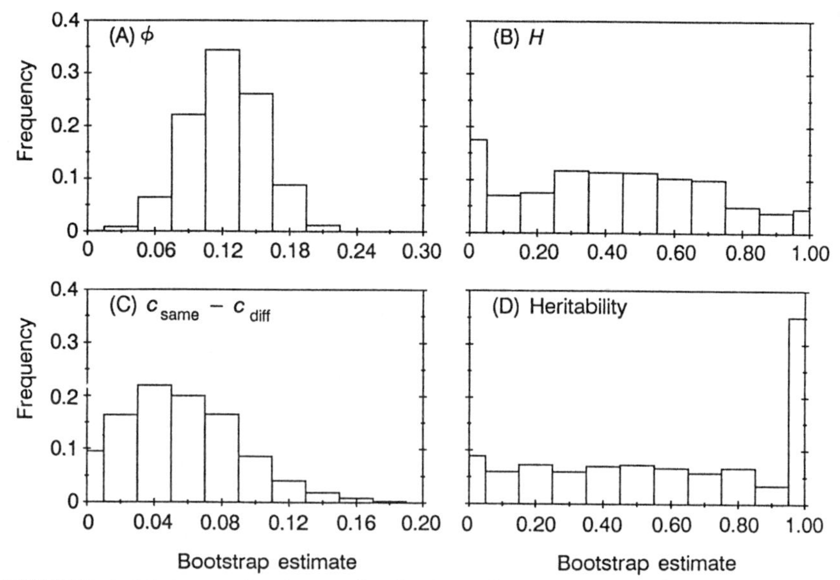

FIGURE 4. Variance of estimates for the *Mimulus guttatus* population, as indicated by the distributions of 1000 bootstrap estimates for (A) ϕ (two-gene coefficient of relatedness, (B) H (correlation of relatedness), (C) $c_{same} - c_{diff}$, and (D) heritability of capsule number.

CONCLUSIONS

Both cases in this chapter are examples of the use of correlations between marker loci and quantitative trait loci to infer quantitative genetic parameters. In the first case, the correlation of homozygosity between loci, which is always present in mixed-mating populations, allows the probability of selfing to be estimated based on homozygosity at a marker locus, and hence allows selfing viability to be estimated. In the second case, the correlation of relationship between loci allows prediction of relatedness at quantitative trait loci based on relatedness at marker loci, and hence allows estimation of heritability. This inference relies on the presence of variation of relatedness among pairs, i.e., some pairs are full sibs, some half-sibs, some unrelated, etc. This variation results in a correlation of relatedness between marker and quantitative trait loci, and is the critical factor to quantify for the estimation of heritability.

Both types of estimation use what was originally termed "gene-identity disequilibrium" by Weir and Cockerham (1969). This type of association can be confused with linkage disequilibrium because both have similar genetic effects. For example, apparent overdominance can be caused by linkage disequilibrium between allozyme alleles and deleterious alleles, or by identity disequilibrium between allozyme loci and heterotic loci. Second, as mentioned earlier, quantitative traits may also be studied using linkage disequilibrium between marker and QT trait loci. Identity disequilibrium involves associations not between alleles, but between *pairs* of alleles and so is a four-gene measure, in contrast to the digenic linkage disequilibrium.

The estimators developed in this chapter are based on very simple models that make numerous assumptions. Any statistical treatment of data poses the conflicting demands of simplicity versus realism (Clegg et al. 1978b). Estimates of quantitative genetic parameters are seldom free of biases introduced by inbreeding, assortative mating, dominance, maternal effects, nonadditive genetic variances, and nonrandom sampling of genotypes from natural populations (Mitchell-Olds and Rutledge 1986). However, the loss in rigor from abandoning artificial experiments is offset by the gain in realism, as the genetic components of characters in the field are estimated entirely free of artificial manipulation.

Because one uses natural levels of gene identity and relationship, one has very little choice over "efficient" experimental designs. One cannot design natural populations with a specified, optimal pattern of relationships. The options available are (1) to increase sample size, either through more loci or (preferably) through more individuals, (2) to choose Mendelian loci with greater polymorphism, and (3) to choose species and populations with moderate levels of inbreeding and/or relatedness (and with maximal potential for variation of inbreeding and/or relationship). The third option can introduce a type of bias, as only species that are nonclonal, show limited dispersal, and have local genetic differentiation can be used. For example, to estimate the viability of selfing, species must show natural selfing rates of at least 10%.

This "marker gene demography" places a premium on the genotypic assay of live plants in the field. To assay genotypes, one needs to collect tissue samples from the field. When adult tissue is unsuitable for electrophoretic assay, genotypes need to be inferred by assay of several progeny. This is laborious and can restrict the sample size, which for quantitative genetic parameters, must be large, much larger than the 89 pairs of the *Mimulus* study.

The derivations of both models were based on conditional probabilities that assigned probabilities of selfing or gene identity to individuals. Individually, these probabilities have extremely high statistical variance. Their use lies in combined, "population level" information about mean levels of relationship and inbreeding in the population, where statistical significance can be attained. In this light, it would be relevant to investigate the inference of parentage of seedlings (Meagher and Thompson 1987) to improve the estimate of H. Estimates of the correlated mating system (Ritland in press-a,b) may also provide clues about H, as the parameters of this model predict H among members of a progeny array.

A central theme of this chapter is that gene identity can form the basis for certain studies in genetic demography. Estimation of relationships is common in human demography (Cannings and Thompson 1981) but the use of these concepts is generally lacking in plant demography. One recent study by the author, an analysis of crossability as a function of distance (Ritland and Ganders 1987b), utilized the correlation of homozygosity between marker loci and fitness loci to estimate gene fixation at fitness loci. Because we cannot see the fixation or identity-by-descent of genes, the reliance on genetical models and the attendant statistical problems are greater in such studies. Despite these problems, the integration of population genetic parameters into demographic studies is worthwhile because of the evolutionary context it brings to demography.

ACKNOWLEDGMENTS

I thank Carol Ritland for encouragement and help with the field study and Tim Prout for comments on the manuscript. This research was supported by a Natural Sciences and Research Council of Canada grant to K.R.

CHAPTER 12

DETECTION AND MEASUREMENT OF SELECTION: GENETIC AND ECOLOGICAL APPROACHES

R. A. Ennos

ABSTRACT

Population and biometrical genetics provide the theoretical framework that is essential for the detection and measurement of selection. Population genetic analysis in particular has revealed the ubiquity, variability, and strength of natural selection in plant populations. In order to exploit such analysis fully, further integration with ecological studies is required. The challenge is to understand how specific genetic differences between individuals are translated, through an interaction with the environment, into differences in fitness. The virtue of this approach is that it could lead to the development of a predictive theory of selection that would be of great benefit in both pure and applied fields.

Selection is the process that occurs when groups of individuals possessing a common set of genetic or genetically based phenotypic attributes show differences in their ability to survive between different life cycle stages and/or to transmit genetic material to the next generation. The experimental investigation of selection is concerned with three main questions:

1. Can selection be detected in present day populations?
2. Can selection be measured in present day populations?
3. Can the causes of selection be understood?

A number of different approaches have been taken to answer these questions. In this chapter I intend to outline each of these approaches, discuss their achievements, highlight their limitations, and suggest a framework for further studies.

GENETIC APPROACHES

Geneticists have developed two families of models that allow us to describe selection in a quantitative manner. The first family of models is concerned with selection of discrete variants that differ as a result of segregation at identifiable genetic loci (Population Genetic models). A second set of models is applicable to selection of variants that differ for traits showing continuous variation (Biometrical Genetic models). It is useful to begin by exploring the ways in which these models conceptualize the process of selection and to illustrate their application in plant populations.

Population genetic models

Population genetic theory is concerned with predicting the dynamics of allele and genotype frequencies by considering the interaction of mutation, migration, genetic drift, the breeding system, and selection. To model selection each genotype is assigned a selective value or fitness that specifies viability or net reproductive success relative to some standard genotype. The link between genotype and viability/reproductive success is direct and unconditional on any other parameter (Figure 1). If the breeding system is known and if we can assume that mutation, migration, and genetic drift have negligible effects on genotypic frequencies, the selective values or fitnesses of genotypes can be estimated from data on genotypic frequencies at different stages of the life cycle, or at the same stage of the life cycle in successive generations (Jain and Allard 1960; Workman and Allard 1964; Allard et al. 1966; Workman and Jain 1966; Prout 1969; Clegg and Allard 1973; Clegg et al. 1978). These estimation techniques, which integrate genetics and demography, have been widely applied to measure selective value and fitness of genotypes, in both predominantly self-pollinated and predominantly outcrossed plant populations. The studies make use of naturally occurring polymorphic marker loci controlling morphological or molecular variation. It is important to note that marker loci are chosen for their genetic resolution, rather than for their ecological significance, which is not directly known.

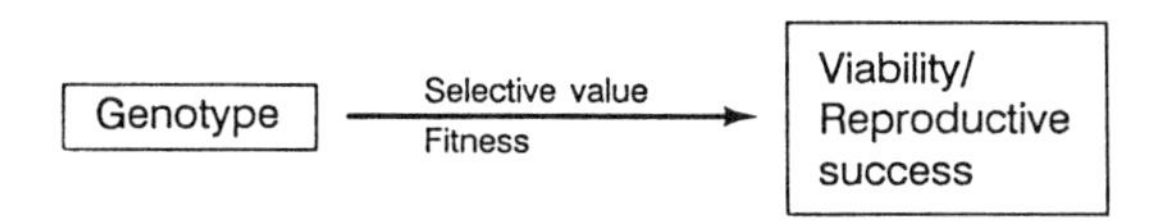

FIGURE 1. Population genetic theory.

From studies of predominantly self-pollinated populations, four general results have emerged: (1) genotypes homozygous for different alleles at the marker loci may differ substantially in fitness in evolving populations; (2) the selective value of a marker locus genotype may differ significantly in different life cycle stages; (3) fitnesses of marker locus genotypes may vary in different seasons; and (4) individuals heterozygous at the marker loci may have fitness values that are twice as large as their homozygous counterparts (Jain and Allard 1960; Allard and Workman 1963; Imam and Allard 1965; Allard et al. 1968, 1972; Marshall and Allard 1970; Hamrick and Allard 1972; Clegg and Allard 1973; Rick et al. 1977; Clegg et al. 1978; Brown 1979).

In order to appreciate the significance of these results, it is necessary to take account of the multilocus structure of these inbreeding populations. Highly significant correlations in allelic state are found among loci in inbreeding populations. This means that although we are ostensibly comparing groups of individuals that differ in genotype at a single marker locus, we are, in reality, measuring differences in fitness between groups that differ at a substantial number of loci. Differences in fitness observed between homozygous marker locus genotypes may be due to the effect of variation at the marker locus itself, but could equally be caused by the cumulative effects of differences at many loci correlated in allelic state.

Careful interpretation of the differences in fitness between heterozygous and homozygous marker locus genotypes is also required. Under extreme inbreeding, correlation in genotypic state among loci will be found, regardless of whether loci are linked or not (Bennett and Binet 1956). Individuals heterozygous at marker loci are likely to be multiply heterozygous. Heterozygotes are also rare genotypes, the product of infrequent outcross events. In these circumstances we can say that the differences in fitness between heterozygous and homozygous genotypes are significant, yet we cannot offer any unequivocal explanation as to how they arise. The observed results could be accounted for by a huge variety of selective effects ranging from single locus overdominance at the marker loci to frequency-dependent selection favoring the rare heterozygous genotype (Harding et al. 1966; Brown 1979).

Similar demographic genetic analysis has been applied in predominantly outcrossing populations practicing mixed selfing and random mating where correlations in allelic and genotype state are likely to be uncommon unless loci are tightly linked (Clegg 1978). Changes in allele frequency over time have rarely been recorded in these populations, but there are many reports of increases in the level of heterozygosity with increasing age of cohorts in both forest trees (Moran and Brown 1980; Farris and Mitton 1984; Yazdani et al. 1985; Cheliak et al. 1985; Plessas and Strauss 1986) and herbs (Schaal and Levin 1976; Burdon et al. 1983).

Two hypotheses have been put forward to account for these observations. The first asserts that increases in heterozygosity result from heterozygous advantage at the marker loci scored (Mitton and Grant 1984). The alternative hypothesis proposes that variation at the marker loci is selectively neutral and that increases in heterozygosity reflect progressive elimination of inbred individuals from the population. Within the fraction of individuals produced by self-pollination in these predominantly outcrossed populations, a positive correlation is expected between marker locus homozygosity and the inbreeding coefficient of an individual. The elimination of inbred individuals will therefore increase heterozygosity at the marker loci even when these loci are under no direct selection (Strauss 1986).

Attempts to distinguish between these two hypotheses have failed to demonstrate the consistent positive correlation between heterozygosity at individual marker loci and components of fitness that would be expected under the overdominance model (Bush et al. 1987). The available evidence suggests that the observed increases in marker locus heterozygosity have resulted from selective elimination of inbred individuals in populations practicing a significant amount of self-pollination (Strauss 1986; Strauss and Libby 1987). However, this process cannot account for the appreciable number of studies that report marker locus heterozygosity levels that exceed those expected under random mating (Ganders et al. 1977; Moran and Brown 1980; Farris and Mitton 1984; Neale and Adams 1985; Plessas and Strauss 1986; Knowles et al. 1987). Further work is required to determine whether overdominance at marker loci plays a role in generating such high levels of heterozygosity.

Summarizing these data for inbreeding and outcrossing species we can say that the application of demographic genetic analysis has proved a powerful technique for detecting and measuring selection in plant populations. It has demonstrated that relative fitness is not constant for any particular genotype, but can vary with the life cycle stage, season, and frequency of the genotype in the population. The approach, however, has limitations. In inbreeding populations where selection has often been detected, differences in fitness cannot readily be ascribed to the effects of particular genetic loci because of the correlated multilocus structure of the genome.

In predominantly outcrossed populations, on the other hand, there may be difficulties in distinguishing between selection on marker loci and selection against inbred genotypes. Another potential weakness of the demographic genetic approach is that while it provides a null hypothesis that we can use to test for significant differences in fitness among genotypes, it does not in general generate a priori predictions about which groups of genotypes are expected to differ in fitness. This criticism can, however, be met in a number of specific situations. These involve instances where genotypes are known to differ in the efficiency with which they transmit genetic material to the next generation. Differences in the rate of genetic transmission are found for chromosomes showing meiotic drive, between hermaphrodites and females in gynodioecious populations, between selfing and outcrossing individuals, and between sexually and asexually reproducing genotypes. In each of these situations, population genetic models can be used not only to indicate which genotypic classes are expected to differ in fitness, but also to predict the direction and magnitude of selection.

One of the best studied cases of meiotic drive involves the B chromosomes of rye. These chromosomes pair among themselves at meiosis and are effectively transmitted to the generative nucleus in the pollen, and the embryo sac in the ovule, thereby doubling their numbers each generation (Jones and Rees 1982). A stable distribution of B chromosomes is nonetheless maintained. This is achieved because there is strong fertility selection against genotypes that possess B chromosomes such that the fertility of plants possessing four or more B chromosomes is effectively zero (Muntzing 1943; Jones and Rees 1967).

In gynodioecious populations females suffer a transmission disadvantage relative to hermaphrodites since they do not contribute genetic material via pollen. Genetic models suggest that females will persist only if they show substantially greater seed production and/or higher offspring fitness than their hermaphrodite counterparts (Ross and Weir 1975; Charlesworth 1985). Thus it should be possible experimentally to detect significant differences in net reproductive rate favoring females, or higher fitness of offspring from females than from hermaphrodites in gynodioecious populations. Studies of a number of gynodioecious species have upheld these predictions (Assoud et al. 1978; Phillip 1980; Van Damme and Van Delden 1984).

A transmission bias favoring selfing genotypes in an outcrossed population was originally noted by Fisher (1941). Populations are expected to remain predominantly outcrossing only if the fitness of selfed progeny is less than half that of outcrossed progeny (Maynard Smith 1978; Lande and Schemske 1985). This prediction has been tested by comparing lifetime reproductive success of selfed and outcrossed progeny from a normally outcrossing population of *Gilia achilleifolia* (Schoen 1983). The fitness of outcrossed progeny was nearly twice that of selfed progeny, in close agreement with genetic predictions. Indirect indications that selection favors outcrossed as opposed to selfed progeny can be derived from analysis of Wright's fixation index at different life cycle stages in outcrossing perennial plants. Consistent decreases in fixation index at progressively later stages of the life cycle may be interpreted as evidence for viability selection for outcrossed progeny (Schaal and Levin 1976; Farris and Mitton 1984; Yazdani et al. 1985; Cheliak et al. 1985; Plessas and Strauss 1986).

The final example concerns differences in transmission efficiency between sexual and asexual genotypes. Genetic considerations indicate that mutants producing seed asexually in a normally outcrossing hermaphrodite population will enjoy a 50% advantage in genetic transmission rates, and this advantage rises to 100% for females in dioecious species (Maynard Smith 1978; Lloyd 1980; Charlesworth 1985). If sexual reproduction is maintained by short-term advantage we expect to be able to detect large fitness advantages for sexually, as opposed to asexually, reproduced offspring in populations that have retained sexuality. This prediction has been tested experimentally in the outbreeding grass *Anthoxanthum odoratum* by planting sexually and asexually derived propagules into their natural population and measuring their net reproductive rates (Antonovics and Ellstrand 1984; Ellstrand and Antonovics 1985). These experiments demonstrated a roughly 2-fold fitness advantage for sexually reproduced propagules, in line with population genetic predictions.

These examples indicate that a purely genetic approach to the study of selection can be more than a descriptive exercise. In prescribed situations where genetic variation affects transmission efficiency, it is possible to make firm predictions about the direction and magnitude of selection favoring a defined set of genotypes. These predictions are amenable to experimental verification. In undertaking these experiments it has been possible to demonstrate selection of considerable magnitude and to quantify this in terms of differences in fitness between genotypes. The experimental results raise a whole series of further questions about how selective differences between genotypes are achieved.

Biometrical genetic models

Biometrical genetic theory deals not with the frequencies of alleles and genotypes, but with the behavior of means and variances of continuously varying characters within a population. The underlying causes of continuous character variation are assumed to be segregation at many unidentified genetic loci together with environmental variation. These two sources of variance, acting together, give the observed phenotypic variation for the character in question. In the absence of selection the means and variances of character distributions are expected to remain unchanged provided there is no change in the environment in which the character is measured (Falconer 1981).

In order to incorporate selection into the biometrical model we note that the additive effects of genotype and environment give rise to a particular phenotypic character value X. We then assume that some function of the phenotypic character value X determines the viability/reproductive success of the individual in the same way that these quantities are directly determined by genotype in the population genetic model (Figure 2). Two modes of phenotypic selection are recognized. Under directional selection, viability/reproductive success is a function of X, and under variance selection it is a function of $(X - \overline{X})^2$, where $\overline{X}$ is the mean character value in the population (Lande and

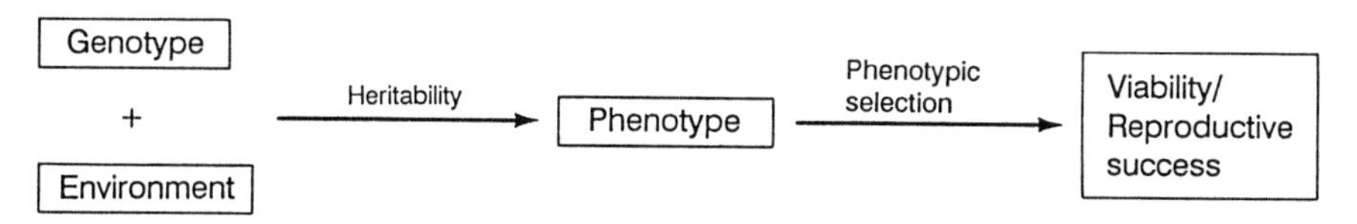

FIGURE 2. Biometrical genetic theory.

Arnold 1983; Arnold and Wade 1984a,b; Endler 1986). Phenotypic selection leads to natural selection only if the heritability of variation for the character is significant (Lande and Arnold 1983).

Two methods can be used to detect and measure selection of continuously varying characters. The simplest involves collecting population samples from successive generations, growing them under identical conditions (which should conform as closely as possible to those from which they have been collected), and comparing the mean and variance of the character distributions. Allard and Jain (1962) used this approach to investigate selection of a wide range of continuously varying characters during the evolution of a barley composite cross-population. Significant changes in mean and variance were found for many characters, attributable to the action of directional and stabilizing selection. Selection of numerous agronomically important characters has since been recorded in other composite cross-populations evolving in the absence of conscious selection (Singh and Johnson 1969; Marshall 1976; Quisenberry et al. 1978; Luedders 1978; Choo et al. 1980) and in genetically variable cultivars of forage crop species (Hayward 1970; Aarssen and Turkington 1985).

The number of situations in which population samples are available from different generations is rather rare. An alternative method for detecting selection is based on the premise that if the two conditions necessary for selection are met (significant heritability of variation and phenotypic selection) then we can conclude that selection is occurring (Lande and Arnold 1983; Arnold and Wade 1984a,b). Realistic measurements of heritability must be made under the environmental conditions in which selection is being studied (Lawrence 1984; Mitchell-Olds and Rutledge 1986). To establish whether phenotypic selection is occurring, the fate of individuals of known phenotype must be recorded in terms of viability/reproductive success. Multiple regression of relative selective value/fitness on character value X and on $(X - \overline{X})^2$ is used to establish the significance of directional and/or variance selection on the phenotype. If X is measured in standard deviation units, the multiple regression coefficients can be regarded as coefficients of directional and variance selection on the phenotype, and their magnitude can be compared among characters (Lande and Arnold 1983; Endler 1986).

There are a number of problems with the interpretation of results arising from these studies. If we detect a significant relationship between selective value/fitness and X and/or $(X - \overline{X})^2$, this does not mean that direct phenotypic selection for this character has occurred. The result could have arisen from phenotypic selection on an environmentally or genetically correlated character (Falconer 1981). Environmental correlations between characters will arise in a heterogeneous environment if environmental differences simultaneously affect the phenotypic value of two characters. Genetic correlations between characters are found not only as a result of linkage disequilibrium between loci (as in the case of single locus variants) but also through the pleiotropic effects of genes that may control variation for more than one character. Genetic correlations are likely to be commonest in inbreeding populations showing high linkage disequilibrium, or where characters are controlled via common developmental pathways, irrespective of the breeding system of the population.

Lande and Arnold (1983) have suggested that the problem of correlated characters may be circumvented by accounting for differences in selective value/fitness in terms of variation not of one, but of many correlated characters using multiple regression techniques. Characters associated with the largest partial regression coefficient are assumed to be those under direct selection, while changes at other characters are presumed to represent correlated responses to selection. This approach suffers from the weakness of all multiple regression techniques, namely that we may fail to measure the selectively important character, and therefore will draw erroneous conclusions from the analysis. It should also be borne in mind that a significant regression of selective value/fitness on character value does not establish a cause–effect relationship between these variables, no matter how much of the variation it explains (Mitchell-Olds and Shaw 1987).

Despite these caveats, regression techniques hold promise for the study of selection in plant populations. Their success will depend on whether significant levels of heritability can be detected for continuously varying characters under field conditions. In the handful of studies conducted so far, heritability values for a range of traits have been found to be much lower when measured under natural conditions than would have been anticipated from trials under standardized conditions (Billington 1985; Mitchell-Olds 1986; Billington et al. 1988). In an integrated study of heritability and phenotypic selection, Kalisz (1986) was able to show significant phenotypic selection for time of germination in a population of *Collinsia verna* but was unable to detect significant heritability for the character; there was no evidence for natural selection on time of germination. These preliminary reports indicate that low heritability of variation for many plant characters under field conditions may limit the usefulness of simple applications of the regression techniques described above for detecting selection in plant populations (Mazer 1987). More data are clearly required.

ECOLOGICAL APPROACHES

Population and biometric genetic theory together provide the framework needed for detecting and measuring selection within and between generations. Simple application of the theory has yielded insights into the rate, variability, and ubiquity of selection in plant populations (Endler 1986). The major limitation of these purely genetic models is that they cannot be used to explain how differences in selective value/fitness of genotypes arise. The reason for this is that they assume a direct link between genotype (population genetics) or phenotype (biometric genetics) and viability/reproductive success (Figures 1 and 2), unconditional on any other parameter. If we wish to gain an understanding of the mechanism of selection these models need to be set within an ecological context. Genetic analysis can then be used to probe the validity of ecologically based models that purport to explain the causes of selection.

In order to develop such ecologically based models it is helpful to remember that the ability of a plant to survive and transmit genetic material to the next generation is determined not by its genotype or phenotype, as the genetic models assume, but by an interaction between its genotype and the environment (Figure 3). The term environment is used here in its broadest sense to encompass all abiotic and biotic factors. The interaction of genotype and environment should therefore provide the focus of attention for ecological approaches to the study of selection. The challenge is to understand how specific genetic differences between individuals are translated through an interaction with the environment into differences in viability/reproductive success.

The importance of an interaction between genotype and environment in determining selective value/fitness has been very clearly demonstrated experimentally using the reciprocal transplant technique. The results of numerous studies have shown that the selective value/fitness of a genotype is not constant, but is highly dependent on the environment in which the genotype is grown (Clausen et al. 1948; Snaydon and Bradshaw 1962; Jain and Bradshaw 1966; Cavers and Harper 1967; Davies and Snaydon 1976; Lovett Doust 1981; Antonovics and Primack 1982; McGraw and Antonovics 1983; Schemske 1984; Schmidt and Levin 1985). Genotypes grown in their home environment almost invariably show higher selective value/fitness than alien genotypes. This is to be expected if, over time, genotypes that interact with the environment to give high net reproductive success have increased in frequency at each site. By looking at separate life cycle stages it has been possible to pinpoint the time at which critical differences in selective value are manifest (McGraw and Antonovics 1983). Wolff (1988) has followed the fate of individuals in a population of *Plantago lanceolata* segregating for marked chromosome segments derived from two contrasting ecotypes. Adaptive differentiation between the two ecotypes is caused by variation at many loci scattered throughout the genome. Variation in performance cannot, therefore, be ascribed to the action of any single genetic difference between populations. For

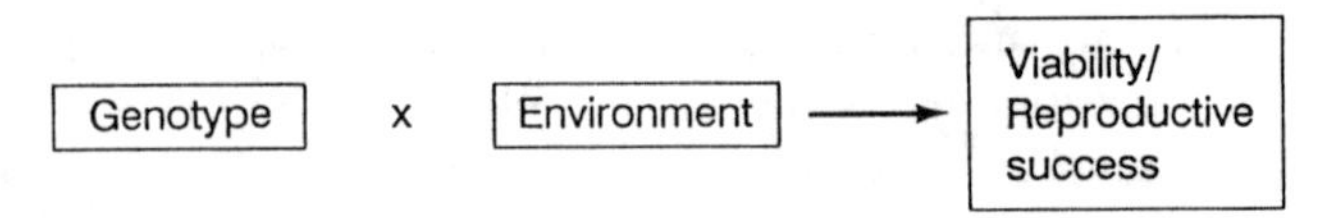

FIGURE 3. Ecological genetic approach.

this reason reciprocal transplant experiments cannot be used to study the question in which we are most interested, namely how specific genetic differences between individuals are translated into differences in selective value/fitness.

In order to establish how specific genetic differences between individuals are translated into differences in fitness via interaction with the environment it is necessary to compare genotypes derived from the same population in which the genetic differences are as far as possible randomized on the genetic background, and in which genetic correlations between characters are minimized. Outbreeding populations provide the best material because correlations in allelic and genotypic state are likely to be minimal among loci, though genetic correlations between continuously varying characters will still be found as a consequence of pleiotropic gene action (Falconer 1981). It is also important to study the effect of specific environmental variables on the fitness of genotypes in defined forms of ecological interaction. The aim should be to dissect the multitude of interactions between genotype and environment occurring within natural populations into their component parts so that the effect of any single ecological interaction on genotypic fitness can be measured, and its importance in the natural population can be assessed.

To provide a framework for investigating these questions we require a simple classification of the types of ecological interaction that are likely to occur within plant populations. The following is a list of the classes of interaction that have been studied to date:

Plant–abiotic environment
Plant–plant
Plant–pollinator
Plant–herbivore
Plant–pathogen

These are all first-order interactions but it must be remembered that higher order interactions will be important too. For instance the outcome of many plant–herbivore interactions may be strongly influenced by variations in the abiotic environment (Rhoades 1983).

If we are adopting this ecological approach to the study of selection the first decision that needs to be made is the type of ecological interaction that we wish to investigate, and the particular system that we will study. Knowledge of the biology of the interaction will suggest forms of genetic variation in the plant population that could affect viability/reproductive success as a consequence of the chosen interaction. Experiments can subsequently be designed to detect and measure selection of these genotypes as a result of the ecological interaction. Hypotheses based on a knowledge of the biology of the interacting system will provide testable predictions about the direction and magnitude of selection anticipated under defined sets of conditions.

This ecological approach differs radically from the purely genetic approaches that have already been described. The starting point is the ecological interaction, which in turn defines the type of genetic variation that is studied. The biological background is then used to provide testable predictions about the direction and rate of selection. The purely genetic approach on the other hand starts with genetic variation and attempts to detect and measure selection for this variation. Causal explanations for any selection detected are often formulated after the results have been obtained. When this happens further experiments will be required to establish the validity of these explanations so as to avoid the trap of "adaptive story telling" (Gould and Lewontin 1979).

The ideas that I have just outlined are not new. Many studies of the effects of specific genotypic differences on viability/reproductive success in defined ecological interactions have already been conducted. A selection of these studies is outlined below.

Plant–abiotic environment

The classic studies of selection occurring within defined ecological interactions have involved work on plants and their edaphic (soil) environment. In order to study selection of this form it is necessary to seek situations in which different edaphic conditions have been imposed on various areas of a common population, usually as a result of man's activities. Population samples may be taken at varying intervals after the imposition of the treatment, or simultaneously from areas that have received different treatments, and compared for changes in genetically controlled characters affecting survival/reproductive success under appropriate edaphic conditions.

Using these approaches Wu et al. (1975) were able to show increases in the level of genetically determined copper tolerance in a population of *Agrostis stolonifera* exposed to copper contaminated soil for increasing periods of time. Davies and Snaydon (1973a,b, 1974) found that samples taken from different areas of an *Anthoxanthum odoratum* population that had received a diversity of fertilizer and liming treatments showed differences in growth response to individual nutrients that were directly related to the level of these nutrients in their source plots. In each case selection had occurred for characters which enhance the viability/reproductive success of genotypes as a result of interaction with the edaphic environment.

Plant–plant

Most workers in this area have concentrated on competitive interactions between plants. Competition may be intra- or interspecific, the importance of each being dependent upon the species structure of the community being studied. Allard and Adams (1969) investigated the effects of genotypic variation on fitness under intraspecific competition. They demonstrated that genotypes sampled from a composite cross-population, in which success in intraspecific competition was likely to be a major determinant of fitness, showed higher reproductive rates when grown with coexisting competitors than when grown in pure stand. Mixtures of genotypes that had no such history of selection under intergenotypic competition showed no evidence of consistent overcompensation. These results provide strong evidence for selection of genotypes that show high fitness under competitive interactions with neighbors. The causes of the observed differences in fitness under intra- and intergenotypic competition are unclear, though it has been suggested that annidation (occupation of slightly different niches by different genotypes) may be involved. This is a hypothesis that could readily be tested by looking at the fitness of genotypes in competition with neighbors that differed in the extent of their niche overlap (Ennos 1985).

Some of the most detailed work on selection under interspecific competition has been carried out on white clover and perennial grasses in permanent pasture. Turkington and Harper (1979) demonstrated that clover genotypes surviving in close association with each of four grass species were those that showed the best vegetative performance in competition with their associated grass. Evans et al. (1985) found that clover genotypes sampled from a range of sites gave higher yields in competition with the ryegrass genotypes from their native pasture than with two standard ryegrass varieties with which they had not previously been grown. Both experiments provide evidence for selection of clover genotypes best able to compete with coexisting species (Aarssen 1983). A more detailed analysis of how genetic differences among clover genotypes are translated into differences in vegetative success under competition with their associated grasses would add greatly to our understanding of the mechanism of selection under interspecific competition.

Plant–pollinator

Differences in fitness among plant genotypes arising from interactions with their pollinators can be studied as long as appropriate genetic variation affecting pollination efficiency is available within the plant population. Flower color variation has been shown to influence the male fitness of hermaphrodite plants through its effect on pollinator behavior. Morphs that are more attractive to pollinators show higher rates of transmission through pollen than their less attractive counterparts (Stanton et al. 1986). Relative reproductive success of self-compatible and self-incompatible morphs may also be affected by an interaction with pollinators. Where pollinators are scarce, selection favoring self-compatible morphs has been demonstrated (Piper et al. 1986). Many further examples of differences in fitness mediated by plant pollinator interactions are likely to be detected if, for instance, genetic variation affecting levels of reward offered to pollinators can be identified within plant populations.

Plant–herbivore

Ecological studies of plant–herbivore interactions are beginning to highlight the enormous impact that herbivores can have on the viability and reproductive success of plants in natural populations (Dirzo 1984). To study the way in which genetic differences between plants can affect their fitness in plant–herbivore interactions it will be simplest to study two-component systems in which some information is already available on forms of genetic variation likely to affect the degree of herbivory suffered by a plant. Genetic variation affecting production of secondary chemicals has proved of enormous value in this respect.

In a model study of this kind Berenbaum et al. (1986) have used biometric genetic analysis to establish that there is significant heritable variation for the concentration of the chemical bergapten in the seeds of wild parsnip and that under herbivory by parsnip webworms significant phenotypic selection favoring plants with high bergapten levels occurs. Conversely, phenotypic selection for low bergapten levels is found in the absence of herbivory. This study illustrates the power of combining an ecological approach to the study of selection with appropriate genetic analysis. Selection can be detected and measured, and the link between genotype and fitness can be understood.

Another plant–herbivore system that has received considerable experimental attention involves the interaction of white clover with the gray field slug in permanent pastures. White clover genotypes able to release hydrogen cyanide when grazed suffer less herbivory by the slug than their acyanogenic counterparts (Angseesing 1974; Dirzo and Harper 1982a). Selection favoring cyanogenic genotypes has been detected under field conditions and this may be most important at the seedling establishment phase (Ennos 1981; Dirzo and Harper 1982b). It is important to remember when dealing with interactions involving two biotic components that genetic change may occur as a result of the ecological interaction in either or both species (Gould 1983). In these circumstances the relative fitness of a plant genotype under herbivory cannot be regarded as constant, but is dependent on the genetic composition of the herbivore population (Edmunds and Alstad 1978; Burgess and Ennos 1987). Selection can be understood only in the context of models which take into account genetic variation in both plant and herbivore populations.

Plant–pathogen

Although the plant pathology literature abounds with studies of plant–pathogen interaction this work has revealed relatively little about selection in plants because the agricultural populations involved tend to be genetically uniform for characters affecting the plants' susceptibility to pathogen attack (Burdon 1987). Composite cross-populations of barley are, however, highly genetically variable and here it has been shown that plant genotypes resistant to local pathogens have increased in frequency since the population was established (Muona et al. 1982; Saghai Maroof et al. 1983).

Studies of selection in natural plant populations as a result of plant–pathogen interactions are as yet in their infancy (Burdon and Jarosz, this volume). Recent work by Parker (1986, 1988) has demonstrated, however, that selection can be detected and measured in these situations. Using populations of the annual legume *Amphicarpaea bracteata* and its obligate pathogen *Synchytrium decipiens* he was able to show that there is genetic variation within the plant population for disease resistance and that strong phenotypic selection for low infection rates occurs within natural populations. The two conditions necessary for natural selection by the pathogen are met.

This situation is, however, more complex than would at first appear for there is genetic variability within the pathogen population affecting its ability to overcome the resistance of the host (Parker 1988). Such a phenomenon is by no means unique for, wherever it has been sought, substantial variability within natural pathogen populations for their ability to overcome the resistance of host plants has been detected (Powers 1980; Harry and Clarke 1986; Ennos and Swales 1988). In order to design critical experiments to detect, measure, and understand selection in plant–pathogen systems a thorough knowledge of genetic variability for resistance in the host and corresponding pathogen variability for ability to overcome resistance will be required.

CONCLUSIONS

Genetic approaches to the study of selection provide the theoretical framework necessary for its detection and measurement. In those cases where genetic variation affects transmission, genetic theory alone can be used to predict the direction and rate of selection expected. However, in many situations a purely genetic approach can neither uncover the underlying causes of selection nor provide explanations for the differences in fitness observed among genotypes (Ennos 1983). On its own, genetic theory provides an incomplete description of selection because it overlooks the importance of interactions between genotype and environment in determining the net reproductive success of individuals.

In order to overcome these deficiencies I believe that we need to adopt an ecologically based approach to the study of selection. Here attention is focused on the interactions that occur between genotypes and their environment. Within specific classes of ecological interaction, genetic analysis can be used to investigate how variation for ecologically relevant characters affects the viability/reproductive success of individuals. In this way an understanding may be gained of how genetic differences are translated into fitness differences in defined ecological situations. If this knowledge can be obtained, a predictive theory of selection can be developed.

To adopt this approach we need to switch the emphasis away from studies of single species toward the study of interacting ecological systems. Genetic analysis can then be used as a tool to detect, measure, and understand selection within these systems. Geneticists may be reluctant to follow this course since it will mean a considerable increase in the complexity of the experimental material with which they work. Genetic change may have to be studied in more than one interacting species and a deeper understanding of the ecological background will be required before genetic studies can be initiated. However, I believe that these difficulties are more than offset by the enormous increase in realism that such an approach is likely to provide. We will have the opportunity to obtain a predictive rather than a purely retrospective view of selection.

From the point of view of crop improvement a predictive theory of selection incorporating both genetic and ecological elements would be of great utility in pinpointing those natural populations in which desirable forms of genetic variation are likely to be located. It would also be important in suggesting new forms of genetic variation that could be incorporated into crop populations and could provide guidance on how they should be deployed. In both basic and applied fields the development of a predictive theory of selection through a fusion of ecological and genetic approaches is long overdue.

ACKNOWLEDGMENTS

I would like to thank the reviewers for their perceptive and moderating comments on an earlier version of this chapter. I am also grateful to the many ecology students whose critical approach to the study of selection has forced me continually to reassess my ideas on the subject.

CHAPTER 13

DISEASE IN MIXED CULTIVARS, COMPOSITES, AND NATURAL PLANT POPULATIONS: SOME EPIDEMIOLOGICAL AND EVOLUTIONARY CONSEQUENCES

J. J. Burdon and A. M. Jarosz

ABSTRACT

Comparisons of host–pathogen interactions in agricultural mixtures and natural plant communities provide different, yet complementary, perspectives on common issues concerning the evolutionary dynamics of these systems. A central question is the role that genetic diversity in the host population plays in reducing the incidence of disease. In the short term this has immediate agricultural importance in the possibility of increased yield; for natural plant populations its importance lies in the genetic structure of the seed population that gives rise to the next generation. In the longer term, the dynamic nature of host–pathogen associations is likely to produce changes in the genetic structure of host and/or pathogen population. In both agricultural and natural populations this has similar evolutionary implications.

Genetics plays a central role in any understanding of the structure of populations and their long-term survival. The selective forces that mold plant populations are many and diverse, and the primary aim of population genetic studies has been to develop a predictive account of the effects of these forces. To date, many studies concentrating on the function of specific genes have approached this problem by making comparisons of genetic diversity within and between different populations. Such investigations have been successful in demonstrating marked discontinuities in the frequency of biochemical characters or morphological traits. Paradoxically though, many of these markers (for example, individual isozyme alleles) may be selectively neutral and the observed patterns of variation presumably reflect differences in the intensity of selective forces acting on other unknown, but linked, parts of the plant's genome.

The selective forces mediating most changes observed in populations have not been determined and an immense gap remains between the ideal of predictive models and reality. In this light, the effects of pathogen pressure—one of the most powerful and ubiquitous selective forces in nature—are relatively easy to establish and consequently deserve greater attention.

While it is true that most plants are resistant to most pathogens, no plant species has broken free of an attendant assemblage of pathogenic organisms. Advanced agricultural varieties, primitive cultivars, land races, and wild relatives of agricultural crops, however distant, are all vulnerable to the depredations of disease. In this chapter we jointly consider host–pathogen interactions from the distinctly different perspectives afforded by agricultural systems and natural plant communities to focus on common issues concerning the evolutionary dynamics of plant and pathogen populations.

HOST–PATHOGEN INTERACTIONS IN NATURAL COMMUNITIES

It is only recently that the importance of disease in natural plant communities has become widely recognized (Burdon 1987a). Diseases in these systems are not just restricted to a few spectacular epidemics like those of Dutch elm disease and *Phytophthora* die-back. Rather, they are an integral part of most communities, the major difference to those in agriculture being in the spatial and temporal scale of disease occurrence. Disease in natural systems is characterized by extreme patchiness. Host populations heavily attacked one year may be free of disease the next, while, for a variety of environmental and genetic reasons, heavily infected and disease-free plants may occur in close association with one another. In all of these situations, pathogen-induced damage has been shown to reduce the reproductive output, survival, and longevity of affected individuals. Disease-induced selection thus favors plants that are resistant. Conversely, host resistance selectively favors the emergence of pathogen races that can overcome that resistance.

The interactions that might be expected to occur between host and pathogen during the development of close coevolutionary associations have been described on many occasions (e.g., Person 1966; Burdon 1982). Essentially these heuristic models see the short-term advantages conferred by novel resistance genes in host populations as leading to the emergence of complementary virulence genes in the pathogen population. Through a series of reciprocal interactions both host and pathogen populations become more diverse and a balance is ultimately achieved in a continuing series of cyclical polymorphisms between specific resistance factors in the host and virulence characters in the pathogen (Person 1966). Mathematical models of genetic aspects of this argument (Jayakar 1970; Leonard 1977) have consistently found that persistent polymorphisms for resistance/susceptibility in the host and avirulence/virulence in the pathogen result only if resistance and virulence have associated fitness costs. Such costs would be reflected in some aspect of the fecundity, longevity, or viability of individuals carrying resistance or virulence genes.

Virtually all empirical knowledge of wild host–pathogen systems is restricted to analyses of the resistance structure of host populations at single points in time. These studies have shown considerable variability in the complexity of individual populations (Burdon 1987b; Parker 1988a) and in the geographic distribution of resistance (Dinoor 1970). In general these results are compatible with those expected on theoretical grounds (Leonard 1984). Unfortunately, very few data are currently available concerning temporal patterns of change in resistance within individual host populations or about any aspect of the structure of pathogen populations (Dinoor and Eshed 1987; Burdon and Jarosz 1988).

HOST–PATHOGEN INTERACTIONS IN AGRICULTURE

The devastating effects that pathogens have on agricultural crop production are apparent both in the occasional pandemics that still ravage crops (e.g., southern corn leaf blight in the United States, 1970) and in the more typically modest losses exacted by epidemics of limited spatial and temporal severity that regularly occur in most crops. In responding to these effects one of the major objectives of plant breeding programs has been the incorporation of effective disease resistance into crop varieties prior to their release.

This has created a multiplicity of host–pathogen "coevolutionary" systems mediated by man. In these systems, short-term ecological changes in the host population are produced by variations in the relative frequency of planting of particular varieties. Longer term genetic changes are produced through the deployment of new resistance genes. Changes in pathogenicity are driven by a combination of the resistance gene deployment strategies adopted by plant breeders and the collective varietal preferences of growers. This is illustrated in the changes that have accompanied the use of the *Mla* 12 gene for mildew resistance in barley in Europe (Wolfe 1984). Shortly after the deployment of the *Mla* 12 gene, the pathogen population overcame its effectiveness, and farmers rapidly adopted newer barley varieties carrying different resistance genes. Over the next few years, the frequency of virulence at the *Va* 12 locus in the pathogen population declined to such a low level that for a year or two newer varieties containing the *Mla* 12 gene were again resistant. Such examples of the rapid response of pathogen populations to changes in the resistance structure of their hosts occur sporadically through the literature (Luig and Watson 1970; Martens et al. 1970). They all demonstrate just how powerful a selective force host populations may be in affecting the genetic structure of their attendant pathogen populations.

MIXTURES—WHERE AGRICULTURAL AND WILD SYSTEMS MEET

The markedly different approach of scientists working with diseases in natural plant populations and those involved with agricultural crops belies the considerable overlap in interests that occurs between them. This overlap is particularly evident in attempts to incorporate a greater level of genetic diversity into agricultural crops through the use of varietal mixtures, multilines, and composite populations. The underlying rationale for the use of mixtures in agriculture has been to try and regain the disease buffering capacity that wild systems are reputed to possess (Browning 1974). It is hardly surprising, therefore, that many of the basic questions concerning both the short- and long-term fate of agricultural mixtures and the workings of natural coevolved systems appear to be the same.

However, the nature of diversity in agricultural mixtures differs markedly from that in natural populations. This fact must temper the desire to improve disease control in agriculture by examining natural systems, and of furthering our knowledge of natural systems by examining agricultural ones. Natural populations are usually small and fragmented, and individual plants are interspersed with a wide range of other nonhost species. Such populations are composed of individuals of variable size and age. Agricultural mixtures, on the other hand, are even-aged, usually far more uniform morphologically, and occur in much larger stands.

Here we ask whether an increase in genetic diversity does in fact lead to reductions in disease and what implications this has for host–pathogen systems in general. There are two levels at which these questions can be addressed. First is the primarily short-term question of whether mixtures actually reduce disease during the life of a crop and whether this is translated into increased reproductive performance. Second, we can look at the longer term fate of mixtures in a genetic sense—that is, how effective is a strategy of applying frequency-dependent selection to a pathogen population likely to be? Will the pathogen population become characterized by the presence of many simple races, each specialized to one component of the mixture, or will the population be dominated by one or a few races capable of attacking most or all of the component lines?

Do mixtures reduce disease and increase yield?

Whether measured indirectly as the number of fungal spores produced (Cournoyer 1970) or directly as numbers of pustules (Wolfe et al. 1976) or rates of disease increase (Leonard 1969a; Burdon and Whitbread 1979), disease levels in mixtures are almost invariably below those occurring in monocultures of their most susceptible components. Moreover, in the large majority of cases this reduction is greater than one of simple proportionality. In a review of over 100 cases, disease levels in two-component mixtures averaged 25% of those occurring in the most susceptible monoculture (Burdon 1987a).

The effects of mixtures on yield in disease-prone environments are considerably more modest. Yield increases are usually only a few percent better than the mean monoculture yield, rarely exceeding 10% and occasionally dipping to a poorer than expected performance (Burdon 1987a). Typically, the yield of disease-affected mixtures has been compared with that of the mean performance of their component monocultures also grown in the presence of disease. Are these, however, the best controls for comparison?

Disease-affected monocultures provide the most relevant comparison with disease-affected mixtures for evolutionary biologists interested in the general question of the long-term fate of natural mixtures growing in disease-prone environments. In these situations, all three populations (two monocultures and the mixture) are likely to suffer disease and interest centers on the performance of the mixture relative to its component lines.

In contrast, from an agricultural point of view the most important yardstick against which the disease-affected mixture should be assessed is the performance of the disease-free mixture and *both* healthy and infected monocultures. The yield of healthy monocultural stands represents the maximum that plant breeders can currently attain. In an economic sense they must be the ultimate target. Remarkably few studies involving disease-affected mixtures have attempted such comparisons. As a result, it is currently not possible to state that the yield advantages shown by disease-affected mixtures over similar monocultures spring totally from their recognized disease reduction abilities rather than possible overyielding attributes of the mixtures themselves. Extensive reviews of the relative performance of disease-free mixtures and their components in a wide range of crops (Marshall 1977; Nitzsche and Hessebach 1983, 1984) indicate that disease-free mixtures frequently although not invariably outyield the mean of their monocultures. These yield increases approach those commonly encountered in diseased mixture trials and may at least partially explain the better performance of mixtures in disease-prone environments. For example, in one experiment in which disease-free controls were used, the yield of healthy wheat mixtures exceeded that of the mean of the monocultures by 19.4% while the yield of infected mixtures was only 17.4% greater than that of the mean of infected monocultures (Jeger et al. 1981).

A further problem that is apparent in current assessments of the agricultural performance of mixtures lies in the use of experimental plots that are too small to reproduce the dynamic nature of the interaction being assessed. In mixtures, as the distance between individuals of the same genotype increases, the production of inoculum and its effective spread from one plant to the next declines and the effective transmission rate of the pathogen falls (see Figure 1). However, the use of small plots juxtaposed with heavily infected pathogen spreader rows negates many of these advantages as such mixtures experience heavier inoculum loads than they would normally generate internally. To avoid this problem mixture plots need to be sufficiently large to ensure that the host and pathogen populations are largely self-contained. Once an initial infection is established, all or nearly all further infections should result from inoculum produced elsewhere within the plot. Relatively little inoculum should enter the plot from outside and relatively little should be lost beyond the edges of the plot.

The potential confounding effects that extraplot pathogen interference can have on the interpretation of mixture yields can be seen in a recent report concerning disease in mixtures of spring barley (Gieffers and Hesselbach 1988). Mixtures occupying small plots (13.5 m^2) showed a 4% yield improvement over the mean of their component monocultures. In contrast, mixtures occupying large 2500–5000 m^2 plots showed a 10% yield improvement. If this

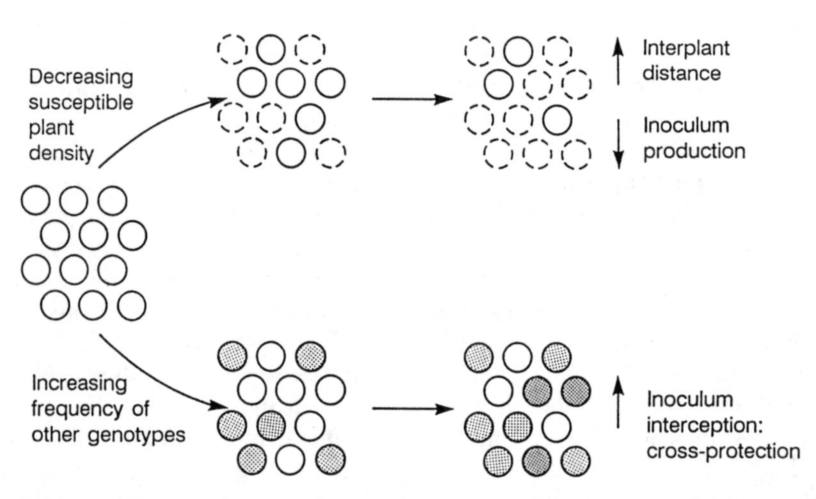

FIGURE 1. The simultaneous changes that occur in the conversion of a pure stand to a mixture of susceptible (unshaded circles) and resistant (shaded circles) individuals (redrawn from Burdon 1987a). Also shown are the ways in which these changes affect the development of disease epidemics by affecting the effective transmission rate of the pathogen.

result is typical of the underestimation that results in experimental plots, then arguments for the greater use of mixtures are strengthened.

Clearly what is needed to resolve questions concerning the short-term yield merits of agricultural mixtures is more careful experimentation including both infected and disease-free mixtures, and using plot sizes that are designed specifically for the host–pathogen interaction under consideration.

Mixtures: implications for natural systems

Experiments with agricultural mixtures stress the performance of the overall population. Despite this, they provide information pertinent to natural populations where the performance of the individual is important. One aspect of mixture experiments that is directly applicable is the effect of plant dispersion. In natural systems, clumped patterns of plant distribution are largely a reflection of propagule and pollen dispersal. However, the interrelationship between such clumped patterns and the occurrence and intensity of disease is markedly affected by a number of epidemiologically related factors including the source of inoculum for the initial disease focus, the dispersal efficiency of the pathogen, and the size of the host patch itself. Experimental work with random and clumped mixtures of resistant and susceptible oats subjected to infection by *Puccinia coronata* has shown that the original distribution of both the susceptible plants and the initial inoculum can play a very important role in determining the ultimate development of disease (Mundt and Leonard 1985). In situations where epidemics developed from a generalized inoculum rain, the clumped mixture had disease levels that were similar to that of a susceptible monoculture while the random mixture displayed much lower disease levels. On the other hand, when epidemics developed from a single inoculum focus, both the clumped and random mixtures had less disease than the susceptible monoculture. In a snap bean-rust system, the severity of epidemics that started from a single focus was affected by the degree to which susceptible plants were clumped (Mundt and Leonard 1986).

In very small host populations, however, even a single heavily infected plant may produce sufficient inoculum to infect all other susceptible individuals rapidly. Once this has happened, the susceptible component of such populations may gain little or no benefit from being mixed with resistant individuals. Exactly this phenomenon has been observed in a study of the epidemiology of *Erysiphe cichoracearum* in experimental populations of wild *Phlox* (Jarosz and Levy 1988) and may explain why natural *Phlox* populations from disease-prone environments tend to be uniformly resistant.

The disease-buffering capacity of mixtures is also affected by the proportion of the mixture that is resistant to the pathogen. In general there appears to be a nonlinear relationship between epidemic rate and the proportion of resistant individuals (Browning and Frey 1969; Leonard 1969a; Burdon and Chilvers 1977). As a result, the pathogen pressure on a particular genotype is determined by (1) the number of virulent pathogen races present, and (2) the proportion of individuals resistant to each of those races. In natural systems, both of these factors vary in time, so that the actual selective pressure exerted by a pathogen will also vary considerably.

THE LONG-TERM CONSEQUENCES OF MIXTURES

In contrast to questions concerning short-term disease and yield levels, knowledge concerning host–pathogen interactions over longer periods of time is very poor. However, a number of general coevolutionary models (Jayakar 1970; Leonard 1977) or ones specifically directed toward the consequences of the use of mixtures in agriculture (Barrett 1980; Marshall and Pryor 1978, 1979) have highlighted the importance of fitness costs associated with resistance in the host and virulence in the pathogen. The existence and magnitude of these fitness costs greatly affect the long-term dynamics of both agricultural and natural host–pathogen systems. Without such costs, pathogen populations should accumulate greater and greater levels of virulence and host populations more and more resistance genes.

The genetic structure of host populations

Attempts to document the existence and magnitude of the fitness costs associated with resistance genes are fraught with difficulty. Like all other genes, those controlling the expression of resistance are an integral part of the genome of a plant and frequently their expression is substantially altered by the genetic background in which they occur (e.g., Dyck and Samborski 1968). Furthermore, the degree to which individual resistance genes are exposed to selection is affected by the breeding system of the plant species in question.

Close inbreeding is a feature of many plants, including most of our major crops, and in such species resistance genes are simply part of multigenic blocks or even whole genomes that are favored or disadvantaged by particular biotic and abiotic environmental conditions (Burdon and Müller 1987; Parker 1988b). In these circumstances, the overall performance of the plant can mask the cost, if any, of individual resistance genes. Only when costs regularly exceed 5–10% are they likely to be recognized as a detrimental feature associated with breeding for disease resistance. This has not been the case and it is, consequently, hardly surprising that the concept of a metabolic cost involved with resistance has attracted little interest from breeders of pure line crop varieties. In the future this will change if molecular biology progresses to the stage of providing raw materials for precise experimental comparisons by inserting individual resistance genes into existing well-adapted backgrounds. In the meantime, however, concerns about the cost of resistance largely remain the domain of those interested in the dynamics of agricultural mixtures and natural populations.

Currently, assessments of the fitness costs associated with resistance will best be made over a number of years in genetically interactive systems. Unfortunately, temporal studies of plant populations are uncommon and those involving a monitoring of resistance are restricted to three. Two of these studies essentially involve mixtures of genetically discreet lines (*Chondrilla juncea*, Burdon et al. 1981; *Avena sativa*, Murphy et al. 1982) and, because of the problems alluded to above, provide no information about the underlying effects of resistance on plant performance in a disease-free environment.

The third study is that by Allard, Webster, and their associates of the evolutionary responses of barley composite cross-populations to *Rhynchosporium secalis* (Jackson et al. 1978; Muona et al. 1982; Saghai-Maroof et al. 1983). For a long time, disease caused by *R. secalis* has been a noticeable feature of these barley populations and extensive investigations have detected substantial temporal changes in the resistance of composite cross II (CCII) to three races of *R. secalis* (Webster et al. 1986). From generation 15 to generation 47 the average frequency of resistance to races 40, 61, and 74 has risen from approximately 0.12 to 0.85. These changes are taken to reflect selection for resistance per se rather than a passive increase resulting from selection acting elsewhere on the genome of this highly inbreeding species (Webster et al. 1986).

Conversely, during the early years of establishment of CCII, *R. secalis* was reportedly very uncommon and by the seventh generation the frequency of resistance to races 40, 61, and 74 was much lower than in the parental lines. Using a knowledge of the resistance present in the original lines Webster et al. (1986) have estimated the frequency of resistance that would have been expected during generations 1 to 13, assuming these resistance alleles were selectively neutral in the disease-free environment then prevailing. These estimated frequency values were substantially higher than those observed in practice. Average selective penalties of 14, 24, and 9% per generation were computed for the alleles associated with resistance to races 40, 61, and 74, respectively!

These values are astonishingly high, presenting a range of enigmatic problems and a number of exciting experimental possibilities. From a strictly practical point of view, if such fitness costs are typical of most resistance genes then some kind of interactive process that substantially reduces their individual effect on yield *must* occur when they are present in combination. Individual barley lines containing the alleles for resistance to both races 40 and 61 of *R. secalis* have been found in CCII (Saghai-Maroof et al. 1983), although the fitness penalty incurred by such individuals is potentially very high (additive model, 38% per generation; multiplicative model, 35%). Similarly, in wheat breeding there has been a tendency to pyramid genes through time. Several varieties grown extensively in the United States, Canada, and Australia have carried large numbers of genes for resistance to stem rust [e.g., Chris (United States), four genes; Selkirk (Canada), six genes; Gatcher (Australia), six genes].

Fitness costs of the magnitude of those reported for resistance in CCII are sufficiently high to allow careful investigation of the relative reproductive performance of homozygous-resistant, homozygous-susceptible, and heterozygous-resistance lines selected from the selfed progeny of heterozygous individuals. Such a comparison based on early generation lines of CCII would provide an independent estimate of the cost of resistance.

In constantly changing situations like those occurring in CCII, the resistance genes favored by selection should be those that counter the pathogen at the lowest possible cost. Furthermore, in the case of particular resistance genes, selection will favor those individuals with genetic backgrounds that minimize the cost. An assessment of the cost of resistance in late-generation lines would consequently provide an interesting comparison with the early generation results. This might give some indication of the interaction between individual resistance genes and the genetic background in which they occur, although, again, the correlated structure of the genome would have a potentially confounding effect on any interpretation.

High costs of resistance also have profound implications for the dynamics of wild plant populations. They are likely to be responsible for massive season-to-season fluctuations in resistance and susceptibility in populations that are tempered only by the frequency and intensity of disease and the longevity of the species and its breeding system. Studies of considerable duration may be necessary to determine even the most general of temporal trends.

Despite the questions raised by the implications of temporal changes in resistance in the barley composite cross-populations, this research is invaluable. It has focused attention on the intriguing but much neglected topic of resistance costs and indicated a variety of ways in which further progress may be made. In addition, this research needs pursuing in populations of outbreeding plants. For example, future experiments might employ deliberately constructed populations of very short life cycle plants (e.g., fast cycling brassicas, Williams and Hill 1986) to test for the confounding effects of correlated changes in the genome.

The genetic structure of pathogen populations

The long-term consequences of host mixtures for the genetic structure of pathogen populations constitute a highly contentious issue. Proponents of the use of mixtures have argued that the frequency-dependent selective effect placed on a pathogen population by the intimate mixing of different host lines each carrying different resistance genes will slow the rate of pathogen evolution. Pathogen populations will be dominated by pathogenically simple races each able to attack only a small proportion of the total mixture. This line of reasoning implies the existence of relatively high fitness costs associated with the possession of virulence. Critics of agricultural mixtures, on the other hand, have claimed that by deploying many different resistance genes, mixtures will enhance the rate of pathogen evolution and foster the rapid emergence of pathogen "superraces" capable of attacking most, if not all, components of the crop. For this argument to be correct, the costs associated with virulence in the pathogen will have to be relatively low or nonexistent. Of course, selection should minimize these costs in the same manner that costs of resistance genes would be minimized in the plant.

To date it has not been possible to reliably measure the costs associated with the possession of virulence in the field. Although interest in mixtures has been increasing, the actual area grown as a proportion of the total of any given crop is still very small. As a consequence, the structure of the pathogen population in any mixture is likely to be strongly affected by the selective pressures placed on the pathogen in surrounding stands of pure lines.

Analogous problems to those encountered in attempts to measure the cost of resistance in inbreeding plants confront most attempts to measure the cost of virulence in pathogens. A large number of pathogens undergo meiotic recombination extremely infrequently. For such pathogens, individual races can be regarded as clonal lines that *may* have very different origins. Comparisons of the reproductive performance of two or more such lines, selected on the basis of virulence differences, may consequently fail to recognize significant differences in other aspects of the genome that affect general fitness. Moreover, even if this pitfall is avoided, it is not clear at what stage in the life cycle or under what conditions selection against virulence is likely to occur. Fecundity selection may have a markedly different effect than viability selection (Marshall et al. 1986), while nothing is known about the effects of virulence on the ability of asexual pathogens to survive through harsh off-season conditions.

Even the experimental design used in most attempts to measure a virulence cost must be questioned. Pathogen races are frequently grown together at high density over a number of asexual generations while changes in the frequency of the different races are monitored. This situation is quite unlike that which normally occurs in both natural and agricultural systems. There, pathogen density is usually too low for individual pustules from different clones to compete directly with one another.

Currently the best way available to examine the cost of virulence is the same as that proposed earlier for directly measuring the cost of resistance in early generations of barley composite cross II. Taking a population of asexual spores derived directly from a cycle of sexual recombination, it is possible to examine the fitness cost associated with particular virulence genes in a randomly reassorted genetic background. Such an investigation could be conducted at either the level of the individual or that of the population. At the population level changes in the frequency of particular virulence genes may be followed over a number of asexual generations. This approach was used by Leonard (1969b) to examine the cost of virulence in *Puccinia graminis avenae*. Over eight successive asexual generations he detected fitness costs ranging between 14 and 46%. However, these measures were made at pathogen densities that are rarely attained under field conditions. While Leonard's work indicates that costs are likely to differ between different virulence genes (Leonard 1969b), this extremely important observation needs to be verified.

The implications that arise from the existence of fitness costs associated with virulence do not stop at the farm gate. Such high costs would select against pathogen races that could overcome large numbers of host resistance genes, favoring instead less virulent races. In turn, this affects patterns of virulence within the pathogen population and has extremely important implications for the long-term dynamics of natural host–pathogen systems. As noted earlier, our knowledge of the structure of natural pathogen populations is extremely limited, but what evidence we do have suggests that even with apparently highly mobile, airborne pathogens, population structures may vary markedly over very short distances. A study of the structure of *Melampsora lini* populations attacking nine populations of *Linum marginale* in the Kosciusko National Park, New South Wales, detected four common races (A, E, K, and N) of the pathogen (Burdon and Jarosz 1988a, and unpublished). Considerable resistance to races E, K, and N was found among the plants within the Park, yet these races attained a frequency of 15% or more at four, six, and eight sites, respectively. In contrast, no resistance to the fourth race (A) was detected, yet this race was common at only two sites. At one of these sites, the *L. marginale* population possessed high levels of resistance to races E and K and here, the pathogen population was dominated by race A (>70% of isolates).

The restricted pattern of distribution of the highly virulent race A is consistent with the hypothesis that there are costs associated with virulence. However, conclusive proof of this has yet to be obtained.

In the long term, these and similar studies will provide vital insights into the functioning of natural pathogen populations. In the meantime, however, it is still necessary to rely on the implications of highly simplified genetic models. One interesting feature of these is the apparent inverse correlation between the ecological complexity of the model and the magnitude of the selective coefficients necessary to prevent domination of the pathogen population by a superrace. In a model incorporating spatial aspects of pathogen spread, Barrett (1980) found that a departure from a simplified view of all spores being distributed at random, to one in which variable proportions reinfect the plant on which they were produced, could result in a pathogen population of only intermediate virulence even when fitness costs were relatively low. This result raises the intriguing possibility that fitness costs associated with virulence that are capable of maintaining a diverse and polymorphic pathogen populations in the wild may be insufficiently high to prevent the evolution and dominance of highly virulent races in the simplified systems of modern agriculture.

CONCLUSIONS

Several questions must be answered to generate a deeper understanding of the evolutionary dynamics of natural plant–pathogen systems and to develop more sophisticated strategies for disease control in agriculture. Aspects of three of these questions—the epidemiology of disease in mixed stands, and the existence and magnitude of the fitness costs associated with resistance in the host and virulence in the pathogen—have been considered in this study.

The adoption of mixtures as a strategy of disease control strategy in agriculture has been slow. While a number of scientific, economic, and sociological factors have contributed to this, the disappointingly small yield increases that have been associated with massive reductions in disease levels in mixtures have done little to encourage their use. However, a clear understanding of the geometry of mixtures and the ways in which they reduce disease (Figure 1) strongly suggests that the yield advantages conferred by mixtures in the presence of disease are likely to be underestimated by the methods currently used to measure them.

Host–pathogen associations are dynamic interactions involving one highly mobile partner and considerable attention needs to be given to experimental design to ensure that this will truly reflect the benefits to be obtained from mixing different plant genotypes. In general, (1) plots will need to be much larger than those used in pure-line breeding trials; (2) the distribution and intensity of initial infections by the pathogen need to reflect more accurately those likely to be encountered in the field; and (3) care needs to be taken to ensure that the correct controls are included in all trials. In an analogous way, the effects of polymorphic host populations on the development of disease in natural plant–pathogen associations are also markedly influenced by interactions between the size of individual host populations and the distribution and number of initial disease foci.

The consequences of fitness costs associated with resistance in the host and virulence in the pathogen cannot be underestimated for either natural or agricultural host–pathogen interactions. While reliable estimates are sparse, models of both these systems require the existence of such fitness costs to prevent pathogen races with many virulence genes from dominating pathogen populations, or the never-ending accumulation of resistance genes in wild host plant populations.

To date the fitness costs involved in the use of resistance genes in agriculture have largely gone unnoticed. However, as the insertion of resistance genes becomes increasingly precise, the existence and importance of these costs will be pushed into greater prominence. This will affect their use in both pure line and mixture systems. The inherent costs of each resistance gene and the possibility that these may vary between different genes (Burdon and Jarosz 1988) obviously will play an important role in decisions concerning their use. Similarly, a deeper understanding of the costs associated with virulence and the possibility that certain combinations of virulence are particularly "expensive" for the pathogen to sustain also has the potential to make a major contribution to the successful control of disease in agriculture. This will especially be the case when combinations are used appropriately in mixtures where disease pressure is already low.

CHAPTER 14

SPATIAL PATTERNS OF GENETIC VARIATION WITHIN PLANT POPULATIONS

B. K. Epperson

ABSTRACT

The spatial pattern or structure of genetic variation within populations is an important part of evolutionary and ecological genetic processes in natural populations of plants. Genetic structure also affects empirical studies of the genetics of natural populations of plants, and is important for sampling populations for gene conservation or breeding purposes. This chapter reviews recent advances in statistical methods that use spatial patterns of genetic variation to characterize genetic structure. A growing number of mathematical and statistical studies are showing how different amounts of dispersal and different types of natural selection change spatial patterns. The chapter reviews theoretical results, and puts forth a hypothesis-testing framework and sampling strategies for using spatial patterns of populations in order to make direct inferences on natural selection and gene flow. This discussion includes novel methods to measure associations between spatial patterns of genetic variation and spatial patterns of microenvironmental heterogeneities. These methods separate the influences of microenvironmental selection from those of limited dispersal on these associations.

INTRODUCTION

Importance of spatial patterns of genetic variation

Spatial patterns of genetic variation within populations figure prominently in evolutionary and ecological processes in natural populations of plants. The spatial distribution of genetic variation in populations was an integral part of Sewall Wright's view of evolution (e.g., Wright 1943, 1978), as well as theories of allopatric or parapatric speciation (e.g., Mayr 1970). Even when the scale of the genetic structure is very fine, structure promotes formation of novel genotypes locally (Wright 1978). Small, partially isolated demes may become sharply differentiated, despite being part of a large ensemble. Spatial structure of genetic variation within populations has immediate ecological–genetic consequences for populations of plants, because it conditions location of genotypes, and survival depends, in large part, on local environmental factors like soil type (Bradshaw 1984) and light intensity. Genetic structure also affects mating when pollen dispersal is limited (Wright 1943).

Spatial genetic structure, per se, is also important for strategies of sampling natural populations, either for gene conservation or breeding purposes (e.g., Marshall and Brown 1975; Brown 1978; Moran and Hopper 1987). Knowledge of structure can improve efficiency of sampling to maximize gene diversity or minimize consanguinity. Structure also affects studies of the genetics of natural populations, and, if ignored, may lead to, for example, misinterpretations of the population breeding system or the selective significance of genotypes. For example, samples collected from areas in which allele frequencies for different loci are spatially correlated produce linkage disequilibrium (D) (Prout 1973; Feldman and Christiansen 1975). Many studies have found evidence of consanguineous outcross matings due to limited pollen dispersal over genetically structured areas. Structure-sensitive single-locus estimates of outcrossing rates are lower than the relatively insensitive multilocus estimates (Ennos and Clegg 1982; Shaw et al. 1981; Shaw and Allard 1982; Ritland 1985; Bijlsma et al. 1986; for an exception, see Epperson and Allard 1984). Structure can cause sample allele frequencies (p) and inbreeding coefficients (F) to differ from reference populations. Moreover, the usual calculations of standard errors of many estimates of parameters (e.g., F, D, p) require the assumption of independence of samples, and this is violated to some degree in samples from structured populations (see, e.g., Cliff and Ord 1981). Thus, in many cases it is important to collect samples either in such a way that spatial correlations are avoided or so that spatial statistics can be used to quantify structure.

Origins of structure

One of the better characterized spatial processes is isolation by distance for selectively neutral genotypes. In this case, mating by proximity results in inbreeding within demes and spatial variation in allele frequencies (see Wright 1943, 1965; also Rohlf and Schnell 1971; Weir and Cockerham 1984). The coefficient of kinship (the probability that two genes each from different individuals are identical by descent) generally decreases exponentially with distances of separation (Malécot 1948, 1973), and so does spatial autocorrelation of allele frequencies (Barbujani 1987). Populations with low to moderate levels of dispersal develop marked patch structure, where patches, or areas that are large homogeneous concentrations of homozygotes, are separated by zones containing mostly heterozygotes (Turner et al. 1982; Sokal and Wartenberg 1983; Sokal et al. 1989). Self-fertilization effectively limits dispersal (Allard 1975) and affects spatial distributions in a manner similar to mating by proximity (Wright 1946).

Microenvironmental selection can create sharp patterns of genetic variation in plant populations (Hamrick and Allard 1972; Allard et al. 1972). Some examples include many plant populations that have considerable dispersal, but are located on and around mine tailings (see review by Bradshaw 1984). This is just one example of how spatial structure and selection can interact to allow populations to track microenvironmental variation genetically.

Studies of structure

The more detailed spatial pattern analyses of large-scale structure have been conducted by Sokal and colleagues on a number of animal species (Sokal and Oden 1978a,b; Sokal and Menozzi 1982; Sokal et al. 1986, 1987; Sokal and Winkler 1987; Sokal 1988). Few studies have used spatial statistics for analysis of fine-scaled structure within populations (Clegg and Epperson 1985; Epperson and Clegg 1986; Epperson and Allard 1989; Dewey and Heywood 1988; Wagner et al. 1989); nonetheless, these studies have demonstrated several advantages of spatial autocorrelation methods. Spatial autocorrelation statistics are particularly informative and, because these methods include all pair comparisons in samples, they have high statistical power.

Structure has theoretical, empirical, and practical importance in plant population genetics. Powerful statistical methods recently have been tied to theoretical results on spatial processes of population genetics. Thus, it is timely to review these methods and their relationship to population genetics. This chapter presents an overview of spatial processes in terms of spatial autocorrelation and other spatial statistics. The first section presents a brief review of the methods, calculation, and interpretation of spatial statistics, including important statistical models that relate spatial patterns of genetic variation to spatial patterns of microenvironmental heterogeneity. The second section reviews theoretical results on the development of spatial patterns under different scenarios of gene flow and natural selection. Temporal correlations are also briefly considered. The final section is a summary of sampling strategies for measuring structure, and further strategies for using spatial pattern analyses to study different forces that act in spatial processes in natural populations.

PATTERN ANALYSIS

A spatial map of a sample of genotypes contains a great amount of information. Simple visual inspection of such maps may be informative but may also be misleading, because visually one may "pick out" patterns or associations of genotypes that are not different from sampling of a truly random distribution. In this section, objective and statistically powerful methods for describing fine-scaled patterns of genetic variation are reviewed, first for spatial samples of individual genotypes and then for collections of allele frequencies in subsamples. These methods include measures of correlations as functions of distance and other types of spatial correlations. Finally, novel methods are outlined for measuring the relationship between spatial distributions of genetic variation and spatial heterogeneity of microenvironments. With these methods, the effects of limited gene flow are separated from the effects of selection in a way that allows spatial patterns to be used to measure selection.

Genotypic data

Spatial statistical analyses for individual genotypes begin with a spatial map of a sample (of size n) genotypes; for incomplete censusing it is convenient to arrange such samples on a regularly spaced grid or lattice. For example, Figure 1 shows a lattice of point samples for a dominant blue versus recessive pink flower color polymorphism from a population of the bee-pollinated common morning glory, *Ipomoea purpurea* (data from Epperson and Clegg 1986).

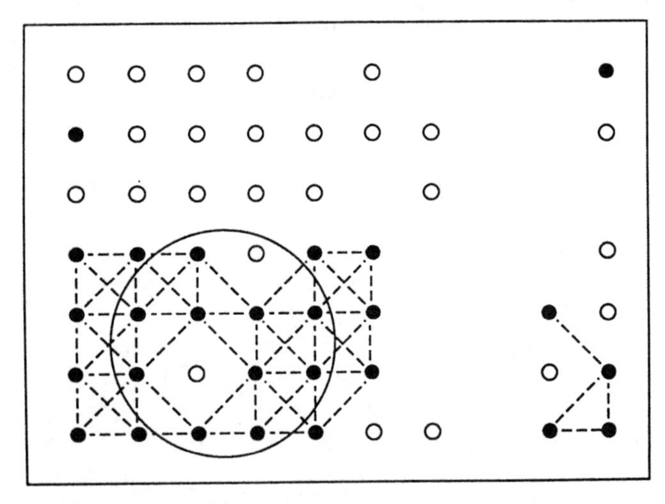

FIGURE 1. Lattice of point samples for a single-locus flower color polymorphism from population number 8312 of *Ipomoea purpurea* (from Epperson and Clegg 1986): ( ) dominant blue morph; ( ) recessive pink morph. Blanks are lattice points that fell in unoccupied areas. Overall, approximately 1 in every 10 plants within the lattice area was sampled. (——) indicates joins between two pink sample types that are separated by a distance, d, of 1 or $\sqrt{2}$ lattice units, i.e., $k - 0.5 \leq d < k + 0.5$, where $k = 1$. The circle has diameter equal to the X-intercept of the SND-correlogram for pink × pink joins (see Figure 2).

ERRATUM

e regret that due to a production error the following captions were complete as printed:

ige 232

GURE 1. Lattice of point samples for a single-locus flower color polymorphism m population number 8312 of *Ipomoea purpurea* (from Epperson and Clegg 1986):) dominant blue morph; (●) recessive pink morph. Blanks are lattice points that fell unoccupied areas. Overall, approximately 1 in every 10 plants within the lattice ea was sampled. (– – –) indicates joins between two pink sample types that are separated by a distance, d, of 1 or $\sqrt{2}$ lattice units, i.e., $k - 0.5 \leq d < k + 0.5$, where $= 1$. The circle has diameter equal to the X-intercept of the SND-correlogram for nk × pink joins (see Figure 2).

ige 234

GURE 2. Correlograms for standard normal deviate (SND) join-count statistics for e sample from population 8312 (see Figure 1): (○) pink × pink joins, (●) ue × blue joins, and (□) pink × blue joins.

own & Clegg: PLANT POPULATION GENETICS, BREEDING, AND GENETIC RESOURCES

nauer Associates, Inc.

Spatial autocorrelation analysis proceeds by adopting criteria to form subsets (or classes) of the $n(n - 1)/2$ total number of pairs of points. The genotypes at both points of a pair define a join. Thus, $n_{ij}(k)$ is the number of joins between genotypes i and j for class k. Usually the criteria for grouping of pairs is based on ranges of the Euclidean distances of separation between pairs of locations. An example of equally ranged distance classes is the case in which each distance class k contains all pairs of points separated by distance d, where $k - 0.5 \leq d < k + 0.5$ lattice units [i.e., the shortest distance between (neighboring) lattice points]. In the example in Figure 1, the number of pink–pink joins for $k = 1$ is 56.

Test statistics for the presence of autocorrelation are usually based on the null hypothesis, H_0, that the sampling distribution of the numbers of joins is that produced by sampling pairs without replacement from the sample genotypes, i.e., that the locations of sample genotypes are randomized. Thus, for any class k the expected number of joins between genotypes i and j, under H_0, is $u_{ii} = Wn_i(n_i - 1)/2n(n - 1)$ and $u_{ij} = Wn_in_j/n(n - 1)$ for $j \neq i$. Here n_i is the number of times genotype i occurs in the sample, and (suppressing k) W is twice the number of joins in total (i.e., between all pairs) for class k (for nonbinary weightings W is the sum of weights between locations l,m and m,l, for all $l \neq m$). The standard errors, $SE_{ij}(k)$, under H_0 are given by Sokal and Oden (1978a) and Cliff and Ord (1981). Under H_0 the test statistic $SND = (n_{ij} - u_{ij})/SE_{ij}$ has an asymptotic standard normal distribution (Cliff and Ord 1981). Spatial patterns of populations can be inferred directly from the sample distributions (Ord 1980), except for patterns with scales smaller than the distance between nearest sample locations. Slightly different expected values are obtained for a different null hypothesis, sampling with replacement (Sokal and Oden 1978a; Cliff and Ord 1981; Ripley 1981). This hypothesis requires that the sample distributions for each point are uniform and known; hence, it is rarely used.

Significance tests based on SNDs and the random hypothesis, H_0, generally have high statistical power (Cliff and Ord 1981), and SNDs for short distances appear to be particularly sensitive to most forms of structure. Isolation by distance causes large positive SNDs for joins between like genotypes at short distances. In the morning glory example, SNDs for pink–pink joins and blue–blue joins are positive for distance classes 1, 2, and 3 (Figure 2). In addition, other null hypotheses could be tested, but this is difficult in practice because the expected standard errors are complex functions of spatial patterns (Cliff et al. 1975). A null hypothesis of random distribution may be unrealistic for some populations; in such cases, SNDs are useful solely as measures of correlations.

Some special problems are encountered in using join counts to infer the structure of distributions of diploid genotypes. Spatial correlations can be quite different among different genotypes. For example, under isolation by distance for a locus with four alleles (A_1, A_2, A_3, A_4), we would expect either

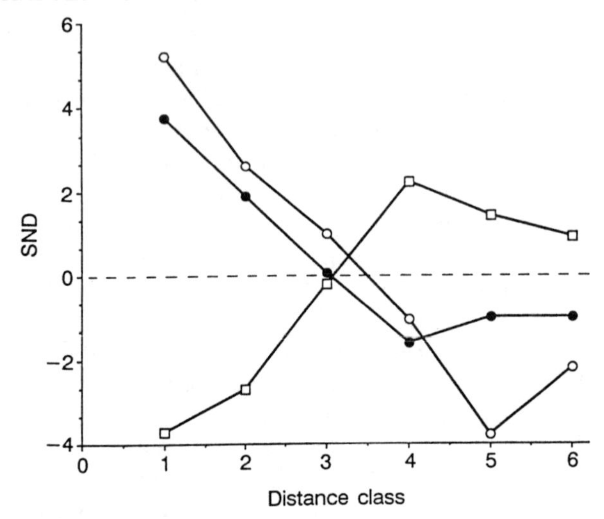

FIGURE 2. Correlograms for standard normal deviate (SND) join-count statistics for the sample from population 8312 (see Figure 1): ( ) pink × pink joins, ( ) blue × blue joins, and ( ) pink × blue joins.

a slight excess or a slight deficit for joins between genotypes $A_1A_2 \times A_1A_1$ at short distances (see Figure 3) and a much larger deficit for joins between $A_1A_1 \times A_2A_2$ or $A_1A_2 \times A_3A_4$. Different types of joins can be combined into the same test statistic. For example, the number of joins between all genotypes that contain a rare allele may be particularly informative, and this can be done by redefining the nominal types. If the desired sum of joins cannot be achieved by redefining types (e.g., sum of joins between like homozygotes), then the expected covariance between types of join under H_0 must be accounted for (Krishna Iyer 1949, 1950). For the total number of joins between unlike types, Cliff and Ord (1981) give the expected number and variance under H_0. The total number of unlike joins is an informative summary measure of the "diversity" as a function of distance (David 1971), and, for diploid genetic data, it is closely related to Jacquard's (1973) measure of genetic distance (probability that two individuals have different genotypes). For data sets from polymorphic populations that nonetheless contain very few heterozygotes, or for haploid data sets (Epperson and Allard 1989), genetic identity bears a direct inverse relationship with the total number of unlike joins.

Additional interpretations of structure can be made from a correlogram, the set of SNDs for all mutually exclusive distance classes. The most interpretable result is where SNDs for joins between like genotypes (e.g., $A_1A_1 \times A_1A_1$) are positive for short distances. The distance (interpolated between distance classes) at which the correlogram becomes negative, referred to as the X-intercept, measures the scale of autocorrelation of each genotype. In particular the X-intercept for joins between like homozygotes is an operational estimate of the diameter of patches of homozygotes (Epperson and Clegg

1986). In the morning glory example, the X-intercept for joins between pink homozygotes is 3.5 (Figure 2), and this roughly corresponds to the width of patches (see Figure 1). In this population, about 1 of 10 individuals were sampled. This suggests that the number of plants in each patch was on the order of 200–400. It must be noted, however, that such estimates are unreliable for small samples. In addition, if the number of joins is highly variable between distance classes, so too will be the power of the SND tests and SNDs for some distance classes will be more likely to have negative values by chance sampling.

Allele frequency data

For spatial statistical analyses of allele frequencies the number of subsamples (quadrats) must be fairly large and spatially arranged either on a lattice or other fairly uniform spacing. The simplest measure of distance between quadrats is the Euclidean distance between quadrat centers. However, if population density is very irregular or "clustered," and long-term gene flow is likely to pass through adjacent intervening clusters, then distances may be measured along connections between quadrats. Different criteria for connectedness include Gabriel connectedness (Gabriel and Sokal 1969; Cliff and Ord 1981; Tobler 1975). Connectedness schemes can also be modified to account for other barriers to gene flow.

If pairs of quadrats are classified into distance classes (using sets of 0 or 1 weights), then the unweighted Moran I-statistic measures the correlation in allele frequencies, p_i, for pairs in the distance class relative to the variation among all locations. Thus, $I = n\Sigma_{ij\epsilon k} Z_i Z_j / W \Sigma_i Z_i^2$, where $Z_i = p_i - \bar{p}$, and $\bar{p}$ is the mean allele frequency of all n quadrats, and W is the sum of weights, or twice the number of pairs of quadrats in class k. Under the randomization hypothesis I has expected value $\mu_1 = -1/(n - 1)$. The variance μ_2 is given for example in Sokal and Oden (1978a) and Cliff and Ord (1981). If the number of quadrats is fairly large then the statistic $(I - \mu_1)/\sqrt{\mu_2}$ has an approximate standard normal distribution (Cliff and Ord 1981) under the randomization hypothesis.

Correlograms of I-statistics for mutually exclusive and exhaustive distance classes can be tested as a whole for significant deviation from a null hypothesis of random distribution (Oden 1984). As yet there are apparently no statistically powerful tests for differences between correlograms from different data sets, for example, frequencies of different genes, but cluster analysis can be used to group correlograms based on the Manhattan distances between them (e.g., Sokal and Wartenberg 1983). As for nominal data, the X-intercept of I correlograms can be used to infer the scale of spatial autocorrelation and isolation by distance (Barbujani 1987). In addition, modified I-statistics can be used to test the fit of observed maps to theoretical maps (Cliff and Ord 1981).

As was alluded to earlier, I-statistics can be based on any specified scheme of weights, w_{ij}, between locations i and j, for all $j \neq i$ ($w_{ii} = 0$), $I = n\Sigma_{i \neq j} w_{ij} Z_i Z_j / (W \Sigma_i Z_i^2)$, where $W = \Sigma_{i \neq j} w_{ij}$. The problem is in specifying the weights. In addition, the Mantel statistic (Mantel 1967; Cliff and Ord 1981) tests the independence of two matrices, one containing values of any specified measure of differences in allele frequencies [e.g., Nei's genetic distance (Nei 1973a)], and a second matrix containing values of any specified measure of spatial distance or relationship (e.g., Euclidean distance, distance-squared, distance along Gabriel connections). Spatial data can also be fitted to and/or tested against several statistical models, such as pure autoregression models and moving average models. Each of these models requires assumptions regarding the form of the spatial autocorrelations; thus, spatial autocorrelation analyses can provide preliminary information for building these models.

Spatial correlations of genetic variation may not be simple functions of distance; one common situation is where dispersal (or microenvironmental selection) has a directional component. For example, pairs of points that are separated primarily along an axis parallel to the predominant direction of dispersal will have higher correlations. One method used to detect directionality classifies pairs of points by their respective locations in subareas defined by sectored concentric rings of an encircled total sample area (Oden and Sokal 1986; Sokal et al. 1986, 1987). Other methods require knowledge of the predominant axis of directional effects. One method computes "distances" as a function of the cosine of the angle formed between the line connecting two points and the directional axis, and the physical distance of separation (Upton and Fingleton 1985). Directionality can also be included in weighted *I*-statistics and the Mantel statistic.

Correlation of genetic variation with environmental heterogeneity

Microenvironmental selection may be detected by measuring correlations between spatial distributions of genetic variation and spatial distributions of microenvironments. However, correlations caused by selection may be confounded with the effects of limited gene flow in several ways. First, mating by proximity produces a patch structure overlay that alone could cause large correlations in small-scaled samples. Gene flow may also blur spatial differences of genetic variation near the boundaries of different environment zones, thereby reducing genetic–environment correlations to statistically nonsignificant levels. Limits on gene flow may produce spatial autocorrelations within an environment, resulting in biased standard errors of estimates of allele frequencies and other parameters within each zone (thus, for example even simple *t* tests between two zones are inflated; Cliff and Ord 1981; Upton and Fingleton 1985).

For allele frequency data, these problems can be addressed by using multiple regression models where the expected values incorporate location- dependent factors (which can be interpreted as being caused by environmental heterogeneity), modified by either an autoregression or autocorrelation component that represents interactions between locations (interpreted as being caused at least in part by mating by proximity). The choice between these models depends on how locations interact through gene flow. Nonetheless, these methods offer several advantages over simpler methods often used in population genetics, because they partition the effects of limited dispersal from those of selection.

In spatial analyses the standard linear regression model expressed in vector format, $\mathbf{Y} = \mathbf{Xb} + \mathbf{e}$, is interpreted as a vector $\mathbf{Y}$ of spatially located observations on genetic values, Y_i, $i = 1,n$, which are a function of an $n \times m$ matrix $\mathbf{X}$ that gives the values of m variates (environmental factors) for each of the n locations. The vector $\mathbf{b}$ gives the coefficients, and the vector $\mathbf{e}$ gives the error terms for all locations. The variates X_i, $i = 1,m$, could represent either known (or at least precisely measured) linearized environmental factor values, or the presence or absence of some environmental factor. For many statistical analyses the e_i are assumed to be identically and independently distributed normal random variables with mean 0 and equal variance σ^2 (i.e., e_i are homoscedastic). If the sampling distribution of each quadrat allele frequency, p_i, is approximately binomially distributed, then the logit transformation will give approximate normality (but the variance will depend on quadrat sample size and p_i); the arcsine transformation will give a distribution that is approximately normal and homoscedastic.

One important aspect of multiple regression is analyses of autocorrelation among the residuals, $\hat{\mathbf{e}} = \mathbf{Y} - \mathbf{X}\hat{\mathbf{b}}$ (where $\hat{\mathbf{b}}$ is the vector of estimates of $\mathbf{b}$), which may indicate either unrecognized variates (Upton and Fingleton 1985) or the influence of gene flow. [*I*-statistics for $\hat{\mathbf{e}}$ have properties similar to those for Z_i (see above), and they are calculated the same way ($\hat{e}_i$ replaces Z_i); however, the expected values under H_0 may differ (Cliff and Ord 1981).] The form of spatial correlations of residuals may distinguish between these two causes. For example, a knowledgeable investigator might inspect a map of the residuals and find associations with another environmental factor.

One way to incorporate statistical interaction effects into regression models is to consider interactions between the Y_i values themselves, by adding an autoregression component. The most tractable model is the linear autoregression–regression model (AR model) with *simultaneous* autoregression. Then in vector format $\mathbf{Y} = \rho\mathbf{WY} + \mathbf{Xb} + \mathbf{e}$, and $\mathbf{W}$ is an $n \times n$ matrix of weights (which measure the relative weights of statistical interactions between the Y_i), and the scalar ρ measures the overall strength of interactions (see Cliff and Ord 1981; Upton and Fingleton 1985). A major problem of fitting AR models is choosing the weights w_{ij}. For population genetic processes, the w_{ij} are not direct measures of gene flow from location i to j, rather they are more complex functions of the spatial process itself. However, even some simple weighting schemes may work reasonably well; for example, where $w_{ij} = 1$ for nearest neighbors and $w_{ij} = 0$ otherwise. Once the weighting matrix is specified, maximum likelihood estimators (MLEs) of $\mathbf{b}$ and ρ can be found, and the MLEs can then be tested for significance (Cliff and Ord 1981; Upton and

Fingleton 1985) using large-sample theory (Rao 1973). Moreover, the residuals of the autoregression–regression model can be investigated—a random distribution of residuals would indicate a good fit.

An alternative model is linear regression with autocorrelated errors (RAE) where the "interaction" between locations is specified in the error term of regression models. The most tractable model is the *simultaneous* model with normality assumptions, for which $\mathbf{Y} = \mathbf{Xb} + \rho\mathbf{We} + \mathbf{u}$, where the elements μ_i are independently distributed normal $(0, \sigma^2)$. Once $\mathbf{W}$ is specified, then the parameters $\mathbf{b}$, σ^2, ρ can be estimated and tested for significance using maximum likelihood methods, and the models can be tested for goodness of fit (see Cliff and Ord 1981; Upton and Fingleton 1985).

To my knowledge, such models have not been applied in plant population genetics. Several detailed examples of the application of RAE and AR models are presented in Upton and Fingleton (1985). One data set is the logit transformed proportions of individuals with blood group A in Irish counties (Hackett and Dawson 1958). In this analysis, the location-dependent factor was a measure of English influence, and represented not an environmental selection agent, but different amounts of immigration from England into the different Irish counties. The autoregression or autocorrelation component measured the effects of local migration between Irish counties. Using relatively simple schemes of weights, Upton and Fingleton (1985) found that both the RAE and AR models fit the data well. Both models showed that interactions between counties were large, and the effects of English influence were also large (see also Sokal and Oden 1978b).

SPATIAL AND TEMPORAL CORRELATIONS IN MICROEVOLUTIONARY PROCESSES

Different forms of selection and different amounts of dispersal together produce different spatial patterns and temporal correlations. This section reviews theoretical results and develops a framework of hypotheses that is amenable to spatial–temporal statistical analyses. Low-to-moderate dispersal produces distinctive spatial patch structures and genetic isolation by distance for selectively neutral loci. In contrast, strong selection in a form where fitness does not depend directly on location produces a nearly random distribution. Different types of microenvironmental selection cause distinct changes in structure. Temporal correlations of genetic variation also differ for different processes for selectively neutral versus adaptive polymorphisms.

Spatial distributions under mating by proximity and neutrality

Monte Carlo simulations show that for selectively neutral loci, severely restricted dispersal results in the development of concentrations or "patches" of homozygotes, up to the exclusion (or near exclusion) of other genotypes

(Sokal and Wartenberg 1983); different patches containing different homozygous genotypes are separated by zones where heterozygotes predominate (Turner et al. 1982). These features are reflected in the spatial statistics of distributions produced in simulated populations (Turner et al. 1982). For example, Figure 3 shows typical SND-correlograms of join counts for Monte Carlo simulations of near-neighbor pollination after 200 generations. In this case, pollen parents were chosen with equal probabilities from the eight nearest neighbors or self (hence, selfing rate is one-ninth), and there was no seed migration. Correlograms for AA × AA and *aa* × *aa* join-counts remain positive for X-intercepts of about 20 lattice units. The correlogram for AA × *aa* is negative at short distances and increases to near zero also at about 20 lattice units. The correlogram for A*a* × A*a* indicates that A*a* genotypes are slightly autocorrelated for short distances, and this suggests that heterozygotes are concentrated into smaller zones that separate homozygous patches. The general shape of SND correlograms for AA × AA, *aa* × *aa*, and AA × *aa* are probably robust to moderate changes in the levels of mating system (see below) but correlograms for AA × A*a*, *aa* × A*a*, and A*a* × A*a* may be expected to differ under slightly different models, especially if the inbreeding

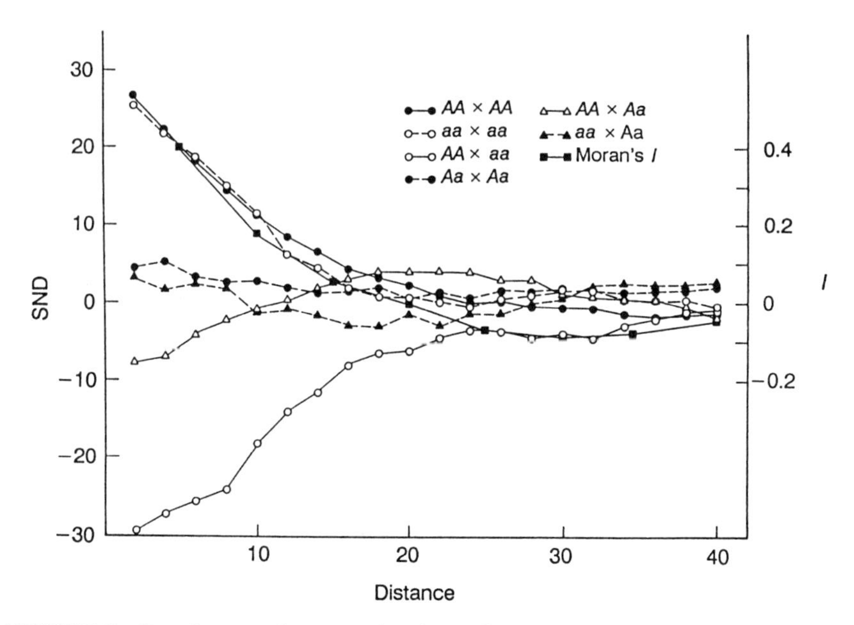

FIGURE 3. Correlograms for a simulated population with near-neighbor pollination (mating system Model 1; see text) at generation 200. Correlograms of SND statistics for each type of join-count (e.g., AA × *aa* for joins between genotypes AA and *aa*). Also, correlogram of *I*-statistics calculated on the 400 quadrats.

coefficient differs. Moreover, SND statistics for the latter three types of joins are likely to be nonsignificant in samples with realistic sizes (see e.g., Epperson and Clegg 1986).

Figure 3 also shows for the same simulation the *I*-correlogram calculated after partitioning the 100 × 100 genotype locations into a new "surface" of 400 allele frequencies in 5 × 5 contiguous quadrats. The X-intercept of the *I*-correlogram is approximately equal to those of the SND correlograms for AA × AA and *aa* × *aa*. This suggests that the X-intercept of the *I*-correlogram circumscribes only the zones of concentrations of homozygotes, and this is supported by visual inspection of the surfaces. The exact relationship of autocorrelation of homozygotes to the X-intercepts of *I*-correlograms may differ somewhat for populations with different dispersal parameters.

The process of patch development in the simulated populations of Sokal and Wartenberg (1983), Sokal et al. (1989), and the author was summarized by periodically partitioning the genotypic surfaces into the same scheme of quadrats mentioned above. Thus, the *I*-correlograms calculated can be directly compared among the various models studied in these papers (see Table 1). My models included, in addition to the model mentioned earlier with pollination from the eight nearest neighbors and self ($N_p = 9$) (Model 1), a model with pollen dispersal with $N_p = 25$ (again, no seed dispersal) (Model

TABLE 1 Dispersal parameters for simulation models of pure isolation by distance.[a]

Model	N_p	N_f
1	9	—[b]
2	25	—[b]
3	9	9
4	25	25
5	49	49
6	81	81
7	121	121
8	10,000	10,000

[a]Pollen parents (or male parents) were chosen with equal probability from the nearest N_p "neighbors" including self. Female parents chosen likewise from the nearest N_f "neighbors" including self.

[b]The female (ovule) parent was always chosen from plant at location.

2). Three different models were included in Sokal and Wartenberg (1983): (1) migration of both female and male parents from nearest 9 neighbors including self ($N_p = 9$; $N_f = 9$) (here denoted Model 3); (2) same as Model 3 but with $N_p = N_f = 25$ (Model 4); and (3) random mating among all 10,000 parents (Model 8). In addition, Sokal et al. (1989) studied three models like Model 3 except with $N_p = N_f = 49$, 81, and 121, respectively, for models 5, 6, and 7. In all of these models except the random mating model (Model 8) substantial autocorrelations of allele frequencies developed within 10 generations. The X-intercepts, as well as the magnitudes of the I values for short distances, increased substantially up to about generation 100, then remained almost constant (on averages) up to termination at generation 200. These and other results suggest that the spatial genetic structure of such populations has stabilized by about the 100th generation (Sokal and Wartenberg 1983).

Figure 4 shows the average correlograms for several replicate simulations of each of the eight models at generation 200 for the first five distance classes (classes were formed similarly in all three studies). This figure shows several differences between the models that were apparent in earlier generations as well. As the level of dispersal increases up to that in Model 5 the X-intercepts (not shown) increase slightly and the values of I increase substantially; as dis-

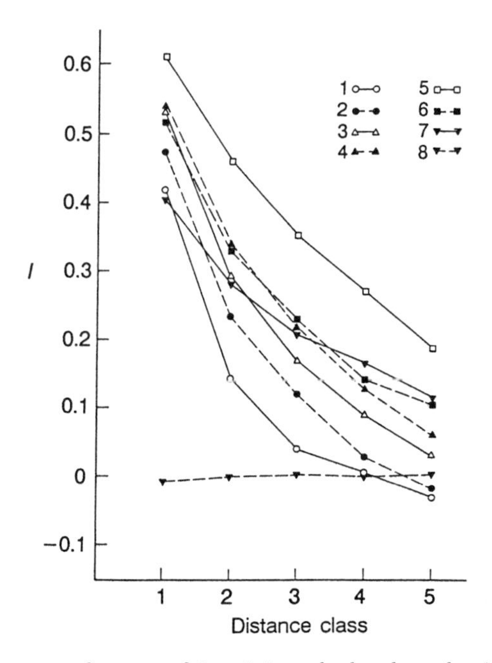

FIGURE 4. Average correlograms of I-statistics calculated on the 400 quadrat system for sets of simulations with different mating systems (Models 1–8) at generation 200. Dispersal is generally greater for larger model numbers. See text for model specifications. Adapted from Sokal and Wartenberg (1983) and Sokal et al. (1989).

persal increases further, the amount of autocorrelation decreases (Sokal et al. 1989). Thus, it appears that as dispersal increases the patches of homozygotes become slightly larger and increasingly diffuse. We expect this trend to continue for even greater levels of dispersal, until there is virtually no identifiable patch structure of homozygotes. This is supported by Wright's (1943) results that showed that the variance of allele frequencies among subpopulations is very small for neighborhood sizes, $N_e > 200$, and is virtually zero for $N_e = 1{,}000$. In addition it is noteworthy that autocorrelation did not differ among simulations with different allele frequencies (Sokal et al. 1989).

The form of spatial distributions produced under pure isolation by distance is characterized by Malécot's (1948) theoretical results on the a priori kinship coefficient, ϕ_{ij} (the probability of identity by descent for two genes, each taken randomly either from different individuals i and j or from different subpopulations i and j, relative to the ancestral founding population). These theoretical results may be particularly important in studies of populations with high levels of gene flow, where homozygotes of neutral loci may not be distributed in sharply differentiated patches. Malécot (1948, 1955) showed that the relation of a priori kinship with distance of separation, d, has the general form

$$\phi(d) = d^{-c}ae^{-bd} \tag{1}$$

The constants a, b, and c are positive. As is discussed below the I-correlogram bears a close relationship to $\phi(d)$, which itself cannot be observed (Barbujani 1987).

The constants a, b, and c of Equation 1 may vary for different specifications of gene flow processes. Most of the theoretical models of kinship by distance under limited gene flow have several features in common, including (1) dispersal is a function of distance alone and is direction invariant; and (2) a parameter m is included as the linearized systematic pressure resulting from random mutation, long-distance migration, or immigration from outside the whole population, and/or selection. The latter feature allows the population to reach equilibrium. The models differ in certain details including (1) infinite or variable alleles (e.g., Nei 1973b) versus fixed alleles (e.g., Kimura and Weiss 1964; Morton 1973a,b), (2) probabilistic (Malécot 1973) versus recurrent or deterministic transitions between generations (Imaizumi et al. 1970), and (3) continuous versus discrete distribution of and migration between subpopulations. Examples include the stepping stone (migration between adjacent subpopulations only) and island (migration equal between all pairs of subpopulations) models.

The results of some early studies suggested that c depended on the number of dimensions of the populational field and on the scale of distances between subpopulations under consideration (Maruyama 1973). However, recursion of various stepping stone models shows that for small distances

$c \cong 0$ for all dimensionalities (Imaizumi et al. 1970) so that Equation 1 is a simple exponential function:

$$\phi(d) = ae^{-bd} \tag{2}$$

The values of a and b generally depend on m and the standard deviation of dispersal distances, σ, and the evolutionary population size N (Imaizumi et al. 1970; Morton 1982). For example, in the model of Malécot (1955) $a \simeq 1/(4\,Nm + 1)$ and $b \simeq (\sqrt{2m})/\sigma$ at equilibrium (Morton 1982).

Morton (1973c) claims that the small probabilities associated with changes in kinship through migration events may be difficult to simulate, unless the paths of descent of all genes are recorded throughout the process. Nonetheless, the spatial patterns of genetic variation produced in the simulations of Sokal and Wartenberg (1983) do fit Equation 2 (Barbujani 1987). Kinship can also be predicted for specific populations for which there are experimental measurements or other data on demography or genealogy (Morton 1973d,e), however the information must be very reliable because errors will affect the predicted $\phi(d)$.

Genetic data can be used to estimate only the *conditional* kinship (i.e., the kinship relative to the extant population). Unlike the a priori kinship, ϕ_{ij}, the conditional kinship, r_{ij}, can be negative. When defined as an average for two subpopulations, i and j, r_{ij} is the covariance of allele frequencies, p_i and p_j (Malécot 1973), and $r_{ij} = (p_i - \bar{p})(p_j - \bar{p})/\bar{p}(1 - \bar{p})$, where $\bar{p}$ is the mean allele frequency. Where r_{ij} is averaged over all pairs of subpopulations within a distance class, Morton and colleagues have claimed in several papers (Morton 1973a,b, 1982; Morton et al. 1968) that

$$r(d) \cong (1 - L)ae^{-bd} + L \tag{3}$$

The parameter L was defined by Morton (1973b) to be the conditional kinship in the limit for large distances within the total region (see also Morton 1973a), and generally L is negative with small absolute value. However, the genetic meaning of L as well as the validity of Equation 3 have been questioned (Harpending 1973, Nei 1973b). Nonetheless, most studies (primarily on human populations) have found a good fit of $r(d)$ to Equation 3 (Morton 1982; Jorde et al. 1982).

Estimates of a, b, and L can be obtained from the function $r(d)$. Moreover, when a specific model is assumed, the demographic parameters N, m, and σ^2 (or neighborhood size N_c, see Crawford 1984) defined above can be estimated from the topological parameters a and b. For example, for the stepping stone model of Morton (1982), approximate estimators are $m = b^2\sigma^2/2$, $N = (1 - a)/2ab^2\sigma^2$, and $4\,Nm = (1 - a)/a$.[1] Tests of fit of genetic distributions to the predicted form of exponential decline are complicated by the

[1]The multiplier 4 was omitted by Morton (1982).

fact that the estimates of a, b, and L are based on the same data (Barbujani 1987), but if there are enough distance classes sufficient degrees of freedom should exist.

Barbujani (1987) showed that if the number of subsamples is large, then the correlograms of Moran I-statistics calculated on the same distance classes, $I(d)$, are approximately equal to $r(d)/F_{ST}$, where F_{ST} is Wright's F statistic between all subpopulations. Thus, the I-correlogram predicted under pure isolation by distance is given by dividing Equation 3 by F_{ST}.

Simulations and probabilistic spatial theories that are directly based on the spatial distributions of genetic variation avoid the problems of dependency of a priori kinship on ancestral populations, and unlike both a priori and *conditional* kinship, the results can be directly compared to empirical studies (Harpending 1973). Join-count statistics, which provide more details on structure in the general case, are not specified by kinship, but require additional descent measures (Yasuda 1968, 1973; Jacquard 1973). It may not be necessary to incorporate into such models other factors ("m") in order that equilibrium be obtained. Spatial autocorrelation stabilizes quickly in many cases under limited gene flow alone. Decreasing rates of change of kinship are also expected for neutral loci (Malécot 1973). Thus, Monte Carlo simulations are particularly useful means to study autocorrelation of genotypes and gene frequencies in isolation by distance and other spatial processes directly.

It may be possible to relate or adapt the large volume of mathematical and statistical models of spatial distributions and space–time processes that have been developed primarily in the fields of geography and physics to the theory on spatial autocorrelation and autoregression of genetic variation. The major obstacle seems to be in relating the stochastic sampling of diploid genotypes during dispersal (e.g., seed dispersal) to the "dispersion" parameters of these models that usually assume normality. However, genetic sampling among subpopulations may approximate binomial sampling; thus, arcsine transformations of genetic samples may approximate normal sampling.

The simplest adaptation may be the use of the covariance-structured autoregression models (Cliff and Ord 1981) to describe statistical interactions between locations at a point in time (Haining 1977, 1979; Moran 1973a,b). (This autoregression model is the one most easily related to spatial autocorrelation.) Another promising model is the moving average model (Haining 1978; Bennett 1979; Cliff and Ord 1981). Studies that have found solutions to spatial stationarity of space–time models include those by Whittle (1954), Bartlett (1971), and Besag (1972).

Probabilistic models and computer simulation studies as well as empirical measures of autoregression and autocorrelation will help considerably in building or adapting space–time models for population genetic processes. The most useful empirical studies are those with spatial data collected from several time periods. Such space–time data can be used to estimate the model parameters that link (e.g., allocation criteria) spatial distributions through time (Haining 1979; Bennett 1979).

Reduction of spatial correlations caused by natural selection with location independent fitnesses

I have conducted simulations in which selection was added to the near neighbor pollination model (see above Model 1). Individuals with genotypes *AA*: *Aa*: *aa* were removed with probabilities 0: $s/2$: s, and were replaced by one of their eight neighbors or by themselves, each with probability one-ninth; thus, selection was directional and fitnesses were nearly independent of locations. A second process was added in which gamete types were replaced by the opposite allele with uniform probability, μ, which is formally equivalent to reversible mutation or effective random immigration, or long-distance migration within the population. Values of μ in these simulations (Table 2) were either 0.001 or 0.01. These large values allowed allele frequency equilibrium to be reached and caused a rapid dynamic of patch structure. They span the range of many realistic levels of immigration and are also reasonable rates for some mutator systems.

The *I*-correlograms (calculated on the same quadrat system described earlier) of simulated populations with strong selection ($s = 0.10$, $\mu \leq 0.01$) are much lower than those for the neutral cases (Figure 5); in particular, the *I*-statistics for short distances were much lower. In contrast, simulated populations with $s = 0.01$ and $\mu = 0.001$ or $\mu = 0.01$ produced *I*-correlograms that were only slightly lower than the corresponding neutral cases. In general, the *I*-correlograms (Figure 5) for different populations with various values of s and μ (Table 2) seem to indicate that selection erodes patches of the disadvantageous allele. The X-intercepts of the correlograms with strong selection showed a reduction of about 20%; however, for samples that have realistic sizes and are scaled properly (see discussion below), sampling errors could

TABLE 2 Parameters of simulations in Figure 5.[a]

Set	N_p	s	μ
1	9	0	0
2	25	0	0
3	9	0.01	0.001
4	9	0	0.001
5	9	0.01	0.01
6	9	0	0.01
7	9	0.10	0.001
8	9	0.10	0.01

[a] N_p, number of pollen parents, including self; s, selection intensity; μ, mutation/migration rate.

easily cause the small values of I-statistics for $k > 1$ in these populations to be negative; thus, as a result empirical estimates of X-intercepts would likely be reduced by 50% or more. Alternatively, spatial autocorrelation in empirical samples from such surfaces might not be significantly different from sampling a randomized sample.

These differences in spatial autocorrelations developed very early in the simulations (by generation 30) and, in general, stability of the genetic structure in the surfaces was apparent by generations 50–100. Among these simulations, spatial autocorrelation at stable phases was again unaffected by either the initial or equilibrium allele frequencies.

These results explain some observed spatial distributions of flower color polymorphisms within each of several populations of morning glory. In these populations, homozygotes for a locus that controls pink versus blue flower color were distributed in large patches (consisting of several hundred plants, e.g., see Figure 1), whereas recessive homozygotes for an unlinked gene that causes white flowers were distributed either in small patches or in a manner not significantly different from random sampling (Epperson and Clegg 1986). White-flowered plants are undervisited by pollinators and as a result probably suffer a substantial reduction in overall fertility (Brown and Clegg 1984; Epperson and Clegg 1987a). There is also evidence that transposable elements

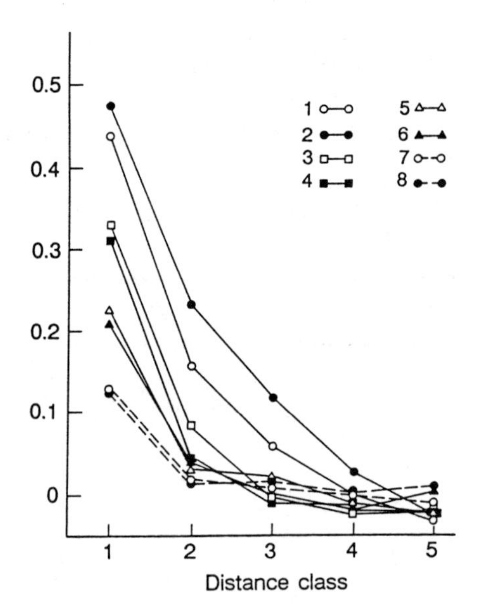

FIGURE 5. Average correlograms of I-statistics calculated on the 400 quadrat system for sets of simulations with different levels of selection and mutation (see Table 1 for specification of sets 1–8).

may produce white flower mutations in high frequencies in these populations (Epperson and Clegg 1987b).

Effects of immigration and mutation on spatial autocorrelations

In the simulation populations with random mutation or effective random immigration with rates $\mu = 0.01$ and $s = 0$ or 0.01 (Table 2), I-correlograms were reduced considerably (Figure 5). In contrast, mutation or immigration had much smaller effects for $\mu = 0.001$. Thus, it appears that either high rates of random mutations or moderate levels of random immigration disrupt or restrict patch structure. A different form of immigration was incorporated into simulations conducted by Sokal et al. (1989). They report results for dispersal Model 5, modified so that locations near one boundary of the populations were allowed to receive immigrants from an outside source (which had differing allele frequencies) each generation (here the average per generation frequency of immigrants was on the order of 0.01). These populations had larger I-statistics for small distance classes, but most strikingly the I-statistics for the longest distance classes were much more negative. Thus, immigration into one margin apparently produced a cline of allele frequencies superimposed onto patch structure.

Spatial correlations under microenvironmental selection

Microenvironmental selection can strongly influence spatial correlations of genetic variation, especially where different microenvironments induce opposing directional selection. How selection influences spatial distributions depends further on several factors, including (1) pattern of the distribution of microenvironmental heterogeneity, (2) scale of microenvironmental heterogeneity, (3) dispersal distances, and (4) strength of selection. Autocorrelation structure can be predicted in several situations in which factors 3 and 4 take extreme values. The expected values of sample distributions are likewise predictable if they are fine scaled and collected over a large area. The situation is more complex when scales of dispersal and microenvironmental heterogeneity are intermediate.

The simplest case is where dispersal is extremely limited and microenvironmental selection is strong; then spatial distributions of genetic variation are determined primarily by the genotypic fitnesses in the different environmental zones. The spatial patterns of environmental heterogeneities may have very irregular shapes. Several relevant points emerge from statistical analyses of artificially generated patterns of spatial data by Sokal and Oden (1978a) and Sokal (1979): (1) X-intercepts of I-correlograms are approximately equal to the average size of zones of spatial data when these zones vary in size; (2) X-intercepts are close to the smaller dimension of irregularly shaped zones; and (3) helical clines ("mountain peaks") and "ridge" clines both produce substantial positive correlations at short distances and negative correlations at longer distances.

When environmental heterogeneity is scaled much smaller than dispersal and spacing (i.e., the inverse of density) of individuals, propagules from each plant will fall into zones with probabilities that are nearly independent of the parents' locations. Thus, the overall average fitnesses of genotypes may be nearly independent of structure, even though the average fitnesses of genotypes may depend on the total amounts of each microenvironmental type.

In cases where dispersal distances and scales of microenvironmental heterogeneity are not extreme, the spatial patterns produced (for genotypes that are subject to induced microenvironmental selection) might vary widely depending on the precise strengths of selection, distributions of environmental heterogeneity, and distances of dispersal. Sokal et al. (1989) conducted simulations with the Model 5 mating system, but with various patterns of microenvironmental selection added. Simulations with a gradient of strengths of directional selection (i.e., always against the same allele) produced *I*-correlograms that differed from neutral simulations primarily in that *I*-statistics for long distances were much more negative, suggesting that a cline is superimposed onto patch structure. Populations with parabolic selection gradients or disjointed zones with different microenvironmental selection produced *I*-statistics (at short distances) that were similar to the neutral case (Sokal et al. 1989). However, in the case of disjointed microenvironmental zones, the zones were similar in size to the patches produced under the neutral model. Clearly, further studies of this type will improve our understanding of spatial distributions of genetic variation produced under different combinations of levels of dispersal and types of microenvironmental selection.

Temporal correlations

Restrictions on gene flow result in correlations among genotypic or allele frequencies at locations through time. In all seven models discussed earlier for neutral loci there were very high correlations among the quadrat allele frequencies over small time lags (i.e., the number of intervening generations), and these decreased steadily as time lags increased. It is worth noting that correlations for the same lag were nearly constant during each simulation run (Sokal and Wartenberg 1983). This, along with observations of the uniformity of *I*-correlograms after 100 to 150 generations, suggests that the genetic structure quickly reached stable phases. Correlations for lags of 150 or more generations had small values, and this suggests that the spatial autocorrelations in the late phases were essentially independent of initial or early conditions.

Simulated populations with greater levels of dispersal produce smaller temporal correlations (Sokal and Wartenberg 1983) indicating that the surfaces of these processes change more rapidly. The reductions are quite large and temporal correlation appears to follow a monotonically decreasing function of the dispersal variance, σ^2, or N_c. Measures of temporal correlations may provide sensitive measures of dispersal parameters in a way that avoids the cumbersome methods of tracking marked individuals.

Simulated populations with near-neighbor pollinations (Model 1) and "locational-independent" selection had much lower temporal correlations. The reductions are greatest (up to 50%) for the shorter time lags and for the simulations with stronger selection ($s = 0.10$). Random mutation or immigration caused somewhat smaller reductions. The reductions caused by location-independent selection or mutations may be expected to be less pronounced in populations that have greater dispersal.

Joint effects of selection and gene dispersal on spatial–temporal distributions of genetic variation

The theoretical results lead to several preliminary generalities regarding how natural selection and dispersal affect spatial–temporal distributions of genetic variation within populations. In populations with low to moderate dispersal ($N_c < 100$) homozygous genotypes of selectively neutral loci quickly aggregate into temporally stable, large (several hundred individuals), and well-defined patches that are surrounded by heterozygotes. The size of patches and the amount of autocorrelation increase with the level of dispersal up to a point; beyond this point patch structure becomes increasingly diffuse or less defined, and spatial autocorrelation decreases (Sokal et al. 1989). Very high levels of dispersal (certainly for $N_c > 1000$) result in essentially random distributions. Temporal correlation decreases with increasing dispersal. These effects of dispersal on spatial and temporal correlations appear to be largely independent of allele frequencies and independent of initial conditions.

Recurrent random mutations or random immigrations or long-distance migrations have relatively little effect on spatial or temporal correlations, unless they occur with high rates (extremely high rates for mutation). However, "directional," recurrent immigration of individuals that have very different allele frequencies can induce clines superimposed onto patch structure, resulting in substantial reductions in the I-statistics for the longest distance classes.

Location-independent selection of moderate strength greatly reduces both temporal and spatial correlations in populations with low dispersal. The effects of microenvironmental selection depend greatly on the strength of selection, and the scale and distributions of microenvironments. If dispersal is very low, then moderate to strong (and relatively large-scaled) microenvironmental selection will essentially determine the spatial distribution of genetic variation. For slightly higher levels of dispersal, lower correlations of genetic variation with environmental heterogeneity will be produced. If dispersal is moderate, then selection gradients will result in clines that are superimposed onto patch structure. It is predicted that with even greater levels of dispersal, the effects of moderate to strong selection on correlograms are probably reduced, and in extreme cases microenvironmental zones may simply be foci of ill-defined concentrations of favored genotypes. It is expected that many forms of large-scaled microenvironmental selection stabilize spatial distributions, resulting in larger temporal correlations.

Clearly, the fitness of the genotypes of a locus could be simultaneously influenced by both location-dependent and -independent factors, and the spatial distribution of genetic variation will clearly depend on the contributions of these two forms of selection.

Different combinations of dispersal patterns, location-independent selection, and spatial patterns of microenvironmental selection values could produce similar *I*-correlograms (but not necessarily similar genotypic distributions), thus join-count statistics are likely to be more informative in complex situations. However, it is less likely that different combinations will also produce the same temporal correlations of allele frequencies.

The consideration of the joint distributions of genotypes at more than one locus adds yet another dimension of complexity to spatial processes of genetic variation. It is expected that for large populations under neutrality, the spatial distributions of genotypes at unlinked loci would show little correlation (Sokal et al. 1986, 1987). In contrast, linkage or selection, especially epistatic selection, can interact with spatial structure in very complex ways. For example epistatic selection of multilocus genotypes and limited dispersal may create linkage disequilibrium. As a consequence, even in cases where the fitnesses of multilocus genotypes do not depend on location, the marginal fitnesses may depend on the joint spatial distribution of genotypes of that locus and distributions for other loci.

EMPIRICAL METHODOLOGY

The best choices of sampling and experimental methods for using spatial statistics and ancillary information to study spatial processes of genetic variation depend on the expected distributions under null hypotheses and under alternative hypotheses that include factors of interest, such as natural selection. The first part of this section details methods of sampling for quantifying isolation by distance and inferring dispersal parameters under neutral theory. The second subsection outlines some empirical methods for inferring selection and dispersal components of spatial processes in certain straightforward and predictive cases.

Sampling strategies for studies of isolation by distance

The multiple-loci data approach offers many advantages for analyses of genetic isolation by distance in populations, using spatial samples, spatial statistics, and neutral theory. Multiple-loci data provide internal controls because all genes at selectively neutral loci should develop very similar spatial autocorrelations (and usually almost independent spatial distributions), regardless of allele frequencies. Excessive differences among single-locus measures would indicate violations of the assumptions. For example, in populations of lodgepole pine, dispersal is great enough that a nearly random distribution is predicted for neutral loci. Thus, even though sample sizes were large, nearly all join-count statistics for genotypes of 14 multiallelic allozyme loci were not significantly different from zero (Epperson and Allard 1989). However, spatial correlations were repeatedly found for the genotypes of a few loci, suggesting that these correlations were caused by selection (see also Wagner et al., in press).

A less desirable alternative to multiple-loci analyses is to take replicated (single locus) samples of the reference population or species. It is also useful to have some knowledge of the general level of dispersal that can serve as a further check on the results of spatial statistical analyses. In addition, efforts should be made to avoid obvious heterogeneities in environment within the sample area.

A critical consideration in planning studies of isolation by distance or patch structure is the scale of sampling. In populations with low to moderate levels of dispersal and fairly uniform densities, homozygotes of neutral loci are aggregated into patches, each containing several hundred individuals within similarly sized areas. If sample locations cover too small an area, and just span boundary zones between two or three patches, then measures of genetic isolation by distance may be inflated, and measures of patch size deflated. Alternatively, if samples span only the core of one patch, then measures of both isolation by distance and patch size may be deflated. The latter situation may not be obvious from inspection of the mapped data. In addition, intentional positioning of samples over identified patch structure violates assumptions required for spatial statistics. On the other hand, spacing of samples can be too large, and important features of structure may be missed. As a rule, sample intensities should be chosen such that roughly at least 5–10 samples will fall within each patch (higher values, say 10–25, will give more distance classes near the X-intercepts and thus more precise measures of patch sizes), and at least 4 to 9 patches within the total area sampled. For a populational area containing 3000 individuals and with patches containing roughly 500 plants, a sample lattice of 150 individuals (here sample "intensity" is 1 in 20), should provide a good picture of structure.

In addition, a large number of sample locations is required for quadrat allele frequency data; thus, the number of individuals per quadrat must usually be small, perhaps 5 to 10. Quadrat samples may be either random samples from within contiguous areas or samples of the nearest k individuals to quadrat centers. (The distances from individuals to points, or the number of individuals per quadrat can be used to estimate density and to test for uniformity of density; see Pielou 1977; Wadley 1954.) Thus, for fixed total numbers of individuals, there is a trade-off between the greater accuracy of location values for quadrat allele frequency data versus better coverage of the area and more detailed information for individual genotypic data.

In addition, sample locations should be as uniformly spaced as possible (unless density is very irregular, see earlier discussion), so that the number of pair comparisons for each distance class is fairly uniform, and thus better estimates of patch sizes are obtained.

Sampling schemes for inferences about complex spatial processes

Methodological procedures are more complex for studies that use spatial data or spatial–temporal data to quantify or test for specific forms of natural selection or other influences on spatial processes, rather than simply to detect aberrant patterns for some loci. An example of situations where deviations can be specified occurs in populations with low dispersal, where sharp patch structure is predicted for neutral loci and selection at loci of interest is thought to be location independent; then reduced autocorrelation is the alternate hypothesis. A case in point is the reduction of patch structure for white-flower genotypes in morning glory populations (see earlier discussion of results of Epperson and Clegg). Alternative hypotheses of specific forms or specific statistical models of spatial distributions are virtually unlimited in number. These tests are achieved through specifications of different classification or weighting schemes for autocorrelation analyses of pair comparisons, or by using specific distance measures for the Mantel statistics (Mantel 1967; Cliff and Ord 1981). Statistical models for analyses of the association of spatial distributions of genetic variation with microenvironmental heterogeneity include regression models with autocorrelated errors and autoregression–regression. Fitting these models requires identification and measurement of important microenvironmental variates and further specifications on the form of interactions between locations due to dispersal. Precise information on these factors should be obtained in order to formulate precise hypothesis frameworks and/or models for estimation and fitting, because excessive numbers of ad hoc tests can lead to spurious statistical significance.

There are further problems in choosing informative sample schemes for more complex situations, because the scales of structure and types of differences in structure under different alternate models may depend in complex ways on the exact scales and forms of microenvironmental heterogeneities and dispersal. In many cases it may be informative to simulate specific processes that incorporate available information before conducting studies of the natural populations of interest. It is expected that as the number of empirical and theoretical studies of the spatial processes of population genetics increases so too will our understanding of space–time processes as well as our knowledge of proper statistical models for spatial data. Spatial autocorrelation analysis can be very useful as an exploratory tool to provide ad hoc, but reasonable procedures for building more specific statistical models for estimation and testing (Cliff and Ord 1981). It would be valid (and not ad hoc) to use the results of spatial autocorrelation on data from one population to build models for another population or for an independent sample from the same population.

More specific suggestions can be made for studies of selection in populations with low dispersal and where correlations are sought between identified, large-scaled microenvironmental heterogeneities and genetic distributions. Clearly, the sample area should include many replicates of each type of microenvironment (or ranges of values) and also be large enough to contain many times more than the number of plants in patches under the neutral theory, in order to avoid spurious correlations of genetic variation with environment. It would also be informative to include, for comparison, suspected neutral loci in such studies. Finally, procedures for collecting and interpreting space–time data are clearly more powerful and may also in some ways be simpler, especially for multiple-loci data, because such data can be used to estimate allocation functions directly.

CONCLUSIONS

The development of spatial statistical methods provides us with new powerful and informative tools for detecting and measuring the structure of genetic variation within populations. Spatial pattern analyses are promising and sometimes necessary parts of studies of processes of population genetics in natural populations. Recently, statistics such as spatial autocorrelation and autoregression for distributions of genetic variation have been directly related to theoretical results on spatial–temporal processes of isolation by distance and natural selection. Such theoretical results will continue to contribute importantly to our understanding of population genetics. A major advantage of theoretical results based on spatial statistics is that they can be compared directly with observations in natural populations, unlike the kinship coefficient. However, spatial patterns may be complex functions of dispersal and natural selection. Hence, the need for caution and the utility of ancillary information and experiments. Similar care should be taken in designing sampling strategies and in choosing hypothesis tests. Theoretical and empirical work is rapidly advancing a framework of hypothesis tests that uses spatial patterns to study gene flow and natural selection. These advances call for increased activity in this important area of plant population genetics.

ACKNOWLEDGMENTS

I thank Spencer Barrett for reviewing this chapter and Patti Fagan for her meticulous wordprocessing of the manuscript. This work was supported in part through NSF Grant BSR-841-8381.

CHAPTER 15

THE GENETICS OF PLANT MIGRATION AND COLONIZATION

Spencer C. H. Barrett and Brian C. Husband

ABSTRACT

Migration and colonization are processes shared by all organisms, yet it is unlikely that the genetic consequences are the same for all. Colonizing episodes will be important in determining population genetic structure when they occur frequently, as in species of ephemeral environments, or when as a result of long-distance dispersal, genetically isolated populations occur. The effects of small populations through founder events or bottlenecks, inbreeding, and strong directional selection in novel environments can all influence population genetic structure depending on the ecology of the species and the scale of colonization. While a theoretical framework for understanding the genetics of migration and colonization is well developed, few studies of plant populations exist that test the predictions of the models. Two particular deficiencies are evident. First, quantitative data on the significant parameters of colonization models, e.g., effective population size and migration rates, are lacking. Second, information on the effects of stochastic processes on quantitative traits is not available for plant populations; yet such traits are likely to be of major importance to survival and reproductive success. Satisfactory explanations for the success or failure of colonizing episodes will most likely come from demographic genetic studies of natural colonization events or from experimental work on artificially established colonies.

Colonization is the establishment and spread of an organism in a region or habitat not previously occupied by that species. On some spatial or temporal scale, it is an integral feature of the population biology of all plants and animals. Therefore, information on the genetic consequences of colonization is essential for understanding the population genetics and evolution of organisms. Information on the genetic structure of colonizing populations can be useful in two ways. It can be interpreted retrospectively to aid in constructing the historical processes of migration and colonization. Second, it can provide insights into the ecological persistence and evolutionary potential of populations once they have entered a new environment. To understand the evolutionary consequences of colonization, it is necessary to know the amounts, kinds, and organization of genetic variation that result from different patterns of colonization. The effects of small populations through founder events or bottlenecks, inbreeding, strong directional selection in novel environments, rapid density-independent population growth, and gene exchange with related taxa can all influence population genetic structure in colonizers depending on the species and scale of colonization.

Colonizing episodes are more likely to be important in determining population genetic structure when they occur frequently, as in many pests, weeds, and early successional species, or when, as a result of long-distance dispersal, genetically isolated populations are produced. In fact, much of the interest in the genetics of colonization has focused on species with particularly well-developed invasive powers whose populations are in a constant state of colonization, extinction, and recolonization. Such "colonizing or invading species" have provided the experimental material used to develop much of our knowledge of the evolutionary process since the modern synthesis. However, in studying the genetics of colonization it is perhaps more useful, as Lewontin (1965) originally pointed out, not to think of colonizing species as a discrete group but to consider the effects of colonizing events for any species. All organisms occur on a continuum in which the frequency of colonization of new territory or habitats varies from high to low. Differences among species in the frequency and importance of colonization for regional persistence largely depend on extrinsic ecological factors, such as the type and distribution of habitats, as well as intrinsic factors associated with the life history and reproductive system of the species in question. The development of a sound theoretical framework for the genetics of colonization necessitates the recognition that models describing the colonization process will vary for species at different positions along the continuum.

While we have made considerable progress in the past two decades in understanding the genetics of plant colonization (reviewed in Brown and Marshall 1981; Barrett and Richardson 1985; Rice and Jain 1985), there are still surprising gaps in our knowledge, particularly in relation to the dynamics of the colonization process. In most empirical studies, patterns of genetic variation are measured at one point in time and inferences are made about the processes that have led to the observed pattern, often without historical information on the populations involved. The absence of chronological genetic studies of colonizing plants severely restricts direct tests of theoretical models of the genetics of migration and colonization.

Understanding the genetics of colonization is important not only to evolutionary studies and population biology but also to agriculture and conservation. The domestication of plants and animals involved many genetic bottlenecks in both space and time and, in most cultigens, this has resulted in an erosion of genetic diversity in comparison with wild relatives (Frankel and Soulé 1981). The accidental or planned introduction of crops and weeds from one continent to another involves many of the same genetic processes that operate during colonizing episodes in wild populations. Knowledge of the migratory history of crop plants can aid in interpreting geographic patterns of genetic diversity and the sampling of germplasm. Efforts to conserve germplasm resources, either by habitat preservation or in collections, requires a knowledge of the effects of small populations on the maintenance of genetic variability and appreciation of the likely effects of prolonged inbreeding. For pest and weed species, information on the genetic diversity present in a particular area, as well as knowledge of the likely source region for a particular invasion, can be of value in devising effective methods of biological control (Marshall et al. 1980). Clearly, both theoretical and empirical studies of migration and colonization can provide basic information to applied biologists whose primary goal is the wise management of genetic resources.

In this chapter, we begin by outlining some simple theoretical concepts and models that describe how colonization affects population genetic structure. Particular attention is given to the effects of stochastic processes on different classes of genetic variation. Because of an overall paucity of empirical data on colonization in plants, most of the data relevant to the models come from species of ephemeral environments. Following a discussion of the evidence for evolutionary changes in mating systems that can accompany colonization, we conclude by suggesting the kinds of genetic studies that could be conducted profitably on plant colonization and where major gaps in our knowledge occur.

GENETICS OF FINITE POPULATIONS

Small populations are a distinct feature of most colonizing events. They are common when a population is first established (e.g., founder events) or to existing populations after disturbance (e.g., population bottlenecks). Often, only a subset of the genetic information present in the source is represented in a single migration or colonizing event. This sampling error causes random fluctuations in allele frequencies called genetic drift (Wright 1969). The genetic theory of finite populations most simply describes the effect of population size and migration on genetic drift and its consequences on population genetic structure. The models assume random mating and no selection or mutation.

Population size

At equilibrium, populations small enough to experience drift will become fixed for one of the alleles at a polymorphic locus and thereby decrease allele richness, and increase inbreeding and population differentiation. The average number of alleles per locus decreases because the probability of a rare allele occurring in a founding population decreases with population size (Nei et al. 1975). If the majority of alleles in the source population are rare, then the average number of alleles in the colonizing population will be most affected by the size of the initial genetic bottleneck, and less by the length of time a population remains small (Sirkkomaa 1983). If a population remains small over long periods, then more common alleles will be lost as random fluctuations alter allele frequencies toward 0 or 1. The frequency of homozygotes increases in small populations as a result of increased mating among relatives. The proportion of heterozygotes as a function of time is given by

$$H_{t+1} = (1 - 1/2N)H_t \quad (1)$$

indicating that the proportionate loss of heterozygosity is expected to be only $1/2N$ each generation and therefore will be substantial only if populations remain small for a number of generations (Nei et al. 1975). Not only will allele richness and heterozygosity decrease within small populations but also, if founding events are repeated in space, the variance among new colonies will depend on the average population size and the time spent at that size. Although the variance among populations increases, the expected average allele frequency among all populations will not deviate from the initial frequency in the source. Where small populations are a significant feature of the colonizing process, theory predicts that populations will exhibit low genetic variability within and a large degree of differentiation among populations. Although the theory of finite populations is well developed, it remains for empiricists to determine its significance for natural populations of plants.

When examining the genetic consequences of colonization, the critical parameter to estimate is population size. A direct count of the number of breeding individuals may be sufficient. However, in most natural populations, the number of breeding individuals does not reflect the degree of drift and inbreeding observed. In other words, the breeding population size is rarely equal to the N of the models described above. Factors that cause this disparity include gene flow over long distances, varying population size, nonrandom mating, unequal sex ratios, and age and size structure (Kimura and Crow 1963; Heywood 1986). These factors violate the assumption of the simple finite population models, that each of the N individuals has an equal probability of contributing gametes to the next generation. A better parameter is N_e, the effective population size, which represents the size of an idealized population, in which each individual contributes equally to the gamete pool, having the same variation in allele frequencies as the observed population (Wright 1931). The effective size can be derived from the actual number of breeding individuals, when factors affecting the breeding structure such as selfing rate, sex ratio, and the distribution of reproductive output are known, or can be inferred from the variance of allele frequencies at neutral loci (Kimura and Crow 1963; Nei and Tajima 1981; Pollak 1983; Crow and Denniston 1988).

Unfortunately, there are few examples where N_e has been estimated and compared to N for plant populations. Jain and Rai (1974) measured the effective population size of subpopulations of *Avena fatua* based on the number of breeding individuals, departures from random mating, and variance in seed output among plants. Since the estimates of N_e varied over the 2 years samples were taken, they calculated the harmonic mean of N_e for each population (Wright 1951). In all cases, the average effective size was less than the average number of individuals per population. Similarly, estimates of N_e based on the distribution of reproductive output within populations of *Papaver* were less than the observed number of reproductive individuals (M. Lawrence, personal communication). While N_e is less than N in these examples, the magnitude and direction of the differences will vary within and among plant species due to their diverse life histories, reproductive systems, and colonization patterns.

Colonization affects genetic structure through its influence on population size; however, there have been relatively few studies of plant colonization from this perspective. Local differentiation has been widely reported for both isozymes (Ennos 1985; Knight and Waller 1987) and life history traits (Antonovics and Primack 1982; Schemske 1984); in many cases drift has been inferred. However, differentiation can often be explained by other factors such as differences in the direction and intensity of selection among sites. To demonstrate the effects of finite population size we must at least measure the effective population size, preferably in past generations, and relate it to both the existing and expected levels of variation. Jain and Rai (1974) compared variation in morphological traits of *Avena fatua* from subpopulations in two orchards. The average effective population size for the orchards differed because of contrasting management practices. Jain and Rai related the changes in allele frequency over 2 years to population size. As predicted by theory, the average gene frequencies in both orchards did not fluctuate significantly from one year to the next, except at one locus. However, over 2 years, the subpopulations became more differentiated, particularly in the orchard with smaller subpopulations. The authors concluded that most changes in allele frequency were due to random drift or sampling errors in the survey.

Drift in small populations is also an important evolutionary force affecting differentiation in style morph diversity among populations of tristylous *Eichhornia paniculata* that inhabit N.E. Brazil. Populations vary in their morph structure, from equal morph frequencies to a single morph. In a sample of 84 populations, morph diversity increased and the variance among populations decreased with increasing plant density. This result is consistent with genetic drift since density is correlated with population size in the species (Barrett et al., in press). Measures of temporal variation in morph evenness reveal a similar pattern. Figure 1 shows changes in morph evenness between 1987 and 1988 in 34 populations of *E. paniculata* in relation to the harmonic mean of population size. Populations deviate away from (+ change) or toward (− change) equal morph frequencies to nearly the same degree (19 versus 15 populations, respectively) and the average change in morph evenness is not significantly different from zero. The largest absolute changes in diversity occur in the smallest populations, particularly those below 100 individuals (Figure 1). The observed patterns indicate that random drift in small populations is a dominant factor affecting genetic structure in *Eichhornia paniculata*.

The inheritance of traits with more obvious ecological significance, such as growth rate or reproductive effort, is not well understood and predicting the effects of small populations is difficult. While there are examples of quantitative characters under relatively simple genetic control (reviewed in Gottlieb 1984), most life history traits are likely to be controlled by many loci. When the genetic variance of polygenic traits is based on additive allele effects, variation should decrease in proportion to $1/2N_e$ after a bottleneck of N_e individuals (Lande 1980). However, if genetic variability is partly the re-

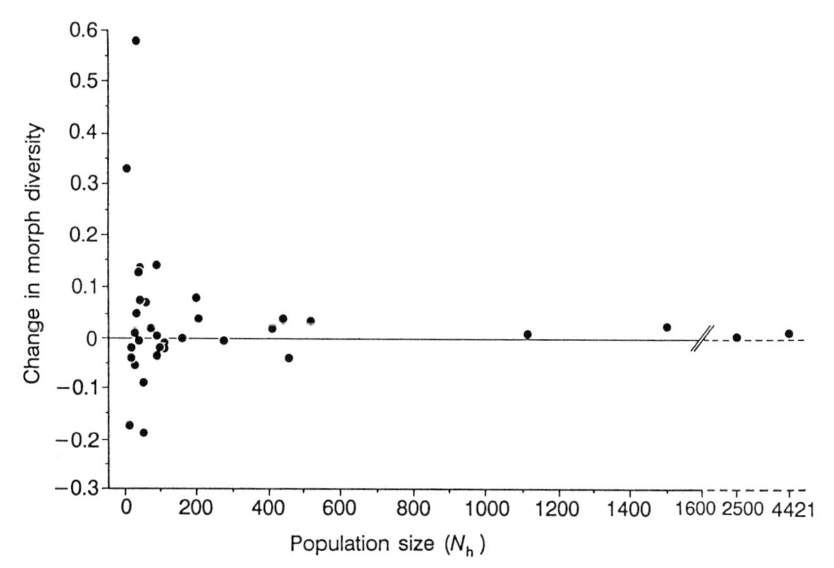

FIGURE 1. Changes in style morph diversity in relation to population size from 1987 to 1988 in 34 Brazilian populations of tristylous *Eichhornia paniculata*. Population size (N_h) is the harmonic mean of the number of reproductive individuals in each year. Style morph diversity was calculated using a modification of Simpson's index (B. C. Husband and S. C. H. Barrett, unpublished data).

sult of nonadditive allele effects (such as epistasis and dominance), the effect of small populations on genetic variation will not be a simple relationship with N. Recent models by Goodnight (1987, 1988) suggest that the additive genetic variance may actually increase, at least temporarily, after a bottleneck, as inbreeding converts nonadditive genetic variation from the donor population to additive genetic variation in the derived population. Figure 2 illustrates such an effect for traits related to fitness, in which 20% of the total variance in the ancestral population is additive and 80% is epistatic. If populations are kept at 16 individuals for 100 generations, the additive genetic variance increases and temporarily exceeds the total genetic variance in the ancestral population. Over the remaining period, the additive genetic variance in the derived population exceeds that in the ancestral population.

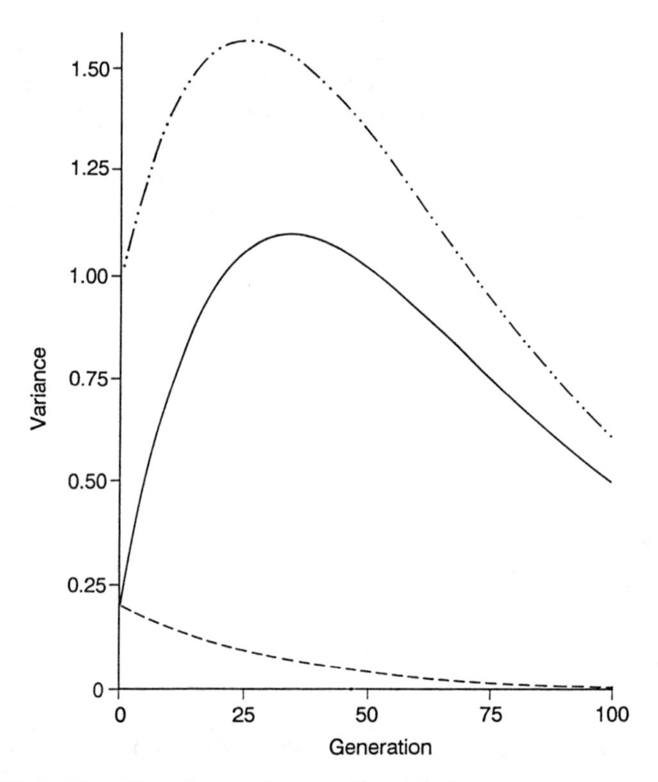

FIGURE 2. The effect of a population bottleneck of intermediate size (16 individuals) on additive genetic variance (solid line) and total genetic variance (dash/dot line) for 100 generations. The contribution of additive genetic variance in the ancestral population to additive genetic variance after the bottleneck is also indicated (dashed line). Twenty percent of the total genetic variance in the ancestral population is additive and 80% is epistatic (after Goodnight 1988).

Goodnight also shows that while epistatic variance changes to additive variation most rapidly in small populations, a greater proportion of the variance converts to additive variation over many generations in populations of moderate size. Aside from laboratory studies on houseflies by Bryant and coworkers (1986a, 1986b), there are few studies designed to test this theory and none is known from the plant literature.

Although less controlled, recent colonization events of known origin can also reveal the effect of bottlenecks on genetic variability in the field. For example, in a study of local colonization, we measured genetic variability in an isolated pair of populations of *Eichhornia paniculata* in N.E. Brazil. In 1987, one population was large and contained all three style morphs; the second population, 3 km west, was small, with only two morphs. Both populations were separated from the remaining concentration of populations by 100–200 km. Based on morph structure, size, and geographic distribution, the large trimorphic population is the most likely source of plants that established the smaller population. While enzyme variability in both populations was low relative to other Brazilian populations (see Glover and Barrett 1987), the variation in the derived population was a reduced subset of the variation in the source population (Table 1). A similar pattern was evident for quantitative

TABLE 1 Comparison of genetic variation in source and derivative populations of *Eichhornia paniculata* in N.E. Brazil.[a]

	B56	B55
Style morph structure	Trimorphic	Dimorphic
Estimated population size		
1987	2000	200
1988	775	0
Outcrossing rate (t)	0.99 ± 0.04	0.60 ± 0.14
Loci polymorphic (%)	20.8	8.3
Average number of alleles	2.3	2.0
H_{obs}	0.054	0.016
Gene diversity (h)	0.062	0.035
F	0.13	0.54
Number of traits with significant family variation	8/15	1/15

[a]The two populations were located in roadside ditches, 3 km apart, 69 km west of Campina Grande, Paraiba State (see Figure 2, Barrett et al., in press). Isozyme data are based on a survey of 24 loci from a sample of 264 plants from B56 and 64 from B55. Measures of quantitative variation in life history traits are based on comparisons of 25 open-pollinated familes of 2 individuals each, per population, grown under uniform glasshouse conditions. Variation among families was significant at the $p<0.05$ level, in a mixed model, hierarchial analysis of variance.

traits. Between-family variation was a significant component of the total variation in the source population for 8 of 15 traits. In the derived population, however, only 1 of 15 traits exhibited significant family variation and in no case did the family component of the variance exceed that in the source. While quantitative genetic variation was apparently reduced by the colonization event, genetic correlations, based on family means between all possible pairs of traits, did not change significantly between the source and derived populations. Although in this example both classes of genetic variability exhibited similar patterns, a lack of congruence between variation in quantitative traits and enzyme loci, as well as few data on the genetic basis of quantitative traits, make it all the more important that experimental studies are conducted on the effect of bottlenecks on genetic variation. If these studies are to be useful to colonization genetics they should examine traits of obvious ecological significance.

Migration

In small populations, migration moderates the random change in allele frequencies due to genetic drift. The degree to which drift occurs will depend on both the rate of migration and the genetic characteristics of the migrants. When the genotypes involved in colonization are at a constant frequency, recurrent migration will oppose the forces of drift and maintain variability within populations. The degree of differentiation among populations that have reached a drift–migration equilibrium is given by

$$V_q = pq/(4Nm + 1) \tag{2}$$

where V_q is the variance of the frequency of allele q, and Nm is the rate of dispersion (Falconer 1981). Thus, a single migration event will have a greater influence on small populations than on large ones. However, because of drift, maintaining a given level of homogeneity requires more migration in small than large populations. In plants, migrants are commonly from neighboring populations and because of the dispersal of multiseeded fruits, are likely to be kin structured. This feature in association with variable migration rates introduces a large stochastic element and increases the potential for random differentiation of populations (Levin 1988).

In many plants species, particularly those of disturbed or ephemeral habitats, the sequence of population bottleneck and expansion, or extinction and recolonization, is repeated continuously over time. Most models of population differentiation, however, assume that each local population lasts indefinitely. Slatkin (1977) showed that without selection and mutation, the effect of local extinctions and recolonizations on genetic structure is complex and depends on the relative strengths of drift and gene flow between populations during recolonization. The impact of gene flow during recolonization depends on whether the migrants that establish new colonies are from the same population, such as the immediate seed bank or the nearest population (the propagule model), or from a random sample of populations, differing in allele frequency (migrant model). In the propagule model, drift outweighs the effect of gene flow, and, therefore, populations differentiate. Under this colonization scheme, genetic differentiation is enhanced most when the number of individuals colonizing an available site is small relative to the number entering extant sites (Wade and McCauley 1988). In the migrant model, a random sample of seeds from a collection of populations can increase variation within and reduce the levels of differentiation between populations, particularly when the number of individuals colonizing available sites is relatively large.

If selection is also considered, the theory of colonization and extinction may be analogous to the nonequilibrium theory of species diversity in communities (Huston 1979). This theory suggests that species diversity is a product of the interacting effects of disturbance and competition, both of which reduce species richness. At intermediate levels of disturbance, competition is interrupted and disturbance is not intense enough to extirpate species. At this point, disturbance is sufficient to maintain the presence of early successional species and thus diversity is higher than at low levels of disturbance (or colonization). If community diversity is comparable to genetic diversity within populations of a species, then, by analogy, variation will be a function of the effects of drift and selection. Selection will be most important in reducing variability in undisturbed populations, while drift causes fixation in frequently disturbed populations. This model predicts that variability within populations will be highest at intermediate levels of disturbance, since the effects of selection and drift are minimized. While apparently no population genetic studies address this issue, community studies on intertidal diversity (Sousa 1979) and Atlantic coast plants (Keddy 1983) appear to support this prediction.

Models of genetic structure in populations experiencing frequent colonization and extinction predict that, under certain circumstances, increasing the frequency of colonization increases the genetic diversity within populations (Wade and McCauley 1988). This is important since we regularly associate colonization and colonizing plants with low genetic diversity within populations. The results may explain why variability in disturbed populations can exceed that in stable populations, as described for *Lupinus succulentus* from California (Harding and Mankinen 1972). Harding and Mankinen found that disturbed populations were more variable at three flower color loci and one seed pigment locus. They suggested that contrasting selection pressures may account for the differences in genetic variation associated with disturbance. An alternate explanation, however, is that interpopulation migration plays a larger role than drift in the colonization dynamics of these populations. Unfortunately, the level of migration between populations is poorly understood and represents one of the least tractable aspects of population genetics (Levin 1988). Not withstanding this difficulty, interpretations of the genetic structure of colonizing populations cannot be made in isolation, but should be viewed in the context of the local distribution and density of populations and their dynamic interrelationships with one another.

MODELS OF COLONIZATION

Measuring and describing the genetic structure of populations, in space and time, pose no difficulties of principle. However, inferring the importance of migration and population size is difficult without knowing the dynamics of colonization. Since so few estimates of plant migration rates exist, conclusions are highly speculative. In light of this deficiency, we consider a number of colonization models, discuss their effects on genetic structure, and then summarize the empirical data available that can be interpreted within this framework. The models involve migration and finite population size, but do not deal with selection.

Continent–island model

The continent–island model is the simplest depiction of colonization, based on Wright's (1940) island model of migration. It assumes unidirectional migration from a relatively large source, with a fixed allele frequency, to small isolated colonies or islands. Wright combined the effects of population size and migration to predict whether island populations will differ from those on the continent. At equilibrium, allele frequencies in small, isolated colonies will differ significantly from their source and, depending on the rate of expansion, may differ from the genetic composition of the initial migrants due to drift (Nei et al. 1975). The distribution of allele frequencies among populations will vary, depending on their size. The frequency of populations fixed for a particular allele will be in proportion to the frequency of the allele in the migrant pool. The continent–island model can be applied to plant colonization involving long-distance dispersal and is particularly relevant to the adventive spread of weeds. Unfortunately there are relatively few examples in which genetic variation has been measured in both the source and colonial (introduced) populations of wide-ranging species and even fewer comparisons of genetic diversity between continental and island populations of plants.

In all cases where island and mainland populations have been compared, insular populations have had reduced or, in extreme cases, no measured genetic diversity (Rick and Fobes 1975; Ledig and Conkle 1983). In the former case, Galapagos Island populations of *Lycopersicon cheesmanii* are sufficiently distinct from their more variable mainland progenitor to warrant separate species status, whereas, in the latter case, both island and mainland populations of *Pinus torreyana* are genetically depauperate, suggesting that the species itself has been subject to an historical bottleneck. Two examples in which a genetic bottleneck has been associated with island colonization involve the heterostylous plants *Turnera ulmifolia* and *Eichhornia paniculata* (Table 2).

TABLE 2 Comparisons of isozyme variation in source and introduced populations of four colonizing species of plants.[a]

Region	Number of populations/loci	PLP	Average number of alleles/polymorphic locus	H_o	H
	Apera spica-venti (Warwick et al. 1987)				
Europe	6/17	0.62	2.5	0.23	0.20
Canada	9/17	0.57	2.5	0.23	0.21
	Echium plantagineum (Burdon and Brown 1986)				
Europe	2/14	0.82	2.6	0.29	0.35[b]
Australia	8/16	0.94	2.7	0.32	0.34[b]
	Eichhornia paniculata (Glover and Barrett 1987)				
Brazil	6/21	0.24	2.2	0.08	0.09
Jamaica	5/21	0.08	2.0	0.02	0.03
	Turnera ulmifolia (Barrett and Shore, in press)				
Latin America	7/14	0.46	2.1	0.11	0.12
Caribbean	16/14	0.20	2.0	0.07	0.04

[a]PLP, proportion of loci that are polymorphic; H_o, observed heterozygosity; H, gene diversity.
[b]Based on polymorphic loci.

Both taxa occur primarily in South America with isolated populations on various Caribbean islands. In the case of *E. paniculata*, island colonization is confounded with a change in mating system from outcrossing to predominant self-fertilization (see below). In contrast, both continental and island populations of *Turnera ulmifolia* var *intermedia* are self-incompatible and outbreeding, but the former populations are diploid whereas the latter are autotetraploid. Since continental autotetraploids of *T. ulmifolia* var *elegans* are highly variable, with populations more diverse than those of diploids, it is unlikely that the reduced diversity of island populations of var. *intermedia* is the result of their autotetraploid origin; more likely it is a direct result of genetic bottlenecks associated with island colonization (Barrett and Shore, in press).

Evidence from investigations of intercontinental migrations of weeds indicates that genetic variation in introduced populations can be lower or higher (reviewed in Brown and Marshall 1981) than populations from the source range. The outbreeding weeds *Apera spica-venti* and *Echium plantagineum* exhibit similar levels of variability in native and introduced populations (Table 2). Both species are native to Europe and have become established in Canada and Australia, respectively. Despite its recent introduction, *Apera* exhibits as much variability in introduced colonies as in European populations. This supports the prediction that the effects of genetic bottlenecks are reduced by factors such as outcrossing, which increase the effective size, N_e. In *E. plantagineum*, the high genetic diversity of Australian populations is in part the result of hybridization among multiple introductions of floral variants used for ornamental purposes. Without information on the source, time, and number of introductions, interpretations of the genetic effects of long-distance colonization will be difficult.

An example in which some historical information is available involves the invasion of two annual barnyard grasses [*Echinochloa microstachya* and *E. oryzoides*] into cultivated rice fields in New South Wales, Australia. Imported Californian rice varieties were used to initiate rice cultivation in New South Wales in 1922 (McIntyre and Barrett 1986). The earliest records of the two barnyard grasses in Australia were from rice fields at Leeton Rice Experiment Station in 1938, the entry point for Californian rice varieties. Comparisons of isozyme variation in North American and Australian populations of the two species indicate a major genetic bottleneck associated with introduction to Australia with Californian rice field populations the most likely source (S. C. H. Barrett and A. H. D. Brown, unpublished). In the case of *E. microstachya*, the predominant Australian genotype could be identified from among the North American sample of populations and occurred in a population from northern California close to Biggs Rice Experiment Station. Historical records suggest this site as the likely exit point for cultivated rice varieties shipped to Australia in the 1920s.

Patterns of genetic differentiation in life history attributes among populations of the two barnyard grass species from the two regions were similar to those obtained for isozyme loci. In the native *E. microstachya*, North American populations were more differentiated than Australian populations with two populations clustered with the Australian sample (Figure 3). Once again this points to northern California as the likely source region for the Australian invasions. Population samples of the crop mimic *E. oryzoides* showed little differentiation between California and Australia. The lack of differentiation may result from the similar cultural conditions for rice growing in the two regions (McIntyre and Barrett 1986) as well as restricted amounts of genetic variations present in the founding stocks.

The continent–island model of colonization is applicable to many examples in the plant literature, although few workers have explicitly tested the relationships between migration and genetic structure as defined by Wright (1969). It is difficult to evaluate the predictions of the model when the appropriate variables are not known. Estimates of the number of introductions, founder size, and levels of migration are required to predict differences between colonial and source populations. While there is increased interest in estimating levels of interpopulation gene flow using the methods of Wright

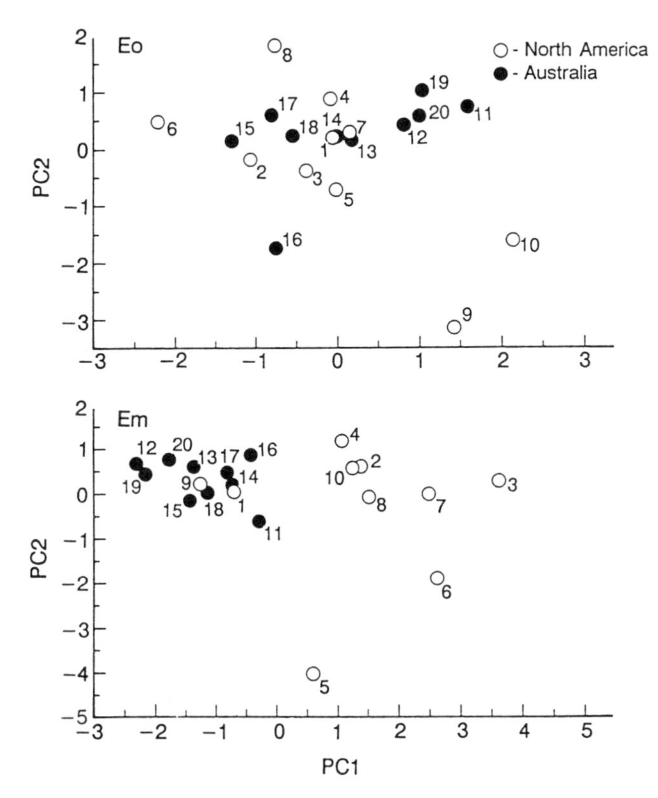

FIGURE 3. Principal components analysis of quantitative variation in 12 life history traits in populations of *Echinochloa oryzoides* (Eo) and *E. microstachya* (Em) from North America (open circles, 1–10) and Australia (closed circles, 11–20). The 20 populations of each species were grown under uniform glasshouse conditions. Note the absence of differentiation between populations from the two regions in *E. oryzoides* and the occurrence of two population of *E. microstachya* (1,9) from northern California that cluster with the Australian sample (S. C. H. Barrett and A. H. D. Brown, unpublished).

(1969) and Slatkin (1985), no studies have as yet examined this in relation to colonization history in plants.

Island model

Island models of colonization assume that migration occurs among subpopulations. In contrast to the continent–island model, migration is multidirectional because the subpopulations are assumed to be of the same effective size. We will describe the botanical evidence for two spatial patterns of population differentiation in the context of the island model. First is the random pattern, in which migration among populations is essentially random. Second is the stepping stone pattern, where migration occurs only between adjacent subpopulations, in one or two dimensions.

The island model has been used as a theoretical framework for interpreting spatial patterns of population structure on a quantitative as well as a qualitative basis. The degree that populations differentiate will depend on the effective population size of each subpopulation. If N_e is small, drift will maintain a random pattern of variation in gene frequencies. In the *Avena* study, discussed earlier, Jain and Rai (1974) compared the observed local differentiation to quantitative predictions of the island model. Using estimates of N_e, migration, and selfing rates, they found that differentiation among subpopulations was consistent with that expected by the island model. That is, the degree of similarity in allele frequencies among subpopulations was not correlated with the distance separating them. Also, the variance in allele frequencies expected based on this model was similar to that observed. Similar spatial patterns of genetic structure have been reported in *Oenothera* (Levin 1975), *Clarkia* (Soltis and Bloom 1986), and *Impatiens* (Knight and Waller 1987). In each case, the genetic distance among populations was not correlated with geographic distance. These species are annual and occur in ephemeral habitats. This suggests that, where habitats are disturbed and extinction and colonization occur frequently, populations rapidly differentiate and a clear source-derivative relationship within a local area may not be maintained.

Not all species from ephemeral environments display a random distribution of alleles among populations. In a recent study of isozyme variation in 12 Jamaican populations of *E. paniculata* sampled throughout the island, genetic distances among populations were significantly correlated with geographic distance (B. C. Husband and S. C. H. Barrett, unpublished). Two spatial patterns are evident from the isozyme survey: (1) central populations on the island are differentiated from populations toward the east and west (Figure 4A,B), and (2) populations in the west are differentiated from those throughout the rest of the island (Figure 4C). These patterns may reflect the initial location of two separate introductions to Jamaica and subsequent diffusion from these points. This hypothesis is supported by both the presence of unique alleles at the loci *Pgi* and *Pgm* and associated differences in floral traits and life history in populations from the western end of the island.

Migration restricted primarily to adjacent populations results in a pattern of genetic variation described as either stepping stone, in patchily distributed species, or clinal, in continuously distributed species. The stepping stone model, first described by Kimura and Weiss (1964), predicts that when migration is restricted in this way, populations will diverge and the correlation among populations will decrease with the distance separating them. A stepping stone model may explain the pattern of variation at neutral loci, for populations that have diffused from a single site of colonization and where

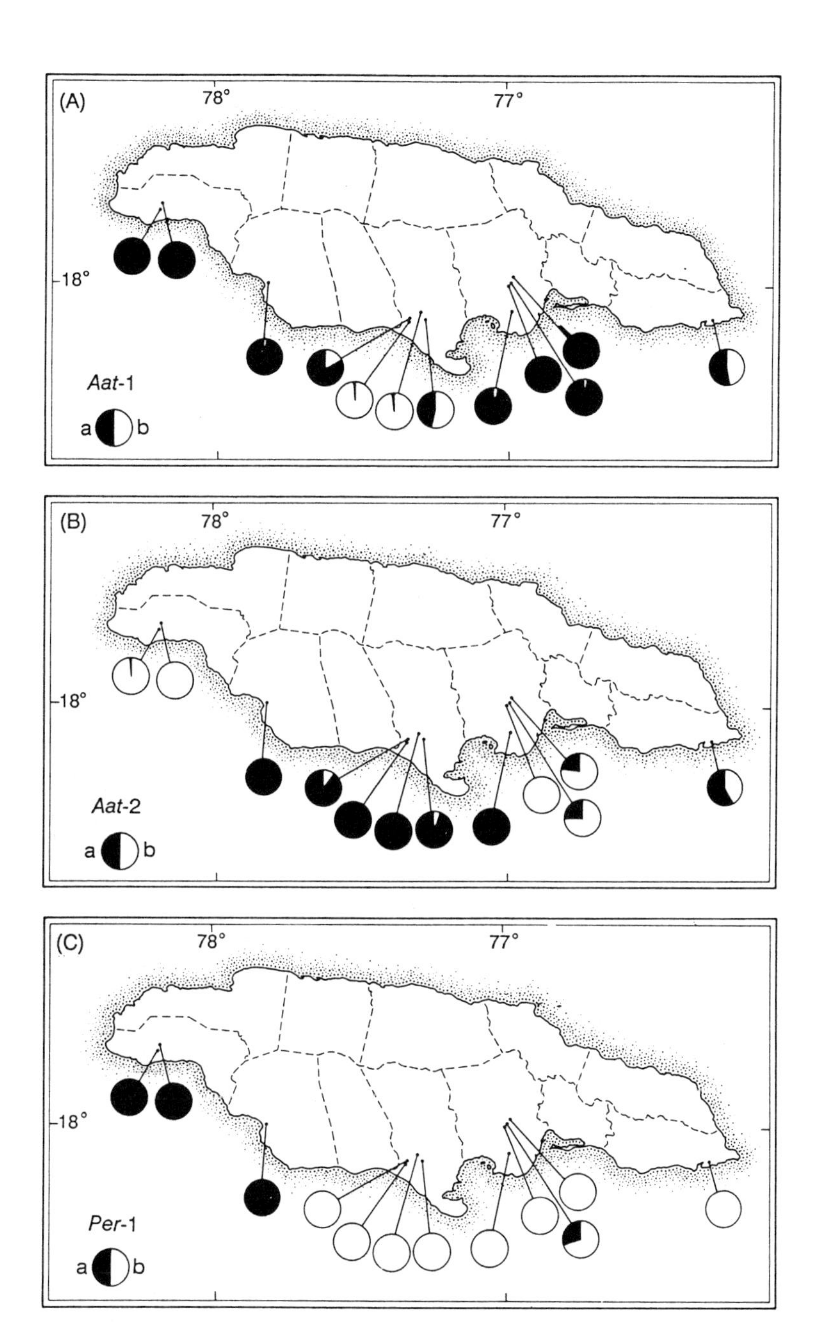

FIGURE 4. Geographic patterns of variability at three isozyme loci in Jamaican populations of *Eichhornia paniculata*. The number of polymorphic loci per population ranged from 0 to 3 of 24 loci screened. A–C, allele frequencies at the loci *Aat*–1, *Aat*-2, and *Per*-1, respectively; a represents the fastest migrating allele and b the slowest (B. C. Husband and S. C. H. Barrett, unpublished).

migration is primarily outward from the source. There are several examples of plant species with populations whose genetic similarity, based on isozymes, is significantly correlated with geographic separation (Lundkvist and Rudin 1977; Yang et al. 1977; Bergmann 1978; Yeh and O'Malley 1980). In contrast to species with random patterns of genetic differentiation (see above), most examples involve tree species from relatively stable habitats. One exception is described by Weber and Stettler (1981), who found a weak but nonsignificant correlation between genetic and geographic distance in *Populus trichocarpa*. Like the annuals that tend to exhibit random variation, *Populus trichocarpa* is an opportunist from ephemeral habitats.

These results suggest that island models of migration may be suitable for predicting the geographic patterns of genetic differentiation throughout the range of a species. Whether a species exhibits regional patterns of differentiation among populations may depend on the direction and frequency of colonization. These are likely to differ among species with contrasting life histories such as long-lived trees and annual plants of ephemeral habitats.

Central–marginal model

The central–marginal model of colonization has provided an important framework for investigating the processes of microevolution (Antonovics 1976). The model assumes that central habitats are environmentally more benign and less isolated than marginal ones. If marginal sites are colonized by migrants from a central source, the model predicts they should be genetically differentiated and less variable than those at the center. Lower genetic variation in marginal populations may be due to genetic drift and/or strong directional selection. The relative importance of these forces is likely to depend on whether central–marginal classifications are based on geographic or ecological criteria.

Empirical studies have not consistently supported predictions of the central–marginal model, regardless of the criteria used to evaluate marginality. Populations in marginal sites may have lower (Farris and Schaal 1983; Silander 1984), similar (Tigerstedt 1973; Levin 1977), or higher (Keeler 1978; Schumaker and Babbel 1980) variability than central sites. These results are not altogether unexpected, however, since most of these studies were conducted across ecological gradients but involved isozyme surveys. Variability at isozyme loci is unlikely to be under strong selection and may be selectively neutral. There are relatively few studies of quantitative genetic variability within central and marginal populations of plants. In *Veronica peregrina* (Linhart 1974) and *Spartina patens* (Silander 1985), the expected decrease in genetic variability in ecologically marginal environments was observed. However, in *Danthonia spicata*, there was no consistent difference between central and marginal populations (Scheiner and Goodnight 1984). These authors suggest that *D. spicata* has only recently colonized the marginal sites they studied and, as a consequence, selection has had little time to operate.

Most comparisons of central and marginal populations lack information on the historical relationships among the populations examined. An exception involves the postglacial migration of *Pinus contorta* in northern Canada (Cwynar and MacDonald 1987). Palynological studies from lakes throughout the northern range of *P. contorta* suggest that the species migrated from southern refugia to the Yukon Territory in the last 12,000 years. Cwynar and MacDonald (1987) compared the date of colonization with genetic diversity, for 42 isozyme loci and several quantitative traits, at 15 locations along a north–south transect through its distribution. The average number of alleles per population decreased from the center to the northern periphery of the range (Figure 5A). However, levels of heterozygosity remained constant. While the effects of selection cannot be completely discounted, the fact that populations with the least genetic diversity are also the youngest supports the hypothesis that the observed central–marginal cline results from drift occurring during the colonization process. Several quantitative traits were also correlated with the time since populations were founded. Cwynar and MacDonald (1987) suggest that selection in peripheral populations on characters conferring greater dispersal ability, such as seed mass and wing loading, may have occurred (Figure 5B). Palynological approaches combined with isozyme studies may prove valuable for determining the colonization history of species for which there is a reliable pollen record.

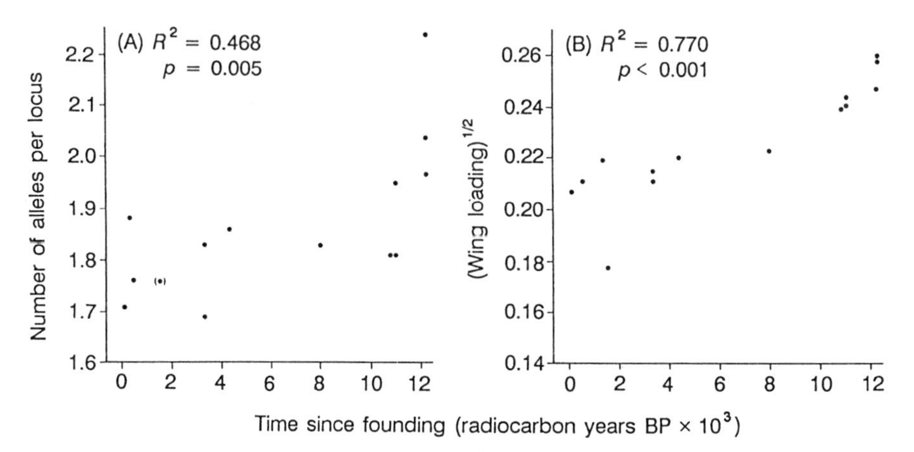

FIGURE 5. The relationship between the time since founding and (A) the number of alleles per locus and (B) the square root of wing loading, an index of seed dispersal, in populations of *Pinus contorta* ssp. *latifolia* from N.W. Canada. Values for time of founding were obtained from palynological records and radiocarbon dating (after Cwynar and MacDonald 1987).

MATING SYSTEMS AND COLONIZATION

Populations colonizing new territory are often confronted with novel environments, particularly after long-distance dispersal. In most cases, the environments are unsuitable and founding colonies are soon extirpated. Unfortunately, we know almost nothing about the causes of failure of most introductions so, for particular colonization models, it is difficult to evaluate the relative importance of factors such as chance, low genetic variation, or difficulties in finding mates. When individuals are preadapted to the new environment, colonizing populations may spread and are then often exposed to a set of conflicting demands. There may be strong selection pressures on the genetic system for phenotypic innovation. This requires mobilization of genetic variability and a flux of new genotypes. It may be achieved by genetic changes in the breeding system through increased levels of outcrossing. However, if, as is often the case during early stages of colonization, population sizes are small and plant density low, mating between individuals may be difficult. Under these circumstances, assured reproduction through self-fertilization or apomixis may be selectively advantageous. The particular solution to the conflicting demands of genetic experimentation on the one hand and assured reproduction on the other may depend largely on the position of a species on the colonization continuum, and the levels and organization of genetic variation in introduced populations.

Among plants characterized by frequent colonizing episodes and severe fluctuations in population size, uniparental reproductive systems tend to predominate (Baker 1955; Allard 1965; Brown and Marshall 1981). A variety of ecological and genetic factors influence the evolution of self-fertilization and apomixis (Jain 1976; Brown and Marshall 1981). Hence it is not clear for most colonizers whether outcrossing systems based on self-incompatibility or dioecism are maladaptive because they impose too severe a constraint on mating during colonization, or because they fail to suppress genotypic diversity and perpetuation of successful genotypes. Experimental studies on the effects of low density on mating success in outbreeding plants would be useful in assessing the role of reproductive assurance in colonizing plants, since it is often argued that, in general, seed set in plants is rarely pollen limited.

An example of the disruptive effects of frequent colonizing episodes and small population size on the maintenance of outcrossing occurs in tristylous *Eichhornia paniculata*. Populations of this species inhabit seasonal pools and ditches in arid N.E. Brazil where the size and life history of populations are largely determined by available moisture (Barrett 1985). Since the region has one of the most unpredictable rainfall regimes in the world, colonization–extinction cycles are a prominent feature of this annual or short-lived perennial aquatic. In N.E. Brazil, *E. paniculata* populations display a wide range of mating systems from large tristylous outcrossing populations that contain high levels of genetic diversity to semihomostylous selfing populations with low levels of genetic polymorphism (Glover and Barrett 1986, 1987). The evolution of selfing in *E. paniculata* is associated with colonization of ecologically and geographically marginal sites and involves changes in population structure from stylar trimorphism through dimorphism to monomorphism. The breakdown of tristyly is occurring in contemporary populations of *E. paniculata* in both N.E. Brazil and Jamaica, enabling examination of the microevolutionary processes responsible for the evolution of self-fertilization (Barrett et al., in press).

Dissolution of the tristylous genetic polymorphism in *E. paniculata* occurs in two stages involving the sequential loss of alleles at the two diallelic loci (*S*,*M*) that govern the inheritance of the polymorphism. Loss of the *S* allele and hence the S morph converts trimorphic populations to dimorphic populations. This process appears to occur by both random and deterministic processes. Stochastic fluctuations in population size and founder events are more likely to result in a loss of the *S* allele from populations than the three remaining alleles at the *S* and *M* loci (Heuch 1980; Barrett et al., in press). In addition, selection mediated by pollinators may also lead to a loss of the S morph from populations. Where specialist long-tongued pollinators are absent, such as in small or isolated populations, the seed set of the S morph suffers, in comparison with the L and M morphs, because of its concealed female reproductive organs. The decline in frequency of the S morph occurs more rapidly when mating patterns change from disassortative to random mating. Such an effect is more likely in small colonizing populations serviced by generalist pollinators (Barrett et al., in press).

Loss of the L morph from dimorphic populations of *E. paniculata* accompanies the spread of genes altering stamen position in the M morph. The genes that are recessive in nature modify the mating system, resulting in high levels of self-fertilization. Their origin in dimorphic populations may be associated with their low density and small size in comparison with trimorphic populations. Recessive genes are more likely to be exposed to selection through inbreeding in small populations than in large outcrossing populations. Once selfing variants arise, they appear to be favored over the L and unmodified M plants through reproductive assurance under conditions of low pollinator service (Barrett et al., in press). However, even where pollinator service is reliable, the automatic selection of the M morph, leading to the origin of floral monomorphism, can occur as a result of mating asymmetries between the morphs. Under this model of mating system evolution, a negative relationship should exist between the frequency of the L morph and of selfing variants of the M morph in dimorphic populations. Surveys of morph frequencies in N.E. Brazil provide evidence for this association (Figure 6). As in many of the cases discussed previously, however, samples collected over space rather than through time can at best give only indirect evidence of the dynamic processes responsible for evolutionary change within populations. This problem is particularly acute in colonizing species where populations are

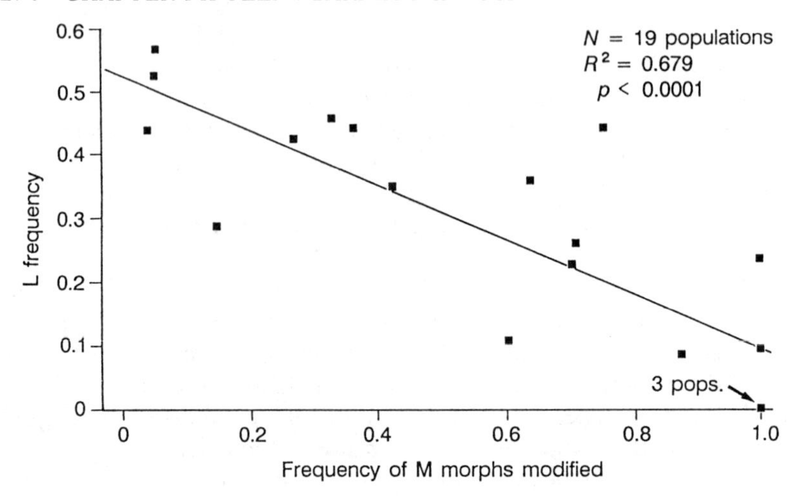

FIGURE 6. The relationship between the frequency of the long-styled morph (L) and the frequency of mid-styled (M) plants that were self-pollinating variants in 19 dimorphic and monomorphic populations of *Eichhornia paniculata* from N.E. Brazil (after Barrett et al., in press).

often short-lived and rarely at equilibrium. Accordingly, it is often difficult to study temporal changes in gene frequencies for any extended period of time and samples in space are likely to be subject to large variances due to genetic drift.

Selfing variants of *E. paniculata* predominate on the island of Jamaica. This is in accord with Baker's Law (Baker 1955, 1967) that establishment following long-distance dispersal favors self-compatible (and in this case autogamous) colonists. Once self-compatible plants have established on islands, however, it is of interest to examine whether their mating systems can evolve in response to new selective pressures. In *E. paniculata*, high levels of self-fertilization are typical of all Jamaican populations that we have examined. This is probably because they occur in highly disturbed, ruderal habitats subject to frequent local extinctions. However, maintenance of high levels of self-fertilization is not typical of all island colonists, particularly those that have undergone ecological diversification. A considerable literature has developed in recent years concerning mating system evolution and patterns of genetic differentiation of island plants (e.g., Lowry and Crawford 1983; Baker and Cox 1984; Helenurm and Ganders 1985; Barrett and Shore 1987; Witter and Carr 1988). While the necessity of self-compatibility and even autogamy for insular establishment is generally recognized, it appears that the autochthonous development of outcrossing mechanisms has occurred in many groups of island plants. This pattern is particularly associated with adaptive radiation into new habitats (Carlquist 1974; Ehrendorfer 1979; Baker and Cox 1984).

A particularly good example of mating system change in island colonists is evident in the genus *Bidens* on the Hawaiian islands (Sun and Ganders 1988). All 19 endemic species are thought to have originated from adaptive radiation of a single ancestral species following long-distance dispersal, probably from the American mainland. All taxa on Hawaii are interfertile but exhibit more morphological and ecological diversity than the remaining 200 species in the genus on five continents. Gynodioecy occurs in nine of the Hawaiian species but is unknown in non-Hawaiian taxa. Genetic and developmental studies indicate that the origin of male sterility is homologous in all Hawaiian *Bidens*, suggesting that gynodioecy most likely evolved during adaptive radiation of an initial colonizer after arrival in the oceanic islands (Sun 1987; Sun and Ganders 1987). The frequencies of females in gynodioecious populations of Hawaiian *Bidens* are positively correlated with the selfing rate of hermaphrodites suggesting that gynodioecy has evolved in response to increased levels of inbreeding in hermaphrodite populations (Sun and Ganders 1986). This is likely to have occurred in *Bidens*, since population sizes of many of the Hawaiian species are extremely small (Helenurm and Ganders 1985) favoring increased inbreeding. Other examples of the autochthonous development of outcrossing mechanisms in island plants include the evolution of dioecism in Hawaiian *Wikstroemia* (Mayer 1987) and the evolution of herkogamy in homostylous *Turnera ulmifolia* var *angustifolia* in the Caribbean (Barrett and Shore 1987).

CONCLUSIONS

In the two decades since the symposium volume on the Genetics of Colonizing Species (Baker and Stebbins 1965), much descriptive information has been amassed on the mating systems and population genetic structure of colonizing plants (reviewed by Brown and Burdon 1987; Barrett and Shore, in press). This has enabled some generalizations to be made concerning the genetic characteristics of species with well-developed colonizing abilities. While stressing that colonizing species are by no means a homogeneous group, Brown and Marshall (1981) identified several recurrent patterns in successful colonizers. Shared features often included fixed heterozygosity through polyploidy, propagation by self-fertilization or asexual means, genetically depauperate populations with respect to isozyme variation, high levels of multilocus association, marked population differentiation, and high levels of phenotypic plasticity. Knowledge of the shared attributes of successful colonizers provides useful information for studies of plant colonization, since it enables us to make educated guesses as to which genetic and ecological features of populations are likely to be important to colonizing success. However, beyond the generalities, there are the specific details of individual colonizing episodes and these are likely to vary with species and environment. Satisfactory explanations for the success or failure of colonizing episodes are most likely to be obtained from detailed observations of natural colonization events or from experimental studies of artificially established colonies.

It is remarkable that, considering the ease with which many plant colonizers can be grown, measured, crossed, and manipulated, so few experimental studies have been conducted on the genetic aspects of plant colonization. Agricultural weed populations in particular offer attractive opportunities for experimental studies since, in comparison with plants of natural communities, they inhabit relatively uniform environments in which the selection pressures are more easily identified and controlled (Barrett 1988). One of the few attempts to examine the genetic consequences of colonization by experimental means are the studies by S. K. Jain and colleagues at Davis. In a series of long-term experiments involving establishment of artificial colonies of known genetic composition in *Trifolium hirtum* and *Limnanthes* spp. they have examined the importance of mating systems and levels of genetic variation for colonizing success (Jain and Martins 1979; Martins and Jain 1979; Jain 1984). Because of the long-term nature of these experiments, few clear answers have been obtained to date, although there is some evidence that variable colonies of *T. hirtum* establish more successfully than less variable colonies. In the future, more widespread use of genetic markers in experimental demographic studies (Ritland, this volume) may enable workers to measure parameters required to test various colonization models and to determine the relative importance of stochastic versus deterministic processes to evolutionary change in colonizing populations.

While experimental studies that involve the founding and manipulation of colonies may enable us to test hypotheses concerned with colonization, there is still need for comparisons of the spatial and temporal patterns of genetic diversity between colonizing and noncolonizing populations within species. As mentioned above, little is known about genetic changes that occur within plant populations through time, and yet these are critical for understanding the effects of population size. In the few studies where investigators have followed genes, genotypes, or phenotypes through time, either within or between seasons, significant changes have been reported and important insights have been obtained into the stages in the life cycle where selection is most intense (Clegg et al. 1978; Gray 1987; Allard 1988). Of particular value in these types of studies would be to examine the patterns of variation exhibited by different classes of genetic variation (e.g., isozymes versus quantitative traits) following colonizing episodes that involve periods of small population size.

Our understanding of the genetic consequences of colonization in natural populations is largely based on the patterns of isozyme variability, as measured by electrophoresis. Yet, levels of isozyme variation may not accurately reflect variability elsewhere in the genome (Giles 1984). Isozymes are well suited to indexing levels of genetic variation and relationship among individuals and among populations (Brown and Burdon 1987). Many quantitative traits have direct effects on survival and reproductive success. Therefore, investigating the effects of colonization on such characters may provide more meaningful insights into the biological significance of colonization and its effect on evolutionary potential subsequent to colonization.

Theories of speciation through genetic bottlenecks are based on the assumption that the genetic architecture of species results from multilocus epistasis (Mayr 1970; Templeton 1980). In these models, speciation occurs through the breakdown and reassembly of this multilocus structure, by founder events. In contrast, models of the effects of small population size within species are often based on single loci with additive gene effects. Although few empirical studies have examined the effects of colonization on multilocus structure, there is some evidence that genetic correlations between quantitative traits are altered by small population size (Mitchell-Olds 1986), inbreeding (Rose 1984), and novel selective pressures (Silander 1984). However, it is not known how often colonizing ability is constrained by negative genetic correlations between fitness components. Future analysis of genetic correlations will likely shed light on the constraints to evolution after colonization as well as allowing a more realistic assessment of speciation models invoking founder effects (Carson and Templeton 1984; Barton and Charlesworth 1984).

From this review, it is clear that the scale and pattern of colonization have profound effects on the organization of genetic variation in plant populations. The genetic theory of finite populations has defined, mathematically, how small populations cause reductions in variation through drift. In addition, simple models of colonization and migration help us to predict the spatial patterns of genetic differentiation among populations. Despite these theoretical foundations, few studies of natural populations exist to evaluate their predictions and many of the critical parameters (e.g., migration rates, effective population sizes) are unmeasured. At best, we can examine the qualitative predictions of the models. Colonization is a process shared by all species. Yet it is unlikely that the genetic implications are the same for all. In order to understand the genetic consequences of migration and colonization for different species, we must acquire better ecological and genetical explanations for the factors regulating their distributions.

ACKNOWLEDGMENTS

We thank J. J. Burdon, L. C. Cwynar, and C. B. Fenster for comments on the manuscript, Elizabeth Campolin for drawing the figures, and the Natural Sciences and Engineering Council of Canada for financial support.

SECTION 3

APPLICATIONS IN PLANT BREEDING AND GENETIC RESOURCES

So far the discussions in this book have centered on the twin pillars of modern plant population genetics. First is the discovery and measurement of genetic diversity among members of populations, now possible at all levels (Section 1). Second are the basic evolutionary forces that most immediately and directly impinge on this diversity. Earlier in this century population genetics began largely with theoretical study of the potential impact of hypothetical evolutionary forces on idealized, discrete genes. Now modern techniques have supplied the tools to uncover such genes and the impact of microevolutionary forces on populations is no longer hypothetical. We now turn to two closely related areas of endeavor where knowledge of the genes and of the forces are major essential ingredients to current and future progress.

In transferring our attention from natural plant populations of the ecological geneticist to the crops of the plant breeder, forest species provide a natural transition. With their long life cycles forest species are at the earliest stages of domestication. Yet with development of clonal systems and the prospect of genetic transformation, forest plantations stand ready for potentially drastic change in their genetic make-up. Plant population genetic research has had a recent profound effect on research in genetic improvement of forests as Muona's chapter shows. Analyses of genetic diversity, of mating patterns, and of gene flow are important because of our growing potential to effect great change to our forests, but changes that will commit resources and strategies in future decades to a much greater extent than in annual crops. The link between continued productivity and conservation requires the most extensive and rigorous knowledge base for making decisions.

Selection has a fundamental role in shaping the genetic structure of plant populations. Selection schemes are also the fundamental transformer of plant breeders' materials. Weber, Qualset, and Wricke survey the whole range of options open to the breeder of self-pollinated crops, from the traditional backcross to the emerging technologies of genetic engineering. Plant breeding is a resource-demanding task, where the basic payoff is the enhanced "value" of the new cultivar (Allard and Hansche 1965). Thus the modeling of selection strategies such as in recurrent selection and the analysis of relative efficiency of varying experimental inputs play a crucial role in optimizing effort.

The very genes or alleles detected by the student of population variation form yet another direct link between population genetics research and plant breeding. This can happen in two ways. First, the variants themselves may have direct effects of economic traits. Second, they may be tags of useful yet undefined genes nearby on the chromosome.

The direct use of alleles is well exemplified by the common bean project described by Bliss, which integrates biochemical techniques, and population genetics in the use of genetic resources for crop improvement. His chapter also addresses the problem of how breeders are to exploit the desirable germplasm presently to be found in an otherwise ill-adapted genetic background. Once the biochemical components of a trait are better understood, the transfer of specific alleles into the breeder's population through conventional crossing is remarkably effective when it can use precise selection criteria.

The importance of the alleles of experimental population genetics as markers is most evident in the maize project described by Stuber, where alleles at the level of isozymes and the restriction fragments offer for the first time a formal genetics (linkage mapping and epistasis analysis) for genuine phenotypic characters of the plant breeder. Given a molecular map of sufficient density, it is possible to identify and locate short regions of the chromosome that account for much of the variation in yield or quality components in a cross. Measurement of the relative contribution from each segment is possible. Contributing segments are thus tagged by molecular variants and can be manipulated much more readily in pedigree breeding. Further the relationship between hybrid response and the degree and nature of genetic divergence can be explored, in the hope of gaining greater predictive power for heterosis. Finally it should be possible to dissect heterosis itself, and analyze the effect of variation in background.

And what of the genetic basis of quantitative traits, the so-called real traits of the plant breeder, and of the student of life history in natural populations? Mayo reviews the present state of plant quantitative genetics, particularly in light of potential molecular approaches. He describes quantitative or biometrical genetics and the plant breeding it supports as mature fields in which new research will be incremental in character. Modeling realistic assumptions about the basic parameters of gene numbers and action spectra will be important in improving mating designs and selection schemes. These tasks will not be removed by the new analytical technology. Indeed the need may be increased as new synthetic genes become available in large number for incorporation into pedigrees, and greater arrays of material arrive in the breeder's field trials for evaluation on the farm. Efficient evaluation will be at a premium.

Finally we come full circle to genetic resources themselves, the concern of the breeder and the biologist. Population genetics theory is clearly a foundation for sound collection, conservation, and use of genetic variation. Yet social and political issues have injected a new, and at times factious, dimension to the discussion of gene resources. In the last chapter Marshall reviews many of the scientific questions that should assist decisions in this area. He ends by challenging us to think of new ways to organize gene banks, which both build on existing active national programs and yet increase international cooperation in a global network. In this way, the precious heritage of the genetic diversity of populations is both safeguarded and made available to all.

CHAPTER 16

POPULATION GENETICS IN FOREST TREE IMPROVEMENT

Outi Muona

ABSTRACT

Research on population genetics of forest trees has concentrated on measuring the level and distribution of genetic variability, in particular at isozyme loci, and on quantifying reproductive patterns using marker loci. Most temperate conifer populations are highly variable at isozyme loci, but there is little differentiation between populations due to efficient gene flow by pollen dispersal. In contrast, adaptive quantitative characters are often more differentiated due to varying selection pressures. Partial selfing occurs in most species, but there is no evidence of inbreeding at the adult stage, as inbred zygotes are eliminated in juvenile life stages. Intense fertility selection takes place during the reproductive cycle in conifers. Reproductive patterns in angiosperm trees are poorly known. The findings have bearing in many areas of tree breeding, e.g., estimation of genetic parameters, design of seed orchards, planning of breeding populations for long-term genetic improvement, and gene conservation.

INTRODUCTION

Tree improvement is a young field of science, especially in relation to the generation time of trees. Most tree improvement programs started less than 50 years ago. Many of the harvested forest tree populations are regenerated naturally, and in most cases even domesticated plantations are just a few generations removed from natural populations (Stern and Roche 1974). As in all breeding, there are two conflicting goals. The first is to maintain the diversity and adaptedness of the natural populations and the second to make rapid genetic improvement (e.g., Cotterill 1984).

As domestication and tree breeding proceed, there will be increasing differentiation between kinds of populations to meet different needs (Namkoong 1984; Cotterill 1984; Ledig 1986; Kang and Nienstaedt 1987). Natural populations of all species will need to be maintained for ecosystem and gene conservation. Breeding populations are designed as the basis of continued long-term tree improvement (Burdon and Namkoong 1983). Material for regenerating production populations will be produced in multiplication populations (e.g., seed orchards) (e.g., Kang and Nienstaedt 1987).

The level of differentiation between these kinds of populations will vary widely between species and cultivation conditions. In particular, the nature of the production populations will differ. In many northern conifers with long generation intervals, production populations in the future may be close to natural populations. They will be grown in a natural environment with a low level of management. Genetic diversity and natural adaptation will be required as protection against environmental and biotic stress. On the opposite extreme, some short rotation species may be grown in intensively managed monoclonal plantations.

Thorough understanding of the distribution of genetic variability and breeding systems of trees, and the evolutionary forces that have shaped them, is a prerequisite for tree breeding. Forest geneticists have emphasized the study of natural populations and such research should continue (Namkoong 1984). This information is needed for planning gene conservation and breeding populations. It also forms a sound basis for deciding on the required genetic constitution of production populations. Mating patterns should be studied for formulating breeding strategies and designing seed orchards.

This chapter focuses on studies on genetic variability and mating patterns in forest trees and discusses their implications for tree breeding. The results will be from natural populations and seed orchards, where experimental work has concentrated in recent years. Much of the research discussed has been stimulated by the use of isozyme marker loci.

GENETIC VARIABILITY IN TREES

Isozyme variability

Genetic variability in trees has been reviewed by Hamrick et al. (1981), Loveless and Hamrick (1984), and Mitton (1983). Gymnosperm trees have been found to contain a higher level of isozyme variation within populations than other groups of organisms (see Hamrick and Godt, this volume). The average gene diversity (H_e) over 20 species was 0.207 (Hamrick et al. 1981), but diversity varied greatly between species, from 0 in *Pinus torreyana* (Ledig and Conkle 1983) to 0.327 in *P. longaeva* (Hiebert and Hamrick 1983). Numerous recent studies have confirmed the high genetic variability of conifers.

Less is known about isozyme variability in angiosperms. The average diversity for nine species of wind-pollinated angiosperm trees representing the genera *Alnus*, *Populus*, and *Quercus* was 0.137, with a range of 0.06–0.42 (Bousquet et al. 1987, references in Hamrick and Loveless 1986; Manos and

Fairbrothers 1987). Hamrick and Loveless (1988) surveyed 16 tropical angiosperm tree species in Panama, which had average H_e = 0.211. Moran and Hopper (1987) gave an average H_e of 0.182 for nine species of eucalypts, while nine Australian species of the genus *Acacia* were somewhat less variable (average H_e = 0.132, with a range from 0.017 to 0.300 (Moran et al. 1989a). The averages for angiosperm trees seem to be slightly higher than for other angiosperm plants, which have an H_e = 0.136 (Loveless and Hamrick 1984). The conclusion of Hamrick et al. (1979) that trees are more variable than other plants is supported by recent data. This is probably due to their longevity, predominant outcrossing, broad range, and potential for extensive gene flow, all of which contribute to large effective population sizes.

Isozyme variability may not always reflect the amount of variability in the rest of the genome. The amount of variability of quantitative characters may be governed more strongly by selection than variability at isozyme loci. Extended bottlenecks may deplete genetic variability at all loci, but when the population size increases again, previous levels of variability may be reached more rapidly in quantitative characters than at enzyme loci (Lande 1988). *Picea omorika* is known as a morphologically uniform species (Burschel 1965), but has considerable variability at enzyme loci (H. Kuittinen and O. Muona, unpublished). *Pinus torreyana* is a species that seems to have lost both isozyme and quantitative variability within populations through bottlenecks (Ledig and Conkle 1983). *Acacia mangium* is a fast-growing multipurpose tree that is being cultivated increasingly in southeast Asia. It contains little variation at isozyme loci throughout its wide range from Indonesia to Australia, with average H_e being just 0.017 (Moran et al. 1989a). Studies of isozyme variation are still the quickest way to obtain a picture of the level and distribution of genetic variability in species undergoing domestication and entering tree breeding programs.

Distribution of variability

Population differentiation is measured by the relative gene differentiation, the proportion of total gene diversity found between populations (G_{ST}, Nei 1973). Wind-pollinated conifers are only slightly differentiated, the average G_{ST} for 23 species being 0.076 (data from Govindaraju 1988). The range in G_{ST} over species is from 0.02 for *Pinus sylvestris* to 1.00 for *Pinus torreyana* (Ledig and Conkle 1983). Data for angiosperm trees are scarce. The average for nine wind-pollinated species of the genera *Alnus*, *Populus*, and *Quercus* was 0.075 (Bousquet et al. 1987; Govindaraju 1988). In animal pollinated populations, widespread eucalypts were found to have G_{ST} values of 0.10–0.12 (Moran and Hopper 1987). In two species of tropical acacias, Australian and New Guinea populations were differentiated with G_{ST} values of 0.09 and 0.18 (Moran et al. 1989b). The few data suggest that insect pollinated trees are more differentiated with respect to allozyme variation than are widespread wind-pollinated species.

Many studies suggest that quantitative traits are much more differentiated between populations than isozymes (e.g., Wheeler and Guries 1982; Falkenhagen 1985). *Pinus sylvestris* is a good example of a species showing such a pattern. It is highly differentiated between latitudes or altitudes with respect to quantitative characters related to climatic constraints. In common garden experiments in Finland, the northernmost populations (70°N) set bud in mid-July, the southernmost (60°N) in mid-September. The correlation of median budset date with latitude was −0.968 (Figure 1, Mikola 1982). Results from transfer experiments in *P. sylvestris* parallel these findings and show that even a short-range transfer of seedlings north or to an altitude higher than their source location results in increased mortality (Eriksson et al. 1980). In contrast, several allozyme studies in *P. sylvestris* have shown little latitudinal differentiation in allelic frequencies in northern Europe (Chung 1981; Gullberg et al. 1985; Muona and Szmidt 1985), with G_{ST} values less than 0.02. In *Pseudotsuga menziesii*, only 1% of variability at isozyme loci was between breeding zones in southwest Oregon, whereas such zones account for a large

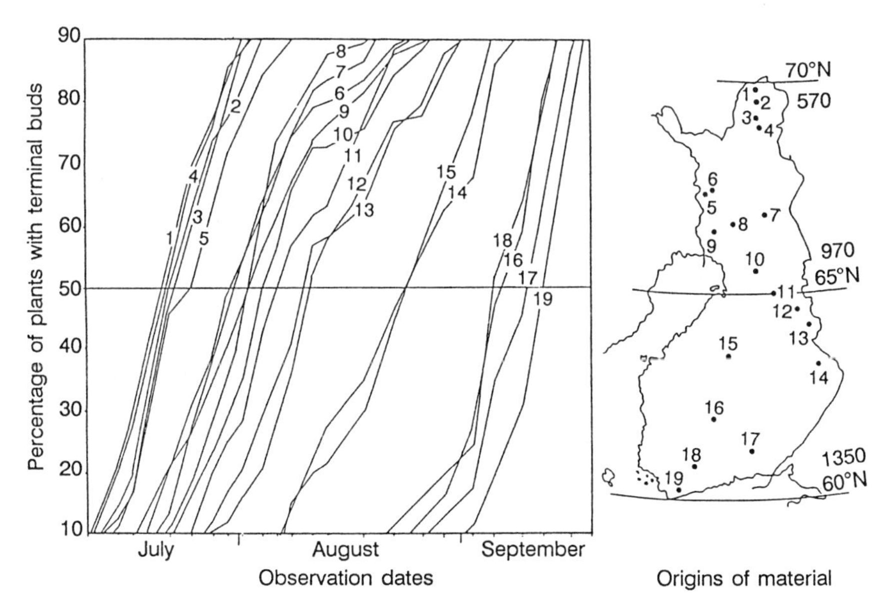

FIGURE 1. Variation in the date of bud set of 1-year-old seedlings of *Pinus sylvestris* in a common garden experiment. The origins of the population samples from within Finland are shown on the map, as well as the temperature sums of the growing season at locations 1, 10, and 19. From Mikola (1982).

proportion of quantitative variability (Merkle and Adams 1987). However, quantitative characters not directly related to climatic constraints, e.g., wood quality, may be less differentiated in many species (Zobel and Talbert 1984).

While quantitative traits often have geographic patterns suggesting adaptive significance, patterns of isozyme variability indicate that isozyme loci are not subject to the same selection pressures. In contrast, the patterns reflect their effective population sizes and gene flow between populations.

Associations between isozyme variability and quantitative characters within populations

Associations between loci are not easily maintained in large predominantly outcrossing populations, except between very tightly linked loci (e.g., Epperson and Allard 1987). As in other outbreeding organisms, gametic disequilibrium between unlinked pairs of loci is expected to be rare in trees (Brown et al. 1975; Muona and Szmidt 1985). However, some reports of associations exist (e.g., Epperson and Allard 1987).

As gametic disequilibrium is not found between pairs of loci, it is not likely that associations will be found between specific marker loci and quantitative characters either. El-Kassaby (1982) failed to find significant associations between isozyme loci and quantitative traits in *Pseudotsuga menziesii*, however, Linhart et al. (1979) reported significant associations between specific alleles and quantitative characters in natural populations of *Pinus ponderosa*. In other plants, useful associations are often generated in crosses between lines that are differentiated with respect to both isozymes and quantitative characters (e.g., Edwards et al. 1987). Disequilibrium between isozyme and quantitative character loci could be used for tree breeding when there is potential for use of hybrids, as, e.g., between *Pinus halepensis* and *P. brutia* (Schiller et al. 1986). Within species, strong associations might be found if a dense map of marker loci with large numbers of alleles is available, and the variation of the quantitative characters concerned is governed by major loci. So far such marker loci have not been used for selection of economically important traits in forest trees (see Adams 1983). This may change as restriction fragment length polymorphisms (RFLPs) become available.

Scope of use of isozyme variation

Isozyme variation provides little information on the pattern of distribution of adaptive quantitative characters, and may not be very useful for describing adaptive patterns of variation in tree species. Thus, decisions on the range of possible seed transfers, on delimiting breeding zones, or on some aspects of sampling for gene conservation would best be based on genetic study of quantitative variability. In most cases it will be difficult to distinguish between seed sources based on isozymes, despite some success with multivariate methods (e.g., Yeh et al. 1985). Isozymes (Wheeler and Guries 1987; Yeh and Arnott 1986) and chloroplast DNA variation (Wagner et al. 1987; Szmidt et al. 1988) have both been used to identify hybrids.

Isozymes can be used for planning many aspects of genetic conservation (see Brown and Moran 1981). They are also suitable for studying genetic change accompanying domestication. The level of variability in natural populations is a standard against which variability in breeding populations, seed orchard crops, or production populations can be compared (e.g., Adams 1981; Szmidt and Muona 1985; Moran and Bell 1987). Isozyme loci are useful markers for studying evolutionary factors that are expected to influence all loci in the genome equally. That is, they are excellent for providing information about reproductive patterns (e.g., outcrossing rates, level of inbreeding) and reduction of variability due to small population size, etc. For these purposes, that isozyme loci seem to be less influenced by selection than many quantitative characters is an advantage. In the rest of this chapter, these kinds of applications will be discussed.

REPRODUCTIVE PATTERNS

Reproductive patterns in a population determine the genetic structure of the progeny generation. Information on mating systems, fertility variation, and migration in natural populations is needed for planning gene conservation programs and for managing breeding and production populations. Outcrossing rates must be known so that losses due to inbreeding depression can be predicted. The family structure of open-pollinated or polycross progenies influences estimation of components of genetic variance. The rate of gene flow is also an important parameter.

Outcrossing rates and inbreeding depression in natural populations

Most tree species are known to express severe inbreeding depression in growth characteristics (e.g., Eriksson et al. 1973a). Selfed *Pinus radiata*, measured at 7 years, grew more slowly, had poorer stem, and were less resistant to pests than outcrossed controls (Wilcox 1983). In three conifers, inbreeding depression in growth ranged from 29 to 36% in the first 10 years (Sorensen and Miles 1982). Severe inbreeding effects also have been found in *Eucalyptus regnans* (Eldridge and Griffin 1983).

To understand the consequences of inbreeding, it is worthwhile to consider the demography of natural tree populations (see Stern and Roche 1974). The thoroughly studied *Pinus sylvestris* is a typical monoecious, wind-pollinated conifer (Sarvas 1962). The proportion of own pollen around the crown of a tree has been estimated at 26%. As no self-incompatibility mechanisms are known, the proportion of self-fertilization is expected to be high. Additional inbreeding may be due to mating between relatives. In each seed, on the average, two archegonia are formed and fertilized, but only one embryo develops to maturity. During embryonic development many of the selfed zygotes die due to homozygosity of embryonic lethal genes. The low viability of selfed seed results in high frequencies of empty seed on complete selfing (Koski 1971). Thus, in wind-pollinated progenies, most selfed zygotes are expected to die before seed maturation. The annual seed crop of about one million seeds/ha still contains some selfs (Koski 1980). However, between the seed stage and the adult stage there is heavy mortality. A stand is usually established with some 50,000 seedlings, of which less than 1000 survive to maturity. Much of this mortality is random, but it also causes genetic changes.

Inbreeding is reflected as increased homozygosity at marker loci. It is not possible to obtain estimates of outcrossing at the zygote stage, except by indirect inference from empty seed frequencies. Outcrossing rates are estimated from germinating embryos of mature seed. Early estimates were based on morphological marker genes (see Squillace 1974 for references), but at present estimates are obtained using isozyme markers (see Brown et al. 1985). Conifers have haploid megagametophytes, which, when analyzed with the embryo (Müller 1977), can distinguish the paternal and maternal contributions to the zygote. This has been used in developing estimation method for conifers (Shaw and Allard 1982a; Neale and Adams 1985a; Ritland and El-Kassaby 1985). Table 1 shows that natural populations of species in the genus *Pinus* have high outcrossing rates (t), except for *Pinus radiata*. The average for the genus *Picea* is 0.88. Outcrossing rates in *Pseudotsuga menziesii* resemble those in pines. *Larix laricina* may have one of the lowest outcrossing rates of conifers. The three species in the genus *Acacia* are predominant outcrossers. Brown et al. (1985) compiled results on 10 eucalypt species, which have intermediate outcrossing rates (mean 0.77).

Part of this apparent selfing ($s = 1 - t$), especially if based on single locus estimates of t, may be due to mating between relatives (Shaw and Allard 1982a; Brown, this volume). There is some evidence of family groups in natural populations of trees based on the distribution of genotypes (see Mitton 1983). Further, studies of empty seed frequencies in crosses between nearby trees suggest that neighbors may be related, e.g., in *Picea glauca* at the half-sib level (Park et al. 1984). Because of the importance of inbreeding in forest trees, the variation in outcrossing rates has been studied in different populations and silvicultural settings. Differences between populations have been found, e.g., between elevations in *Abies balsamea* (Neale and Adams 1985b). Low population density was associated with lower outcrossing rates in *Picea engelmannii* and *Abies lasiocarpa* (Shea 1987) and *Larix laricina* (Knowles et al. 1987). Temporal variation in outcrossing rates also has been described, e.g., in *Eucalyptus stellulata* (see Brown et al. 1985).

Direct estimates of outcrossing can be made only with family structured data. In population samples, observed and expected heterozygosity can be compared for monitoring levels of inbreeding. Many of the studies listed in Table 1 have shown that partial selfing results in positive fixation indices at the seed stage, e.g., in *Pseudotsuga menziesii* (Shaw and Allard 1982b) and *Pinus sylvestris* (Yazdani et al. 1985b). In contrast, adult genotypic frequencies have shown no evidence of inbreeding in *Eucalyptus delegatensis* (Moran and Brown 1980), *Pseudotsuga menziesii* (Shaw and Allard 1982b), *Sequoiadendron giganteum* (Fins and Libby 1982), *Picea glauca* (King et al. 1984), *Abies balsamea* (Neale and Adams 1985b), *Liriodendron tulipifera* (Brotschol et al. 1986), and *Larix laricina* (Knowles et al. 1987). Apparently, inbred seedlings do not survive to the adult stage.

The stage of elimination of inbreds has been studied in only a few species. In species that retain unopened fruits for several seasons, some elimination occurs on the tree, as in *Eucalyptus stellulata* (see Brown et al. 1985) and *Pinus banksiana* (Cheliak et al. 1985a; Snyder et al. 1985). In *P. sylvestris*, homozygote excess has been found to be eliminated in a naturally regenerated stand by the age of 10 years (Yazdani et al. 1985b; see also Muona et al. 1987, 1988; see, however, Tigerstedt et al. 1982). Early selection against inbreds also has been demonstrated in *P. ponderosa* (Farris and Mitton 1984). In natural populations, inbreeding is of little concern, as most inbreds are probably eliminated before the adult stage as a part of natural mortality.

The timing of inbreeding depression effects is of great concern in forestry. When forests are artificially regenerated, seedlings are produced in nurseries where environmental conditions are often optimized and mortality minimized. In such conditions inbred seedlings survive better than in the harsh conditions of natural populations (see Sorensen and Miles 1982 for discussion). In many species, most of the inbreds are fortunately eliminated before the mature seed stage. Thus, assuming most inbred seeds resulting from wind pollination survive, Sorensen and Miles (1982) estimated that losses due to inbreeding depression in artificially regenerated plantations of *Pseudotsuga menziesii* would be as low as 5%. In species with higher self-fertility, the problem could be far more serious. *Eucalyptus regnans* has an outcrossing rate of 0.74 in natural populations (Moran et al. 1989c), and it has considerable inbreeding depression in growth. Even after heavy culling in the nursery, enough inbreds may be outplanted to significantly decrease productivity (Eldridge and Griffin 1983).

Outcrossing rates in seed orchards

Most of the genetically improved seed is produced in seed orchards, where selected trees are assumed to reproduce as a panmictic unit (Faulkner 1975). Such seed should not be inbred, because of the potential losses in productivity due to inbreeding when stands are artificially regenerated. Outcrossing rates in seed orchards could be lower than those in natural populations because of repetition of genotypes in clonal orchards, or because of increased

TABLE 1 Estimates of outcrossing rates (t) in natural populations of trees based on isozyme markers.[a]

Species	t	Number of populations	Method	Sources
Abies balsamea	0.89	4	M	Neale and Adams (1985b)
Abies lasiocarpa	0.89	1	M	Shea (1987)
Peudotsuga menziesii	0.90	8	M	Shaw and Allard (1982a)
	0.90	3		El-Kassaby et al. (1981)
	0.98	2	M	Neale and Adams (1985a)
	0.95[b]	2	M	Neale and Adams (1985a)
Picea abies	0.88	1	U	Müller (1977)
	0.88		U	Lundkvist (1979)
	0.83	1	M	O. Muona, L. Paule, and A. Szmidt (unpublished)
Picea glauca	0.98	1		Cheliak et al. (1985a)
	0.91[c]	1		King et al. (1984)
Picea engelmannii	0.87	1	M	Shea (1987)
Larix laricina	0.73	5	M	Knowles et al. (1987)
Pinus monticola	0.98	1	M	El-Kassaby et al. (1987)
Pinus sylvestris	0.94	1	U	Müller (1977)
	0.81[b]	1	U	Rudin et al. (1977)
	0.89	1	U	Rudin et al. (1986)
	0.88[b]	1	U	Yazdani et al. (1985a)
	0.94	1	S	Muona et al. (1987)
	0.95	3	M	Muona and Harju (in press)
Pinus ponderosa	0.96	1	S	Mitton et al. (1977)
	0.96	1	S	Mitton et al. (1981)

mating between relatives in seedling seed orchards. On the other hand, the family structure of natural stands is broken up in seed orchards, which may reduce the probability of inbreeding. So far the results suggest that outcrossing rates in seed orchards and natural stands are fairly similar in *Pseudotsuga menziesii* (Shaw and Allard 1982a; Ritland and El-Kassaby 1985), *Pinus sylvestris* (Müller-Starck 1982; Rudin et al. 1986 and references therein), and *P. taeda* (Adams and Joly 1980; Friedman and Adams 1985a). In some cases, outcrossing rates in seed orchards have been considerably higher than in natural stands. The average outcrossing rate for *Pinus radiata* in natural stands was 0.74 (Table 1), but the seed orchard estimate was 0.90 (Moran et al. 1980). Similarly, *Eucalyptus regnans* had an outcrossing rate of 0.74 in a natural stand, but on the average 0.91 in a seed orchard (Moran et al. 1989c). This decrease in selfing may be due to the lack of family structure in the orchard as compared to natural stands.

TABLE 1 (Continued)

Species	*t*	Number of populations	Method	Sources
Pinus jeffreyi	0.93	5	M	Furnier and Adams (1986)
Pinus rigida	0.95			Guries and Ledig (1982)
Pinus banksiana	0.88	1	M	Cheliak et al. (1985b)
	0.88	1	M	Snyder et al. (1985)
Pinus contorta	0.99	2	M	Epperson and Allard (1984)
	0.95	3	M	Perry and Dancik (1986)
Pinus radiata	0.74	18	M	G. F. Moran, J. C. Bell, and O. Muona (unpublished)
Liriodendron tulipifera	0.65	3	S	Brotschol et al. (1986)
Pithecellobium pedicellare	0.95	1	M	O'Malley and Bawa (1987)
Acacia auriculiformis	0.93	2	M	Moran et al. (1989b)
Acacia crassicarpa	0.96	2	M	Moran et al. (1989b)
Acacia melanoxylon	0.90	2	M	O. Muona, G. F. Moran, and J. C. Bell (unpublished)
Eucalyptus, 10 species	0.77			Brown et al. (1985)

[a]U, rare allele; S, single locus; M, multilocus.
[b]Shelterwood, seed tree stand.
[c]Seed production area.

Outcrossing rates may vary between parts of the seed crop. Lower parts of the crown have more selfing than higher parts in *Pseudotsuga menziesii* (e.g., Shaw and Allard 1982a) and *Pinus sylvestris* (Shen et al. 1981). El-Kassaby and Ritland (1986a) have shown that the amount of pollen migration from outside the orchard and phenological differences between clones also influence outcrossing rates in *Pseudotsuga menziesii*. Despite high average outcrossing, some individual trees have been found to have fairly low rates of outcrossing (e.g., Shaw and Allard 1982a; Ritland and El-Kassaby 1985). Much of the variation between individual trees is probably genetic, due to differences in self-fertility or in pollen production (for discussion on *Pinus sylvestris*, see Koski and Muona 1986).

Estimating inbreeding levels of single trees

Estimating the level of inbreeding for single trees is difficult if parentage is not known (Strauss 1986). However, especially when trees are to be vegetatively propagated, it is important that they are not inbred individuals. This problem concerns young trees, which are the main material for vegetative propagation. Among older trees, inbreds have been eliminated by natural selection. In many trees, a positive relationship has been found between isozyme locus heterozygosity and growth rate (see Mitton and Grant 1984). The interpretation of this finding is still debated. Bush et al. (1987) suggested that in *Pinus rigida* the relationship is due to overdominance at specific loci. However, the overdominant loci seem to differ among different populations. Strauss and Libby (1987) studied *Pinus radiata* collected at an early stage, and found a positive relationship between growth rate and heterozygosity for the trees that had lower or average levels of heterozygous loci. They interpreted the results as reflections of inbreeding depression in the genotypes with low heterozygosity. Strauss (1986) found a weak positive relationship between growth and heterozygosity within both inbred and outcrossed trees of *P. attenuata*. Further study is needed to elucidate the mechanisms responsible for the relationships.

Correlated matings

In previous sections the consequences of partial selfing and other inbreeding were discussed mainly with respect to inbreeding depression. Here, deviations from random outcrossing are considered with respect to the assumption that open pollinated progenies or progenies from polycrosses are half-sibs, as is assumed in models for estimating components of genetic variance.

Partial selfing Partial selfing and random mating gives rise to progenies that consist of selfs and half-sibs. For characters that do not display inbreeding depression, this should result in upwardly biased heritability estimates, if a half-sib relationship is assumed (Squillace 1974). However, estimates are frequently obtained for characters that show inbreeding depression. Sorensen and White (1988) found that in *Pseudotsuga menziesii* the inbreds among wind-pollinated progeny gave rise to large within-family components of variance for height, and in fact resulted in significantly reduced estimates of heritabilities compared to estimates from controlled crosses.

A low number of males A low number of trees fathering the progeny also violates the assumptions of these models (see Sorensen and White 1988). Under this circumstance, there will be a mixture of half-sib and full-sib relatives in open-pollinated progenies. Such paternal correlation results in overestimation of heritabilities. Schoen (1985) found little correlation between seeds within cones of *Picea glauca*. Cheliak et al. (1985b) have, however, suggested that the effective size of the male population in natural populations of *Picea glauca* may be quite small. In insect pollinated plants, the pollen load may be largely due to the most recently visited flower. For this reason, Schoen and Clegg (1984) have developed a one pollen parent mating model, where they assume that all progeny from a maternal plant are sired by just one father. The most favorable situation for correlated matings occurs when pollen grains are not distributed singly, but as composite pollen grains, polyads. Species of the genus *Acacia* have polyads, the number of pollen grains per polyad corresponding closely to the number of ovules per flower (Kenrick and Knox 1982). In most cases pollinated flowers have just one polyad per stigma. This suggests that all offspring in a pod may have just one father. Inflorescenses contain a large number of flowers. Flowers within an inflorescence and a group of inflorescenses (raceme) flower nearly simultaneously. This kind of flower and fruiting structure indicates that there may be a hierarchy of genetic relationships between seeds. O. Muona, G. F. Moran, and J. C. Bell (unpublished) analyzed the probability of multiple paternity within pods, and the probability of two pods sharing fathers by using isozyme markers. Outcrosses were not random, instead there was a hierarchy of relationships. The probability of sharing a father for seeds within a pod was about 0.76, and for pairs of pods within a tree at large 0.25. This represents a major deviation from the random pollen pool assumption, and could result in biased estimates of outcrossing rates if the mixed mating model were used, and inflated variance components estimates, if not corrected for. It is easy to eliminate the family structure within a tree by mixing seeds collected from different parts of the tree. There will still be some full sibs among the progeny, but the influence of correlated matings will be substantially reduced. This phenomenon may be of more evolutionary than tree breeding interest, but it is important to be aware of it. Correlated matings due to polyads may occur in the approximately 1000 species of *Acacia*, of which many are being domesticated and entering tree breeding programs. Furthermore, correlated matings may occur in other insect pollinated trees.

Polycrosses In an attempt to avoid problems with open-pollinated families, even pollen mixtures from many trees (>10) are used in polycrosses. Polycrosses are frequently used in breeding programs to estimate genetic parameters. Studies in *Pinus radiata* (Moran and Griffin 1985) and *Picea abies* (Schoen and Cheliak 1987) have demonstrated that, contrary to assumption, males may be unequally represented in the progeny. This may be mainly due to viability differences during embryonic development between the progeny of different trees. Fowler (1987) has suggested that the problem may be minimized in many conifer species by using a sufficiently large number of potential fathers in polycrosses. In angiosperm trees, pollen competition during pollen tube growth would result in similar effects (e.g., Stephenson and Bertin 1983).

Fertility variation and sexual asymmetry

Variation in seed set of plants is well known, but male fertility has been more difficult to study, and has received attention only recently. If there are genetic differences between the male or female fertility classes, genotypic frequencies will be altered in the progeny. Even if the variation were entirely environmental, it may have indirect genetic effects through a reduction in effective population size (e.g., Stern and Gregorius 1972). In seed orchards differential fertility may lead to unbalanced representation of different parental clones.

Fertility differences in conifers can be approximated by counting female strobili or cones, and in many species by counting male strobili (Eriksson et al. 1973b). This implies the assumption that trees do not differ in pollen viability or seed or pollen production per strobilus. In wind-pollinated species it is likely that the production of male pollen is linearly related to reproductive success. Schoen and Stewart (1986) have provided evidence for this in a *Picea glauca* seed orchard. Direct genetic estimates of male reproductive success are difficult to make in natural populations, because current methods are for use in small populations. Such estimation was possible in a small natural stand of *Pinus radiata* on Guadelupe Island off the coast of Mexico, where one population consists of just 14 trees. Half of the trees produced unique gametes, and, among those seven trees, wide variation in the number of pollen gametes was present in viable seed. The coefficient of variation of male reproductive success for this group of trees was 0.8, causing a considerable reduction in the male effective population size, from 14 to 9 (G. F. Moran, J. C. Bell, and O. Muona, unpublished).

Variation in fertility among seed orchard clones has been demonstrated in several conifer species, e.g., *Pinus sylvestris* (Jonsson et al. 1976) and *Picea abies* (Eriksson et al. 1973b). An example of variation among 28 clones in a *Pinus sylvestris* seed orchard in Finland is shown for the 1985 flowering season in Figure 2. Both male and female fertility varied extensively. This was due primarily to differential strobilus production per ramet (coefficient of variation 1.08), but also partly to uneven replication of clones.

Hermaphrodite plants need not be sexually symmetrical. Clear examples of sexual asymmetry, based on strobilus counts, have been reported for seed orchard clones of *Pinus sylvestris* (Ross 1984). Müller-Starck and Ziehe (1984) have shown, using marker genes, that gamete contributions in mature seed are also asymmetrical. Some clones function predominantly as males and others as females. The sexual function of individual clones may change from year to year. Such asymmetry renders many genetic parameters frequency dependent (Ross 1984; Gregorius et al. 1987). Fertility variation may also influence outcrossing rates. Shea (1987) found that in *Picea engelmannii* and *Abies lasiocarpa*, the heaviest male strobilus producers had the lowest outcrossing rates.

As Figure 2 shows, the clones in the Finnish *Pinus sylvestris* seed orchard are sexually asymmetrical. Fertility variation increases the probability of self-fertilization. If all clones had an equal probability of contributing gametes, the inbreeding effective size at fertilization would be $N_e = 28$. Due to fertility variation and sexual asymmetry, the inbreeding effective size is reduced to

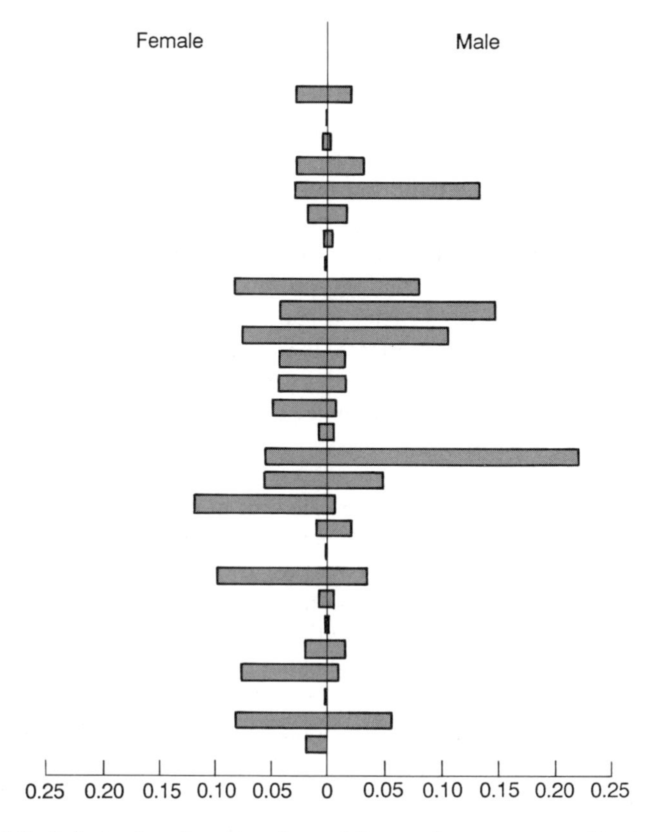

FIGURE 2. Relative female and male strobilus production among 28 clones in a *Pinus sylvestris* seed orchard in Finland in 1985.

18.5 (Muona and Harju, in press). The ratio of effective to actual population sizes observed in a Finnish seed orchard (18.5/28 = 0.66) resembles ratios found in plantations of *P. sylvestris* by Stern and Gregorius (1972). In species where flowering is less regular, such as *Picea abies*, the ratio of effective to actual size may be far lower, perhaps as low as 0.15 to 0.35 in years with poor cone crops (D. Lindgren, personal communication).

Fertility variation may also have directional effects on the genetic composition of the progeny. Several studies show that flowering and growth characteristics are negatively correlated (e.g., Nikkanen and Velling 1987). The progeny of abundantly flowering clones grow more poorly than average.

Pollen and seed movement

Pollen and seed movement influence mating patterns within populations and gene flow between populations. Both have been studied by direct measurements and by genetic methods (Levin and Kerster 1974).

Dispersal within populations Limited dispersal of seeds may lead to grouping of genotypes within populations (e.g., Furnier et al. 1987). Koski (1980) reported that about half of the fertilizing pollen of *Pinus sylvestris* comes from further than 50 m from the maternal tree. Genetic markers have been used in seed tree stands of *P. sylvestris* (Yazdani et al. 1985a) and seed orchards (Shen et al. 1981; Müller-Starck 1982). These data suggest that pollen movement results in fairly random mating within populations or seed orchards. Gene flow in insect pollinated plants is supposedly quite limited (Levin and Kerster 1974). However, W. T. Adams (personal communication) found that in a seed orchard of *Eucalyptus regnans*, where spacing between trees was 7 m, the average distance between the maternal tree and the pollen contributor was 42 m.

Gene flow between populations Gene flow in trees is mainly due to pollen dispersal. Measurements of efficient pollen dispersal between populations (Koski 1970) and the limited differentiation of populations at isozyme loci suggest that gene flow between populations is extensive. This may cause problems in tree breeding. Gene conservation areas and breeding populations should be isolated from production populations (Ledig 1986). An essential requirement for seed orchards is that they should be isolated from unimproved stands, as fertilization by pollen from outside sources reduces expected genetic gains. Smith and Adams (1983), Friedman and Adams (1985b), and El-Kassaby and Ritland (1986b) have developed methods for estimating the proportion of background pollen in seed orchards. Friedman and Adams (1985b) found that 36% of the fertilizing pollen in a *Pinus taeda* seed orchard came from background stands. When seed orchards are located outside the area of origin of the clones, pollen contamination can have more serious consequences. In Finland seed orchards containing northern clones (latitudes 66°N to 69°N) are situated far south (lat. 62°N) of the origin of the clones in order to promote flowering (Sarvas 1970). Background pollination will result in crosses between northern and southern seed sources. Such offspring will be poorly adapted to northern growing conditions and will therefore not be usable for reforestation in the zone of intended use (northern latitudes). The northern clones were expected to flower somewhat earlier, on average, than the local stands, because of a decreased temperature sum requirement, providing some measure of isolation. This isolation has, however, not proved sufficient, as the orchards containing northern clones may receive about a third of their pollen from outside (Harju et al. 1987; A. Harju and O. Muona, unpublished). Observations on the high level of background pollination have led to the development of methods to reduce it, e.g., by delaying flowering in *Pseudotsuga menziesii* orchards by water-spray cooling (see El-Kassaby and

Ritland 1986a). Nevertheless, pollen contamination may remain the most serious single problem in wind-pollinated seed orchards.

CONCLUSIONS

We have obtained a fairly good picture of the evolutionary forces that shape the genotypic composition of natural populations of forest trees, especially those of temperate conifers. A balance between selection and gene flow due to pollen dispersal results in high differentiation between populations for adaptive quantitative characters, while variability at isozyme loci remains largely undifferentiated. Partial selfing occurs in many species, but viability selection eliminates inbreds during juvenile life stages. During the reproductive cycle, intense fertility selection takes place. Variation in reproductive success between individuals results in lowered local effective population sizes. These evolutionary factors are relevant not only in natural but also in artificial populations.

Population genetic methods are used in many areas of tree breeding, as has been discussed above in detail. When estimating genetic components from crosses or designing seed orchards, breeders make assumptions about the genetic composition of the expected progeny. Use of genetic markers has made it possible to check these assumptions. Studies of reproductive patterns have revealed many deviations from the desired reproductive patterns in seed orchards. In New Zealand, wind-pollinated seed from *Pinus radiata* seed orchards result in trees that grow 10–19% better than genetically nonimproved bulk lots. Progeny from control-pollinated orchards formed from the best clones grow 28% better than bulk lots (Anonymous 1987). These differences may be partly due to deviations from the optimum mating system in the wind-pollinated orchard. Once such problems are known, ways can be found to correct them.

At present, plans are being made for gene conservation (e.g., Ledig 1986) and for long-term breeding (e.g., Cotterill 1984). Information on distribution of variability in natural populations and on the evolutionary forces influencing it form a good basis for this work.

Current knowledge on distribution of variability in trees is concentrated on quantitative characters and biochemical marker loci in temperate conifers. Little is known about angiosperm trees. Work on variation at the DNA level in trees has only just started (e.g., Wagner et al. 1987; Szmidt et al. 1988).

So far tree breeders have hardly been able to use marker genes in selection for important quantitative traits. However, if crosses are designed carefully, if a large number of DNA markers are available, and the quantitative character concerned is governed by major loci, this approach may be useful in breeding programs.

Many studies (Table 1) have documented that trees are predominantly outcrossing. However, the genetic basis of variation in outcrossing rates is poorly known. Research on other aspects of the reproductive cycle, e.g., fertility selection (Schoen and Stewart 1987) or correlated matings should continue. Highly variable molecular markers may make more detailed studies of reproductive systems feasible.

ACKNOWLEDGMENTS

Anni Harju, Katri Kärkkänen, Päivi Karvonen, and Sirpa Filppula have participated in the research and given valuable help in the preparation of this chapter. Drs. W. T. Adams, Veikko Koski, Jaakko Lumme, Jyrki Muona, and P. M. A. Tigerstedt provided helpful comments on earlier drafts of the chapter. The Department of Forest Genetics, Finnish Forest Research Institute has provided flowering data from seed orchards. Our research has been financially supported by the Academy of Finland and the Foundation for Natural Resources in Finland.

CHAPTER 17

SELECTION STRATEGIES FOR THE IMPROVEMENT OF AUTOGAMOUS SPECIES

W. E. Weber, C. O. Qualset, and G. Wricke

ABSTRACT

The breeding of autogamous crop plants involves the creation of genetically variable populations for selection, and population improvement and variety development using selection strategies such as pedigree selection or recurrent selection. In many breeding programs for autogamous species the genetic base is smaller than in outcrossing species, since selfing leads to the isolation of lines within the population. Therefore broadening the genetic base is very important.

After a survey of different breeding strategies, procedures for one- and two-stage optimization in pedigree selection are discussed. For recurrent selection programs, the published experimental results on various crops are summarized. The merit of recurrent selection is investigated more efficiently by computer simulation. This helps to optimize the inputs into programs that may last many years. Finally, the use of doubled haploids is sometimes less desirable than that of single-seed descent as a breeding scheme.

Selection is a powerful force in changing the mean of a population for fitness and other traits in plant populations. Natural populations and landraces of crops are almost always genetically variable for many traits, thus explaining the observed responses when these genetic resources are subjected to artificial selection in controlled culture conditions. Plant breeders and their farmer predecessors adopted empirical methods for maintaining and improving populations. By now a large arsenal of selection strategies is available to the plant breeder. These methods differ from crop to crop depending on the mating system, mode of propagation, length of life cycle, nature of the traits under selection, and desired genetic structure of the end-point cultivar (Allard 1960). This chapter is restricted to autogamous crop species and points out some of the alternative selection strategies and approaches to allocating resources for successful and efficient plant breeding.

Schnell (1982) provided a useful framework for discussing the breeding of autogamous species by focusing on factors relating to the types of cultivars under development by plant breeders. We have expanded his framework in Table 1 to emphasize the importance of the genetic structure of the population of plants in the various types of cultivars. Clonal and synthetic cultivars are not common in autogamous seed-producing species. Hybrid cultivars are important in only a few autogamous crops (e.g., sorghum), and their production depends on subvention of the mating and reproductive system by male-sterility genes or synthetic gametocides. By introducing male-sterility genes into autogamous populations an approach to random mating is possible, but the frequency of male sterility alleles declines rapidly without means to ensure maintenance of those alleles. Harland barley was a cultivar developed from a composite-cross population that had a low frequency of male-sterility genes (Suneson 1968). Such a cultivar is a heterogeneous population composed of a mixture of homozygous and heterozygous plants and thus does not fit clearly into the framework of Table 1.

GENERATING GENETICALLY VARIABLE POPULATIONS

The breeder is faced with important decisions about how to develop a source population from which selection can be effective. Selection within landraces was historically important in the development of new cultivars, but now landraces are used primarily as donors for genes to improve an adapted cultivar. In general terms, breeders emphasize adapted cultivars for improvement and will reach outside the adapted gene pool for specific traits. Thus populations developed for selection are of two principal types, which are developed by hybridization of adapted × adapted or adapted × exotic germplasm. Plant breeding has been remarkably successful in developing commercial cultivars from intensive intermating and selection within adapted germplasm. However, adapted germplasm may have specific deficiencies such as disease resistance, or it may appear to have reached an improvement plateau. Selection strategies must take into account the need for introducing exotic germplasm into the gene pools for cultivar development. Most comprehensive breeding programs have concurrent selection-for-cultivar development subprograms with complementary genetic enhancement subprograms to provide new source materials for cultivar development. Jensen's diallel selective mating program is a good illustration of an integration of these two functions (Jensen 1988).

The usual means of creating populations in adapted (A) autogamous genetic resources include single crosses $A_i \times A_j$, three-way crosses $(A_i \times A_j) \times A_k$, double-crosses $(A_i \times A_j) \times (A_k \times A_l)$, backcrosses

TABLE 1. Characteristics of cultivars which influence choice of selection strategies for cultivar improvement.[a]

Characteristics and requirements	Genetic structure (and type)				
	Homogenous			Heterogeneous	
	Homozygous	Heterozygous		Homozygous	Heterozygous
	(Pure line)	(Hybrid)	(Clone)	(Blend or bulk)	(Synthetic)
Is heterosis a positive performance factor?	No	Yes	Yes	No	Yes
Is the cultivar a single genotype?	Yes	Yes	Yes	No	No
Is the cultivar reproduced directly by seeds?	Yes	No	No	Yes[b]	Yes[c]
Is the cultivar reproduced by clonal propagation?	No	No	Yes	No	No
Is the cultivar reproduced by hybridization?	No	Yes	No	No	Yes[c]

[a]Adapted from Schnell (1982).
[b]Component lines may be propagated in isolation and composited each generation or after a limited number of self-reproducing generations.
[c]Component lines may be propagated in isolation and combined by hybridization for production of seed of the cultivar, not a common practice for autogamous crops.

$(A_i \times A_j) \times A_j$, and population crosses $(A_i \times A_j)$. Similar mating schemes are used to introduce exotic (E) genetic resources but with a conscious effort to introduce rapidly, and at high frequency, genes from the adapted materials, as for example backcrosses $(A_i \times E) \times A_i$, three-way crosses $(A_i \times E) \times A_j$, or double crosses $(A_i \times A_j) \times (A_k \times E)$. An important point is how exotic genetic resources are chosen for use in genetic enhancement. Such materials may have very specific traits that are highly heritable and easily identified in screening of genetic resource collections. In other cases, the traits of interest in exotic materials may be complexly inherited, environmentally sensitive, or difficult to detect. This introduces a whole new strategic problem for plant breeders who may have access to very large genetic resource collections. This problem is further explored in Marshall (this volume), but in passing we mention that large collections may be subdivided by geographic origins or by statistical cluster analysis of indicator traits (Spagnoletti Zeuli and Qualset 1987). Accessions, once selected, can be mated to an adapted gene pool or to individual cultivars to enhance selection for the traits of interest and for immediate creation of gene pools with a high frequency of “adapted” genes (Qualset 1979).

The plant breeder has special problems in the use of exotic genetic resources with cross species barriers. Many difficult interspecific gene transfers have been made, but often undesirable traits are linked to the gene(s) being transferred. For this reason, and because of the inordinate amount of time required for most interspecific gene transfers, plant breeders are anxiously awaiting the perfection of new methods for isolation of genes and routine transformation of these genes into adapted germplasm. Hence we have included gene transformation among the breeding strategies of Table 2. While the methodologies for molecular-based gene transfer are advancing rapidly, questions about the stability and expression of transformed genes in a new genetic background and about the transfer of multigenic traits from exotic germplasm remain to be addressed. The breeder's tool box may contain new ways to transfer genes, but traditional selection and evaluation methods will be required to develop cultivars having complex agricultural adaptation and end-use traits.

Finally, genetic variability on which selection can be applied can be generated without hybridization by the application of chemical and physical mutagens. Such variation has been used effectively in developing new cultivars. Sigurbjörnsson (1983) listed 126 cultivars in more than 25 crop species that have been selected from mutagenized cultivars or populations and released for commercialization. An additional 82 cultivars have been developed using a mutation-derived line as a parent. The regeneration of plants from tissue culture can release genetic variability (Meredith 1984; Orton 1984). In some instances stress factors, such as mineral nutrients or organic compounds added to the culture medium, have led to the recovery of heritable variants. The regenerated plants are known as somaclones and the variation among them, either genetic or epigenetic, has been termed somaclonal variation. Aneuploidy or chromosomal rearrangements are often associated with somaclonal variation, but the nature of the variation released via tissue culture is not completely clear (Orton 1984).

BREEDING STRATEGIES

Breeding systems for the improvement of autogamous species have the ultimate goal of producing homozygous lines to be used as single pure-line cultivars, as populations of lines in multiline cultivars, or as parents in an F_1 hybrid cultivar. Various mating designs have been devised, as outlined in the previous section, that are appropriate to the end-point goal. Table 2 lists nine categories of breeding strategies. These include the traditionally used and highly successful methods, and gene transformation, which is an emerging breeding methodology. This last strategy will require that gene cloning and gene integration into the crop plant DNA be rather routine.

Table 2 does not distinguish between breeding for cultivar improvement and for development of unique germplasm to be used later in breeding. The

TABLE 2. Typical breeding strategies for autogamous crops.[a]

Breeding strategy	Generation											
	1	2	3	4	5	6	7	8	9	10	11	12
Backcross (BC)	Cross	BC1	Sel	BC2	BC3	Sel	BC4	BC5	Sel	Sel	Eval	Eval
Pedigree (P)	Cross	F_1, self	Sel	Sel	Sel	Eval	Eval	Eval				
Doubled haploids (DH)	Cross	Anther culture	Double chromosome	Bulk	Sel	Sel	Sel	Eval	Eval	Eval		
Single seed descent (SSD)	Cross	Bulk[b]	Bulk[b]	Bulk[b]	Sel	Sel	Sel	Eval	Eval	Eval		
Bulk, random (RB)	Cross	Bulk[c]	Bulk[c]	Bulk[c]	Sel	Sel	Sel	Eval	Eval	Eval	Eval	
Bulk, selected (SB)	Cross	Sel, bulk[d]	Sel, bulk[d]	Sel, bulk[d]	Sel, bulk[d]	Sel	Sel	Eval	Eval	Eval		
Recurrent selection (RS)	Cross	Intermate	Sel	Intermate	Sel	Intermate	Sel	Intermate	Sel	Sel	Eval	Eval
Induced mutation (IM)	Mutation treatment	IM, self	Sel	Sel	Sel	Eval	Eval	Eval				
Gene transformation (GT)	Gene insertion	GT, self	Sel	Sel	Sel	Eval	Eval	Eval				

[a]Self-pollination is assumed in all generations not designated for crossing, backcrossing, or intermating. The number of generations for selection and evaluation is arbitrary and varies depending on population size, the traits under selection, desired degree of homozygosity, and stability of the selected traits. Sel refers to selection of single plant and advancement as progeny lines; Eval refers to comparative trials for evaluation.
[b]A single seed is selected from each plant, then all seeds are bulked and planted for the next generation.
[c]All plants are harvested en masse; natural selection may occur.
[d]Single plants having desirable traits are selected and bulked to produce the next generation.

methods for these goals are the same, but differ in the intensity of selection for uniformity (homozygosity) and extent of evaluation for agricultural adaptation. The table is not meant to project conclusions about relative efficiencies of the various strategies. Selection for uniformity and homozygosity was usually allotted three generations and a similar number for evaluation. Obviously, the breeder adjusts the protocol according to the nature of gene expression and other factors. If the most rapid strategy to produce a new cultivar is needed then induced mutation (IM), gene transformation (GT), and pedigree (P) are indicated. IM suffers from a lack of assurance that desired alleles will be mutated and GT is not readily usable. The backcross (BC) strategy is quite efficient (Briggs and Allard 1953) and can be contracted in number of required generations by producing large numbers of BC plants each generation and deferring selection until after the final BC is made. This method produces homozygous lines that are congenic to the recurrent parent whose performance level sets the upper limit on the new cultivar except for the value added by the newly transferred character.

The bulk population strategies, as described in Table 2, including random (RB) and selected (SB) bulks and single-seed descent (SSD), are used to generate homozygous lines with a minimum of effort. Both RB and SSD can be used under conditions of accelerated plant development to reduce generation turnover time. After a high degree of homozygosity is achieved (say F_6), individual progeny lines are extracted from the populations and handled more or less as a pedigree selection program and single-genotype homozygous lines are produced for ultimate use as cultivars. Recurrent selection (RS) is a member of the family of bulk-population breeding strategies. Its distinguishing feature is deliberate intermating among selected individuals or progenies for several cycles of selection. After the final intermating cycle the population may be distributed as improved germplasm to be used in further breeding or for extraction of individual lines for evaluation as new cultivars. RS will be discussed in some detail later in this chapter.

Two important points need to be emphasized here. The first is that the practical breeder adopts a combination of breeding strategies to meet the challenges and opportunities as they arise. Jensen's (1988) diallel selective mating system is a strategy that identifies several of the options that a breeder can use simultaneously, including multiple matings to generate variability, selection to identify parents for the next cycle of mating, and selection at any stage to exploit the potential agricultural value of the population as cultivars. Thus, breeders desire flexibility in management of breeding populations. Such a system has been used in wheat breeding in California for a number of years (Qualset and Vogt 1980; reviewed in Jensen 1988).

The second point to emphasize is that breeding for long-term improvement in crops involves a form of recurrent selection. Sometimes the interval between cycles is extremely long because the parents selected for intermating to begin a new cycle are often cultivars that have been demonstrated to be agriculturally fit and widely used in commercial production. Breeders tend to select the best products of their own and other breeding programs as source materials for the next round or cycle of selection. This has been recorded in the long-term wheat breeding efforts in Sweden (MacKey and Qualset 1986). The highly successful international wheat breeding program at the International Center for Wheat and Maize Improvement (CIMMYT) is another example. This program is one of recurrent selection and relies heavily on the intermating of selected materials from within the program with occasional infusion of exotic germplasm.

SELECTION CRITERIA

The goals for a breeding program dictate the plant characters chosen as selection criteria and it is the nature of these characters that has a large bearing on the overall breeding strategy. Direct selection for the characters needed is most desirable and is very effective when the characters have high heritability. High heritability is often associated with single genes or at least a small number of genes so elaborate mating and selection schemes are not necessary. Later in this chapter we consider selection strategies for characters that are multigenically controlled and rather strongly influenced by environment.

Characters closely linked genetically to the targeted selection goals can provide efficient aids to selection if they are easily observed and simply inherited. Stuber (this volume) and Stuber et al. (1987) consider this situation in some detail. For indirect selection to be effective for multigenic traits, marker genes distributed throughout the genome are needed and, of course, appropriate marker alleles must be introduced into a breeding population. Early work on *Drosophila* (e.g., Thoday 1961) suggested that quantitative trait genes were distributed throughout the genome. Fasoulas and Allard (1962) used congenic lines of Atlas barley differing in two independent loci (orange lemma and smooth awn) and found several quantitative characters associated with these loci that a priori had no physiological connection to the quantitative traits. The Atlas barley work was extended to an additional locus for blue aleurone and two loci for awn length (Qualset et al. 1965; Qualset and Schaller 1966). Additive and epistatic associations were identified with these loci that could directly affect the quantitative trait (photosynthesis and seed size related to awn length), but additional relationships were observed that were not obviously related.

Congenic lines have a limitation in developing indirect selection criteria because the results may be restricted to a single background genotype and not generally useful to the plant breeder. The breeder may be more interested in associations of marker genes with quantitative traits when the associations are established in a variable background genotype. Random recombinant inbred lines derived from the crosses or backcrosses of inbred lines are useful in this regard (Wehrhahn and Allard 1965). Recent work by Payne et al. (1979)

showed an association of specific seed storage protein alleles with industrial quality of wheat. This association has been confirmed with random recombinant inbred lines (Qualset et al., unpublished).

SELECTION RESPONSE—PEDIGREE SYSTEM

It is obvious that the biological and agronomic characteristics of each crop species have a large bearing on the breeding strategy adopted. A common dominating factor in all breeding programs is selection response. The breeder would like to have guidelines for prediction of response, i.e., the potential success of the program, so that allocation of time and resources will be optimized to give a sufficient return on investment. This subject has been under theoretical and experimental investigation for several decades and the reader is referred to Falconer (1981), Mayo (1986), Hallauer (1981), Hallauer and Miranda (1981), Baker (1984, 1986), and Wricke and Weber (1986) for development of the entire topic.

With respect to autogamous species, we present an approach following Wricke and Weber (1986) that illustrates how single-trait selection response is influenced by the number of crosses and progenies within crosses and how multistage selection can be adapted to the breeders' advantage.

Optimum number and size of crosses

The breeder has access to a large number of parental lines, but the capacity to raise cross progenies is restricted. The use of many crosses could cover a broad genetic base, or, alternatively, only a few crosses might be made with carefully selected parents. Yonezawa and Yamagata (1978) and Weber (1979) using a risk function discovered that the risk of raising no good genotypes is minimized if the breeder makes as many crosses as possible. A consequence of such a strategy would be to raise only a small number of progenies from each cross.

The situation is changed by selection, since the good genotypes must also be detected by the breeder. In a pedigree system the results from the whole cross and from individual lines can be used for selection. To derive the optimum number and size of crosses, the following equation for the prediction of the selection response (R) in its standardized form is used:

$$R = i\,\mathrm{Cov}(p,g) = i\rho_{pg}$$

We define p as the phenotypic value of the selected unit and g as the expected genotypic value of a line; i is the standardized selection intensity. If the correlation ρ_{pg} between the phenotypic and genotypic values is zero, no response will occur. The phenotypic variance of p, σ_p^2, contains the genetic variance σ_g^2, the genotype × environment interaction σ_{ge}^2, and the nongenetic variance σ_e^2. The influences of σ_{ge}^2 and σ_e^2 can be reduced by increasing the number of environments and number of replications and by appropriate blocking of crosses and families within the experimental area. For the following analyses σ_{ge}^2 and σ_e^2 are lumped into one variance σ^2, as the breeder can control the size of this component by the means just mentioned. For one-stage selection there are $N = n_c n_s$ test units with n_c crosses and n_s progenies within each cross. Generally, n_s is not constant for all crosses, but here we will assume constant n_s.

The phenotypic value of a test unit p_{jk} can be described by the linear model:

$$p_{jk} = \mu + g_{jk} + e = \mu + c_j + s_{jk} + e$$

where the genetic effect g_{jk} is due to an effect contributed by each cross c_j and lines or progenies within each cross s_{jk}. The corresponding variance components are σ_c^2 and σ_s^2. The breeder has two sources of information contributing to selection response, c_j and s_{jk}, and can construct the index

$$I_k = bp_{j\cdot} + p_{jk}$$

based on the phenotypic values of the cross mean and progenies. Given n_c and n_s, the optimum index weight b is

$$b = \frac{n_s \sigma_c^2 \sigma^2}{\sigma_s^2 (n_s \sigma_c^2 + \sigma_s^2 + \sigma^2)}$$

The selection intensity i is dependent on the number of progenies within a cross (n_s) because of the intraclass correlation among progenies within crosses. Thus maximation of R is done by trial and error (Weber 1982).

The ratios σ_c^2/σ_s^2 and σ^2/σ_s^2 must be known. Estimates for some traits in oat and peas are given in Table 3. The ratios varied much between different traits. The response R can be examined for a range of the ratios σ_c^2/σ_s^2 and σ^2/σ_s^2 and the optimum number of crosses and progenies can be computed. Table 4 shows some results. The optimum curves for response to selection are always flat. When the nongenetic variance is small ($\sigma^2/\sigma_s^2 = 1$) then $n_s=1$ is nearly as good as the optimum value. When heritability is much lower ($\sigma^2/\sigma_s^2 = 10$) then n_s becomes far more critical.

This approach can also be used to optimize the number of lines from single crosses at various generations. In this case σ_c^2 is the variance component among F_j-derived progenies from a cross and σ_s^2 is the variance component among the progenies obtained from each F_j, i.e., $F_{k(j)}$. For example, consider selection from among F_5 lines derived by selection from F_2 and F_3 lines ($F_{3(2)}$). The variance among lines is $\sigma_c^2 = \sigma_A^2 + \sigma_{AA}^2 + \cdots$ and the variance among progenies within the lines is $\sigma_s^2 = 0.5\,\sigma_A^2 + 1.25\,\sigma_{AA}^2 + \cdots$. Only additive and additive $\times$ additive epistatic effects are included. This is reasonable because in the F_5 test generation, dominance

TABLE 3 Estimates for variance components between and within crosses.

Character and crop	σ_c^2/σ_s^2	σ^2/σ_s^2
Oats (Weber and Wricke 1981)		
Yield	1.2	1.6
Plant height	6.9	3.0
Peas (Weber 1976, 1980)		
Flowering time	3.9	0.8
First flowering node	4.8	1.4
Plant height	6.1	0.8
Pods/node	1.2	3.0
1000 kernel weight	1.4	0.9

and additive × dominance effects are expected to be very small. Then for a pure additive model $\sigma_c^2/\sigma_s^2 = 2$ and for a model with epistasis ($\sigma_A^2 = 1$ and $\sigma_{AA}^2 = 2$), the ratio $\sigma_c^2/\sigma_s^2 = 1$ and Table 4 can be used to examine selection response for optimum of F_j- and $F_{k(j)}$-derived progenies. In general, epistasis reduces the ratio σ_c^2/σ_s^2 and shifts the optimum toward larger values of n_s, but the influence is weak.

Multistage selection

It is especially convenient in autogamous crops to produce selfed progenies, thus providing a family structure so that families, as for example a population

TABLE 4 Optimum number (n_c), size of crosses (n_s) and response to selection (R) relative to $n_s = 1$; $N = n_c n_s$.

		$\sigma^2/\sigma_s^2 = 1$				$\sigma^2/\sigma_s^2 = 10$			
N	$\sigma_c^2/\sigma_s^2 = 1$	n_c	n_s	n_c/n_s	***R*** **(%)**	n_c	n_s	n_c/n_s	***R*** **(%)**
120	1	30	4	7.5	101	6	20	0.33	134
	2	60	2	30	101	10	12	0.83	133
	5	60	2	30	100	20	6	3.33	123
1200	1	50	24	2.1	105	12	100	0.08	165
	2	80	15	5.3	104	20	60	0.33	161
	5	150	8	18.7	102	40	30	1.33	141

of F_2-derived F_4 lines, can be evaluated. Selection can be practiced on F_2-derived F_3 lines and in the next generation on a family of F_2-derived F_4 lines, each family tracing to a selected F_2-derived F_3 line. Alternatively, a nonselected population of F_2-derived F_3 lines can be advanced. In the F_4, selection can be made first on the family group and then on the individual lines within each family. Thus, there are two forms of selection. In the first case selection occurs in two crop cycles (two stages) and in the second, selection occurs in one crop cycle (one stage). The former is more desirable for the breeder because it avoids growing a large fraction of unselected, presumably undesirable lines. The latter case is appropriate when selection cannot be practiced in one generation as during single-seed descent in a growth chamber or in off-season seed increase generations.

As an example, consider F_2-derived progenies being evaluated in the F_4 in year 1 and as F_2-derived progenies evaluated in the F_5 in year 2. There are n_2 progenies from the F_2-derived F_4 generation in the first year; after selection in the first year there are n_{21} F_4 lines left. In the next year F_3-derived lines already exist, but only those tracing back to a selected family are tested. Thus, with n_3 lines for each selected family, the total number of tested progenies in 2 years is

$$N = n_2 + n_{21}n_3$$

To optimize the number of lines (families) and lines within families, N is a constant.

For $n_3 = 1$ the problem is reduced to the specific case of two-stage selection considered by Cochran (1951). He recognized this as a form of independent culling and developed the methodology to estimate selection response using the same concepts as for multitrait selection. Wricke and Weber (1986) discussed multistage selection in detail using the Cochran approach and compared it with an alternate method developed by Finney (1958). Utz and Schnell (1973) extended this approach and included a third stage and genotype × environment interaction.

When n_3 exceeds 1, no exact solution exists for the optimization, but Weber (1981) and Utz (1981) independently proposed the following method. First, perform selection to identify the best F_2-derived family and estimate the selection response with Cochran's method, then select within this family the best F_3-derived line and add the selection responses. The Cochran approach is based on infinite population size so the expected responses to selection cannot be compared directly with selection in one crop cycle only. Simulations were done by Weber (1984) and some of the results on optimum numbers n_2, n_{21}, and n_3 defined above are given in Table 5. The results are based on the restriction that one out of N progenies were selected; N = 100 in this example. Four combinations of genetic and environmental variances were selected. The major difference in conditions was a 10-fold difference in environmental variance σ^2. When σ^2 is large more progenies are selected

TABLE 5. Multistage selection for F_2- and F_3-derived progenies ($N = 100$).

σ_A^2	σ_{AA}^2	σ^2	n_2	n_{21}	n_3	***R* (2-stage)/ *R* (1-stage)**
2	0	1	60	2	20	118
0	1	1	40	3	20	120
2	0	10	51	7	7	131
0	1	10	44	8	7	125

(n_{21}) at the first stage than with small σ^2. Response to selection was compared as the ratio of two-stage response to one-stage response. Two-stage response was clearly superior in all four situations. Of great interest to the breeder is the result that two-stage selection response was greater than one-stage response when $\sigma^2 = 10$ than when $\sigma^2 = 1$, but more selection is delayed to the second year. Large epistatic variances (rows 2 and 4) also dictate that the initial selection intensity on F_3 progenies should be reduced and more intense selection be practiced in the next generation.

SELECTION RESPONSE—RECURRENT SELECTION SYSTEM

Recurrent selection in autogamous crops has been generally viewed as a method of population improvement. In a few cases, most notably in tobacco (Matzinger et al. 1977), several cycles have been completed and selection responses evaluated. Good responses to selection have been found, but the rates are variable (Table 6) depending on the trait under selection, the genetic diversity of the base population, the degree of environmental influences on the selection unit, and the population size. It is desirable to advance from one cycle to the next as rapidly as possible. Thus, only a limited number of generations of selfing and evaluation is completed between intermatings of selected parents. When this is done, the evaluated progenies may be highly heterozygous and heterogeneous and large genotype × environment effects may adversely influence the phenotypic values of parents selected for the next cycle.

This was illustrated by Vanselow (in press) using oats, who found low correlations for grain yield between replicates in the same location and year (Table 7). Kernel weight and plant height showed higher correlations. The response on grain yield after one cycle was nearly the same for direct selection on grain yield alone as for selection on the index GY × KW, but only the selection on the index could improve grain yield and kernel weight simultaneously (Table 8).

TABLE 6. Some recurrent selection programs in autogamous species.

Crop	Initial material	Number of cycles	Selected character	Gain/cycle (%)	Source
Pearl millet	158 lines and ms forms	3	Yield	2.5	Doggett (1972)
Tobacco	8-line hybrid	5	Height, low	−4.9	Matzinger et al. (1977)
			Leaf number	7.0	
Tobacco	8 lines	3	Leaf weight	7.2	Gupton (1981)
		5	Alkaloid content	5.2	
Cotton	2 lines	4	Yield	9.9	Miller and Rawlings (1967)
Soybeans	10 lines	3	Yield	5.4	Kenworthy and Brim (1979)
Soybeans	40 lines	3	Yield, early	4.3	Sumarno and Fehr (1982)
			Yield, midseason	0.0	
			Yield, late	1.2	
Soybeans	3 lines	4	Oil content	5.5	Burton et al. (1983)
Barley	2 lines	3	Sr content, high	7.4	Byrne and Rasmusson (1974)
			Sr content, low	−12.2	
Barley	7 lines	1	Yield	0.0	Patel et al. (1985)
Barley	7 lines	1	Yield	39.8	Delogu et al. (1988)
Wheat	12 lines	2	Protein	7.8	McNeal et al. (1978)
Wheat	10 lines	4	Kernel weight	5.8	Busch and Kofoid (1982)
Wheat	2 lines	3	Sr content, high	12.4	Byrne and Rasmusson (1974)
			Sr content, low	−11.2	
Oats	12 lines	3	Yield	3.8	Payne et al. (1986)
Oats	4 lines	1	Yield	2.5	Vanselow (in press)

TABLE 7. Correlation coefficients in spring oats.[a]

Type	Yield	1000 kernel weight	Plant height
Between replications, same year, same location			
F_2	0.137	0.414	—
F_3	0.183	0.409	0.292
F_4	0.087	0.609	0.167
Between locations, same year			
F_2	0.124	0.588	—
F_3	0.108	0.611	0.292
F_4	−0.000	0.646	0.201
Between generations (mean over locations)			
F_2–F_3	0.075	0.686	0.262
F_2–F_4	0.077	0.657	0.256
F_3–F_4	0.131	0.804	0.401

[a]Vanselow (unpublished).

Simulation of recurrent selection by Monte Carlo methods has provided some conceptual guidance in the design of RS programs. Several details can be varied: the genetic model (variance components, number of loci, linkage), the number of parents at each cycle, and the structure of a single cycle. Just one cycle was simulated by Bliss and Gates (1968) and Bailey and Comstock (1976). The potential of an RS program can be judged only from the simulation of several cycles. Such a study was conducted by Silvela and Diez-Barra (1985). Fouilloux (1980) analyzed a two-parent scheme with two alleles per locus, using the binomial distribution. No nongenetic variance was super-

TABLE 8. Selection response (per cent) after one cycle of recurrent selection in oat.[a]

Character	Grain yield (GY)			Kernel weight (KW)		
Generation of selection	F_2	F_2	F_3	F_2	F_2	F_3
Test generation	F_3	F_4	F_4	F_3	F_3	F_4
Selection on GY	2.5	0.5	2.5	1.2	0.0	−2.1
Selection on index = GY*KW	2.1	1.6	3.0	2.1	1.0	1.7

[a]Vanselow (in press).

TABLE 9. Percentage fixed and segregating loci after five cycles of recurrent selection for 50 loci (*d/a* = degree of dominance).

Status	Selection on heterozygotes		Selection on homozygotes
	d/a = o	*d/a* = 0.6	
Favorable allele fixed	68.4	64.6	78.0
Unfavorable allele fixed	4.4	2.4	16.8
Segregating loci	27.2	33.0	5.2

imposed and homozygous lines (SSD or DH were advanced for the next cycle. Weber (1983) extended this approach to include heterozygotes. Table 9 gives the proportion of fixed or segregating loci after five cycles. Selection of heterozygotes is preferable because fewer loci are fixed for the unfavorable allele. Progress is slower but this is compensated by shorter cycle length since homozygotes do not have to be produced. Some dominance gene action in the desired direction increases the selection response. As a practical matter this type of dominance is useful if selection can be done after only one generation of selfing.

Vanselow (unpublished) simulated RS with 2, 4, 8, or 16 initial parents from which 1, 2, 4, or 8 crosses were made to initiate the selection program. The same numbers of parents and crosses were used to initiate the next cycle of selection. Table 10 shows some of the results after 10, 20, and 30 cycles of RS when single-trait selection was applied. The conclusions from this study were as follows:

1. Genetic variability was not exhausted quickly, even when the number of parents was very small. For rapid response a low to medium number of parents would be optimal. In a long-term recurrent selection program, genetic variability is exhausted by more parents, but this does not happen until after 10–20 cycles. Silvela and Diez-Barra (1985) obtained similar results.
2. The main advantage of more parents lies in a larger degree of safety. The simulations showed that the selection response varied considerably between runs (see standard deviations in Table 10). A useful limit is the difference D = E − S, or the expected response minus the standard deviation. For an approximate normal distribution a single response will exceed D in about 85% of the cases. From Table 10 one can see that for 10 loci and h^2 = 0.2 a shift from less to more parents is preferred, since for more parents the standard deviation is reduced. This effect increased at lower values of heritability.

TABLE 10. Expected (E), standard deviation (S), and difference (D = E − S) of response to selection (percent of maximum response) for recurrent selection schemes with 2, 4, 8, and 16 parents, 10 loci, and $h^2 = 0.2$.[a]

	10 cycles			20 cycles			30 cycles		
Parents	E	S	D	E	S	D	E	S	D
2	48	22	26	66	23	43	70	24	46
4	44	16	28	68	16	52	88	15	73
8	36	12	24	64	13	51	90	10	80
16	24	9	15	44	11	33	78	10	68

[a]Data from Vanselow (unpublished).

3. Multiplying the number of crosses by four was approximately as efficient as halving the error variance. Halving of the error variance means an increase in the number of experimental units and therefore is often connected with an additional generation of seed multiplication. However, four times as many crosses requires much labor, especially in a crop like oats, which is hard to hybridize.

With a mechanism to induce male sterility the species can be temporarily an outcrossing one. Chemical gametocides can provide a very useful role in RS. Genetic male sterility has been proposed as an aid to RS by Brim and Stuber (1973) for recessive male sterility and by Knapp and Cox (1988) for dominant male sterility.

SELECTION OF DOUBLED HAPLOIDS

Chase (1952) developed doubled haploids and proposed this method for rapid production of inbred lines of maize. This method can be used in some autogamous species like barley, wheat, or tobacco. It is expected that the technical possibilities will increase, so that in the future this method can be used by the breeder for several crops. In autogamous species doubling haploids could serve as a rapid method to fix already selected lines. In this connection one must consider whether complete homozygosity and uniformity is a desired goal (Allard and Hansche 1963). Some observations in practical plant breeding lead to arguments against such a perfect pure line. Another point is that completely pure lines cannot be reached if the technique itself creates a mutation pressure. In tobacco, Burk and Matzinger (1976) observed differences between inbred lines and doubled haploid lines extracted from them. In wheat and barley, Snape et al. (1988) took doubled haploids from homozygous lines and again doubled haploids with the *Hordeum bulbosum* hybridization technique. Variation at the first step may be caused by cryptic variation already existing in the lines. Variation at the second step would be an indication of mutation pressure. This was observed only in hexaploid wheat, but not in diploid barley.

Doubling haploids quickly fixes lines as fully homozygous after the initial cross with low probability of recombination, especially if F_1-gametes are already doubled. The method is comparable to bulk breeding and SSD. In a theoretical study Griffing (1975) came to the conclusion that the selection of doubled haploids can increase the response to selection remarkably. But some important questions are still open. Of the thousands of potential gametes, only a few regenerate to doubled haploid plants. Thus strong selection of unknown mode could operate. In SSD plants are grown under conditions of close competition with other genotypes, creating another type of selection pressure. In bulk breeding natural selection can take place (Allard and Jain 1962). It is not known how far the selection pressure in each of these methods supports or impedes the breeder.

The time saved with doubling haploids is not as large as first assumed. With SSD three or more generations can be grown in 1 year in many species. For doubled haploids Snape et al. (1986) estimated about 13 months in wheat with the bulbosum and anther culture techniques from sowing the F_1 until harvest of doubled haploid seed. The authors estimated that about 50–80 doubled haploid plant per person per week can be produced.

CONCLUSION

A large arsenal of selection strategies is at the disposal of a breeder of autogamous plant species. But each crop raises its own problems. Some crops produce only a few seeds so that multiplication is slow, for example, contrast tobacco with oats. There are large differences in the outcrossing rate, and some crops like rapeseed and field beans are partially allogamous. In other crops the genetic basis is extremely small in advanced material.

Generally, self-pollination leads to more or less isolated homozygous lines, and recombination must be forced by the breeder. The primary purpose of recurrent selection programs is to increase the recombination frequency.

In most cases genetic variability is created from crosses between few lines. Exotic material is used only to introduce genes not available in advanced material. In the future, routine gene transfer can be of great importance to overcome crossing barriers between species and to make use of genes from other species in a breeding program. Another advantage of such a method may be the precise insertion of only a few specific alien genes.

Pedigree and bulk methods are the main strategies for variety development. Optimization procedures need efficient estimates of population parameters like components of variance and number of genes. Gene mapping may help here. For many crops homozygous lines can be developed easily by rapid continuous selfing (SSD). Thus, development of homozygous lines via doubled haploids is not in every case the quicker method. Future experimental work should also concentrate on exact comparisons between these two procedures. At the moment not enough experience exists on the amount and nature of variation introduced by *in vitro* methods.

Generally only genotypes that prove their potential under field conditions will become good varieties. This test phase at many places over several years is an essential and time-consuming part of all breeding programs.

CHAPTER 18

UTILIZATION OF GENETIC RESOURCES FOR CROP IMPROVEMENT: THE COMMON BEAN

Fredrick A. Bliss

ABSTRACT

The greatest difficulties experienced by plant breeders in the utilization of unadapted germplasm for improving quantitative traits are predicting the genetic potential of raw germplasm, transferring specific favorable alleles especially without deleterious linked genes, and describing the differential expression of favorable alleles in diverse environments and different genetic backgrounds. The usual ways of overcoming these constraints include replicated testing, testcrosses, and multiple backcrosses to an adapted cultivar. Additional selection criteria based on biochemical traits that are less sensitive to confounding by nongenetic factors have been used in common bean to provide greater selection efficiency for improving seed protein concentration and quality, resistance to Mexican bean weevil, and increased dinitrogen fixation potential.

Plant breeders have used exotic germplasm extensively in cultivar improvement for highly heritable traits controlled by one or a few genes. In contrast, unadapted germplasm has been used sparingly for improving quantitative traits that show low heritability and complex genetic control. Overall, only a small portion of the plant materials held in either *in situ* collections or gene banks has been utilized for crop improvement (Duvick 1984). The collection, maintenance, and evaluation of germplasm are receiving increased emphasis; yet if these materials are to realize their full potential in crop improvement, they must be utilized more fully for improving complex as well as simply inherited traits. The greatest value for understanding the genetic control of a quantitative trait occurs when the superior alleles at the segregating loci can be identified and then used for many subsequent transfers into an array of adapted lines.

The domesticated types of common bean (*Phaseolus vulgaris* L.) include garden bean (snap bean, French bean), of which the immature fruits (pods) are eaten, and dry beans (field beans), of which the mature, dry seeds are consumed after suitable preparation. The immature pods are an important green vegetable that provides trace minerals, vitamin A, and dietary fiber, and the mature seeds contain from 18 to 25% protein.

Inadequate grain supplies limit consumption even where beans are important. Production is constrained because of low yields, poor response to inputs such as nitrogen fertilizer and irrigation, and susceptibility to diseases and pests. While the development of high-yielding cereal cultivars has resulted in increased productivity of those crops, yields of the grain legumes, including beans, have increased very little.

Bean production is often relegated to poor soils because of their low-yielding potential compared to more attractive, high-yielding cereals. Despite the potential for symbiotic nitrogen fixation, beans are considered a poor N_2-fixing crop and the plants often lack enough nitrogen to produce high yields of protein-rich seeds.

While there has been substantial progress toward breeding disease-resistant bean cultivars, destruction by insects, particularly of the stored grains, is extensive. Consequently, the inability to store seeds from one harvest to the next results in limited supplies and high prices prior to harvest, and large market supplies and low prices after harvest.

Beans are a major source of dietary protein, especially where animal protein is scarce, such as in parts of Central, South, and North America, East Africa, and Asia. In Rwanda, for example, the per capita consumption is 40 kg/year (Lamb and Hardman 1985). Although the seeds are moderately rich in total protein, the protein nutritional value is low because the protein is difficult to digest, there are inadequate amounts of the sulfur-containing amino acids, and the seeds contain antinutritional compounds. Because of these constraints, the actual contribution of beans to dietary needs is considerably less than their potential value. There is a need for increased bean yields, improved protein nutritional properties, increased ability to fix atmospheric N_2 efficiently, and greater disease and insect resistance. The analyses of pedigrees (National Academy of Sciences 1972) and of phaseolin protein patterns of wild and cultivated beans (Gepts et al. 1986) provide evidence that the genetic base of commercial cultivars of both snap and dry beans is quite narrow. The germplasm used for bean breeding has consisted primarily of cultivated materials and only recently have wild forms been used.

As the need to expand the genetic base of breeding populations increases, more unadapted germplasm will likely be used. These materials may include cultivars that are well adapted to other conditions, wild and weedy conspecific relatives, and different species. While crossability greatly influences the transfer of genes for all traits, there are additional difficulties associated with utilization of germplasm for the improvement of quantitative traits. These include (1) predicting the genetic potential, (2) transferring favorable alleles without concomitant deleterious traits, and (3) describing the differential expression of favorable alleles in diverse environments and different genetic backgrounds. These problems are common to many crops and a wide range of different quantitative traits, and although specific solutions may differ for each trait and crop combination, there are some useful general approaches to the problems encountered. Studies in common bean provide examples applicable to situations of a similar nature in other crops.

THE GENETIC POTENTIAL OF UNADAPTED GERMPLASM

Several factors contribute to difficulties in determining the genetic potential of unadapted germplasm. These include the confounding of favorable alleles by other plant traits, low heritability of the trait undergoing improvement, and genetic control by multiple genes that show different magnitudes of either direct or indirect effects on trait expression. If the difficulties posed by these factors can be overcome, the genetic potential of the donor stocks are more likely to be realized. Suitable methods for predicting the genetic potential include performance per se, progeny tests and combining ability analysis, and exploiting associations between markers and favorable alleles.

Overcoming factors that obscure the genetic potential

Confounding of favorable alleles by other plant traits Often the effects, and hence the potential value, of favorable alleles are confounded with expression of other plant traits. With biological N_2 fixation, the expression of favorable host alleles could be influenced by plant growth habit. In assessing common bean landraces for differences in N_2 fixation potential, Graham (1981) concluded that in most instances, determinate bush beans (type I plant habit) fixed small amounts of atmospheric N_2. Among the accessions studied, "Puebla 152" from Mexico, which has an indeterminate, prostrate (Type III) plant habit, and "Cargamento" from Colombia, which has an indeterminate, climbing (Type IV) plant habit, showed the greatest N_2 fixation potential. We were interested in improving N_2 fixation of white navy bean cultivars that are early maturing, determinate bush beans (type I). Thus, it was important to determine whether the high N_2 fixation of Puebla 152 and Cargamento was due to genes controlling N_2 fixation directly (e.g., nodulation characteristics such as number of nodules and nodule efficiency) or the result of other traits, such as greater photosynthetic capacity conferred by large plant type and later maturity. Presumably favorable alleles affecting N_2 fixation directly could be transferred to type I plants, whereas if increased N_2 fixation was due to greater leaf area and larger plant type, it would be difficult to produce a high N_2-fixing, determinate, type I navy bean by hybridization and selection.

This question was resolved by evaluating a population of inbred backcross lines resulting from the cross Sanilac × Puebla 152, for plant type and N_2 fixation. The critical observation relative to cultivar improvement was the occurrence of progenies with plant type and days to maturity similar to Sanilac, but which also fixed more N_2 (Table 1) (McFerson 1983; St. Clair et al. 1988). Since these BC_2S_4 progeny lines did not fix as much N_2 as Puebla 152, perhaps several genes contribute to N_2 fixation and/or other traits were also important. Subsequently, F_3 families from crosses among selected progeny lines were identified that fix levels of N_2 comparable to Puebla 152 (St. Clair 1986).

Low heritability For characters with low heritability, the phenotypic expression is not a consistent reflection of the genotype. Although traits that show low heritability often are controlled polygenically, low heritability is a consequence of complex gene expression and a relatively large contribution of nongenetic variation rather than gene number per se. Also because of low heritability, it is difficult to determine whether the same intensity of trait expression shown in a poorly adapted donor can be transferred to a cultivar. Progeny tests and repeated observations are required to discriminate accurately between superior and inferior genotypes. More time is required to transfer the trait and greater resources must be devoted to identifying favorable alleles at the locus or loci involved. Increasing the heritability by better

TABLE 1. Dinitrogen fixation parameters at maturity of Sanilac, Puebla 152, and five BC_2S_4 progenies.[a]

Line	Plant type	Days to maturity	Plant N from fixation	
			mg/plant	% of Total
Puebla 152	III	114	790	54
24–17	III	107	519	45
24–21	I	93	219	23
24–48	I	91	191	21
24–55	I+	100	240	26
24–65	I	105	260	27
Sanilac	I	96	48	7
LSD (0.05)			109	8

[a]Hancock, Wisconsin Agric. Expt. Station, mean of 1984 and 1985.

identification of the biochemical basis of the trait will contribute to more effective use of unadapted germplasm. An example of how this is possible involves insect resistance in common bean.

Two bruchid insects are major pests of beans, and when uncontrolled they quickly destroy large amounts of stored grain. Physical and chemical control measures are difficult, dangerous, and expensive; therefore biological control through the use of resistant cultivars is an attractive alternative method. After finding no usable levels of resistance among the world collection of cultivated bean accessions, scientists at the Centro International de Agricultura Tropical (CIAT) screened accessions of wild beans collected in Mexico using insect feeding trials to assess susceptibility. Among those accessions, they found seeds with promising levels of resistance (Schoonhoven et al. 1983).

A standard approach to breeding for resistance was begun at CIAT by crossing susceptible bean cultivars with resistant wild accessions. When resistance was measured by insect feeding trials, it was difficult to determine the inheritance of resistance and resistant cultivars were not easy to develop. The heritability of resistance based on the criterion of insect feeding damage was low because much of the phenotypic variability was due to nongenetic factors.

Subsequently the basis of resistance to the Mexican bean weevil, *Zabrotes subfasciatus*, was shown to be relatively simple. A major seed protein, arcelin, unique to wild beans and not found in cultivars is responsible for resistance (Osborn et al. 1988). Arcelin expression is inherited as a single Mendelian unit, with presence of arcelin being dominant to absence (Romero Andreas et al. 1986), and multiple alleles accounting for at least four different electrophoretic forms (Osborn et al. 1986). Resistant cultivars are being developed by a standard backcross procedure in which single seeds are scored phenotypically for arcelin presence using either SDS-PAGE or detection with antibodies (Harmsen et al. 1988). By using a more definitive selection criterion (e.g., electrophoretic separation or immunological reaction) rather than insect feeding damage to discriminate between resistant and susceptible seeds, the heritability of resistance was effectively raised.

Multiple genes and differential magnitudes of direct and indirect gene effects on trait expression One of several simplifying assumptions often made when studying quantitative traits is that they are controlled by many genes each having small and equal effects. For all traits this assumption probably is not valid. In several cases, segregation at a few loci accounts for much of the genetic variation seen in populations resulting from biparental crosses. Also, it appears that gene effects usually are not of equal magnitude.

When multiple genes are responsible for the genetic variability it is useful to know how many are segregating and the relative magnitude of individual gene effects. If the gene effects are unequal, it is most advantageous to first transfer those with the largest effects to the adapted parent. Depending on the need and difficulty, it may not be worthwhile to transfer the remaining "minor" genes. If gene effects are of similar magnitude, transfer of each is of nearly equal value and no priority need be given unless there is an unfavorable linkage to secondary effects.

Seed protein concentration is a quantitative trait usually considered to be controlled by many genes with small effects. Thus, inheritance is often studied using classical biometric procedures. To simplify breeding for increased protein concentration and modified protein quality in beans we have used an approach that relies on the manipulation of specific genes that affect the quantitative expression of protein traits. These genes have been identified in germplasm ranging from cultivated, but poorly adapted snap and dry beans to wild *Phaseolus vulgaris* plants and the related species *P. coccineus*. Different loci show either direct or indirect effects on the expression of phaseolin protein, the major constituent protein of bean seeds, with correlated changes in the amount and concentration (percentage) of total seed protein also resulting.

There is limited variability for phaseolin type among cultivated beans in which the S-, T-, and C-type phaseolins are predominant (Brown et al. 1981b; Gepts et al. 1986), but a wide array of variants in wild *P. vulgaris* (Romero Andreas and Bliss 1985; Gepts and Bliss 1986; Gepts et al. 1986). The snap bean cultivar "BBL240" reportedly contains elevated concentrations of total protein and methionine (Kelly 1971). The T^{240}-type phaseolin allele was backcrossed into the standard genotype Sanilac (which contains S-type phaseolin). Field-grown seeds containing T^{240} phaseolin had 1 to 2 percentage points more total protein and no changes in seed size or yield compared to S-type sib lines (Hartana 1983) and Sanilac, and T^{240}-type seeds contained more phaseolin per seed than Sanilac (Table 2). The T^{240} phaseolin allele can be transferred into standard cultivars using electrophoresis to identify the superior allele, to produce an increase in phaseolin and total protein.

Two phaseolin alleles (T- and C-types) from cultivated beans and two (M1- and M2-types) from wild Mexican beans were backcrossed into standard lines to allow comparisons of different allelic effects on phaseolin expression. The protein concentrations (%) of the backcross-derived lines were higher than Sanilac in all cases (Table 3). In lines with the T or C allele, the protein increase was due to more phaseolin and nonphaseolin protein per seed, while in the lines derived from crosses to wild beans (M1 and M2 alleles) the increased concentration resulted from smaller seed size due to less nonprotein components (e.g., starch) (Hartana 1986).

Knowledge of whether genes produce direct or indirect effects on the trait of interest is important not only for choosing efficient ways to improve the primary trait but also for determining possible correlated responses. The potential value of new germplasm may be limited if there are deleterious effects associated with otherwise favorable alleles. Substantial amounts of seed lectins (phytohemagglutins) are present in most beans, but about 10% of the domesticated and wild beans are lectin deficient (Brucher 1968). The structural

TABLE 2. Comparison of the backcross-derived line, 809598–3^{T2+0}, to the navy bean cultivar Sanilac for protein expression traits in field-grown seeds produced at Hancock (H) and Arlington (A), Wisconsin averaged over 2 years.[a]

	Trait							
	Seed yield		Seed weight		Protein (%)		Phaseolin (%)	
Line	H	A	H	A	H	A	H	A
Sanilac	168	158	188	165	19	24	9	13
809598–3^{T2+0}	174	203	198	201	21	25	12	15
Significance	NS	NS	NS	*b*	*b*	NS	*b*	*b*

	Trait (mg/seed)							
	Protein		Nonprotein		Phaseolin		Nonphaseolin	
Line	H	A	H	A	H	A	H	A
Sanilac	36	39	152	126	17	21	18	19
809598–3^{T2+0}	41	48	158	152	23	30	18	18
Significance	*b*	*b*	NS	*b*	*b*	*b*	NS	NS

[a]Hartana (1986).
[b]0.05 level of significance.

genes encoding seed lectins are inherited independently of the phaseolin multigene family (Brown et al. 1981a), and different electrophoretic forms of lectin are controlled by a series of multiple alleles. In addition to influencing the amount of lectin, the *Lec* alleles display indirect effects on the quantitative expression of phaseolin protein. When the lec$^-$ phenotype was compared to the Lec$^+$ in the standard Sanilac background and in combination with four different phaseolin types in backcross-derived lines, more phaseolin was present in lec$^-$ lines than in the Lec$^+$ counterparts in all cases (Table 3).

By identifying superior alleles at loci that either directly or indirectly affect phaseolin, it is possible to increase phaseolin and concomitantly the concentration of total seed protein. This provides a simpler strategy than using standard biometric procedures that have sometimes resulted in an undesirable correlated response such as decreased yield. When plants containing variant alleles controlling either phaseolin type or lectin expression were analyzed in their original state, the potential effects of quantitative expression of phaseolin and total protein were not evident. After evaluation in an adapted background, it was clear that transfer of superior alleles into cultivated materials should produce further increases in protein expression.

TABLE 3 Seed characteristics of backcross-derived lines (Lec^+ and lec^-) with different phaseolin types (T, C, M1, and M2) compared to the standard cv. Sanilac and the near-isogenic, lectin-deficient (lec^-) line derived from Sanilac.[a]

Line	Seed yield (g/10 plants)	Protein (%)	Constituent weight (mg/seed)				
			Total	Nonprotein	Protein	Phaseolin	Nonphaseolin
Sanilac	123	21.4	169	133	36	16	20
Sanilac (lec^-)	88	24.5**	159	120	39	21**	18
"T"BC_2S_2	111	24.1**	172	131	41**	18*	23*
"T"BC_2S_2(lec^-)	109	24.4**	192*	145	47**	26**	21
"C"BC_2S_2	120	23.1**	172	132	40*	18*	22
"C"BC_2S_2(lec^-)	144	24.0**	191*	145	46**	26**	20
"M1"BC_2S_2	112	23.9**	146**	111*	35	14	21
"M1"BC_2S_2(lec^-)	92	24.5**	148	112*	36	19*	17*
"M2"BC_2S_2	113	23.4**	97**	74**	23**	9**	14**
"M2"BC_2S_2(lec^-)	135	25.1**	109**	82**	27**	14*	13**

[a]Hartana (1986).
*, **0.05 and 0.01 levels of significance for comparisons to Sanilac.

Methods for predicting genetic potential

Performance per se Sometimes phenotypic value can be used to predict accurately the genotypic or breeding value of individuals or lines being considered as potential parents for the improvement of pure lines or intermating populations. If genotype × environment interactions are relatively small and additive genetic variance is the major constituent of the total genetic variance, performance per se is a good indicator of breeding value. When heritability is low, adequate replication is needed to discriminate between superior and less-valuable potential parents.

In common beans, N_2 fixation is a trait with low heritability and no major genes affecting nodulation have been identified. Although nodule number is a determinant of N_2 fixation, it has not been shown to affect the amount of fixation in field-grown plants. We conducted an experiment recently to determine whether increased nodule number was an indicator of greater susceptibility to rhizobial infection, and whether selection for increased nodule number resulted in greater N_2 fixation in plants challenged by populations of competitive, effective rhizobia. Black bean lines which fixed the highest levels of N_2 in the field also had the most nodules when grown in the growth room. Significant estimates of general combining ability (GCA), but not specific combining ability (SCA), for nodule number were found upon analysis of F_1 plants, with the lines having the most nodules per se also showing the highest GCA values. Substantial gain from two cycles of recurrent mass selection based on nodule number of individual plants was realized. The pedigrees of the selected plants of the C_3 cycle contained the greatest contribution from the original parents with high nodule number per se and high GCA, and the poorest parents made no contribution to the C_3 cycle. Even though N_2 fixation is a complex trait with low heritability, performance per se was a useful criterion for identifying superior parental genotypes able to transmit the high nodule number trait to the offspring.

Progeny tests and combining ability analysis Performance per se may not be a good indicator of breeding value because of low heritability, large G × E interactions, complex genetic control, and contributions of nonadditive genetic variance to total genotypic variance. Furthermore, it may be desirable to develop inbreds for use as parents of commercial F_1 hybrids rather than as pure lines. In such cases, identification of superior parents based on progeny tests and combining ability rather than performance per se will be appropriate. It is necessary also to choose appropriate testers and elite lines into which the putative favorable alleles are to be transferred.

Association between markers and favorable alleles The use of molecular markers provides an efficient, effective means of identifying superior single seeds or families of seeds. The favorable genotype can be identified from the biochemical phenotype, thus providing a superior alternative to either performance per se or progeny testing for certain trait. Enzyme polymorphisms linked to genes affecting traits that are difficult to assess accurately have been useful for identifying lines with superior breeding potential. Selection for the favorable marker allele rather than for trait expression per se can be made (e.g., Tanksley et al. 1982). However, use of isozymes is limited by insufficient segregating loci and lack of linkages to favorable trait alleles for many of the important economic traits. Genes that control complex traits with low heritability can be identified by high-density genomic mapping using restriction fragment length polymorphisms (RFLPs) (Stuber, this volume). Linkages with favorable alleles can then be sought.

TRANSFER OF FAVORABLE ALLELES CONTROLLING QUANTITATIVE TRAIT EXPRESSION

The second difficulty in using unadapted germplasm relates to the transfer of favorable alleles. Usually it is desired to combine them into a single inbred line by pyramiding or to improve the population mean performance by recurrent selection. In either case, transfer is presumed to occur because of selection of individuals or families with the highest expression of the desirable trait even though favorable alleles cannot be identified specifically. If favorable alleles can be identified, transfer can be facilitated by basing the selection criterion on allele presence rather than quantitative expression.

Selection for electrophoretic variants related to trait expression

Phaseolin seed protein Bean lines with variant alleles at loci that either directly or indirectly affect the expression of phaseolin were combined to form a population for recurrent selection in which the primary objective was to increase phaseolin percentage (mg phaseolin/100 mg total protein) (Delaney 1988). Two strategies were compared. In a more conventional approach, recurrent selection among S_1 families was based on phaseolin percentage estimated by rocket immunoelectrophoresis (RIE). The alternative approach, marker based selection, involved selecting individual seeds having at each locus, the favorable allele that produces more phaseolin, then recombining to produce the subsequent generation.

One generation of selection for favorable markers produced an increase in percentage phaseolin (mg phaseolin/100 mg protein) equal to the gain realized from three generations of recurrent selection for phaseolin percentage (Table 4). Based on the total cumulative value expected from combining the effects of individual favorable alleles and superior genotypes (Table 5), it should be

TABLE 4. Comparison of cycle means for several agronomic characters and seed and protein fractions for two methods of selection for increased percentage phaseolin.[a]

Cycle	Maturity (days)	Seed yield (g/plot)[b]	Seed wt. (mg/seed)	Percentage	
				Protein	Phaseolin
Parents	104.5 c[c]	220.6 c	206.5 bc	22.9 d	9.2 d
		SELECTION FOR % PHASEOLIN (PPS)			
C_1	115.9 a	250.8 b	209.5 bc	23.7 c	12.2 c
C_2	113.2 ab	253.3 b	204.3 b	24.0 bc	13.2 b
C_3	111.4 b	244.3 b	210.4 b	24.6 a	14.2 a
		SELECTION FOR MARKERS (MS)			
C_1	116.4 a	265.6 a	226.8 a	24.3 ab	13.9 a

Cycle	mg phaseolin / 100 mg protein	Seed fractions		Protein fractions	
		Protein (mg/seed)	Nonprotein (mg/seed)	Phaseolin (mg/seed)	Nonphaseolin (mg/seed)
Parents	39.5 d	47.3 d	159.2 bc	19.1 e	28.1 a
		SELECTION FOR % PHASEOLIN (PPS)			
C_1	51.3 c	49.6 c	159.9 b	25.4 d	24.2 b
C_2	55.0 b	49.1 cd	155.2 c	26.9 c	22.1 c
C_3	57.6 a	51.7 b	158.7 bc	29.8 b	22.0 c
		SELECTION FOR MARKERS (MS)			
C_1	57.1 a	55.2 a	171.7 a	31.5 a	23.7 b

[a]Delaney (1988).
[b]Means were adjusted for number of plants by covariance analysis.
[c]Mean separation in columns by Bonferroni pairwise *t* tests.

possible to increase phaseolin to about 16.5%, considerably higher than the gain achieved thus far with either selection for phaseolin percentage of favorable marker alleles (Table 4). The fact that parents 2–4–1 and 6–30 contain large amounts of phaseolin even though they also have some less favorable alleles, e.g., *Lec*$^+$, and the S-type phaseolin, suggests that additional loci undetected by the available markers have a substantial effect on phaseolin quantity. The rapid selection response using easily identified favorable alleles demonstrates the efficiency of this approach for short-term improvement of quantitative traits.

TABLE 5. Effects of favorable alleles on phaseolin expression at loci controlling phaseolin (phas$^-$, Phas^{T,C}), lectin (lec$^-$), and arcelin (arc$^-$) and of superior lines, 2–4–1 and 6–30.[a]

Line	Marker	% Phaseolin	Effect
2–4–1[b]	S	12.93	2.57
6–30[b]	S	11.78	1.42
PP11–37	C	11.75	1.39
809598–3	T	11.44	1.08
L12–56	S and lec$^-$	12.12	1.76
Sanilac	S	10.36	0
Sarcl–7	C and Arc 1$^+$	4.40	−5.92
BMC 3522	phas$^-$ and lec$^-$	0	−10.36

[a]Delaney (1988).
[b]These lines carry unknown gene(s) for higher percentage phaseolin.

Bruchid weevil resistance Knowledge of the relationship between arcelin protein and bruchid resistance was used to facilitate the selection of resistant bean lines. It is difficult to measure accurately the level of resistance using insect feeding damage as the selection criterion. The demands of feeding trials and the variable and sometimes unpredictable nature of the insects may lead to phenotypic misclassification. Pure line seeds (U.W. 325) derived from a single seed of PI 325690 which contained Arcelin-1 protein were tested at CIAT and found to be weevil resistant (Hallman, personal communication). Using SDS-PAGE to identify arcelin-containing seeds, a standard backcross procedure was used to transfer the Arc^1 allele into cultivated dry beans. The backcross-derived lines, developed with no on-going selection for resistance during backcrossing, were resistant to *Zabrotes subfasciatus* in feeding trials conducted at CIAT (Table 6). Either electrophoretic separation or double diffusion assays have been used to identify arcelin-containing seeds to facilitate transferring different Arc$^+$ alleles into commercial bean lines.

Another way to facilitate the transfer of favorable alleles that control quantitative trait expression is to establish linkages between molecular markers and loci or chromosome segments associated with favored trait expression. Naturally occurring enzyme variation has been used, e.g., in tomato, for mapping quantitative trait loci (Tanksley et al. 1982) and to facilitate selection for nematode resistance (Rick and Fobes 1974). RFLPs were used to select for increased soluble solids in tomato fruit (Osborn et al. 1987; Tanksley and Hewitt 1988).

Expression of major genes in inbred backcross lines

When studying quantitative traits, segregation of controlling genes is usually deduced from the analyses of means and variances of segregating populations. Often the only discernible effects are the observed changes in the mean and variance of the quantitative trait. Line 6–30, one of the parents intercrossed to form the base population for recurrent selection (see previous section), carries neither of the alleles for increased phaseolin at either the *Lec* or *Pha* locus, but has seeds with considerably more phaseolin than Sanilac, the recurrent parent used to develop line 6–30.

A population of BC_2S_3 lines resulting from an initial cross of Sanilac × 15R–148 followed by two backcrosses to Sanilac [BC_2 and three generations of selfing with single seed descent (S_3] was developed following the procedure of Wehrhahn and Allard (1965). The donor parent used to produce line 6–30, 15R–148, has high percentage seed protein and high seed yield but poor adaptation to temperate conditions. It has small, hard, red seeds, a weak plant type and is susceptible to many diseases, which clearly make it a poor parent in its current form. Based on analyses of seeds from 54 field-grown BC_2S_3 and BC_2S_4 lines, three lines, 6–17, 6–34, and 6–30, were identified that had significantly higher seed protein concentration then either the recurrent parent or the other progenies (Sullivan and Bliss 1983b). Since the increased protein concentration was due to more phaseolin protein per seed and the three lines formed a discrete group within the population of lines, the authors concluded that one or a few genes from 15R–148 were responsible for expression of enhanced phaseolin (Sullivan and Bliss 1983a).

As several authors (e.g., Baker 1978; Multize and Baker 1985) have pointed out, it is not possible to determine conclusively the number of genes controlling a quantitative trait using the inbred backcross approach if discrete segregation is not observed. However, this method can be used to facilitate

TABLE 6. Resistance to *Zabrotes subfasciatus* in a bean line selected for arcelin protein (Arc^+) compared to a sib line without arcelin (arc^-) and the resistance and susceptable checks.[a]

Line	Arcelin	Eggs/seed	Progeny/female	Emergence %	Days to emergence	Wt./adult (g × 10^{-3})	Rating
859446–67	+	5.6 b[b]	3.1 b	7.8 c	46.4 b	0.8 c	R
859446–59	–	5.5 b	37.7 a	95.4 a	32.8 c	1.3 b	S
G12952 (ck)	+	1.8 c	0.4 c	3.0 d	50.0 a	0.7 d	R
Calima (ck)	–	6.6 a	38.3 a	80.5 b	33.1 c	1.4 a	S

[a]Weevil feeding trials conducted at CIAT (1986).
[b]Means in the same column followed by the same letter are not significantly different at the 0.05 level.

the transfer of genes controlling expression of a quantitative trait into a desirable genetic background. If no backcross lines with the intensity of expression of the donor are recovered, the best lines can be intercrossed to recombine the favorable alleles into one or several superior genotypes (Owens et al. 1985; St. Clair 1986).

DIFFERENTIAL EXPRESSION OF FAVORABLE ALLELES

Alleles that have been used most widely for improved performance of economic crops are those that have a well-defined and predictable expression. In contrast, it is more difficult to use alleles that display unreliable expression due either to unknown modifier genes in different genetic backgrounds or to environmental factors such as location effects and random events occurring between different years. A large differential response of a gene or genotype to fluctuating environmental factors is undesirable particularly if the response is unpredictable.

Effect of genetic background

The effect of genetic background is important in several respects. The cultivars used as recurrent parents into which new variability is being introduced may have different merits. Although the populations of progeny lines that result from the introgression of new genetic variability usually are not of equal value, they also may not reflect the same relative merits of the original cultivar used as the recurrent parent.

Improved populations for increased dinitrogen fixation Two widely grown black bean cultivars of subtropical origin, Porrillo Sintético and Jamapa, were chosen as potential parents in which to improve N_2 fixation ability. Using the high-N_2-fixing donor, Puebla 152, as the source of increased genetic variability, two populations of inbred backcross lines were developed and analyzed in field trials using indirect measures of N_2 fixation potential. The population derived from Porrillo Sintético had the greater number of progeny lines with higher N_2 fixation potential than the recurrent parent (McFerson 1983). Based on results from a separate study, in which the combining abilities of Porrillo Sintético and Jamapa for N_2 fixation were compared using top cross progeny performance, Miranda (1987) also concluded that more gain from selection would be realized in the population derived from Porrillo Sintético than from Jamapa (Table 7). Even though the two cultivars differ little in their own N_2 fixation potential, they showed different values as parents when genes were introgressed from a common donor parent.

Effect of lectin alleles with different phaseolin types The allelic state of the *Lec* locus has an indirect effect on the quantity of phaseolin produced (Os-

TABLE 7 Mean performance of testcross populations of F_3 families, predictions of gain from selection, and population mean after selection for total seed N in common bean: 1985.[a]

Testcross	Testcross population mean seed N (kgN/ha)	G_s[b] (kgN/ha)	Predicted population mean (kgN/ha)
BTS × T[c]	80 a[d]	7	87
UW21–38 × T	66 b	7	73
UW21–58 × T	77 a	7	84
UW24–4 × T	75 a	9	84
Porrillo Sintetico × T	77 a	10	87
Jamapa × T	61 c	10	71

[a]Miranda (1987).
[b]Gain from selection of the top 20% within each testcross.
[c]T, tester = UW24–21; BTS, Black Turtle Soup; UW 21–38, UW 21–58, and UW 24–4, breeding lines.
[d]Values not followed by the same letter are significantly different at the 0.05 level of probability.

born et al. 1985). Since phaseolin type also affects the amount of phaseolin expression, the *lec*$^-$ allele was introduced into backcross-derived bean lines having different phaseolin types in the standard Sanilac background. In combination with all phaseolin types, the lec/lec seeds (lec$^-$) showed increased phaseolin of from 5 to 8 mg/seed, which corresponded to increased concentrations (%) of both total and phaseolin protein (Table 3). There was little evidence of differential expression of the *lec*$^-$ allele and the genetic background effect appeared to be minimal.

Estimation of the differential effects of different genetic backgrounds on allele expression is usually made from backcross-derived lines. Even when multiple backcrosses are made, care must be taken to distinguish specific gene effects from those of tightly linked genes or pleiotropic effects. The ability to produce transformed plants by the introduction of gene-specific DNA will allow greater precision in the study of different genetic backgrounds and modifier genes on single gene expression.

Environmental effects on the utility of favorable alleles

The presence of arcelin protein in bean seeds provides a high level of resistance to the Mexican bean weevil, *Zabrotes subfasciatus* (Boheman) (Osborn et al. 1988), but a second important insect, *Acanthosceledes obtectus* (Say), the common bean weevil, reacts somewhat differently to arcelin. In addition to having different life cycle characteristics, the geographic distribution of the two bruchid insects are quite different. Z. *subfasciatus* is found at higher temperatures, while A. *obtectus* is limited to areas with cooler temperatures, since high temperatures produce sterility in the males. Because of the differences in the geographic distribution and the fact that each insect reacts differently to the arcelin protein, the utility of the different arcelin alleles will be different. Optimal use of the wild germplasm for developing bruchid-resistant bean cultivars will depend on understanding the environmental and biological factors that influence the reaction of each insect to the different *Arc+* alleles.

The availability of different arcelin types offers the opportunity to study the effects of the different alleles on the level and stability of resistance to one or both bruchids. Several strategies for cultivar development are being explored and the following questions are being addressed. (1) Is it necessary that all seeds contain arcelin (resistance) to produce an economically useful resistance level, or will mixtures of seeds (some without arcelin) suffice? (2) If high levels of resistance are imposed will this resistance be overcome by the insects? (3) Will seed mixtures of different arcelin types provide a stable level of resistance and still produce an economically useful resistance level? Answers to these questions will influence the further use of unadapted germplasm for improvement of this and other traits.

CONCLUSIONS

Considerable genetic variability exists in the gene pools of most cultivated crops. Although breeders have used plant introductions extensively to improve qualitative traits, the use of unadapted germplasm as a source of increased genetic variability for quantitative trait improvement has been limited. The major constraints involve predicting the genetic potential of untested germplasm, transferring specific favorable alleles with unwanted linked loci, and predicting the expression of favorable alleles in diverse environments and different genetic backgrounds.

In common bean, extensive interspecific hybridization has been made with *P. coccineus*, but relatively few crosses have been made with *P. acutifolius* (Waines et al. 1988) and there has been only one report of a successful cross with *P. lunatus* (Honma and Hecht 1959). In only a few cases have there been successful crosses with nondomesticated *Phaseolus* species.

Although fertile F_1 plants can be obtained easily from crosses between common bean cultivars and wild *P. vulgaris*, the recent transfer of bruchid resistance into common bean cultivars represents the first use of wild germplasm. Many naturally occurring variant alleles that alter the expression of phaseolin protein are found in both cultivated and wild accessions of *P. vulgaris* and in *P. coccineus*. The transfer of these alleles into adapted bean cultivars has been used to increase seed protein percentage, an important quantitative trait of bean. Although no specific genes with large affects that alter nitrogen fixation have been found in common bean, selection for increased N_2 fixation has been successful in segregating populations of inbred backcross lines for which a poorly adapted landrace cultivar was the high-fixing donor parent.

The constraints to greater utilization of unadapted bean germplasm have been minimized by using replicated family testing, making test crosses to a common parent, and transferring putative favorable alleles into the genetic background of an adapted cultivar. Precise selection criteria based on biochemical traits that are affected less by nongenetic factors that alter expression have been employed to increase selection efficiency for complex traits that show low heritability.

CHAPTER 19

MOLECULAR MARKERS IN THE MANIPULATION OF QUANTITATIVE CHARACTERS

Charles W. Stuber

ABSTRACT

Molecular markers (isozymes and RFLPs) provide a powerful approach for investigating the genetic bases of variation in quantitative traits. Recent evidence shows that these markers should be useful for manipulating these complexly inherited traits. Extensive sets of mapped molecular loci have been compiled and documented in several plant taxa. Investigations in a number of plant populations have demonstrated significant associations of these markers with quantitatively inherited traits. In maize, studies have shown that isozyme and RFLP markers were effective for identifying and locating quantitative trait loci (QTLs). For most of the traits studied, QTLs were found throughout the genome, but the magnitudes of the effects associated with individual QTLs varied greatly.

Effective manipulation of quantitative characters first requires detailed knowledge of their genetic bases. For most characters, such as grain yield in plants, this knowledge has been difficult to obtain because the inheritance is complex, usually involving numerous genetic factors that frequently interact with environmental effects. Until recently, investigations of quantitative variation have relied largely on biometric procedures, which characterize, *en masse*, the activity of the genetic factors involved in the expression of this variation. The use of mapped genetic markers now provides a powerful approach for studying quantitative traits and for locating and manipulating individual genetic factors associated with these traits. Although markers associated with the expression of morphological traits have had some limited usage, molecular markers [isozymes and restriction fragment length polymorphisms (RFLPs)] are far superior for these applications.

GENERAL CONCEPTS FOR USING MAPPED GENETIC MARKERS TO INVESTIGATE AND MANIPULATE QUANTITATIVE TRAITS

Applicable theory for using mapped genetic markers, such as isozyme loci and RFLPs, to identify, locate, and manipulate quantitative trait loci (QTLs) has been documented by several investigators (Jayakar 1970; Mather and Jinks 1971; McMillan and Robertson 1974; Soller and Plotkin-Hazan 1977; Tanksley et al. 1982; Soller and Beckmann 1983; Edwards et al. 1987; Lebowitz et al. 1987). The concept is based on the fact that the marker locus identifies or "marks" the chromosomal segment in its vicinity, and allows that segment to be followed through various genetic manipulations. Alternative homologous chromosomal segments, which have alternative alleles at the marker locus, may be replicated repeatedly in different individuals (or lines) and compared for their effects on the expression of quantitative traits while other chromosomal regions in the same individuals and the environmental factors affecting these traits are permitted to vary at random.

As an example, consider the segregating generation produced from self-fertilizing or intermating hybrid plants derived from the cross of two homozygous lines. By following the transmission of each chromosomal segment with identifiable genetic markers the entire genome can be surveyed, segment by segment, for genetic factors associated with the variation of any desired quantitative character; and the effects contributed by individual chromosomal regions can be quantified. With adequate markers, theoretically, then, it should be possible to construct a detailed map showing the location of all major genes associated with the quantitative trait and to describe their individual and interactive effects.

Although the expression of traits such as disease resistance, male sterility, and self-incompatibility may be due to a single discrete major gene, frequently several loci are involved. The traditional backcrossing method for transferring alleles affecting such traits is relatively simple if only a single locus affects the trait. If the trait is conditioned by several loci, and if the desired alleles are dominant, traditional backcrossing methods are very inefficient for transferring genes. Marker-facilitated backcrossing should be very effective, however, for these "semiqualitative" or "quasiquantitative" traits for "pyramiding" several dominant genes (such as for disease resistance) into a single line (Brown et al. 1988).

In tomato a rare allele of the acid phosphatase locus, *Aps-1*, is associated with an allele for nematode resistance at the *Mi* locus (Rick and Fobes 1974; Medina-Filho 1980). This association has greatly benefited breeding for nematode resistance through marker-facilitated backcrossing. Tomato breeders in the United States and Europe quickly adopted this technique in their breeding programs (Tanksley 1983).

Tight linkage (probably less than 5 cM) is required for "tagging" or manipulating desired loci with single markers. The level of recombination with looser linkage is so great that the benefits of using marker loci to manipulate alleles at desired loci are largely negated. However, if tagging can be accomplished with flanking markers, gene manipulations can be highly successful with more loosely linked markers by simultaneous selection for the markers bracketing the gene(s) of interest. Tanksley (1983) reported that it would be necessary to have markers spaced at 20 cM intervals throughout the genome in order to tag any gene of interest with a selection fidelity of 99%. It is unlikely that this spacing of isozyme markers throughout the genome will be achieved in many crop plants. However, this level of saturation and distribution of RFLP markers is nearly attained in crops such as maize and tomato (Helentjaris et al. 1985, 1988; Tanksley and Hewitt 1988).

ATTRIBUTES OF MOLECULAR MARKERS FOR QUANTITATIVE TRAIT STUDY AND MANIPULATION

As stated earlier, morphological markers have had limited use for quantitative trait applications (e.g., Qualset et al. 1965). However, markers with neutral effects, such as isozymes and RFLPs, have numerous inherent properties that allow them to be used very effectively for characterizing quantitative trait variation and for manipulating both quantitative and "semiqualitative" traits.

As outlined by Tanksley (1983) and Stuber (in press), molecular markers, such as isozymes and RFLPs, are preferable to morphological trait markers for most genetic applications because of the following: (1) Alleles at most molecular marker loci are usually codominant, therefore all possible marker genotypes can be distinguished in segregating populations. With morphological traits, dominant-recessive interactions frequently prevent distinguishing all genotypes. (2) For molecular markers, genotypes can usually be determined at the whole plant, tissue, or cellular levels. For most morphological markers, genotypes may be ascertained only at the whole plant level and, frequently, the mature plant is required. (3) For many plant species, a number of naturally occurring alleles is available at most molecular marker loci. For morphological traits, distinguishable alleles often must be induced through mutagenesis. (4) For molecular marker loci, specific alleles very rarely show deleterious effects. Alleles at loci affecting morphological traits often show deleterious phenotypic effects. (5) Unfavorable interactions frequently occur among loci encoding morphological traits, which limits the number of segregating markers that can be used in a specific population or generation. The number of molecular markers that can be monitored in a single population is usually limited only to those polymorphic markers for which assay procedures are available or by the number that can be handled with the facilities and resources available to the investigator.

For several plant taxa, extensive sets of mapped molecular loci have been compiled and documented. For isozyme markers, these taxa include maize (Goodman et al. 1980; Goodman and Stuber 1983; Wendel et al. 1986, 1988), tomato (Tanksley 1983; Tanksley and Rick 1980), wheat (Hart 1983), and pine (Conkle 1981). Isozyme genotypes at 22 loci for 406 publicly available inbred lines of maize provide an extensive information base for selecting lines with different marker-loci genotypes in this species (Stuber and Goodman 1983). Although RFLP markers are now being used in a number of plant taxa, the maps developed in maize and tomato (Helentjaris et al. 1985, 1988; Tanksley and Hewitt 1988) are probably more extensive than for any other plants. Landry et al. (1987) have reported the construction of a linkage map in lettuce using 53 genetic markers which included 41 RFLP loci.

ATTRIBUTES OF POPULATIONS AMENABLE TO MOLECULAR MARKER APPLICATIONS FOR QUANTITATIVE TRAITS

A number of attributes concerned with the linkage relationships between marker loci and adjacent QTLs determine the effectiveness of using molecular markers for identifying, mapping, and manipulating those factors affecting quantitative traits. These attributes include (1) the number of segregating marker loci available in the population or material of interest, (2) the distribution or uniformity of spacing of the marker loci, and (3) the level of linkage disequilibrium in the population. If only a few marker loci are available for the investigation, a population such as the F_2 derived from the cross of two homozygous lines may be preferred because linkage disequilibrium is maximized in this generation. Although an F_2 is advantageous for detecting QTLs with a minimum number of markers, large genomic regions would probably be represented by specific marker loci in this generation (Hanson 1959). Thus, there is a high probability that genotypic classes at an individual marker locus may be reflecting the effects of multiple QTLs. Desirable attributes of a population that should be useful for investigating quantitative traits would include closely spaced, uniformly distributed, marker loci (i.e., about every 10 cM) and sufficient recombinational generations so that the size of the genomic region associated with each marker allele has been reduced significantly from that found in an F_2 population.

The kinds of populations that may be used for determination of associations of marker genotypes with quantitative trait expression include F_2 populations, backcrosses, topcrosses, and recombinant inbred line populations. The traditional approach involves computation of mean quantitative trait values (for individuals, lines, families, etc.) for each marker class. These mean values are then compared using a statistical measure such as an F test or t

test. If the quantitative trait means differ significantly, this is interpreted as evidence that the marker locus is linked to a QTL(s) that influences the expression of the quantitative trait. The significance of associations between marker loci and a QTL diminishes relative to the true effect of the QTL as the distance (recombination frequency) between the marker locus and the QTL increases (Edwards et al. 1987). Interval mapping using LOD scores may increase the efficiency of QTL mapping, however (Lander and Botstein 1989).

Populations that have been subjected to a number of generations of recurrent selection have been used by several researchers to evaluate associations of marker genotypes with quantitative traits by monitoring marker allele frequency changes over selection cycles (Stuber et al. 1980; Pollak et al. 1984; Kahler 1985). However, the large number of recombination generations in such populations will likely diminish the probability of detecting meaningful associations unless specific marker loci are tightly linked to QTLs affecting the trait of interest.

ASSOCIATIONS OF MOLECULAR MARKERS WITH QUANTITATIVE TRAITS IN DIVERGENT AND/OR RANDOMLY MATING POPULATIONS

Using isozyme markers, Clegg and Allard (1972) analyzed genetic variability among populations of slender oat (*Avena barbata*) in California and they reported two complementary five-locus allozyme complexes whose distributions were closely associated with environment. Further studies in these populations by Hamrick and Allard (1975) were conducted to separate environmental effects from genetic effects for several quantitative characters (plant height, number of tillers, number of seeds, and two measures of maturity). Plants representing these two five-locus enzyme genotypes differed genetically with respect to four of the five quantitative characters studied.

Price et al. (1985) evaluated associations of several isozyme and two morphological markers with herbicide response in 10 populations of slender wild oat and six populations of wild oat (*Avena fatua*). The herbicide barban (4-chloro-2-butynyl-3-chlorophenylcarbamate) was used for this research. Herbicide resistance showed significant variation and was significantly associated with the isozyme and morphological markers in these oat populations.

Several investigations have showed significant associations of isozyme marker loci with quantitative characters in interspecific crosses of tomato (*Lycopersicon* spp.). Twelve isozyme loci were used by Tanksley et al. (1982) to locate factors influencing four quantitative traits in a backcross population of 400 plants derived from *L. esculentum* and *L. penneli*. Of the 48 possible comparisons of marker loci with quantitative trait expression, 27 showed significance, and a minimum of five QTLs were detected per trait. In another study in an interspecific cross of *Lycopersicon*, Vallejos and Tanksley (1983) reported linkages between 11 segregating enzyme loci and genetic factors responsible for cold tolerance. A minimum of three QTLs responsible for growth at low temperatures was detected. Weller et al. (1988) examined marker-locus associations with 18 quantitative traits in an interspecific F_2 population involving *L. pimpinellifolium* and *L. esculentum*. Of the 180 possible comparisons among 10 markers and 18 traits, 85 were significant.

Both isozyme and RFLP markers were used by Tanksley and Hewitt (1988) to identify introgressed chromosomal segments of the wild species *Lycopersicon chmielewskii* in cultivated tomato. Of three detected introgressed segments, two were associated with an increase in soluble solids, however, the effect of one segment was dependent on genetic background. Both of these segments were associated with deleterious characters including lower fruit yield, smaller fruit, and an increase in fruit pH.

The effectiveness of molecular markers for identifying and locating QTLs and for the detailed genetic investigation of quantitative variation in suitably marked populations was demonstrated quite conclusively in these tomato studies. However, the dangers of establishing breeding programs based on linkages of markers with quantitative traits until the quantitative genes have been tested in several genetic backgrounds, and have been evaluated for associated effects on other characters of agronomic or economic importance, was dramatically illustrated in the studies by Tanksley and Hewitt (1988).

Several investigations have been conducted in maize in which frequency changes at a large number of isozyme loci were monitored over different cycles of long-term recurrent selection experiments. Changes of allelic frequencies at eight isozyme loci were statistically significant and greater than would be expected from genetic drift acting alone in several long-term maize selection experiments conducted in North Carolina. These changes also were highly correlated with changes in the populations due to selection for increased grain yield (Stuber and Moll 1972; Stuber et al. 1980). Associations between isozyme marker loci and several agronomic and morphological traits in maize also have been reported for a number of selection experiments conducted by Pollak et al. (1984) and Kahler (1985). Results from these studies provided the impetus for more intensive studies with the use of marker loci in maize as discussed in the following sections.

IDENTIFYING AND LOCATING QUANTITATIVE TRAIT LOCI IN F_2 POPULATIONS OF MAIZE

Edwards et al. (1987) and Stuber et al. (1987) conducted extensive investigations to identify and locate QTLs in two large F_2 populations (about 1800 plants each). The two F_2 populations, "COTX" derived from inbred lines CO159 and Tx303 and "CMT" derived from lines T232 and CM37, showed segregation at 15 and 18 isozyme loci, respectively, plus loci coding two easily scoreable morphological traits. These isozyme and morphological marker loci are distributed on 9 of the 10 maize chromosomes and are located within about 20 cM of 40 to 45% of the genome. Measurements recorded on individual plants in these investigations included weights, dimensions, and counts of vegetative and reproductive plant parts as well as silking and pollen shedding dates and were used to generate 82 quantitative variables.

For these 82 quantitative variables, trait expressions were found to be significantly associated with marker-locus genotypes in 830 (60%) of the 1394 possible comparisons in COTX and in 1079 (66%) of the 1640 comparisons in CMT. The number of marker loci significantly associated with factors influencing the expression of each trait averaged 10.2 (of 17) and 13.8 (of 20) in COTX and CMT, respectively, and QTLs were identified that influenced every trait measured in each population.

For many of the traits studied, individual marker loci accounted for relatively small proportions of the total phenotypic variation. However, differences between mean phenotypic values of homozygous classes at some marker loci occasionally were greater than 16% of the population mean for the trait. For example, in CMT grain weight differences between the two homozygous classes for four unlinked loci (*Dia1*, *Pgd2*, *Amp3*, and *Mdh1*) were each slightly more than 20 grams per plant (16% of the mean grain weight in this population). However, these loci, individually, accounted for only about 4–5% of the total phenotypic variation for this trait.

Appropriate contrasts among trait mean values of marker-locus genotypic classes showed primarily dominance or overdominance types of gene action for grain weight and ear length. Additive gene action was largely implicated for several yield-related traits such as ear number, kernel row number, and second ear weight (Stuber et al. 1987). There was little evidence for digenic epistatic gene action for the traits evaluated in these two F_2 populations. Comparisons of mean grain weight of plants with varying numbers of heterozygous marker loci showed that heterozygosity was highly associated with the expression of this trait (Edwards et al. 1987). This would be expected because the gene action for grain weight was largely in the dominance and overdominance classes. A similar relationship was found by Kahler and Wehrhahn (1986) in an F_2 population derived from Wf9 and Pa405.

Investigations in several smaller F_2 maize populations have corroborated the findings in the COTX and CMT populations, viz that segregating isozyme marker loci were effective in identifying chromosomal regions affecting a wide array of phenotypic characteristics in maize. Six F_2 populations derived from elite inbred line crosses (B73 × Tx303, B73 × T232, Mo17 × Tx303, Mo17 × T232, Oh43 × Tx303, and Oh43 × T232) were evaluated for marker-locus associations with quantitative traits (B. Bateman, M. D. Edwards, and C. W. Stuber, unpublished). Although only about 500 plants were evaluated in each population, the percentage of significant associations agreed with the percentage found in COTX and CMT after accounting for the reduced population size.

The studies by Bateman, Edwards, and Stuber showed that QTLs affecting most of the traits measured were generally distributed throughout the genome, which also agreed with the COTX and CMT studies. In the eight populations evaluated, however, certain chromosomal regions appeared to contribute a greater effect than others to trait expression. Major factors affecting the expression of grain weight were associated with *Mdh4*, *Adh1*, and *Phi1* on chromosome 1L; *Dia1* on chromosome 2S; *Mdh3* and *Pgd2* on chromosome 3L; *Amp3*, *Mdh5*, and *Pgm2* on chromosome 5S; *Idh1* on chromosome 8L; and *Acp1* on chromosome 9S. It is likely that major factors affecting grain weight also are present in regions of the genome that were devoid of marker loci in these studies.

Kahler and Wehrhahn (1986) reported that 42% of the associations between eight isozyme loci and 11 quantitative traits were significant in a study of 460 F_2 plants derived from the single cross, Wf9 x Pa405. Although the number of marker loci was limited in their study, the three chromosomal regions that showed significant effects in the expression of grain yield also showed small, but significant, associations in at least one of the eight F_2 populations discussed previously.

RELATIONSHIPS BETWEEN HYBRID RESPONSE AND MARKER DIVERSITY

Development of inbred lines with superior combining ability (i.e., superior heterosis) in hybrid combinations is costly and requires considerable time, particularly for field testing. A number of investigations have been conducted to determine whether genetic markers, such as isozyme loci, could be used as predictors of hybrid performance in maize. Hunter and Kannenberg (1971) reported a correlation of only 0.09 between isozyme diversity at 11 enzyme loci and grain yield in single cross hybrids derived from 15 inbred lines of maize. In a study of hybrids derived from eight inbred lines, using eight enzyme marker loci, Heidrich-Sobrinho and Cordeiro (1975) found a correlation of 0.23 between specific combining ability and isozyme diversity and a correlation of 0.72 between general combining ability and isozyme diversity. In a similar study involving seven inbred lines of maize and four isozyme systems, Gonella and Peterson (1978) reported a correlation of −0.42 between yield and percentage relatedness based on isozyme diversity. Correlations between hybrid yield and diversity of lines also were low and varied with environments in an investigation of 26 inbred lines of maize and 15 isozyme systems reported by Hadjinov et al. (1982). Peng et al. (1988) reported no association between hybrid grain yield and isozyme diversity in a study of 75 F_1 rice hybrids and six enzyme loci.

Frei et al. (1986a) reported results from a somewhat different study of 114 single-cross hybrids to assess the value of isozyme markers for predicting hybrid yield performance among maize inbreds in which line pairs for specific hybrids were classified into similar and dissimilar groups based on 21 isozyme loci. These isozyme diversity groups were then further divided into similar and dissimilar pedigree classes based on commonality of pedigree background between line pairs. Grain yield of hybrids in the dissimilar isozyme group was significantly higher (10%) than in the similar isozyme group. Hybrids in the dissimilar pedigree class, however, yielded about 37% more than in the similar pedigree class. From this study, it was concluded that although isozyme marker dissimilarity was significantly associated with higher grain yield in single-cross hybrids, the useful predictive value of these markers would be largely limited to lines with similar pedigrees.

In another maize study, Lamkey et al. (1987) evaluated single-cross hybrids among 24 high-yielding and 21 low-yielding lines selected from a group of 247 inbred lines derived from single-seed descent in the Iowa Stiff Stalk Synthetic (SSS) population. Their objective was to determine whether diversity at 11 isozyme loci could be used to predict hybrid performance of maize in single crosses generated from a group of lines derived from the SSS population. Comparisons of high × high, high × low, and low × low hybrids of the high- and low-yielding lines indicated that allelic differences at the 11 enzyme loci did not reflect hybrid performance.

Results from the above studies suggested that isozyme genotypes would have limited value for predicting hybrid performance in maize and rice. Several confounding factors, however, may have contributed to the somewhat disappointing outcomes of these investigations. With the small number of enzyme loci assayed in most of the studies, only a fraction of the genome would be effectively marked; consequently only a small proportion of the genetic factors affecting the quantitative traits would be effectively sampled. More importantly, alleles at most isozyme loci probably do not affect the phenotypic expression of the quantitative trait directly; rather they likely serve only as markers for adjacent (linked) segments of the chromosome on which they are located. Therefore, to be useful as predictors of hybrid performance, the effects of QTLs linked with specific marker alleles must be known.

In addition, if the lines were derived from a randomly mated population, or comprise some subset of publicly available inbred lines, then the alleles present at the linked QTLs might be expected to be somewhat random, i.e., near linkage equilibrium. For marker-facilitated procedures to be successful for predictive or selective purposes for complexly inherited traits, such as grain yield, the genome should be well saturated with markers and/or a high level of linkage disequilibrium must be present. Even for more simply inherited traits, effective marker-facilitated selection requires that the target gene(s) be tightly linked with the marker locus or that the target gene(s) be bracketed (within about 20 cM) with marker loci.

MANIPULATIONS OF QUANTITATIVE TRAITS IN MAIZE USING MARKER-FACILITATED PROCEDURES

Based on the results of the earlier studies in maize (Stuber et al. 1980) in which isozyme allelic frequencies were monitored in long-term recurrent selection populations, it was hypothesized that manipulation of allelic frequencies at selected isozyme loci might produce significant responses in the correlated quantitative traits. Experiments to evaluate this hypothesis, in an open-pollinated maize population, indicated that selections based solely on manipulations of allelic frequencies at seven enzyme loci significantly increased grain yield and ear number, highly correlated traits (Stuber et al. 1982). In a similar study conducted in a population generated from a composite of elite inbred lines, marker-facilitated selection responses were about equal to the responses found for phenotypic selection conducted in the same population (Frei et al. 1986b). Because these populations had been subjected to several generations of random mating, the level of linkage disequilibrium between marker loci and QTLs would have been markedly reduced. This undoubtedly limited the effectiveness of the marker loci for manipulating the associated quantitative traits. The limited positive results, nevertheless, did provide the impetus to pursue further investigations into the value of using molecular markers for studying and manipulating quantitatively inherited traits.

A more definitive study in the use of marker loci for the manipulation of quantitative traits, such as grain yield in maize, was reported by Stuber and Edwards (1986). Results from the COTX and CMT F_2 population studies discussed previously in this chapter were used as the basis for marker-facilitated selection. In the first study, selections were made among the F_2 plants grown in the earlier studies, and evaluations of selection response were made on progeny (half-sibs) of these open-pollinated plants. In the second study, selfed progeny from a different sample of F_2 plants from the same two populations were evaluated. Mass selection, based solely on the phenotypic expression of each F_2 plant, also was conducted for comparison with the marker-facilitated responses.

In each of the marker-facilitated selection studies, a breeding value was determined for each marker for each plant and for each trait being manipulated. This value was equal to one-half the difference between the quantitative means of the homozygous classes for that locus. A value of zero was given to each heterozygote because its progeny would segregate and, thus, would not contribute to a change in gene frequency of the population. Breeding values were totaled over 15 isozyme marker loci to derive a composite breeding score for each plant for each trait selected.

Based on these composite scores, individual plants were chosen to provide divergent selection classes (positive and negative) for several traits and combinations of traits, including grain yield, ear height, and ear number. In the open-pollinated F_2 population studies, for each selection criterion kernels from 37 selected plants were bulked to plant in the evaluation trials. This gave a selection intensity of about 2%. In the selfed-progeny studies, selection intensities were 17% and 30% in COTX and CMT, respectively.

Results from the selections designed to manipulate grain yield in the open-pollinated F_2 populations are shown in Tables 1 and 2. For genotypic selection in the COTX population, the mean for the increased-yield entry was about 12% greater than the mean for the decreased-yield entry. The same comparison showed a 16% difference for the phenotypic (mass) selection. Changes for the correlated traits, ear height and ear number, were in the same direction as yield and percentage differences were similar in size.

Marker-facilitated manipulations for grain yield were more effective in CMT than in COTX (Table 2). This was expected because the marker-locus genotypes accounted for twice as much variation for grain yield in CMT as in COTX. For genotypic selection in the CMT population the mean of the increased-yield entry was 40% greater than the mean for the decreased-yield entry. Also, selection for increased yield showed nearly 20% gain over the check population mean. The response to phenotypic selection for increased yield was similar to genotypic selection, however, phenotypic selection for decreased yield did not differ from the check means. The correlated responses for ear height and ear number in CMT were similar to those found for COTX.

TABLE 1. Means for three traits following divergent genotypic (marker-facilitated) and phenotypic (mass) selection designed to manipulate grain yield in (CO159 × Tx303)F_2 randomly mated populations.

	Traits		
Selection criterion	**Grain yield (g/plant)**	**Ear height (cm)**	**Ear number**
GRAIN YIELD			
Genotypic—increase	138.9	75.0	1.40
Genotypic—decrease	123.9	61.9	1.24
Phenotypic—increase	143.2	79.8	1.43
Phenotypic—decrease	123.0	69.7	1.25
CHECKS—NO SELECTION			
(CO159 × Tx303)F_2	137.9	66.2	1.26
(CO159 × Tx303)F_2—randomly mated	133.0	70.0	1.31
$SE_{\bar{d}}$[a]	5.81	1.96	0.04

[a]Standard error of mean difference.

TABLE 2. Means for three traits following divergent genotypic (marker-facilitated) and phenotypic (mass) selection designed to manipulate grain yield in (T232 × CM37)F_2 randomly mated population.

Selection criterion	Traits		
	Grain yield (g/plant)	Ear height (cm)	Ear number
GRAIN YIELD			
Genotypic—increase	151.2	73.5	1.48
Genotypic—decrease	107.7	47.1	1.20
Phenotypic—increase	151.7	68.5	1.43
Phenotypic—decrease	122.4	57.8	1.28
CHECKS—NO SELECTION			
(T232 × CM37)F_2	120.1	55.1	1.30
(T232 × CM37)F_2—randomly mated	127.2	59.2	1.35
$SE_{\bar{d}}$[a]	6.36	2.16	0.04

[a]Standard error of mean difference.

In the selfed-progeny evaluation studies, divergent selection for grain yield also was more effective in CMT than in COTX. In CMT the mean of the increased-yield entry was about 31% greater than for the decreased-yield mean. The same contrast was about 17% in COTX.

The monitoring of marker-locus allelic frequencies was possible in these selection studies because the marker genotypes were known for each selected plant. In the CMT population, frequencies of the alleles at the 15 marker loci averaged 0.38 greater in the population selected for increased yield than in the one selected for decreased yield (M. D. Edwards and C. W. Stuber, unpublished). For individual loci, differences between frequencies ranged from 0.02 to 0.73. Loci with the greatest differences were those that accounted for the largest proportion of the phenotypic variation for grain yield. In the populations generated from phenotypic (mass) selection for grain yield, frequency differences between populations selected for increased and decreased yield averaged 0.13, about one-third the 0.38 found for genotypic selection. Thus, the similar responses from genotypic and phenotypic selection on grain yield resulted from 3-fold greater changes for loci linked to the marker loci in the populations derived from marker-facilitated selection than in those derived from phenotypic selection. Therefore, loci in unmarked regions of the genome, which would receive no selection pressure from genotypic selection, undoubtedly contributed to the phenotypic selection response. The results for the COTX population were similar to those for CMT, but magnitudes of frequency changes were less in COTX.

These results indicate that marker-facilitated selection (based on 15 isozyme marker loci that represented about 30–40% of the genome) was as effective as phenotypic selection, which, presumably, involves the entire genome. If the entire genome was effectively marked with uniformly distributed loci, a significant increase in the effectiveness of marker-facilitated selection would be expected. These results also suggest that marker-facilitated techniques should be effective for line-to-line transfers of desired factors at multiple loci for improving complexly inherited traits such as grain yield. Effects and success of such transfers will likely vary in different genetic backgrounds.

DISSECTION OF HETEROSIS IN A SINGLE-CROSS HYBRID OF ELITE MAIZE LINES

Heterosis is a basic phenomenon in the expression of many quantitatively inherited traits and has been widely exploited in both plant and animal species. The genetic basis of this phenomenon, however, is not well understood and has been dealt with theoretically in general terms of dominance and overdominance (and occasionally, epistasis), and only in terms of average effects over the entire genome. Attempts have been made to explain heterosis in terms of biochemical or physiological effects, although most have shown negative results or have not been repeatable by other researchers.

In collaboration with Dr. Tim Helentjaris and colleagues at Native Plant Institute, Salt Lake City, Utah, our maize research project at Raleigh, NC, has undertaken an extensive project to use molecular markers to identify and locate genetic factors (QTLs) that significantly contribute to the heterotic response in the hybrid of two elite maize lines, B73 and Mo17. The study will evaluate the effects and types of gene action associated with identified genetic factors. A second objective is to identify and locate genetic factors in two other elite inbred lines, Tx303 and Oh43, that can be used to enhance the heterotic response of the elite hybrid, B73 × Mo17. A third part of the study aims to determine whether the factors detected show consistent effects across environments.

A diagram of the experimental procedures is shown in Figure 1. The two lines, B73 and Mo17, were crossed and, then, through two selfing generations, 264 F_3 lines were developed. A single plant in each line was backcrossed to each of the parental lines and selfed to provide progenies for genotyping (at 12 isozyme and 65 RFLP loci) each of the 264 plants. Progenies of each of the 528 backcrosses were evaluated in six field environments, four in North Carolina, one in Iowa, and one in Illinois. The quantitative traits measured included grain yield, ear number, plant height, ear height, ear leaf area, days to pollen shed, and lodging.

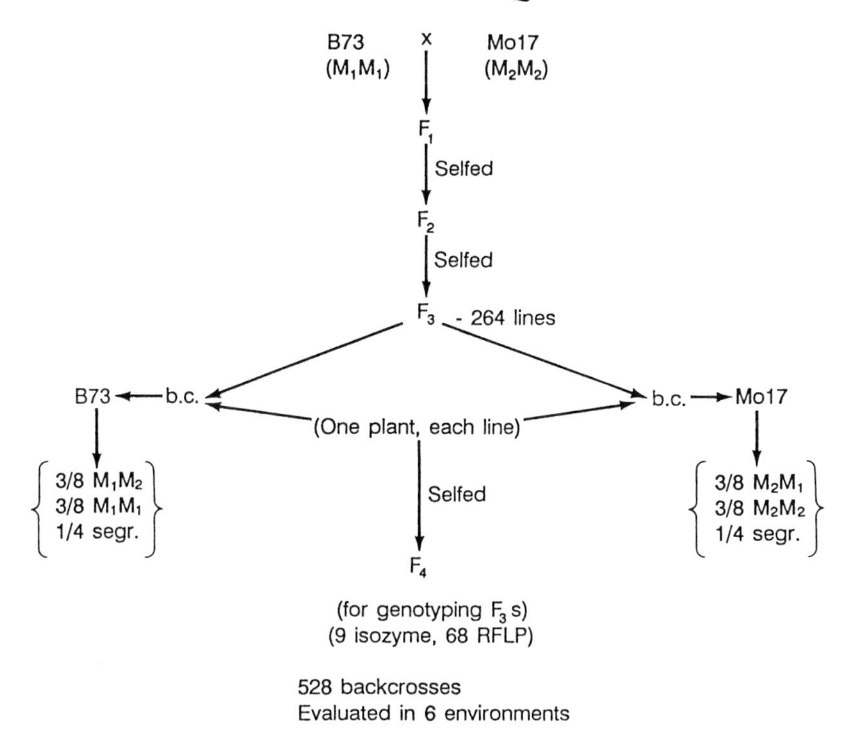

FIGURE 1. Diagram of experimental procedures for development of B73 × Mo17 F_3 lines and backcrosses.

Experimental procedures for the second objective are shown in Figure 2. The two lines, Oh43 and Tx303, were crossed and, then, selfed two generations, as above, to generate 216 F_3 lines. A single plant in each line was testcrossed to both B73 and Mo17 and selfed to provide progeny for marker genotyping each of the 216 plants. Progenies of each of the 432 testcrosses were field evaluated in the same environments as the backcrosses and the same quantitative traits were measured.

The trait means of various genotypic classes for each marker locus were compared to test whether that marker significantly affects trait expression in the cross, B73 × Mo17, and whether the gene action was additive or dominant. Further tests were made for factors in Tx303 and Oh43 that might affect trait expression in B73 × Mo17.

Chromosomal segments were identified having factors affecting trait expression in the cross, B73 × Mo17. Figure 3 summarizes these results for grain yield graphically. For this trait, the factors with the greatest effects were located on chromosome arms 1L, 2S, 3L, 4L, 5S, and 5L. Locations of the major factors affecting grain yield showed a number of similarities with those

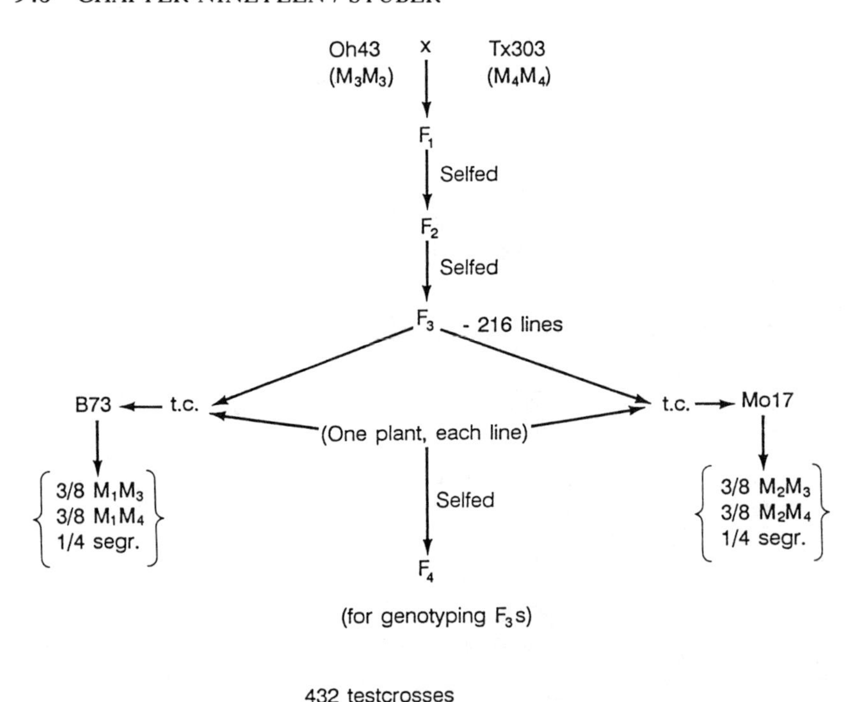

FIGURE 2. Diagram of experimental procedures for development of Oh43 × Tx303 F_3 lines and testcrosses to B73 and Mo17.

found in the earlier F_2 population studies. Chromosomal segments also were identified in Tx303 and Oh43 that would be expected to enhance the B73 × Mo17 hybrid response.

CONCLUSIONS

Investigations in several different plant populations have demonstrated quite conclusively the effectiveness of molecular markers for identifying and locating QTLs and for the detailed genetic investigation of variation in quantitative traits in suitably marked populations. For more precise mapping of QTLs, additional recombination will be required in most populations studied to break up blocks of linked loci. Results from studies in maize and other plant taxa should provide the impetus for additional avenues of research for the quantitative geneticist and plant breeder. For example, multiple-trait associations with genomic regions are complex and studies are necessary to determine whether these associations can be explained by pleitropy or by groups of linked factors. In addition, the stability of these identified factors when trans-

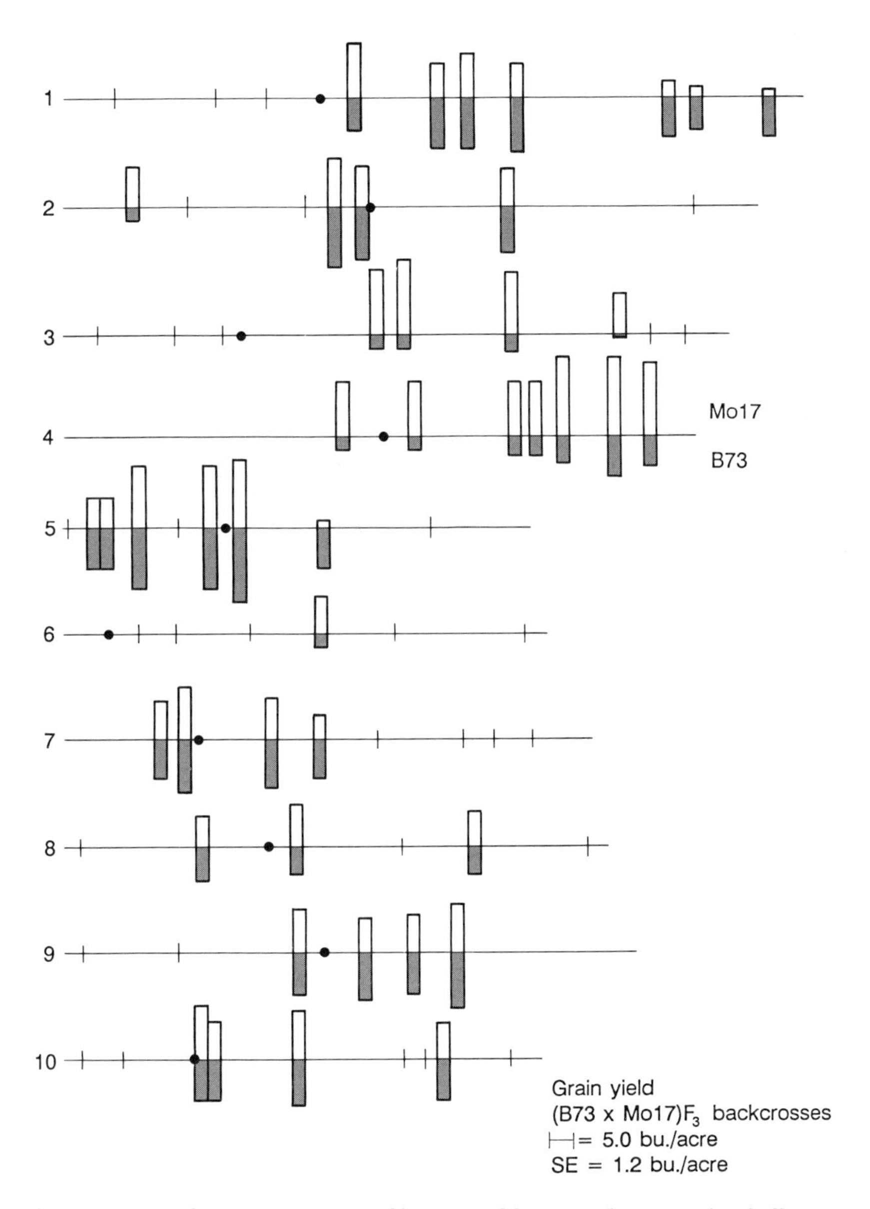

FIGURE 3. Graphic representation of location of factors and magnitude of effects on grain yield in B73 × Mo17 hybrid. As indicated at chromosome 4, the shaded bars at the bottom of each chromosome represent the location and size of the effects when the backcross was made to B73, i.e., the effect of substituting an Mo17 "allele" for a B73 "allele." Likewise, the bars above represent the effect of substituting a B73 "allele" for an Mo17 "allele" in the backcross to Mo17.

ferred to other genetic backgrounds and when evaluated in varying environments requires investigation. Tanksley and Hewitt (1988) dramatically illustrated the dangers of establishing breeding programs based on associations of markers with quantitative traits until the identified genetic factors have been evaluated in appropriate genetic backgrounds. Our maize research program in Raleigh, North Carolina, is currently addressing several of the areas discussed above.

ACKNOWLEDGMENTS

I thank Marlin Edwards, Jonathan Wendel, Major Goodman, Brenda Batemen, and Tim Helentjaris (and his staff) for their collaborative efforts in much of the research summarized here. I also thank Dianne Beattie, Sylvia Boles, Wayne Dillard, and Elizabeth Terrell for their technical assistance.

The research represents a joint contribution from the USDA Agricultural Research Service and the North Carolina Agricultural Research Service, North Carolina State University, Raleigh, North Carolina. The investigations were supported in part by USDA Competitive Grants 83–CRCR–1–1273 and 86–CRCR–1–2030 and in part by National Institutes of Health Research Grant GM 11546 from the National Institute of General Medical Sciences of the United States.

CHAPTER 20

PLANT QUANTITATIVE GENETICS

O. Mayo

ABSTRACT

Quantitative genetics and plant breeding are mature fields of enquiry. Advances will mostly be incremental, despite the promise of recombinant DNA technology. Three aspects of theoretical quantitative genetics (maintenance of variability, genetical control of outbreeding, heterosis) and four aspects of plant breeding (diallel crossing, disease resistance, crossbreeding systems, breeding *Vicia faba*) are discussed, to illustrate both incremental changes and the place of "genetic engineering."

Novel results on the genetic control of outbreeding illustrate the properties of a system of self-incompatibility determined by many genes. In small populations, these genes can be maintained by quite low rates of mutation, despite extremely high rates of ineffective (i.e., incompatible) pollination. Results of a competition diallel are presented and discussed.

The novel results presented here as well as the general reviews of certain areas of work illustrate the limits of quantitative genetics, as an essentially statistical methodology, in resolving the genetic determination of traits at the molecular level.

INTRODUCTION

Quantitative genetics rests on the solid foundation of Mendelian inheritance, but because it deals with continuously variable traits, it uses the methods of classical statistics rather than other areas of genetics. That is, it relies on the facts, demonstrated by Fisher in 1918, that if a trait is determined by the summation of many small genetic and environmental factors interacting in development, it will have a Gaussian distribution, and that the variance in the trait may be partitioned into components corresponding to the terms in this linear model. This very strength has guided the improvement through breeding of crop plants and domestic animals, and added to the understanding of evolution and human genetics. It is, however, a limitation when one seeks to move past the phenotype to the genotype, past the genotype to the genes, and back again.

At one level of analysis, therefore, both quantitative genetics and plant breeding are relatively well developed, the former as an aspect of applied statistics and the latter as an aspect of applied biology. In both cases, improvements are likely to be incremental within the existing frameworks. Recombinant DNA technology has the promise to revolutionize plant breeding. Indeed, it has had this promise for more than a decade, but expectations remain high:

> Biotechnology is set to deliver the biggest jolt to farm economics since Joseph invented buffer stocks. Over the past 40 years, economies of scale, farm machinery, improved crop strains and pesticides have made farms in the developed world 1–2% more efficient on average each year. This is the welcome reason that granaries and cold stores are full of food that nobody will buy. The unwelcome one is the way farmers have persuaded governments to favour them with price-support schemes once designed to beget self-sufficiency in food. Now a battery of biotechnology products—new veterinary drugs, pest-resistant or frost-resistant crops, new fertilisers and genetically engineered animals—is about to provide rapid increases in farm yields of 20% or more. Already creaking, the we-pay-as-you-reap approach to farm subsidy cannot survive this new fecundity. (*The Economist* 16 April, 1988, p. 17)

It is certain that plant breeding will indeed change radically as a result of genetic engineering, but I believe that those of us who are still above ground and active in 20 years time will find that plant breeding practice has not changed out of all recognition. So much of the activity depends on the environment, and the scale of agriculture is so vast that cereals like wheat, rice, maize, barley, and oats will still be grown on a large scale in roughly the same parts of the world as they are grown now. Even where deserts have been made to bloom, they will be only a small part of the land area devoted to crop production.

In this chapter, I shall first examine some of the incremental improvements in quantitative genetics, and then consider the place of genetic engineering in plant improvement, in the light of what I have said about quantitative genetics.

THEORETICAL QUANTITATIVE GENETICS

Quantitative genetics is a vast field, and much of it is already well developed. This is evident in the Proceedings of the Second International Conference of Quantitative Genetics (Weir et al. 1988). In this section, three topics of current interest are discussed: the maintenance of variability, the genetic determination of breeding systems, and genetic interaction and heterosis.

Maintenance of variation

Knowledge of the mechanisms of variability is fundamental to the understanding of how to use this variability. A balance between mutation and selection may lead to unfixable nonadditive genetic variation in traits, just as different patterns of dominance and epistasis will result in different responses to selection for heterosis.

The topic has long been controversial, in that models that yield Gaussian trait distributions under stabilizing selection (which is known to be widespread) require that new variation be generated by mutation at a rapid rate. In contrast, if more modest rates of mutation are the norm, they would appear not to generate the observed patterns of variation (Lande 1976; Turelli 1984). However, a resolution of the theoretical difficulties may be reached through the work of Bürger et al. (in press), as follows from some of their main results.

We have the following definitions:

σ^2 = mutational variance/gene
σ^2 = variance/gene
$V_s = \bar{\omega}^2 + 1$ = total variance in fitness w
w^2 = intensity of selection
$w(G) = e^{-G^2/(2V_s)}$ where G is the genotypic deviation from the optimum
n = number of genes
N_e = effective population size
μ = mutation rate/gene
$V_E = 1$
V_G = total genetic variance
V_P = total phenotypic variance

The form of $w(G)$ is an assumption as well as a definition, but as Tachida and Cockerham (1988) have argued, many inverse square functions will yield similar results. Equilibrium variance in an infinite population is approximately

$$2n\sqrt{\mu\alpha^2 V_s} \tag{1}$$

with an upper bound approximately

$$4n\mu V_s \tag{2}$$

(The approximation is strictly an upper bound only for the haploid case.) If population size is finite, and there is no selection, we obtain the long established result

$$4N_e n\mu E(a^2) \tag{3}$$

where a is a random variable denoting the effects of mutant alleles, i.e., $E(a^2)$ replaces α^2 in the infinite population case (Clayton and Robertson 1955)

If there is selection as well as mutation in the finite population, Bürger et al. show that a good approximation is given by

$$4n\mu V_s/[1 + V_s/(N_e\alpha^2)]$$

To derive a Gaussian distribution of breeding values, as is approximately the case, one requires that $n\mu \geq 0.005$. Estimates range from 0.0028 to 0.0067, but there are also suggestions from recent work that μ may not be constant (Cox et al. 1987).

I now consider some of the problems of this model.

When selection is weak ($V_s = 100$), heterozygosity is always predicted to be higher than it would be for neutrality, and predicted to be of the order of 5–10%. Observed values of heterozygosity are compatible with this prediction on the basis of weak selection, as they are with those for other models of selection.

Heritability in the narrow sense has the expectation

$$E[h^2] = \frac{4n\mu}{4n\mu + (N_e\alpha^2)^{-1} + V_s^{-1}}$$

in the absence of linkage, dominance, and epistasis. Hence, virtually any value is compatible with realistic values of n, μ, N_e, etc. Furthermore, this value is scaled to $V_E = 1$, and, as shown by Bull (1987), a given level of V_P may be maintained by selection which influences V_E or V_G. For example, germination is a trait with very large V_P, and Bull cites evidence that this is almost entirely environmental in origin.

Dominance can increase variance above the value given by (3) by a factor of two at most, usually less, and overdominance can also increase the variance above (3) (Lynch and Hill 1986), and these have not been explicitly included, though they are of course to be found in all populations (Mitton and Grant 1984; Strauss and Libby 1987). Further, the sketch presented ignores linkage, the effects of which will increase with the magnitude of the mutational effects (Hill and Keightley 1988).

Problems arise in estimating V_s, n, etc. Indeed, n is one of the most difficult parameters to begin to assess. Suppose that an organism has 10,000 structural genes and that about 10% are polymorphic at any time in a given population. Then suppose that only 10 influence any trait (Lande 1981; Mayo and Hopkins 1985). These can be chosen in $\binom{1000}{10}$ ways, i.e., more than 10^{20} ways. Thus, the genome can easily provide sufficient genes for any number of traits. Bürger et al. (in press) conclude that there are three possible ways to explain natural heritabilities of 0.25 or more: phenotypic neutrality, very large numbers of genes with small effects, or a few hypervariable pleiotropic genes. The third possibility is the only one allowing pleiotropy, yet given the different patterns of interaction possible in the determination of diverse traits by overlapping sets of genes, the phenotypes may range from apparent close association to apparent independence. As argued by Turelli (1985), testing predictions about pleiotropy requires joint estimation of mutation and selection parameters for all traits under consideration, and this is particularly difficult, especially where the form of the estimate depends on the unknown mode of gene action. Possibly the theory of Kacser (see Keightley and Kacser 1987) will provide a resolution of this difficulty, but not in a manner directly usable by the plant breeder. For example, consider a branched pathway leading to two outputs, almost the simplest possible model for two traits that might be jointly selected. In this model, use of anything other than the products of the two branches as quantitative traits requires a doubling of the number of parameters that needs to be estimated. In addition, because this very important methodology uses a model at the level of the gene product, not the gene, it does not incorporate complications that may well yield "polygenic" effects, such as intron variation (Murai et al. 1983; Rosahl et al. 1987; Stone et al. 1985) or pseudogene variation (Drouin and Dover 1987).

Tachida and Cockerham (1988) have pointed out that when the phenotypic variance fitness in the model of Bürger et al. is partitioned into its components, it is given by

$$V_A + V_D + V_{AA} + V_E$$

i.e., there are no higher order interactions with Tachida and Cockerham's quadratic fitness loss model. If this is so, the relative magnitude of V_A, V_D, and V_{AA} would allow discrimination between the "high mutation" and "low mutation" models of mutation-selection balance. What few data are available suggest that V_A is much greater than V_D and V_{AA}. Such a result is more compatible with the low mutation rate model, i.e., the non-Gaussian (or "House of Cards": Kingman 1978) model, rather than that set out above. However, this line of argument does not allow for the problem of pleiotropy mentioned above, since it extrapolates from data on only one component of fitness.

Recent statistical investigations of selection in natural populations are being carried out with much more statistical sophistication, using methods originally developed largely for animal breeding (Falconer 1981; Lande 1979; Shaw 1987; Mitchell-Olds and Shaw 1987). They will allow much more precise understanding of microevolution, especially joint changes in correlated traits subject to natural selection. However, to the extent that they rest on linear models, variance component estimation, etc., they will not permit the dissection of genetical variation in fitness into components directly related to the underlying genes.

For all of these reasons, it is hard to accept that the approach of "cloning quantitative trait loci by insertional mutagenesis" (Soller and Beckman 1987) will be highly rewarding in the near future. Curiously, it may be that some very complex quantitative traits will respond to molecular methods more rapidly than some simpler traits. Consider flavor in wine. This is very complex, and special methods of examining it have been developed over many years. However, in the vineyard we can have a range of desirable clones of a given variety and the certainty that improvement by conventional breeding will be very slow and difficult. The target gene approach may work here: to modify a single gene in a good clone will not require a cross and three or four backcrosses: it may require no crosses. If the relevant genes lie "in the 'twilight zone' between major genes and polygenes" (Allard 1956), the problem will be more difficult but soluble by the same methods.

Genetic determination of outbreeding

In general, for breeding purposes, plants are characterized as self-pollinated or cross-pollinated, and substantial bodies of theory have been developed to deal with selection and prediction at these two extremes (Allard 1960). Of course it has long been recognized that they are simply extremes, and that plant reproductive systems are not actually so simple. Indeed, the success of hybrid breeding in inbreeding species has shown that the theory of heterosis is inadequate or defective. I shall discuss one of the reasons for this further in the next section.

Considerable theory has been developed to evaluate mixed crossing systems, i.e., mixtures of selfing and random outcrossing. It is not clear that they describe satisfactorily the vast numbers of species which fall into this category. Such species include many important crop plants, such as *Vicia faba* and *Medicago sativa*, and economically significant weeds, such as *Avena barbata*.

In *V. faba*, crossbred individuals are likely to be more self-fertile than inbreds, a phenomenon called heterotic autofertility. Genotypic differences exist in autofertility, which is the ability of flowers to fertilize themselves without intervention by pollinators. In typical crops of slightly improved field beans, it was shown many years ago that about one-third of the plants are the direct products of cross-fertilization (Fyfe and Bailey 1951). In current improved varieties, the degree of autofertility varies between 10 and 50% (Stoddard 1986a). Heterotic autofertility, which is very important in improving productivity, arises from improved self-pollination. It is enhanced by an increased production of ovules per flower, greater fertilization of basal ovules, produced abortion of apical ovules, and greater maturation of ovules overall (Stoddard 1986b). Thus, there are genotypic differences in the ability to self-pollinate, but these are enhanced by increased heterozygosity. While the breeder can take advantage of this situation, such a breeding system is unstable from an evolutionary view point. It has long been recognized that any system that involves a mixture of selfing and outcrossing will, in the presence of genetic variation for the degree of selfing, move rapidly toward self-fertilization unless there is very strong selection against it (see Mayo and Leach 1987 for references). Accordingly, the persistence of such systems is a matter of speculation and controversy (see, e.g., Schemske and Lande 1987).

It therefore seemed appropriate to evaluate a particular species that appeared to have some of these properties. Darwin (1876) demonstrated that borage *Borago officinalis* ($2n = 16$) was a self-fertile species that suffered limited inbreeding depression. In 1971, Crowe published results to support the suggestion that the reproductive system was an incompletely self-incompatible one, polygenically determined. While multilocus gametophytic and sporophytic self-incompatibility systems have been described (see Richards 1986 for review), species with these systems have all been highly or completely self-sterile, i.e., there has been very little breakdown or imprecision in the system. This is clearly not the case with borage.

Accordingly, we have assayed 20 different isozyme systems, and found none to display variation in 12 self-sown garden plants (C. R. Leach and O. Mayo, unpublished). In larger numbers of plants obtained from several seedsmen, variation has been detected at only two loci. Such low levels of genetic variation are unexpected for a plant claimed to be an outbreeder. In a 12 × 12 diallel of the garden plants, we have found h^2 for seed set to be no more than 0.05. It could be argued that as seedset is a fitness-related trait, it should have low heritability. However, dominance variance was also low, given that specific combining ability variance was insignificant. If there is indeed polygenic self-incompatibility system maintained by embryo abortion, this should have been displayed as nonadditive genetic variation. In lucerne, for example, two cycles of selection for self-fertility doubled self-fertility and for self-sterility increased this slightly, suggesting substantial genetical variation for the trait (Busbice et al. 1975).

At the same time, a multilocus system was devised to allow a substantial proportion of selfing. For simplicity in modeling, the model had 10 loci, with two alleles per locus and complete additivity, and with the probability of successful fertilization proportional to the genetic distance between the pollen and style, each organ contributing equally. That is, there are 21 possible stylar genotypes that may be assigned values $g_s = 0, 1, 2, \ldots, 20$ and 11 possible pollen genotypes that may be assigned values $g_p = 0, 2, 4, \ldots, 20$, i.e., pollen and style are made equal in contribution. Then incompatibility occurs if $|g_s - g_p|/20 < c$, where c represented different degrees of stringency. The least stringent condition is $c = 0$, where any pollination is effective, the most stringent less than $c = 0.5$, at which point only one generation of effective pollination would be possible. The system also permits system breakdown, either on crossing or selfing. (This was allowed to occur at random, with a frequency of 0.1, 0.2, etc., independently of the self-incompatibility system.) Simulation of small populations showed that if $c = 0.1$ these systems are about as unstable as gametophytic self-incompatibility (e.g., Mayo 1966, 1971). That is to say, if some pollinations are effective regardless of the degree of difference between pollen and style, the 10 loci will be fixed by random drift within a finite period of time, in the absence of very substantial mutation. In gametophytic self-incompatibility, a high mutation rate is necessary to maintain a large number of alleles, but three alleles form a quasiequilibrial system that has almost no chance of dying out except in the very long term. However, if c is as high as 0.25, the system modeled here is very stable, and the mutation rate necessary to maintain it is quite low. Tables 1 and 2 show the results of some of the simulation trials which we have carried out.

What is particularly notable is that a system that provides a very large number of specificities will, if a threshold effect of the kind modeled here exists, have to support a rate of ineffective (incompatible) pollination that may be very high indeed. The case of borage perhaps supports this view, in that each borage flower can set 0, 1, 2, 3, or 4 seeds, and the mean in open pollination in the field is little above 2, i.e., despite much cross-pollination seed-set is not maximal. This does not occur because of lack of pollinator activity. Darwin (1876) noted that it was visited by more bees than any other species that he had observed. Our limited observations of bee activity agree with Darwin's, hardly surprisingly. Thus, the model is compatible with borage's seed-set but not with its genetic variation, which appears to be what would be expected of a largely self-pollinated plant.

Genetic interaction and heterosis

The work of Crow (1952) established dominance as a sufficient explanation for the bulk of economically useful heterosis. However, single gene heterosis is a demonstrable phenomenon (see, e.g., Section 9.1 of Mayo 1986). Furthermore, as was pointed out by Arunachalam (1977) and again by Minvielle (1987), heterosis in any two-population cross can arise in both F_1 and F_2, either by linkage disequilibrium if gene frequencies in the two populations are

TABLE 1. Numbers of genes fixed in 20,000 generations in a population of size 100, with stringency $c = 0.1$ and mutation rate $\mu = 4 \times 10^{-6}$.[a]

Breakdown rate	Number of genes fixed	Proportion of incompatible pollinations
0.1	4.5 ± 0.4	0.35 ± 0.01
0.2	4.5 ± 0.5	0.31 ± 0.01
0.3	5.5 ± 0.5	0.31 ± 0.01
0.4	5.2 ± 0.5	0.27 ± 0.01
0.5	6.2 ± 0.5	0.24 ± 0.01

[a]Means of four replicates of each trial ± standard error

TABLE 2 Mutation rate necessary to maintain segregation of the diallelic genes.

Population size	Stringency c	Breakdown rate	Proportion of incompatable pollination	Mutation rate ($\times 10^{-4}$)	Number of trials
100	0.1	0.1	0.29 ± 0.01	2.96 ± 0.64	4
		0.2	0.26 ± 0.01	3.08 ± 0.64	4
		0.3	0.24 ± 0.01	3.35 ± 0.64	4
		0.4	0.22 ± 0.01	3.38 ± 0.64	4
		0.5	0.16 ± 0.01	5.02 ± 0.45	4
	0.25	0.1	0.66 ± 0.01	0.00 ± 0.25	5
		0.2	0.61 ± 0.01	0.29 ± 0.25	5
		0.3	0.54 ± 0.01	0.26 ± 0.25	5
		0.4	0.47 ± 0.01	0.81 ± 0.25	5
		0.5	0.40 ± 0.01	1.83 ± 0.18	5
250	0.1	0.1	0.28 ± 0.01	0.01 ± 0.12	4
		0.2	0.25 ± 0.01	0.02 ± 0.12	4
		0.3	0.21 ± 0.01	0.08 ± 0.12	4
		0.4	0.17 ± 0.01	0.25 ± 0.12	4
		0.5	0.16 ± 0.00	0.16 ± 0.08	4
	0.25	0.1	0.66 ± 0.01	0 ± 0	4
		0.2	0.59 ± 0.01	0 ± 0	4
		0.3	0.52 ± 0.01	0 ± 0	4
		0.4	0.46 ± 0.01	0 ± 0	4
		0.5	0.38 ± 0.00	0 ± 0	4

equal, or through unequal gene frequencies. (Thus, it results from epistasis rather than dominance.) Table 3 shows the genotypic values used by these workers. A major point to note is that the simpler of the two models requires the estimation of five parameters (a, b, k, n and the frequency of recombination between A and B) apart from gene frequencies (Mayo in press). Thus, the experimental detection of two-gene heterosis will be difficult, except where it results from the combination of known major genes. Hence, such models are dubiously relevant only until a complete, fine-grained genetic map is available for any given target species.

Such maps are still relatively rare, being expensive and difficult to construct. As they are developed, they tend to reveal complications in recombination, e.g., different distance estimates from DNA markers compared with

TABLE 3. Multiplicative and additive two-locus interactions that yield heterosis in the absence of dominance.

	B_1B_1	B_1B_2	B_2B_2
		Arunachalam (1977)	
A_1A_1	$a_A + a_B + i_{AB}$	$a_A + d_B + j_{AB}$	$a_A - a_B - i_{AB}$
A_1A_2	$d_A + a_B + j_{BA}$	$d_A + d_B + l_{AB}$	$d_A - a_B - j_{BA}$
A_2A_2	$-a_A + a_B - i_{AB}$	$-a_A + d_B - j_{AB}$	$-a_A - a_B + i_{AB}$
		Minvielle (1987)	
A_1A_1	ka	$\frac{1}{2}k(a + b)$	kb
A_1A_2	$\frac{1}{2}(k + n)a$	$\frac{1}{4}(k + n)(a + b)$	$\frac{1}{2}(k + n)b$
A_2A_2	na	$\frac{1}{2}n(a + b)$	nb

other markers. (See Landry et al. 1987, for discussion of the first attempt to develop a linkage map for lettuce.) When such maps are available, they will allow the investigation of substantial heterotic interactions, but it should be noted that mapping of additive "quantitative trait loci" is a difficult exercise (McMillan and Robertson 1974), and mapping nonadditive genes will be even more difficult.

APPLIED QUANTITATIVE GENETICS (PLANT BREEDING)

Current field crop breeding practice is highly effective, but for this very reason it achieves only modest annual improvements in the major well-developed crop plants. Similarly, the methodology is improved by small increments, not great leaps. It is perhaps worth noting that when a new idea is reported, e.g., the use of tiller survival as a criterion for selection in wheat (Kulshrestha and Chowdhury 1987), it is often closely related to an old one, in this example being the converse, in a sense, of the high sowing density uniculm approach of Donald (1968). And the breeding methods cannot differ greatly for inbred cereals, apart from the use of haploids (Kulshrestha and Chowdhury's modified pedigree method may be traced back to that of Breakwell and Hutton 1939).

Two developments provide examples of how plant breeding has become a mature technology.

First, consider the very extensive applications, following the work of Lande (1979) and others, of the methods of quantitative genetics to natural populations, to investigate questions of natural selection and evolution. It was very evident at the Second International Conference on Quantitative Genetics (Weir et al. 1988) that this is a major area of growth in ecology and genetics. In course of time, the same obstacles will be met that have limited progress in plant breeding: the strong assumptions of linearity and normality, the imprecision of higher order statistics, and the difficulty of interpreting multivariate measures. For the moment it is a lively area of research, but it uses methods that have reached their limits in plant breeding, without adding to them.

The second example is the overelaboration of the diallel cross. Quite apart from its extensive statistical problems (see, e.g., Baker 1978; Wricke and Weber 1986), the diallel has a range of problems of interpretation when it is taken to a new area, e.g., grafting (Lefort and Legisle 1977). What should one make of reciprocal grafts, given the purpose of grafts in viticulture?

A diallel may be used for investigating competition between plants as well as genic combination within plants or physiological combination between stock and scion. A. J. Rathjen, T. W. Hancock, G. N. Wilkinson, and I carried out a competition diallel in wheat at two sites. Six genotypes were used, two important breeders' lines and four varieties widely used in South Australia at the time (1979), with four complete replicates at each site. The genotypic performances are shown in Table 4. The analysis of variance is shown in Table 5. The components of variance were uninformative, as is frequently the case (Table 6); despite the significance of GCA effects, the variance component does not differ significantly from zero. Table 4 gives clues to the breeder as to the type of genotype likely to interact well (e.g., Oxley with MKR/111/2) or badly (e.g., Shortim with Gamenya) at the two sites

TABLE 4. Yields of diallel at two sites, shown as deviations from site means.[a]

	Gamenya	Halberd	Oxley	Warimba	Shortim	MKR/111/2
Gamenya	−81.5 55.9	−21.7	−65.0	15.3	−96.0	50.5
Halberd	−14.9	28.5 −142.9	−24.5	18.8	−48.7	48.0
Oxley	−83.9	−9.4	−2.2 −10.6	−5.2	−74.2	89.8
Warimba	11.1	29.4	−71.4	51.8 91.6	−29.2	21.5
Shortim	−31.6	−61.1	−6.1	−7.9	−49.0 13.9	−7.5
MKR/111/2	−10.1	7.1	129.1	136.6	37.4	180.3 49.9

[a]Upper triangle: yields at Charlick, mean 260.0; lower triangle: yields at Mortlock, mean 699.1.

TABLE 5. Analysis of variance for a six genotype competition diallel at two sites.

Source of variation	Degrees of freedom	Sums of squares	Mean squares	Variance ratios
Mean	1	38,635,233	38,635,233	
Site	1	8,100,868	8,100,868	1144.35
Replicates within sites	6	95,838	15,973	2.26
General competitive ability (GCA)	5	408,532	81,706	11.54
Specific competitive ability (SCA)	15	101,614	6,774	0.96
Site × GCA interaction	5	70,617	14,123	2.00
Error	135	955,726	7,079	
Total	168	48,368,429		

used. The analysis of variance simply tells us, in this case, that we are considering real effects.

To illustrate how the methodology of quantitative genetics may or may not be altered by genetic engineering, I shall consider three examples.

Disease resistance

Disease resistance has been assessed to be the subject of more than a quarter of all of the plant breeding effort worldwide. A limited part of this activity relates to quantitative genetics. The main reason for this is the difficulty of

TABLE 6. Estimated variance components for diallel analysis of Table 5.

Site	96,100 ± 79,740
Replicate	420 ± 380
GCA	1,060 ± 690
SCA	0
GCA × site	220 ± 240
Error	7,080 ± 860

working with quantitative resistance under experimental conditions (Van der Plank 1963). Multifactorial resistance, almost by definition, will require a quantitative examination of the dynamics of pathogen spread and growth on different host phenotypes, in contrast to the relatively clear-cut qualitative responses that can be obtained in a gene-for-gene system, where corresponding to a gene for resistance versus susceptibility in the host there is a gene for virulence versus avirulence in the pathogen. For these reasons, the most widely used method has been the incorporation of different alleles at species-specific resistance-susceptibility loci, for which virulent alleles are absent or rare in the pathogen. Also important currently is the incorporation of resistance genes into crop plants from closely related species, a method of course having more general applications. These methods have been in use for some 50 years, and because of the precision of the gene-for-gene interaction, and the difficulty of selecting for quantitatively determined resistance or tolerance, the latter process has been largely neglected (see Mayo 1986 for a review). The qualitative patterns of interaction of a gene-for-gene system may of course be converted into quantitative patterns by appropriate use of a multivariate technique such as principal component analysis (Lebeda and Jendoulek, 1987). However, classification and selection are two different problems. Where the simplest or most practicable measure of resistance is a readily measured quantitative trait, like time from infection to a given level of symptom occurrence, as with Fiji disease in sugar cane (Stevenson et al. 1972), truncation selection will be the initial breeding method of choice, but such cases seem to be rare.

Several novel methods of obtaining disease or pest resistance have been developed or proposed recently. First, there is the possibility of enhanced expression of genes that are expressed in response to pathogenic organisms (Collinge and Slusarenko 1987). Such enhanced expression could come about through traditional truncation selection, but is more likely to be achieved by direct gene manipulation. Second, there are methods based on incorporation of a pathogen's genetic material such as satellite RNA (Gerlach et al. 1987). By their nature, these methods require direct gene manipulation. Third, there is the incorporation of completely unrelated protective genes. The first example is the use of the gene for the toxin from *Bacillus thuringiensis*, which when expressed in plants may protect usefully against larval leaf-eating insects (Vaeck et al. 1987). This too requires DNA manipulation.

Recent work on characterization of quantitative resistance highlights the difficulties of experimentation. For example, Leonard and Mundt (1984) have shown that, in order to determine how host plants respond to pathogens, one needs a satisfactory model for pathogen development, which will require estimates of such factors as the latent period (from inoculation to sporulation), the time to peak spore production, the time to cessation of sporulation, and the total reproduction per generation. Since a proper understanding of the epidemiology of each host-pathogen interaction must help in long-term, stable disease control, these extensive investigations have value for other reasons, but they will remain slower and more laborious than manipulation of major genes.

In summary, quantitative genetics has rarely had a place in resistance breeding, and if genetic engineering is successful in resistance breeding, it will limit the future use of quantitative genetics. This rather negative conclusion is drawn despite the obvious advantages of multifactorial resistance, viz. its slower rate of breakdown and its broader range of responses.

Crossbreeding systems

Breeding systems have been developed for crossbreeding, which are of three basic types: line crossing systems, cyclic systems, and panmictic (synthetic, composite) systems. All breeding systems are aimed to maximise heterozygosity and hence heterosis (on the assumption of a linear relationship between the two). Most of the theory developed for plants uses the assumption that the initial lines are completely homozygous, except where this is impossible, as with fully effective self-incompatibility. It is therefore of interest that evidence has recently been presented for the timber tree *Pinus radiata* (Strauss and Libby 1987) that heterosis is quadratically related to the level of heterozygosity, with an intermediate optimum, i.e., very high levels of heterozygosity are associated with yield (growth) depression. This evidence confirms much earlier work which established a quadratic relationship between heterozygosity and heterosis in maize (Moll et al. 1965). If this were widely true, why is line crossing following extreme inbreeding and line selection so effective? Presumably the answer lies in the selection among lines, since lines homozygous for the rare alleles hypothesized by Strauss and Libby to be deleterious as heterozygotes would certainly be culled. But this is another example of how defective is our understanding of heterosis.

Carmon et al. (1956) presented derivations of expected phenotypic means for four breeding systems: random mating, single-crosses, double-crosses, and rotational crossbreeding involving 2, 3, and 4 lines. They showed that no system was uniformly superior, but that "rotational breeding offers an even greater advantage to breeders than the crossing of several breeds or lines followed by random mating." However, composite populations are widely used, especially where there is no annual planting, as in forage crops.

After that work, Cockerham (1961) noted that multiple crosses were not as widely used as might have been expected from the posited advantages of better productivity of three-way and double crosses. This is probably for the same reason as composites and synthetics are not both recommended and used widely: the production costs of the seed and the more frequent resowing of perennials exceed the expected yield increases, with single-cross responses greater than the multiple crosses.

Breeding *Vicia faba*

Field beans have a breeding system that is a mixture of selfing and outcrossing not determined in any simple genetical manner. Pollination, seed set, and yield are greatly enhanced by insects, either natural populations of bumblebees or honeybees (Stoddard 1986c). Because of the genetic variation in autopollination ability, selection for or against this trait is possible, but given that there is a substantial normal proportion of outcrossing, selection for yield will automatically influence this trait.

Breeding aims may have changed in recent times as a result of the widespread introduction of Plant Variety Rights, and this fact will influence breeding methods. However, because field beans, though grown very widely as domestic crops, have been increasing in importance as cash crops only recently (Bond 1987), methods based both on line breeding and on cross-breeding can be carried out successfully.

The early calculations of Cockerham, referred to above, did not allow for a mixed reproductive system like that of *V. faba*. This has been provided by Busbice (1969, 1970) and Wright (1977). However, none of these approaches allows for the extreme variability in the level of outcrossing (Stoddard 1986c), given the difficulty of keeping pollination constant. (Wright noted that substantial differences in both autofertility and heterosis would not alter the equilibrial level of inbreeding greatly, but these were systematic differences, not unpredictable fluctuations.)

There appear accordingly to be strong arguments in favor of the development of entirely self-pollinating and entirely cross-pollinating lines, so that these may be used in more predictable systems, once the genetics of the breeding system is understood.

The recommendations in section 9.4.2.2 of Mayo (1986) may be too sanguine. That is, it appears likely that for many crops, not just *Vicia faba*, the long-term development of successful synthetics will depend on knowing much more than just line performance and mean GCA.

CONCLUSIONS

Quantitative genetics is a highly successful methodology at its appropriate level of resolution. It will continue to be the basis for plant breeding for the forseeable future. However, further insight into the genetic architecture of quantitative traits (modes of gene action, numbers of influential genes, etc.) will come from the measured genotype approach (e.g., Boerwinkle and Sing 1987; Mayo et al. 1969), whereby the contributions of particular major genes are evaluated, once these have been mapped and identified. Wittgenstein (91922) wrote "That Newtonian physics *can* be used to describe the world, tells us nothing about the world. But this *does* tell us something in that it can be used to describe the world *in the way in which we do in fact use it.*"

Whether this is true of Newtonian physics or not, it seems to be applicable to the present state of quantitative genetics.

ACKNOWLEDGMENTS

I thank C. R. Leach, T. W. Hancock, A. J. Rathjen, and G. N. Wilkinson for permission to mention unpublished results, S. Newman and W. J. Pitchford for some diallel analyses, and R. Bürger, C. R. Leach, D. R. Marshall, S. Newman, F. Stoddard, C. W. Stuber, and W. E. Weber for comments on the manuscript.

CHAPTER 21

CROP GENETIC RESOURCES: CURRENT AND EMERGING ISSUES

D. R. Marshall

ABSTRACT

The conservation of plant genetic resources has, in recent years, attracted growing public and scientific interest and support. There is an increasing awareness of the crucial importance of plant genetic resources to human well-being and broad agreement on the need to conserve these resources for use now and in the future. There is, however, far less agreement on how best to meet this need. Some of the controversial issues are predominantly economic and political, such as who should pay for conservation programs and who should own or control the conserved resources, while other issues are predominantly scientific. This chapter examines five of the latter. The first is the question of future conservation priorities and whether the current emphasis on the major crops and their near relatives should be continued or the range of priority species should be broadened to include other species such as minor crops, tropical fruits, and medicinal plants. The second is the question of the most appropriate method of conservation for landrace varieties of crops and populations of their near relatives—*in situ* in the environment in which they evolved or *ex situ* in gene banks. The third is the question of optimal field sampling strategies for genetic resource conservation, and the types of genetic variants, if any, deserve priority in such programs. The fourth issue is the increasingly vexing question of the growing size of collections and gene bank management. Here emphasis is given to the role of bulk hybrid populations and core collections in improving the management efficiency of collections. The final issue considered is the deficiencies in the current global network of gene banks and how these may be rectified to develop a more comprehensive, integrated, and useful group of collections.

INTRODUCTION

The "genetic resources movement" has had a short but remarkably successful history. Initiated in the early 1960s, this movement clearly identified the need for the conservation of crop genetic resources and strongly promoted action programs to meet this need (Wilkes 1983; Williams 1984; Frankel 1985a,b, 1986). As a result there has been a dramatic increase in the number of gene banks in the world with facilities for the medium to long-term storage of seed especially in developing countries (Plucknett et al. 1983; TAC 1986). There has also been a concomitant increase in the number of samples stored in these gene banks (Lawrence 1984; Holden 1984). Overall it would appear that more germplasm is now readily available to breeders and there is less risk of its loss than at any previous time (Brown et al. 1989).

The development of a supportive organizational framework at the national, regional, and international levels has underpinned this growth of collections. At the national level, many countries have developed genetic resources programs to strengthen and improve their conservation efforts over the last 20 years. At a regional level, the effort has been more sporadic and generally less successful, but with a number of notable exceptions including the IBPGR Regional Committee for South East Asia and the UNDP/IBPGR European Co-operative Program for the Conservation and Exchange of Crop Genetic Resources (IBPGR, 1987, 1988). At the international level, the International Agricultural Research Centres, of which 10 are directly concerned with germplasm conservation, have assumed a significant role. The International Board of Plant Genetic Resources (IBPGR) established in 1974 with a mandate to promote an international network of genetic resources centers to further the collection, conservation, documentation, evaluation, and use of plant germplasm, has been of particular importance.

The growth in collections has also been underpinned by the continued development of the scientific background and methodology of genetic resources conservation. This has involved research on seed (Roberts and Ellis 1984; Roberts et al. 1984) and *in vitro* (Withers 1989) conservation technologies as well as germplasm acquisition, multiplication, maintenance, and regeneration. With respect to the latter issues, the critical importance of plant population genetics in developing effective and efficient genetic resources programs was quickly recognized (Allard 1970a,b) and research in this field continues to be a high priority (IBPGR 1988).

However, despite the progress made over the last 20 years, crop genetic resources conservation has recently been the focus of unprecedented controversy and debate. The controversial issues are many and varied and include

1. The size and adequacy of existing collections and priorities for future collecting.

2. The ownership of, and access to, current collections and access to material still in the field.
3. The safety of germplasm stored *ex situ* in gene banks relative to that conserved *in situ*.
4. The future impact of genetic engineering and biotechnology on genetic resources conservation and use.

Many of these issues have been discussed in detail in recent books and articles (McIvor and Bray 1983; Mooney 1983; Holden and Williams 1984; Plucknett et al. 1987; Brown et al. 1989; Kloppenburg 1988). The aim of the present chapter is to highlight several current and emerging issues that relate to the central theme of this volume.

FUTURE CONSERVATION PRIORITIES

The genetic resources of crop plants can be functionally divided into four categories (Frankel and Bennett 1970):

1. Advanced varieties in current use and bred varieties no longer in commercial use;
2. Primitive "folk" varieties or "land races" of traditional prescientific agriculture;
3. Wild or weed relatives of crop plants and wild species of actual or potential use in crop breeding or as new crops; and
4. Genetic stocks such as mutations, cytogenetic stocks (translocation, inversion, and addition lines), and linkage testers.

Until recently the major emphasis in genetic resources programs was on the landrace varieties of the important food crops that could be conserved *ex situ* as dried frozen seed. A major reason was the pivotal role that landrace varieties have played in the development of scientific agriculture. They are the antecedents of all modern varieties. Another was their potential value as sources of variation for future plant breeding and the fact that they were often under imminent threat of extinction.

This sharp focus of effort on an important well-defined and finite group of species and populations has undoubtedly been a major factor in the successes achieved to date. In recent years, however, it has become increasingly obvious that the time for collecting landraces is past, or passing quickly, in most of the major crops because they have either disappeared or been collected (Frankel 1988). Collection of landraces in these species is now restricted to filling obvious and important gaps in collections and it is expected that this "gap filling" will be completed over the next few years (Williams 1984). This has forced both national and international programs to reassess their future collecting needs and priorities. From this reassessment process a variety of viewpoints have emerged.

One viewpoint is that future collecting should be very restricted and limited to emergency situations where valuable germplasm is in immediate danger of loss (for example, because of a development project) or to fill critically important gaps in collections. Instead, available resources should be devoted to rationalizing already collected germplasm and improving its accessibility and use. Proponents of this viewpoint argue that with the flood of new material into gene banks since the mid-1970s some collections have grown so large that they inhibit efficient use. However, whether existing collections require extensive rationalization, how this should be done, and whether it should replace further collecting are the subjects of continuing debate. The methodology of collection rationalization is discussed in a later section.

A second viewpoint is that further collecting is not only required, but often urgently required, for many regionally or nationally important species used for food, fiber, pharmaceuticals, or fuel. Proponents of this viewpoint argue that too much emphasis has been given to the major temperate food crops in framing genetic resources policies and there are many locally important food crops in the subtropics and tropics where collections are nonexistent, small, or poorly representative of the species. Examples that have been considered by various authors to warrant increased attention are

1. Tropical and subtropical fruits of southeast Asia, and central and south America;
2. The unique high elevation crops of the Andean region that include *Chenopodium*, *Amaranthus*, *Oxalis*, *Ullucus*, and *Tropaeolum* species;
3. Traditional medicinal plants of south and southeast Asia and many tropical areas; and
4. The leafy vegetables used worldwide in subsistence agriculture.

However, there are difficulties in determining priorities for many of these species. Take, as an example, the minor tropical leafy vegetables. Each represents a unique resource and each historically has been, and in some cases still is, important to local farmers. Further, some are under threat from alternative species, land uses, or life-styles (Van Sloten 1984). These same factors led to programs for conservation of the landrace varieties of the major food crops. It does not necessarily follow that the same effort should be devoted to all crops. The landraces of the major food crops are a potentially useful in current and future plant breeding programs. In contrast, many of the minor leafy vegetables are not now, and have never been, subjected to scientific breeding. Why then should they be conserved in a genetic resources program? Because they may be useful as introduced plants in some regions other than the area of current use? This is possible but does not appear a compelling argument. As sources of genes for future genetic engineering? This also seems too remote to justify a genetic resources collection. Of course samples of such species should be retained as a historical reference. However, this would involve maintenance of a modest number of samples of such species in botanic gardens or on "living history farms" (Woods 1987) rather than full-scale representative collections in gene banks.

A third viewpoint is that further collection is required and the emphasis should be on the major food crops but shifted from the landraces to the wild and weed relatives of these crops. The arguments for further collection of the wild relatives of the crops are the poor representation of this group of species in germplasm collections, the large stores of genetic variation they often contain, and their potential value in plant breeding. However, the development of efficient and effective conservation programs will often be far more difficult for these species than for the primitive crop cultivars. The reasons include the number of wild and weed relatives of the major crops, their often confused taxonomic status, the limited information available on their ecogeographic distribution, the difficulties in the collection, maintenance, and regeneration of wild species because of seed dispersal and variation in seed ripening and dormancy among individual plants and populations. However, a more critical problem is the definition of the species of long-term interest. Until recently, this was relatively easy. Harlan and de Wet (1971) divided crop plants and their near relatives into three categories:

1. The *primary* gene pool, consisting of the cultivated species and those relatives that are readily crossable and between which gene transfer is relatively simple;
2. The *secondary* gene pool, including all biological species that will cross with the crop but where gene transfer is difficult;
3. The *tertiary* gene pool, including those species that can be crossed with the crop only with difficulty and where gene transfer is impossible, or extremely difficult and requiring radical techniques.

The value of this system, based on crossability and ease of gene transfer, was its direct relevance to plant breeding. However, developing molecular genetic technologies may make this system obsolete. The prospect in the foreseeable future is that the potential gene pool of any species will include *all* other life forms. Why then should related species, and in particular crossable related species, of crops be targeted for special treatment in genetic resources programs? Should not all species be given equal weight? Perhaps bacteria should be given greater emphasis than at present as a source of exotic genes. The isolation of individual genes is, and will continue to be for some time, easier from bacteria than from higher plants.

Consideration of the impact of new biotechnologies leads to a fourth viewpoint—that the emphasis in genetic resource programs should be shifted from field collected germplasm to laboratory generated genetic stocks. Such stocks have great potential in the identification, location, manipulation, and isolation of genes in the major crops. For example, a mutant gene that radically alters sexual reproduction may be of no direct value to plant breeders. However, if such a mutant allows molecular geneticists to identify and isolate particular genes controlling reproduction, and, hence, to construct variants useful to breeders by direct DNA manipulation, then it is of great indirect value. To date, genetic stock collections have served different purposes and required different inputs from germplasm collections. However, in the future, greater integration between genetic stock and germplasm collections in maintenance, management, and use is likely to occur.

CONSERVATION METHODOLOGY

The best means of conserving crop plant genetic resources—*in situ* in the environment in which they evolved or *ex situ* in gene banks—has been, and remains, a contentious issue. *In situ* conservation, because it offers the prospect that plant populations would be maintained with a high degree of genetic integrity, has invariably been the method of first choice (Frankel 1970). However, *in situ* conservation is not always feasible so that samples must be collected and conserved *ex situ*.

Traditional varieties or landraces

In particular, *in situ* conservation is impractical for landrace crop populations. Farms, and more particularly farming systems with their human component, cannot be conserved in reserves like natural ecosystems. As a consequence, *ex situ* conservation in gene banks, as dried frozen seed where this is possible, and in field collections where it is not, has been generally regarded as the only pragmatic means of conserving primitive varieties.

However, this view has been persistently questioned by some authors. First, it is argued that the collection and conservation of traditional varieties in gene banks mean that the control of these varieties is taken out of the hands of those whose fathers and forefathers developed them and transferred to others, usually in developed countries. Second, it is suggested that the conservation of germplasm as dried frozen seed is not as safe as has commonly been supposed. The reason is that gene banks are often poorly funded, and, as a result, their operating standards fall below the minimum necessary to ensure the long-term safety of the samples in their care. Third, it is pointed out that "static" preservation in gene banks preserves only the existing variation captured in a relatively small number of samples. Further, there is likely to be loss of genetic variation in time; the better the operating standards the slower the loss, but even in the best banks some loss is probably inevitable. Gene bank samples are thus seen as a limited and decaying resource. This problem is perceived to be most acute for genetic variations in disease and pest resistance. Genetic variability for resistance in landrace varieties developed in response to the continued coevolution of the crop and its parasites. *Ex situ* conservation of landraces halts natural adaptive evolution in the host. But evolution in parasites can continue uninhibited. Sceptics of the value of *ex situ* gene banks suggest breeders will end up with limited and diminishing sources of resistance in the crop to counter potentially unlimited and evolving sources of virulence in parasites (Marshall and Brown 1981).

Consequently, it has periodically been proposed that landraces should be conserved *in situ* by local agricultural officers, or, preferably, by local farmers paid to grow them in the traditional way (Bennett 1968; Mooney 1983; Barnes-McConnell 1987; Oldfield and Alcorn 1987). Such an *in situ* conservation program it is suggested would be complementary to, and overcome the major deficiencies of, the current *ex situ* conservation effort. However, such proposals have not received wide support. One reason is the sheer size of an effective program of *in situ* conservation for landrace populations and the organizational, administrative, and funding problems that this would pose. For the major field crops there were thousands, and, in some case, tens of thousands of landraces in existence. Even if each farmer grew landrace varieties of several crops in the traditional way, many thousands of farmers in all countries of Asia, Africa, the Americas, and the Middle East would still need to be recruited to conserve a representative sample of landraces. Another reason is that alternative, and more pragmatic, solutions are available to the important issues of ownership and control of landrace germplasm and the deficiencies in *ex situ* gene banks.

However, the failure of proposals for *in situ* conservation of landraces to be implemented on a wide scale means that the perceived need for a "dynamic" conservation system to facilitate the continuing evolution of the crop gene pool (Frankel and Soule 1981) remains unmet. Another suggested means of meeting this need is via "mass reservoirs" (composite crosses or bulk hybrid populations) grown at a range of sites. This concept was originally proposed by Simmonds (1962), based in part on the earlier studies of Suneson (1956) and his colleagues at the University of California, Davis, and has been the focus of considerable discussion in the context of germplasm conservation (Allard 1970b, 1977; Frankel 1970; Marshall and Brown 1975; Marshall 1977; Frankel and Soule 1981; Jana and Acharya 1981; Jana and Khangura 1986). However, use of such populations as a supplemental "dynamic" means of germplasm conservation remains controversial.

Two outstanding questions are: Is there a pressing need for continuing evolution to provide newly generated resistance and other genes for future plant breeders? If so, how does this need vary among species?

It is difficult to argue against the need for new sources of disease and pest resistance, if in the future the resources available in our crops will be fixed at present levels but those in their major pests will be free to evolve. How pressing this need is, given the substantial numbers of samples in existing collections, is difficult to assess. However, for most crops and diseases it is unlikely that breeders will run out of useful resistances in the foreseeable future. With respect to the second question, it is clear that need varies greatly among crops. For crops with an associated cross-compatible and coevolving weedy taxon that has spread with cultivation of the crop, the associated weeds will provide ample opportunities for future evolution (Burdon 1987). The most obvious examples are the hexaploid weedy oats—*Avena sativa* ssp. *fatua*, *sterilis*, or *ludoviciana*—which have spread into all major cereal growing areas of the world, occupy an impressive range of habitats, are subject to an equally impressive range of diseases and pests, and regularly hybridize with the cultivated crop. Other examples include *Oryza sativa* ssp. *spontanea*, *Hordeum vulgare* ssp. *spontaneum*, *Sorghum arundinaceum*, *Triticum turgidum* var. *dicoccoides*, *Zea mays* ssp. *mexicana*, and *Glycine soja*. These species currently provide large evolving gene pools for rice, barley, sorghum, wheat, maize, and soybeans, respectively, although the long-term survival of some of these weedy populations may be in doubt (Harlan 1984).

Crop plants in which commercial cultivars are genetically heterogeneous populations and regenerate naturally, or are repeatedly harvested as seed and resown, are another group that appears to have ample capacity for continuing evolution in response to changing pest and pathogen populations. This group includes many partially, or completely outcrossing crops, such as cereal rye and faba beans, pasture plants, such are ryegrass, phalaris, and white clover, and forest trees, such as radiata pine and eucalypts. The continued use of a range of genetically heterogeneous cultivars of such species over the major areas where they are grown commercially would appear to offer greater opportunities for continuing evolution than the growing of a limited number bulk hybrid populations on experiment farms.

The remaining species of greatest concern in terms of continuing evolution are those where the commercial cultivars are generally genetically homogeneous (pure line varieties, F_1 hybrids, or asexual clones) and where cross-compatible wild or weedy species are unknown, limited in their distribution, or under threat of extinction. Examples include peas, lentils, and cotton. If it is accepted that for this group of plants, lack of opportunities for adaptive evolution may severely restrict the level of useful variability for disease and pest resistance available to future plant breeders, the critical question becomes whether mass reservoirs or composite crosses are the best means of overcoming this problem. Will the cultivation of genetically heterogeneous populations generated from diverse sets of parents and grown in areas where specified pests and pathogens are endemic provide adequate potential for the evolution of new resistance sources?

It is difficult to provide a definitive answer to this question for several reasons, not the least of which is the problem of assessing future breeding needs. However, two points can be made. First, as noted by Marshall and Brown (1981), the potential of mass reservoirs as sources of new and effective resistance genes is likely to be restricted because of the limited number and size of the populations that could be grown. This is especially true when mass reservoirs are compared to landrace populations that were large in number, rich in diversity, and often collectively grown over very large areas. Second, it remains to be established that the generation of new sources of variation for disease and pest resistance in naturally infected mass reservoirs is the most cost-effective means of meeting potential future needs. Direct generation of new variants by mutagenesis, if and when such variants were needed, is one possible alternative. Genetic engineering may provide other attractive alternatives in the future.

It is this lack of compelling evidence in most crop species for dynamic methods of conservation in general, and for mass reservoirs in particular, which has hindered their adoption and use.

Wild and weed relatives

In contrast, *in situ* conservation is widely regarded as the method of choice for the preservation of the wild and weed relatives of crop plants (Frankel 1970, 1974; Jain 1975; Frankel and Soule 1981; Ingram and Williams 1984). Representative samples would also be stored *ex situ* to promote ready access for study and use. The main reason for this contrast is that *in situ* conservation seeks to maintain self-perpetuating populations in natural or agricultural ecosystems and this is appropriate for wild species but not for cultivars. Other reasons include the fact that the number of wild and weed relatives of crops of interest to breeders is large, and these species are often substantially more difficult to conserve *ex situ* than cultivated plants.

Not only has it been widely agreed that *in situ* conservation is the method of choice for the preservation of wild gene pools, but also that the conservation of the wild relatives of crop plants is a priority need. As a result considerable effort has gone into elucidating the scientific principles on which strategies for the genetic conservation of the wild and weedy relatives of crop plants should be based (Jain 1975; Frankel and Soule 1981; Ingram and Williams 1984; IBPGR 1985). Clearly nature conservation and *in situ* genetic resources conservation have many important issues in common. Both are concerned with population extinction and its avoidance, adaptive evolution and its genetic basis, and the long-term stability of population size and species abundance in plant communities. However, programs of nature conservation and genetic resource conservation differ substantially in emphasis. Nature conservation programs are aimed at the level of ecosystems, biomes, or communities. They may also give particular attention to threatened or near extinct species that are generally held to be of special aesthetic, social, or scientific interest. Less emphasis is given to the maintenance of intraspecific genetic diversity and population structure of individual species. In the case of *in situ* conservation of the wild and weed relatives of crop plants, the genetic composition and structure of populations in reserves are of primary importance. For these species the aim is not just to preserve sufficient variability to ensure long-term survival and continued evolution but, if possible, to ensure the preservation of representative populations throughout their natural geographic range. In view of this requirement for the conservation of a representative samples of existing intraspecific variability, it has usually been argued that *in situ* conservation programs for wild crop relatives must involve two steps (Jain 1975; Ingram and Williams 1984; IBPGR 1985):

1. The close integration of genetic resources and nature conservation programs. The conservation of plant communities and ecosystems has attracted widespread support. Many countries have allocated, or are planning to allocate, substantial areas to conservation as national parks, nature reserves, or other forms of protected land use. Where these reserves contain wild relatives of crop plants every effort should be made to ensure that the location, design, and management of these protected areas reflect the special conservation needs of these species and provide for future access (collection and use).
2. The establishment of special genetic reserves (Jain 1975) or gene parks (Browning et al. 1979) in areas not covered by other conservation programs. Ecogeographic surveys are necessary to determine what portion of each species gene pool occurs inside and outside established reserves and, hence, the optimal locations of additional new "genetic" reserves. (IBPGR 1985).

Steps to improve the coordination between agencies concerned with *in situ* conservation including the International Board for Plant Genetic Resources (IBPGR), the International Union for Conservation of Nature and National Resources (IUCN), United Nations Environment Program (UNEP), Man and the Biosphere Program—UNESCO (MAB), and the Food and Agriculture Organisation of United Nations (FAO) have been taken (IBPGR 1987). However, this effort has yet to yield significant gains in terms of alterations in design and management of existing reserves to facilitate the *in situ* conservation of wild relatives of crop plants. Similarly, new genetic reserves purely for the preservation of crop wild relatives have been prepared, but only a few, for example, the reserve for perennial diploid teosinte in Jalisco, Mexico, have been established. However, surveys of several species are in progress or have been completed which can be used as a guide for their *in situ* conservation (IBPGR 1988).

It has become increasingly clear that progress in the establishment of soundly based *in situ* conservation programs for crop relatives will be difficult and time consuming. It is far easier in most countries to establish *ex situ* gene banks than *in situ* reserves because the latter involve the permanent alienation of land with no obvious immediate economic benefit to local peoples. As the world population continues to increase, the establishment of specialist genetic reserves is likely to become even more difficult. It would seem reasonable to suggest therefore that establishment of genetic reserves should be attempted only for species where *in situ* conservation is the only effective means of conservation. This would include species where *ex situ* conservation is still problematic and that are endangered or threatened with extinction. For the others, and this would now appear to be the great majority, *ex situ* conservation would increasingly appear to be the only pragmatic option.

To this end the wild relatives of crop plants can be classified into three broad groups:

1. Species, or segments of a species gene pool, that are endangered or threatened. Examples include *Zea diploperennis* (Iltis et al. 1979), *Lupinus princei* and *L. digitatus* (W. A. Cowling, personal communication), and *Glycine latrobeana* (A. H. D. Brown, personal communication). Where these species can be conserved effectively *ex situ* this would be method of choice. Where this is difficult or impossible then the very substantial effort required to establish a special genetic reserve, or reserves, would be warranted.
2. Species that are unlikely to survive in the long term unless protected but that are under no immediate threat. Many wild relatives of crop plants fall into this category. For these species, it should be possible to develop an appropriate conservation strategy based on balance between *ex situ* or *in situ* conservation.
3. Species that are adapted to man disturbed habitats and are in no danger of extinction. For these species collection is required only to ensure ready availability of material to breeders and research biologists and should perhaps be done only by the user.

Such a classification should ensure that appropriate effort is devoted to those species in class 1 in the short term while appropriate planning is put in place to minimize the number of species that move from class 2 to class 1.

SAMPLING STRATEGIES

The question of the most appropriate field sampling strategies for landrace varieties and the wild and weed relatives of crop plants has also generated considerable debate in recent years. Prior to the emergence of the "genetic resources movement" and its concern for germplasm conservation, plant collectors were often interested only in gathering a specific sample of germplasm for immediate evaluation and use. Germplasm that was found to be of no immediate value was often discarded because it was assumed that a wealth of material could readily be obtained from the same area, or other areas, if it were ever needed.

The increasing erosion of crop genetic resources over the last half century has resulted in a marked change in attitude to plant collecting. The emphasis has shifted from the collection of specific genes, populations, or ecotypes for immediate use to the collection of a representative sample of the total existing variability for conservation and use, both now and in the future. Further, since the total number of samples collected for conservation is generally restricted by both time and resources, sampling efficiency has received increased attention. Efficient sampling is of greatest importance for plant populations that are threatened by extinction or occur in remote areas on difficult terrain so they are likely to be sampled only once. In these cases plant explorers carry significant responsibilities because it is their decisions that determine what will be available in the future (Marshall and Brown 1981).

Allard (1970a) was the first to consider the question of optimum sampling strategies in the new era of collection for conservation. He stressed that most species contain remarkable stores of genetic variation and hence millions, or even hundreds of millions, of different genotypes. As only a fraction of the existing genotypes can be collected, he emphasized that the problem facing the plant explorer was how to collect the maximum amount of useful variability while keeping the number of samples within practical limits. Based on studies of genetic variation in populations of the wild oat, A. *sterilis* ssp. *fatua*, in California, he suggested that a collection of a single panicle, or approximately 10 seeds, from 200–300 plants in each of 500 populations covering the full range of its distribution, would capture most of the significant variation in local populations of this species.

However, Allard's (1970) recommendations with respect to sample sizes were substantially in excess of what had been recommended, and used in practice, previously (Whyte 1958). This led Marshall and Brown (1975) to reexamine the issue of sampling strategies. These authors argued that if the aim of plant exploration is to maximize the amount of useful variation captured, then the development of a general quantitative theory required the definition of (1) an appropriate measure of genetic variation, the parameter to be maximized, and (2) that portion of the variation in populations that is potentially most useful and that is therefore to have priority in sampling. They also suggested that in the context of genetic conservation the best measure of genetic diversity is the number of alleles per locus (n_a) at qualitative marker loci such as those governing simply inherited morphological traits, or isozymes, or DNA markers. This parameter provides a direct measure of the variation at each locus within and between populations unlike measures based on variance in quantitative characters that measure only the proportion of genetic variation expressed phenotypically and confound the effects of individual loci. Further, as an ideal, genetic conservationists are interested in preserving at least one copy of each allele in a population irrespective of its frequency. The number of allelic variants captured in a sample compared to the number in the sampled population is therefore a critical index of sampling efficiency for genetic conservation (Marshall and Brown 1981).

Marshall and Brown (1975) also suggested that since the aim of genetic conservation is to ensure the current and future availability of adequate germplasm, and since these needs are diverse and unpredictable, no particular group of genes can be regarded as generally more useful and deserving of a higher priority in sampling. Rather the aim should be to preserve as many variants as possible at each locus. The issue of defining "useful" variation that deserves priority in collecting reduces, under this argument, to discriminating among the alleles that occur at each locus. This led Marshall and Brown (1975) to make two arbitrary divisions of alleles at each locus based on their frequency of occurrence within and between populations, viz. those that are common (frequency >5%) and those that are rare (frequency <5%) and those that are widespread occurring in many populations, or local, and occurring in one or few adjacent populations. Of the four classes of alleles generated by these subdivisions they suggested that it was the locally common alleles that should receive priority in sampling. In support of this viewpoint they argued that widespread common alleles are inevitably collected regardless of sampling strategy. Further, the probability of including alleles that are rare in any given population but occur in many populations in a sample, depends largely on the total number of plants collected and is, therefore, also largely independent of sampling strategy. Finally, alleles that are rare in any given population and also localized in their occurrence will be difficult to collect, and presumably represent recent mutants or deleterious genes maintained at low frequency and, therefore, of less interest to breeders. In contrast, locally common alleles will also be more difficult to collect than their widespread counterparts. However they are presumably adaptive and selectively maintained in populations. They also presumably represent adaptations to significant local variations in the biotic or abiotic environment, and are therefore likely to be of special interest to plant breeders.

Marshall and Brown (1975), on the basis of these arguments, suggested that the aim of exploration for genetic conservation should be the collection of at least one copy of each common allele (frequency >0.05) in populations of the target species. If the collector has no prior knowledge of the distribution of alleles in the target species, which is most commonly the case, they suggested that the optimum strategy to achieve this objective would be to collect a bulk sample of 50 to 100 individuals from as many sites, and covering as broad a range of environments, as possible. The aim is to spend the minimum time and effort at each site consistent with collection of all common alleles, and to visit as many different and diverse sites as possible to maximize the possibility of collecting new adaptive variants. Marshall and Brown (1975) also pointed out that with information on the distribution of alleles within and between populations it is possible to develop more sophisticated sampling strategies, a point reinforced by Oka (1975) and more recently Yonezawa (1985). Marshall and Brown (1983) detailed modifications to the above procedures which can be used if problems in the field prevent the use of optimum sampling strategy.

However, the approach advocated by Marshall and Brown (1975) has been criticized on several grounds. One of these has been the use of the number alleles per locus (n_a) as the primary measure of genetic variation and sampling efficiency (Witcombe and Gilani 1979; Bogyo et al. 1980). This criticism is based on two arguments. One is that the sorts of marker genes used as a basis for measuring genetic diversity in this way, until recently morphological variants and enzyme polymorphisms, are of marginal interest to plant breeders. The latter, primarily, utilize germplasm resources to improve agronomically useful, quantitatively varying, characters. Marshall and Brown (1981) answered this objection by pointing out that while they used the data available from studies to assess the patterns of allelic diversity in plant populations, this did not mean that the sampling technique applied only to such marker genes. Rather it applied to all loci including those governing quantitative traits. Further, it could also be applied to combinations of genes or coadapted gene complexes, which in this context could be regarded as "alleles" at a "supergene."

The second argument is that patterns of variation within and between populations may differ for marker loci and quantitative traits. On the basis of such a difference, Bogyo et al. (1980) recommended that the "sampling strategy should be different for characters which are inherited quantitatively from those which are controlled by one or two genes." However, the fact that different classes of characters, or different classes of loci, may show different patterns of variation in nature, is neither unexpected nor unusual (e.g., Rick et al. 1977; Kahler et al. 1980; Giles and Edwards 1983; Smith et al. 1984). Critics of the use of the number of alleles as a measure of genetic diversity suggest that in such situations greater weight should be given to variation in quantitative traits in framing sampling strategies. However, for the great majority of plant collectors such arguments are irrelevant because they completely lack information on the population genetic structure of their target species. In such cases it would appear reasonable to suggest that the most effective strategy would be to spend only sufficient time at each site to ensure the collection of common genes, and to visit as many sites as possible, so as to maximize the opportunity of sampling those populations that are highly variable for each class of character or gene.

If information is available on the population genetic structure of the target species, and this may more often be the case in the future if greater emphasis is given to two stage sampling procedures (Jain 1975; Chapman 1989), then collectors are faced with the decision as to the relative weights to give to various classes of loci in deciding sampling patterns. However, this need not necessarily be an onerous task, even when the patterns of variation for different classes of alleles are markedly discordant. One simple approach would be to give equal weight to each class of characters and ensure that the most variable populations of each class are sampled. This approach is prudent because the patterns of variation for the sample of loci surveyed in any given class may not necessarily be representative of all loci in that class. For example, the patterns of variation for leaf esterase in barley may not necessarily agree with the patterns of variation at less polymorphic loci governing enzyme makers (Nevo et al. 1979). By sampling the most variable populations for each class of loci it should be possible to maximize the average number of common alleles collected over all loci.

Another recurring criticism of the approach of Marshall and Brown (1975) comes from those who favor biased or selective sampling of rare phenotypic variants at each site. This strategy was first advocated by Bennett (1970) who suggested collecting a random sample of 500 plants from each population and enriching that with a selected sample of rare (frequency <0.01) variants. Marshall and Brown (1975, 1981) argued against the selective sampling of rare variants on the grounds that it is time consuming, reduces the effort that can be devoted to the collection of locally common alleles, and may lead to the collection of diseased specimens that appear as rare phenotypic variants. However, this issue has recently been raised again by Ford-Lloyd and Jackson (1986). They argued that collectors should selectively sample plants that are phenotypically distinguishable," in terms of a morphological character or the manifestation of a physiological character, such as drought tolerance or disease resistance," and that these should be retained as a separate sample.

The case put forward by Ford-Lloyd and Jackson (1986) contains several significant flaws. The first is that it is impossible to select in the field individual plants from a variable population that are genetically superior for quantitatively varying traits such as drought resistance. There is ample evidence in the plant breeding literature that the selection of individual plants from variable populations for such traits is largely ineffective (Shebeski 1967). Second, although in time, collectors could assemble a representative collection of rare morphological variants from each population, the utility of such variants in breeding programs is open to serious question. In wild barley populations, for example, traits such as black glumes and hooded awns occur as rare variants in most populations. However, they are rare because they usually reduce fitness. Perhaps the question to ask proponents of the selective sampling of rare morphological forms is what breeders are expected to do with such a collection of genetic cripples, especially if the same or similar sets are collected from many populations. There are very few examples in the literature where rare variants collected on plant exploration missions (as distinct from field stations, and this is a crucial distinction) have proved to be useful in breeding.

Overall it would appear that the strategy proposed by Marshall and Brown (1975) has proved to be effective in practice and is applicable to a wide range of species.

GENE BANK MANAGEMENT AND RATIONALIZATION

Another contentious issue that has arisen over the last 5 years is the size of collections and the impact this has on their use. It has been argued that many collections are now so large and diffuse that they deter effective management and rational use by breeders (Frankel 1984; Frankel and Brown 1984; Holden 1984). Further, with increased emphasis on the collection of wild and weed relatives of crop plants and the continuing generation of elite new gene combinations by plant breeders, there is little doubt that already large collections will grow progressively larger. The problems this poses for curators, charged with characterizing, monitoring, and periodically multiplying these collections, are also increasing.

One obvious means of improving the efficiency of gene banks would be to identify and eliminate, as far as possible, redundant, duplicate accessions within and between collections. Frankel and Soule (1981) and Marshall and Brown (1981) have discussed possible approaches to the identification of duplicate or near-identical accessions in gene banks. While the identification of duplicates may be tedious it is conceptually easy. The difficult problem is to decide what to do with the duplicate accessions identified—discard, retain reduced numbers or amounts, or bulk. Brown in Frankel and Soule (1981) examined the genetic consequences of several such options using simple probability models and concluded that simply discarding samples would rarely, if ever, be an appropriate action. However, this was a first and limited study and more research is needed in this area if effective and reliable management strategies are to be developed.

It should be emphasized that even if all duplicates within collection were eliminated, the gains in efficiency from reduction in entry number are likely to be modest for many collections. Greater gains could perhaps be made by reducing redundancy among collections to the minimum necessary for insurance against catastrophe. However, this is far more difficult to achieve—in part, because it requires a high level of cooperation and coordination among national as well as international gene banks, and, in part, because it has to be extremely difficult to persuade curators to reduce the size of collections even when material is freely available from other gene banks. The reasons are (1) the size of collections is seen to be of importance in terms of prestige and funding, (2) many countries have strict quarantine laws and accessions are retained to avoid the risk, inconvenience, and cost of repeated quarantine, and (3) where countries are heavily dependent economically on particular crops, germplasm collections of those species acquire strategic importance and larger national collections are seen as insurance against an uncertain future, especially in the light of the growing controversy surrounding the ownership and control of crop genetic resources.

The limited scope for reduction in the size of collections due to the elimination of redundancy within gene banks, and the slow progress in reducing redundancy among gene banks, has led to alternate radical proposals to reduce the workloan of curators and to improve the management of collections. Two such radical proposals are considered below.

Conservation via bulk populations

One suggested means of rationalizing collections to improve management efficiency is by the use of bulk populations, composite crosses, or mass reservoirs. The role of such populations in facilitating continuing evolution in crop gene pools was discussed earlier. Here we are concerned with a potential role in reducing the size of collections to more manageable proportions.

As noted earlier the use of mass reservoirs in genetic conservations was first proposed by Simmonds (1962). He argued that mass reservoirs were a relatively cheap method of maintaining an extensive range of germplasm in a dynamic system of direct relevance to breeders in contrast to "museum" collections, which were expensive to maintain and an often rapidly wasting resource. Allard (1970a,b) supported this view and cited an example of continued effective maintenance of lima bean germplasm, in part, by the development of bulk populations. However, in the same volume, Frankel (1970) questioned the role of mass reservoirs in genetic conservation because of the irreversible loss of genetic integrity of individual accessions and the potential loss of variation associated with adaptive evolution of such populations. This issue has been discussed periodically in the literature with the proponents of bulk populations emphasizing the remarkable stores of genetic variation they contain, even after 40 or more generations, and critics of such populations emphasizing the equally remarkable loss of genetic variation during the early generations of their cultivation (Marshall and Brown 1975; Allard 1977; Marshall 1977; Frankel and Soule 1981; Jana and Acharya 1981). Nevertheless, in reality there has been little interest in using mass reservoirs in genetic conservation because the rapid growth in the number and capacity of low-temperature seed storage facilities over the last two decades has meant that all material collected could be stored readily as individual accessions.

However, the continual growth of collections has refocused attention on the genetic conservation potential of bulk populations (Jana and Khangura 1986). While there is little doubt that the maintenance of individual accessions, or preferably individual plants within accessions, as distinct entities is the most effective method of preserving genes and genotypes, this may become increasingly impractical. As a result the creation of bulks or composites from accessions with similar ecological, agronomic, and genetic backgrounds may be unavoidable (Singh and Williams 1984).

If this is the case the development and maintenance of such composites will require special attention. In the past, mass reservoirs have been regarded as serving three purposes—an adjunct to plant breeding, a means of facilitating continuing evolution in crop gene pools, and genetic conservation. To a major degree these purposes, particularly breeding with its radical alterations in gene and genotype frequencies and conservation, are in conflict. However, this conflict can easily be resolved by establishing separate populations for each purpose. Since one of the perceived problems with the use of bulk populations in conservation is the loss of genes during their development and early stages of mass propagation, it is evident that populations developed for conservation purposes should be regenerated infrequently to minimize potential losses. Clear guidelines also need to be developed for (1) the number of plants per variable parental population to be included in a composite with a view to ensuring what has been collected is not lost during the bulking process, (2) the number of plants that need to be grown each cycle of regeneration, and the procedures to be used, to ensure maximum retention of variability in such bulks, and (3) the size of sample that needs to be sent to users to ensure they receive a reasonable sample of the population variation. If bulk populations were developed and maintained solely for gene conservation, they would undoubtedly be more acceptable to gene bank curators as a management tool in the future than they appear to have been in the past.

Core collections

The second radical approach, proposed a Frankel (1984), was the subdivision of the accessions in large gene banks into a *core collection*, which would include with minimum redundancy the genetic diversity of a crop species and its relatives, and a *reserve collection* that would include all other material. This approach was developed further by Frankel and Brown (1984) and Frankel (1986). More recently, Brown (1989, in press) has outlined procedures to develop a representative and diverse core collection using information on the origin and nature of the accessions.

Using the sampling theory of selectively neutral alleles, Brown (1989) argued that a core should consist of about 10% of the collection, up to a maximum of about 3000 accessions, for each species. At this level of sampling the core will generally contain over 70% of the alleles in the whole collection. Brown (1989) also argued that, as a general rule, a fixed proportion of 10% is more useful than a fixed number upper limit.

Brown (in press) examined alternate procedures for choosing core entries. He showed that stratified sampling, where the collection is first divided into nonoverlapping groups and a sample is taken from each group, is generally more efficient in establishing a core than simple random sampling, where accessions are chosen from the whole collection at random. He also examined three options for deciding the number of samples from each nonoverlapping group to include in the core. These were a constant strategy (an equal number of accessions from each defined group), a proportional strategy (a fixed fraction of accessions is included for each group), and a logarithmic strategy (number of accessions included is proportional to the logarithm of the size of the group). Of these, the logarithmic and proportional strategies gave broadly similar results and were better than the constant strategy.

The aim of the hierarchical subdivision of collections into core and reserve components is to facilitate use, and, in particular, to provide efficient access to the whole collection (Brown 1989). Consider a breeder faced with a new virulent race of a pest or pathogen. The breeder will obviously search for resistance in readily available material, usually a limited working collection. If no resistance is found in the working collection, the next step is to screen material from an appropriate gene bank collection. If that gene bank has a well-defined core, this would be screened first. If no effective resistance is located in the core, the breeder knows that resistance is relatively rare. He is faced with a substantial problem and the prospect of screening the other 90% of the collection. If resistance is found in the core, then he can use that resistance immediately in his breeding program as well as screen additional accessions, from the same geographic area, held in the reserve collection.

The core collection concept also has advantages for curators. It is envisaged that more seed of the core collection would be kept on hand packaged ready for distribution to meet generalized seed requests. It is also envisaged that the core would receive priority in evaluation and characterization, so that in time many more characters would be evaluated at more locations on core samples than on reserve samples. In this way curators could better use a limited budget to promote the distribution of information and material and hence to facilitate use of the collection.

The principal disadvantage of the core concept is the possibility, or indeed probability, that the reserve collection will erode away and disappear from neglect, or, alternatively, will be seen by hard-nosed administrators as of less value and, therefore, dispensible in the interests of economy. To minimize this disadvantage Frankel and Brown (1984) and Brown (1989) have stressed that the reserve collection must remain an important and integral part of the collection. Under the core concept the reserve collection serves at least two important functions: (1) as a back up collection to be screened if needed variation is not found in the core collection, and (2) as a source of additional diversity where many different genes are required for the same trait, as in some breeding strategies for disease and pest resistance (Marshall 1977). Nevertheless, there are many who feel that once collections are stratified, they will be more vulnerable to dismemberment regardless of how strong a case is presented for the maintenance of the reserve collection.

However, the core concept has had, and will continue to have, a significant impact at a conceptual level beyond its practical implications. In particular, it forces us to question whether size is a problem because collections are too large to use, or too large to maintain, or both. This is a critical distinction. If use is the essential problem, this points to a need for a core collection. If maintenance is the essential problem, this points to a need to reduce redundancy and accession number by bulking. If both are a problem, then combined programs involving development of a core plus bulking of reserve entries may be necessary (Brown 1989).

THE GLOBAL NETWORK OF GENE BANKS

The establishment of a globally coordinated and cooperative network of genetic resources centers has been a major goal of the genetic resources movement since its inception (Frankel and Bennett 1970; Frankel 1975). The motivation for the development of such a network was that few countries, if any, could individually conserve all the germplasm that they may need, and further, it was only through a cooperative global effort that all economically important plants could be adequately conserved.

The FAO Panel of Experts and others (Frankel 1975, 1986) developed detailed plans for a global network of genetic resource centers. The proposed network was centered around a modest number of 8–10 *base collections* that would be concerned solely with long-term conservation. It was envisaged that the base collections would include strategically placed regional gene banks in areas rich in genetic diversity, the crop-specific International Agricultural Research Centres and significant national collections (e.g., Seed Storage Laboratory, Fort Collins, Colorado and Vavilov Institute of Plant Industry, Leningrad, USSR). It was also envisaged that some of the participating gene banks would be established *de novo* while others would be developed by strengthening already established institutes. The scheme was to be centered in, and coordinated by, FAO in Rome.

Associated with each base collection was to be a series of *active collections* that would be responsible for medium-term storage, regeneration, distribution of seed to users, evaluation, and documentation. If the base and active collections were not in the same institute, close links between them were to be mandatory. Breeders working collections per se were not included as part of the system (Frankel 1975). Under this proposal, the base collections would hold for long-term safekeeping all unique accessions of all important crop species. Surveys would be done to identify unique accessions in collections throughout the world, and these would be increased by the active collections for distribution and supplied to the base collections for conservation. The base collection would arrange replicate storage in at least one other base collection to minimize the risk of catastrophic loss.

This proposal was never implemented by FAO. This task logically passed to IBPGR after its establishment in 1974. However, IBPGR had insufficient funds to foster the development of 8–10 new genetic resources centers in areas rich in crop diversity. Further, the concept of cooperative regional programs and centers often proved to be politically unworkable. As a pragmatic alternative, IBPGR sought to establish a global network of base collections with the cooperation of national and international centers and institutes already holding significant global or regional collections. Details of the requirements for host institutes to join this network are given in the 1986 IBPGR Annual Report (pp. 27–28) and a list of designated base collections was published in the 1987 IBPGR Annual Report (pp. 29–32).

The present network falls significantly short of the ideal system and the FAO Panel of Experts proposals. In particular it covers only base collections and does not yet include the associated active collections (although nearly all the base collections have active collections associated with them in the same institution). For many species, significant unique material in national collections is not included in any designated base collection. Often this material is not readily available to breeders internationally simply because they do not know it exists. In some cases it is at substantial risk of loss. Procedures for identifying such material and including it in a base collection have not been established. Indeed, communications among base collections, the linking element of any network, are poor. Communications between base collections and active collections are often nonexistent.

Clearly the plan put forward by the FAO Panel of Experts has proved unworkable. Occasionally the concepts of a series of international genetic resources centers in the major centres of diversity, or a single large international gene bank in a neutral country, are recycled but to date have failed to attract significant support. Rather, more and more countries are developing national genetic resources programs and establishing facilities for the long-term conservation of their own important germplasm.

Equally clearly, the time has come to replace impractical past plans and philosophies with new proposals compatible with today's political and financial realities. Some of the main features of a revamped scheme would be as follows:

1. The global network of gene banks should include all significant national, regional, and international collections, especially those that contain any material not duplicated elsewhere. Particular emphasis should be given to the involving national collections because logically as the primary source of most germplasm they will form the backbone of the network.
2. The concept of a few large centralized base collections should be discarded. Rather each collection should have the responsibility for the long-term conservation of the germplasm unique to that collection. In particular, national gene banks should have primary responsibility for, and control of, their unique or important germplasm.
3. The concept of separate base and active collections should also be discarded. Rather there should be a single network of collections, all of which assume the responsibilities of an active collection and selected ones of which would assume responsibility for the long-term conservation of specified set of germplasm. Of course, long-term storage facilities may be physically located at different site to the active collection component. The important point is that active collections should have associated base collections rather than vice versa and should assume responsibility for coordination among gene banks.
4. National gene banks should be fully involved in policy making and prac-

tical arrangements for coordination at an international level. International gene banks would still play a critically important role but as an adjunct to the national gene banks.

5. Participation would be voluntary. However, the functionality of the network would clearly require participants to agree to a code of collaboration that included making material freely available and arranging for replication of unique material, for safety, in at least one other long-term storage facility.

It is difficult to envisage how a scheme based on national collections could function other than on a crop by crop basis. Most national collections contain germplasm of only a limited number of crops, generally those in which the country has indigenous resources or those that are most important to its agriculture. Further, plant breeders tend to specialize in particular crops.

Given the number of gene banks in existence, to adopt a new philosophy and reorganize accordingly will be a large, long-term, and difficult task. However, since the previous philosophy has not led to effective action in 18 years there seems to be ample justification for fundamental changes. If the next effort to establish a global network of gene banks is to be more successful, then cooperation and goodwill, active participation by gene bank curators, and a clear focus on identifying and meeting mutual needs will be essential.

ACKNOWLEDGMENTS

The author thanks Drs. O. H. Frankel, A. McCusker, D. H. van Sloten, and J. T. Williams for their comments and discussion.

Literature Cited

The numbers in brackets that follow each entry identify the chapter(s) in which the reference is cited.

Aarssen, L. W., 1983 Ecological combining ability and competitive combining ability in plants: Towards a general evolutionary theory of coexistence in systems of competition. Am. Nat. 122: 707–731. [12]

Aarssen, L. W., and R. Turkington, 1985 Intraspecific diversity in natural populations of *Holcus lanatus, Lolium perenne*, and *Trifolium repens* from four different aged pastures. J. Ecol. 73: 869–886. [12]

Adams, W. T., 1981 Population genetics and gene conservation in Pacific Northwest conifers. pp. 401–415. In: *Evolution Today, Proc. Second International Congress Systematics and Evolutionary Biology*, Edited by G. G. E. Scudder and J. L. Reveal. Hunt Institute for Botanical Documentation, Pittsburgh. [16]

Adams, W. T., 1983 Application of isozymes in tree breeding. pp. 401–415. In: *Isozymes in plant genetics and breeding*, Edited by S. D. Tanksley and T. J. Orton. Elsevier, Amsterdam. [16]

Adams, W. T., and R. J. Joly, 1980 Allozyme studies in loblolly pine seed orchards: Clonal variation and frequency of progeny due to self-fertilization. Silv. Genet. 29: 1–4. [16]

Adriaanse, A., W. Klop and J. E. Robers, 1969 Characterization of *Phaseolus vulgaris* cultivars by their electrophoretic patterns. J. Sci. Food Agric. 20: 647–650. [4]

Albertini, A. M., M. Hofer, M. P. Calos and J. H. Miller, 1982 On the formation of spontaneous deletions: The importance of short sequence homologies in the generation of large deletions. Cell 29: 319–328. [5]

Aldrich, J., B. Cherney and E. Merlin, 1986a Sequence of the chloroplast-encoded *psbA* gene for the Q_B polypeptide of alfalfa. Nucl. Acids Res. 14: 9537. [8]

Aldrich, J., B. Cherney, E. Merlin, L. Christopherson and C. Williams, 1986b Sequence of the chloroplast-encoded *psbA* gene for the Q_B polypeptide of petunia. Nucl. Acids Res. 14: 9536. [8]

Allard, R. W., 1956 Biometrical approach to plant breeding. Brookhaven Symp. Biol. 9: 69–84. [20]

Allard, R. W., 1960 *Principles of Plant Breeding*. John Wiley, New York. [17, 20]

Allard, R. W., 1965 Genetic systems associated with colonizing ability in predominantly self-pollinated species. pp. 49–75. In: *The Genetics of Colonizing Species*, Edited by H. G. Baker and G. L. Stebbins. Academic Press, NY. [15]

Allard, R. W., 1970a Population structure and sampling methods. pp. 97–107. In: *Genetic Resources in Plants—their Exploration and Conservation*, Edited by O. H. Frankel and E. Bennett. Blackwell, Oxford and Edinburgh. [21]

Allard, R. W., 1970b Problems of maintenance. pp. 491–494. In: *Genetic Resources in Plants—their Exploration and Conservation*, Edited by O. H. Frankel and E. Bennett. Blackwell, Oxford. [21]

Allard, R. W., 1975 The mating system and microevolution. Genetics 79: 115–126. [4, 9, 14]

Allard, R. W., 1977 Coadaptation in plant populations. pp. 223–231. In: *Genetic Diversity in Plants*, Edited by A. Muhammed, R. Askel, and R. C. von Borstel. Plenum, New York. [21]

Allard, R. W., 1988 Genetic changes associated with the evolution of adaptedness in cultivated plants and their wild progenitors. J. Hered. 79: 225–238. [4, 5, 6, 15]

Allard, R. W., and J. Adams, 1969 Population studies in predominantly self-pollinating species XII. Intergenotypic competition and population structure in barley and wheat. Am. Nat. 103: 621–645. [12]

Allard, R. W., and P. E. Hansche, 1963 Population and biometrical genetics in plant breeding. *Genet. Today* 3: 665–679. Proc. XI. Int. Congress of Genetics, The Hague, The Netherlands. [17]

Allard, R. W., and S. K. Jain, 1962 Population studies in predominantly self-pollinated species. II. Analysis of quantitative genetic changes in a bulk-hybrid population of barley. Evolution 16: 90–101. [12, 17]

Allard, R. W., and P. L. Workman, 1963 Populations studies in predominantly self-pollinated species. IV. Seasonal fluctuations in estimated values of genetic parameters in lima bean populations. Evolution 17: 470–480. [9, 11, 12]

Allard, R. W., J. Harding and C. Wehrhahn, 1966 The estimation and use of selective values in predicting population change. Heredity 21: 547–563. [12]

Allard, R. W., S. K. Jain and P. L. Workman, 1968 The genetics of inbreeding populations. Adv. Genet. 14: 55–131. [9, 12]

Allard, R. W., A. L. Kahler and B. S. Weir. 1972 The effect of selection on esterase allozymes in a barley population. Genetics 73: 489–503. [2, 12]

Allard, R. W., G. R. Babbel, M. T. Clegg and A. L. Kahler, 1972 Evidence for coadaptation in *Avena barbata*. Proc. Natl. Acad. Sci. U.S.A. 69: 3043–3048. [3, 10, 14]

Ambros, V., and H. R. Horvitz, 1984 Heterochronic mutants of the nematode *Caenorhabdtitis elegans*. Science 226: 409–416. [8]

Anderson, S. M., and J. F. McDonald, 1983 Biochemical and molecular analysis of naturally occurring *Adh* variants in *Drosophila melanogaster.* Proc. Natl. Acad. Sci. U.S.A. 80: 4798–4802. [6]

Angeesing, J. P. A., 1974 Selective eating of the acyanogenic form of *Trifolium repens*. Heredity 32: 73–83. [12]

Anonymous, 1987 Which radiata pine seed should you use? What's new in forest research 157. Rotorua, New Zealand. [16]

Antonovics, J., 1976 The nature of limits to natural selection. Ann. Miss. Bot. Gard. 63: 224–247. [15]

Antonovics, J., and N. C. Ellstrand, 1984 Experimental studies of the evolutionary significance of sexual reproduction. I. A test of the frequency-dependent hypothesis. Evolution 38: 103–115. [12]

Antonovics, J., and R. B. Primack, 1982 Experimental ecological genetics in Plantago. VI. The demography of seedling transplants of *P. lanceolata* L. J. Ecol. 70: 55–75. [12, 15]

Antonovics, J., 1968 Evolution in closely adjacent plant populations. V. Evolution of self-fertility. Heredity 23: 219–238. [9]

Appels, R., and J. Dvořák, 1982a The wheat ribosomal DNA spacer region: Its structure and variation in populations and among species. Theor. Appl. Genet. 63: 337–348. [5]

Appels, R., and J. Dvořák, 1982b Relative rates of divergence of spacer and gene sequences within the rDNA region of species in the Triticeae: Implications for the maintenance of homogeneity of a repeated gene family. Theor. Appl. Genet. 63: 361–365. [5]

Appels, R., W. L. Gerlach, E. S. Dennis, H. Swift and W. J. Peacock, 1980 Molecular and chromosomal organization of DNA sequences coding for the ribosomal RNAs in cereals. Chromosoma 78: 293–311. [5]

Apuya, N. R., B. L. Frazier, P. Keim, E. J. Roth and K. G. Clark, 1988 Restriction fragment length polymorphisms as genetic markers in soybean, *Glycine max* (L.) merrill. Theor. Appl. Genet. 75: 889–901. [6]

Aquadro, C. F., S. F. Desse, M. M. Bland, C. H. Langley and C. C. Laurie-Ahlberg, 1986 Molecular population genetics of the alcohol dehydrogenase region of *Drosophila melanogaster.* Genetics 114: 1165–1190. [6]

Aquadro, C. F., K. M. Lado and W. A. Noon, 1988 The *rosy* region of *Drosophila melanogaster* and *Drosophila simulans*. I. Contrasting levels of naturally occurring DNA restriction map variation and divergence. Genetics 119: 875–888. [6]

Arnheim, N., 1983 Concerted evolution of multigene families. pp. 38–61. In: *Evolution of Genes and Proteins*. Edited M. Nei and R. K. Koehn. Sinauer, Sunderland, Mass. [5]

Arnheim, N., M. Krystal, R. Schmickel, G. Wilson, O. Ryder and E. Zimmer, 1980 Molecular evidence for genetic exchanges among ribosomal chromosomes in man and apes. Proc. Natl. Acad. Sci. U.S.A. 77: 7323–7327. [6]

Arnold, S. J., and M. J. Wade, 1984a On the measurement of natural and sexual selection: Theory. Evolution 38: 709–719. [11, 12]

Arnold, S. J., and M. J. Wade, 1984b On the measurement of natural and sexual selection: Applications. Evolution 38: 720–734. [12]

Arunachalam, V., 1977 Heterosis for characters governed by two genes. J. Genet. 63: 15–24. [20]
Asker, S., 1980 Gametophytic apomixis: Elements and genetic regulation. Hereditas 93: 277–293. [9]
Asmussen, M. A., and M. T. Clegg, 1981 Dynamics of the linkage disequilibrium function under models of gene-frequency hitchhiking. Genetics 99: 337–356. [6]
Asmussen, M. A., and M. T. Clegg, 1982 Rates of decay of linkage disequilibrium under two-locus models of selection. J. Math. Biol. 14: 37–70. [10]
Assouad, M. W., B. Dommee, R. Lumaret and G. Valdeyron, 1978 Reproductive capacities in the sexual forms of the gynodioecious species *Thymus vulgaris*. Bot. J. Linn. Soc. 77: 29–39. [12]
Autran, J. C., and A. Bourdet, 1975 L'identification des variétés de blé: Etablissement d'un tableau général de détermination fondé sur le diagramme électrophorétique des glaidines du grain. Ann. Amél. Plantes 25: 227–301. [4]
Avery, P. J., and W. G. Hill, 1979 Distribution of linkage disequilibrium with selection and finite population size. Genet. Res. 33: 29–48. [10]

Bailey, N. T. J. 1961 *Introduction to the Mathematical Theory of Genetic Linkage*. Oxford University Press, Oxford. [2]
Bailey, T. B., and R. E. Comstock 1976 Linkage and the synthesis of better genotypes in self-fertilizing species. Crop Sci. 16: 363–370. [17]
Baker, H. G., 1955 Self-compatibility and establishment after "long-distance" dispersal. Evolution 9: 347–348. [15]
Baker, H. G., 1967 Support for Baker's Law—as a rule. Evolution 21: 853–856. [15]
Baker, H. G., and P. A. Cox, 1984 Further thoughts on dioecism and islands. Ann. Miss. Bot. Gard. 71: 244–253. [15]
Baker, H. G., and G. L. Stebbins (eds.), 1965 *The Genetics of Colonizing Species* Academic Press, New York. [15]
Baker, J. J., Maynard-Smith and C. Strobeck, 1975 Genetic polymorphism in the bladder campion, *Silene maritima*. Biochem. Genet. 13: 393–410. [10]
Baker, R. J., 1978a Evaluation of the inbred-backcross method for studying the genetics of continuous variation. Can. J. Plant Sci. 58: 7–12. [18]
Baker, R. J., 1978b Issues in diallel analysis. Crop Sci. 18: 533–536. [20]
Baker, R. J., 1984 Quantitative genetic principles in plant breeding. pp. 147–176. In: *Gene Manipulation in Plant Improvement*, Edited by J. P. Gustafson. Plenum, New York. [17]
Baker, R. J., 1986 *Selection Indices in Plant Breeding*. CRC Press, Boca Ration, Fl. [17]
Banks, J. A., and C. W. Birky, 1985 Chloroplast diversity is low in a wild plant, *Lupinus texensis*. Proc. Natl. Acad. Sci. U.S.A. 82: 6950–6954. [6]
Barbujani, G., 1987 Autocorrelation of gene frequencies under isolation by distance. Genetics 117: 777–782. [14]
Barker, R. F., N. P. Harberd, M. G. Jarvis and R. B. Flavell, 1988 Structure and evolution of the intergenic region in a ribosomal DNA repeat unit of wheat. J. Mol. Biol. 201: 1–17. [5]
Barnes-McConnell, P., 1987 Keepers of the trust. Diversity 12:22–24. [21]
Barratt, D. H., 1980 Cultivar identification of *Vicia faba* by SDS-PAGE of seed globulins. J. Sci. Food Agric. 31: 813–819. [4]
Barrett, J. A., 1980 Pathogen evolution in multilines and varietal mixtures. J. Plant Dis. Prot. 87: 383–396. [13]
Barrett, S. C. H., 1985 Floral trimorphism and monomorphism in continental and island populations of *Eichhornia paniculata*. Biol. J. Linn. Soc. 25: 41–60. [15]
Barrett, S. C. H., 1988 Genetics and evolution of agricultural weeds. pp 57–75. In: *Weed Management in Agroecosystems: Ecological Approaches*, Edited by M. A. Altieri and M. Liebman. CRC Press, Boca Raton, Fl. [15]
Barrett, S. C. H., and B. J. Richardson 1985 Genetic attributes of invading species. pp. 21–33. In: *Ecology of Biological Invasions*, Edited by R. H. Groves and J. J. Burdon. Academy of Science, Canberra. [15]
Barrett, S. C. H., and J. S. Shore 1987 Variation and evolution of breeding systems in the *Turnera ulmifolia* L. complex (Turneraceae). Evolution 41: 340–354. [15]

Barrett, S. C. H., and J. S. Shore In press Isozyme variation in colonizing plants. In: *Isozymes in Plant Biology,* Edited by D. Soltis and P. Soltis. Dioscorides Press. [15]
Barrett, S. C. H., M. T. Morgan and B. C. Husband In press The dissolution of a complex genetic polymorphism: The evolution of self-fertilization in *Eichhornia paniculata* (Pontederiaceae). Evolution. [15]
Barros, M. D. C., and T. A. Dyer, 1988 Atrazine resistance in the grass *Poa annua* is due to single base change in the chloroplast gene for the D1 protein of photosystem II. Theor. Appl. Genet. 75: 610–616. [8]
Bartlett, M. S., 1971 Physical nearest-neighbor models and non-linear time-series. J. Appl. Prob. 8: 222–232. [14]
Barton, H. H., and B. Charlesworth, 1984 Genetic revolutions, founder effects and speciation. Annu. Rev. Ecol. Syst. 15: 133–164. [15]
Bekele, E., 1983 The neutralist-selectionist debate and estimates of allozyme multilocus structure in conservation genetics of the primitive land races of Ethiopian barley. Hereditas 99: 73–88. [10]
Bennett, E., 1968 *Record of the FAO/IBP Technical Conference on the Exploration, Utilisation and Conservation of Plant Genetic Resources,* 1967. FAO, Rome. [21]
Bennett, E., 1970 Tactics of plant exploration. pp. 157–179. In: *Genetic Resources in Plants—their Exploration and Conservation,* Edited by O. H. Frankel and E. Bennett. Blackwell, Oxford and Edinburgh. [21]
Bennett, J. H., and F. E. Binet, 1956. Association between Mendelian factors with mixed selfing and random mating. Heredity 10: 51–56. [12]
Bennett, R. J., 1979 *Spatial Time Series.* Pion, London. [14]
Berenbaum, M. R., A. R. Zangerl and J. K. Nitao, 1986 Constraints on chemical coevolution: Wild parsnips and the parsnip webworm. Evolution 40: 1215–1228. [12]
Bergmann, F., 1978 The allelic distribution at an acid phosphatase locus in Norway spruce (*Picea abies*) along similar climatic gradients. Theor. Appl. Genet. 52: 57–64. [15]
Bernatzky, R., and S. D. Tanksley, 1986 Toward a saturated linkage map in tomato based on isozymes and random cDNA sequences. Genetics 112: 887–898. [6]
Besag, J. E., 1972 Nearest-neighbor systems and the auto-logistic model for binary data. J. Roy. Stat. Soc. Ser. B. 34: 75–83. [14]
Bijlsma, R., R. W. Allard and A. L. Kahler, 1986 Nonrandom mating in an open-pollinated maize population. Genetics 112: 669–680. [14]
Billington, H. L., 1985 Population ecology and genetics of *Holcus lanatus.* Ph.D. thesis, University of Liverpool. [12]
Billington, H. L., A. M. Mortimer and T. McNeilly, 1988. Divergence and genetic structure in adjacent grass populations. I. Quantitative genetics. Evolution 42: 1267–1277. [12]
Birky, C. W., 1988 Evolution and variation in plant chloroplast and mitochondrial genomes. pp. 23–53. In: *Plant Evolutionary Biology,* Edited by L. D, Gottlieb and S. K. Jain. Chapman & Hall, London. [6]
Birky, C. W., and R. V. Skavaril, 1976 Maintenance of genetic homogeneity in systems with multiple genomes. Genet. Res., Cambr. 27: 249–265. [5]
Bishop, J. A., and M. E. Korn, 1969 Natural selection and cyanogenesis in white clover. Heredity 24: 423–430. [11]
Bjorkman, P. J., M. A. Saper, B. Samraoui, W. S. Bennett, J. L. Strominger and D. C. Wiley, 1987 The foreign antigen binding site and T cell recognition regions of class I histocompatibility antigens. Nature (London). 329: 512–518. [8]
Blagrove, R. J., G. G. Lilley and T. J. V. Higgins, 1984 Macrozin, a novel storage globulin from seeds of *Macrozamia communis* L. Johnson. Aust. J. Plant Physiol. 11: 69–77. [4]
Bliss, F. A., and C. E. Gates, 1968 Directional selection in simulated populations of self-pollinated plants. Aust. J. Biol. Sci. 21: 705–719. [17]
Blyden, E. R., and J. C. Gray, 1986 The molecular basis of triazine herbicide resistance in *Senecio vulgaris* L. Biochem. Soc. Trans. 14: 62 [8]
Bodmer, M., and M. Ashburner, 1984 Conservation and changes in the DNA sequences coding for alcohol dehydrogenase in sibling species of *Drosophila.* Nature (London). 309: 425–430. [6]
Bodmer, W. F., 1972 Evolutionary significance of the HL-A System. Nature (London). 237: 139–146. [8]

Boerwinkle, E., and C. F. Sing, 1987 The use of measured genotype information in the analysis of quantitative phenotypes in man III. Simultaneous estimation of the frequencies and effects of the *apolipoprotein E* polymorphism and residual polygenetic effects on cholesterol, betalipoprotein and triglyceride levels. Ann. Hum. Genet. 51: 211–226. [20]

Bogyo, T. P., E. Porceddu and P. Perrino, 1980 Analysis of sampling strategies for collecting genetic material. Econ. Bot. 34: 160–174. [21]

Bond, D. A., 1987 Recent developments in breeding field beans (*Vicia faba* L.). Plant Breeding 99: 1–26. [20]

Borroto, K., and L. Dure, III, 1987 The globulin seed storage proteins of flowering plants are derived from two ancestral genes. Plant Mol. Biol. 8: 112–131. [4]

Bousquet, J., W. M. Cheliak and M. Lalonde, 1987 Genetic differentiation among 22 mature populations of green alder (*Alnus crispa*) in central Quebec. Can. J. For. Res. 17: 219–227. [16]

Bradshaw, A. D., 1984 Ecological significance of genetic variation between populations. pp. 213–228. In: *Perspectives on Plant Population Ecology*, Edited by R. Dirzo and J. Sarukhan. Sinauer, Sunderland, Mass. [14]

Breakwell, E. S., and E. M. Hutton, 1939 Cereal breeding and variety trials, at Roseworthy College, 1937–38. J. Agric. South Aust. 42: 632–641. [20]

Brettell, R. I. S., M. A. Pallotta, J. P. Gustafson and R. Appels, 1986 Variation at the *Nor* loci in triticale derived from tissue culture. Theor. Appl. Genet. 71: 637–643. [5]

Briggs, F. N., and R. W. Allard, 1953 The current status of the backcross method of plant breeding. Agron J. 45: 131–138. [17]

Brim,C. A., and C. W. Stuber, 1973 Application of genetic male sterility to recurrent selection schemes in soybeans. Crop Sci. 13: 528–530. [17]

Britten, R. J., and D. E. Kohne, 1968 Repeated sequences in DNA. Science 161: 529–540. [5]

Brotschol, J. V., J. H. Roberds and G. Namkoong, 1986 Allozyme variation among North Carolina populations of *Liriodendron tulipifera* L. Silv. Genet. 35: 131–138. [16]

Brown, A. H. D., 1975 Sample sizes required to detect linkage disequilibrium between two or three loci. Theor. Pop. Biol. 8: 184–201. [10]

Brown, A. H. D., 1978 Isozymes, plant population genetic structure and genetic conservation. Theor. Appl. Genet. 52: 145–157. [14]

Brown, A. H. D., 1979 Enzyme polymorphism in plant populations. Theor. Pop. Biol. 15: 1–42. [3, 9, 10, 11, 12]

Brown, A. H. D., 1989 The case for core collections. pp. 136–156. In: *The Use of Plant Genetic Resources*, Edited by A. H. D. Brown, O. H. Frankel, D. R. Marshall, and J. T. Williams. Cambridge University Press, Cambridge. [21]

Brown, A. H. D., In press Core collections: A practical approach to genetic resources management. Proceedings of International Genetics Congress, Toronto. Genome. [21]

Brown, A. H. D., and L. Albrecht, 1980 Variable outcrossing and the genetic structure of predominantly self-pollinated species. J. Theor. Biol. 82: 591–707. [9]

Brown, A. H. D., and R. W. Allard, 1970 Estimation of the mating system in open-pollinated maize populations using isozyme polymorphisms. Genetics 66: 133–145. [9, 11]

Brown, A. H. D., and R. W. Allard, 1971 Effects of reciprocal recurrent selection for yield on isozyme polymorphisms in maize (*Zea mays* L.). Crop Sci. 11: 888–893. [10]

Brown, A. H. D., and J. J. Burdon, 1987 Mating systems and colonizing success in plants. pp. 115–131. In: *Colonization, Succession and Stability*, Edited by A. J. Gray, M. J. Crawley and P. J. Edwards. Blackwell Scientific Publ., Oxford. [3, 15]

Brown, A. H. D., and M. W. Feldman, 1981 Population structure of multilocus associations. Proc. Natl. Acad. Sci. U.S.A. 78: 5913–5916. [10]

Brown, A. H. D., and D. R. Marshall, 1981 Evolutionary changes accompanying colonization in plants. pp. 351–363. In: *Evolution Today*, Proc. II Int. Congress Syst. Evol. Biol. Edited by G. C. E. Scudder and J. L. Reveal. [9, 15]

Brown, A. H. D., and G. F. Moran, 1981 Isozymes and the genetic resources of forest trees. pp. 1–10. In: *Isozymes in North American Forest Trees and Forest Insects*, Edited by M. T. Conkle. Pac. SW Forest and Range Expt. Station Technical Report #48. [16]

Brown, A. H. D., and J. Munday, 1982 Population genetic structure and optimal sampling of land races of barley from Iran. Genetica 58: 85–96. [3]

Brown, A. H. D., and B. S. Weir. 1983 Measuring genetic variability in plant populations. pp.

219–239. In: *Isozymes in Plant Genetics and Breeding, Part* A, Edited by S. D. Tanksley and T. J. Orton. Elsevier, Amsterdam. [2]

Brown, A. H. D., A. C. Matheson and K. G. Eldridge, 1975 Estimation of the mating system for *Eucalyptus obliqua* L'Herit. using allozyme polymorphisms. Aust. J. Bot. 23: 931–949. [10, 16]

Brown, A. H. D., E. Nevo and D. Zohary, 1977 Association of alleles at esterase loci in wild barley, *Hordeum spontaneum* L. Nature (London). 268: 430–431. [10]

Brown, A. H. D., M. W. Feldman and E. Nevo, 1980 Multi-locus structure of natural populations of *Hordeum spontaneum*. Genetics 96: 523–536. [3, 10]

Brown, A. H. D., S. C. H. Barrett and G. F. Moran, 1985 Mating system estimation in forest trees: models, methods and meanings. pp. 32–49 In: *Population Genetics in Forestry*, Edited by H. R. Gregorius. Springer-Verlag, Berlin. [9, 16]

Brown, A. H. D., J. E. Grant and R. Pullen, 1986 Outcrossing and paternity in *Glycine argyrea* by paired fruit analysis. Biol. J. Linn. Soc. 29: 283–294. [9]

Brown, A. H. D., J. Munday and R. N. Oram, 1988 Use of isozyme-marked segments from wild barley (*Hordeum spontaneum*) in barley breeding. Plant Breeding 100: 280–288. [19]

Brown, A. H. D., O. H. Frankel, D. R. Marshall and J. T. Williams, (1989) *The Use of Plant Genetic Resources*. Cambridge University Press, Cambridge. [21]

Brown, A. H. D., J. J. Burdon and A. M. Jarosz, In press. Isozyme analysis of plant mating systems. In: *Isozymes in Plant Biology*, Edited by D. Soltis and P. Soltis. Dioscorides Press. [9]

Brown, B. A., and M. T. Clegg, 1984 Influence of flower color polymorphism on genetic transmission in a natural population of the common morning glory, *Ipomoea purpurea*. Evolution 38: 796–803. [14]

Brown, J. H., T. Jardetzky, M. A. Saper, B. Samraoui, P. J. Bjorkman and D. C. Wiley, 1988 A hypothetical model of the foreign antigen binding site of class II histocompatibility molecules. Nature (London). 332: 845–850. [8]

Brown, J. W. S., F. A. Bliss and T. C. Hall, 1981a Linkage relationships between genes controlling seed proteins in French Bean.Theor. Appl. Genet. 60: 251–259. [4, 18]

Brown, J. W. S., Y. Ma, F. A. Bliss and T. C. Hall, 1981b Genetic variation in the subunits of globulin-1 storage protein of French bean. Theor. Appl. Genet. 59: 83–88. [4, 18]

Brown, J. W. S., T. C. Osborn, F. A. Bliss and T. C. Hall, 1981c Genetic variation in the subunits of the globulin-2 and albumin seed proteins of French bean. Theor. Appl. Genet. 60: 245–251. [4]

Brown, J. W. S., J. R. McFerson, F. A. Bliss and T. C. Hall, 1982 Genetic divergence among commercial classes of *Phaseolus vulgaris* in relation to phaseolin pattern. HortSci. 17: 752–754. [4]

Browning, J. A., 1974 Relevance of knowledge about natural ecosystems to development of pest management programs for agro-ecosystems. Proc. Amer. Phytopath. Soc. 1: 191–199. [13]

Browning, J. A., and K. J. Frey, 1969 Multiline cultivars as a means of disease control. Annu. Rev. Phytopathol. 7: 355–382. [13]

Browning, J. A., K. J. Frey, M. E. McDaniel, M. D. Simons and I. Wahl, 1979 The biologic of using multilines to buffer pathogen populations and prevent disease loss. Indian J. Genet. Plant Breed. 39: 2–9. Discussion pp. 105–106. [21]

Brucher, O., 1968 Absence of phytohemagglutinin in wild and cultivated beans from South America. Proc. Trop. Region Am. Soc. Hort. Sci. 12: 68–85. [18]

Bryant, E. H., S. A. McCommas and L. M. Combs, 1986a The effect of an experimental bottleneck upon quantitative genetic variation in the housefly. Genetics 114: 1191–1211. [15]

Bryant, E. H., L. M. Combs and S. A. McCommas, 1986b Morphometric differentiation among experimental lines of the housefly in relation to a bottleneck. Genetics 114: 1213–1223. [15]

Bull, J. J., 1987 Evolution of phenotypic variance. Evolution 41: 303–315. [20]

Bulmer, M. G., 1971 The effect of selection on genetic variability. Am. Nat. 105: 201–211. [10]

Burdon, J. J., 1982 The effect of fungal pathogens on plant communities. pp. 99–112. In: *The Plant Community as a Working Mechanism*, Edited by E. I. Newman. Blackwell, Oxford. [13]

Burdon, J. J., 1987a *Diseases and Plant Population Biology*. Cambridge University Press, Cambridge. [12, 13, 21]

Burdon, J. J., 1987b Phenotypic and genetic patterns of resistance to the pathogen *Phakopsora pachyrhizi* in populations of *Glycine canescens*. Oecologia 73: 257–267. [13]

Burdon, J. J., and A. H. D. Brown, 1986 Population genetics of *Echium plantagineum* L.—Target weed for biological control. Aust. J. Biol. Sci. 30: 369–378. [3, 15]
Burdon, J. J., and G. A. Chilvers, 1977 Controlled environment experiments on epidemic rates of barley mildew in different mixtures of barley and wheat. Oecologia 28: 141–146. [13]
Burdon, J. J. and A. M. Jarosz, 1988a The ecological genetics of plant-pathogen interactions in natural communities. Phil. Trans. Royal Soc. Ser. B. 321: 349–363. [13]
Burdon, J. J., and A. M. Jarosz, 1988b Wild relatives as sources of disease resistance. pp. 280–296. In: *Use of Plant Genetic Resources*, Edited by A. H. D. Brown, O. H. Frankel, D. R. Marshall, and J. T. Williams. Cambridge University Press, Cambridge. [13]
Burdon, J. J., and W. J. Müller, 1987 Measuring the cost of resistance to *Puccinia coronata* Cda in *Avena fatua* L. J. Appl. Ecol. 24: 191–200. [13]
Burdon, J. J., and R. Whitbread, 1979 Rates of increase of barley mildew in mixed strands of barley and wheat. J. Appl. Ecol. 16: 253–258. [13]
Burdon, J. J., R. H. Groves and J. M. Cullen, 1981 The impact of biological control on the distribution and abundance of *Chondrilla juncea* in south-eastern Australia. J. Appl. Ecol. 18: 957–966. [13]
Burdon, J. J., D. R. Marshall and A. H. D. Brown, 1983 Demographic and genetic changes in populations of *Echium plantagineum*. J. Ecol. 71: 667–679. [12]
Burdon, J. J., A. M. Jarosz and A. H. D. Brown, 1988 Temporal patterns of reproduction and outcrossing in weedy populations of *Echium plantagineum*. Biol. J. Linn Soc. 34: 81–92. [9]
Burdon, R. D., and G. Namkoong, 1983 Short note: Multiple populations and sublines. Silv. Genet. 32: 221–222. [16]
Bürger, R., G. P. Wagner and F. Stettinger, In press How much heritable variation can be maintained in finite populations by a mutation selection balance? Evolution. [20]
Burgess, R. S. L., and R. A. Ennos, 1987 Selective grazing of acyanogenic white clover: Variation in behaviour among populations of the slug *Deroceras reticulatum*. Oecologia 73: 432–435. [12]
Burk, L. G., and D. F. Matzinger, 1976 Variation among anther-derived doubled haploids from an inbred line of tobacco. J. Hered. 67: 381–384. [17]
Burnouf, T., and R. Bourriquet, 1980 Glutenin subunits of genetically related European hexaploid wheat cultivars, their relation to bread-making quality. Theor. Appl. Genet. 50: 107–111. [4]
Burschel, P., 1965 Die Omorikafichte. Forstarchiv 36: 113–131. [16]
Burton, J. M., R. F. Wilson, and C. A. Brim, 1983 Recurrent selection in soybean. IV. Selection for increased oleic acid percentage in seed oil. Crop Sci. 23: 744–747. [17]
Busbice, T. W., 1969 Inbreeding in synthetic varieties. Crop Sci. 9: 601–604. [20]
Busbice, T. W., 1970 Predicting yield of synthetic varieties. Crop Sci. 10: 265–269. [20]
Busbice, T. H., R. Y. Gurgic and H. B. Collins, 1975 Effect of selection for self-fertility and self-sterility in alfalfa and related characters. Crop Sci. 15: 471–475. [20]
Busch, R. H., and K. Kofoid, 1982 Recurrent selection for kernel weight in spring wheat. Crop Sci. 23: 744–747. [17]
Bush, R. M., P. E. Smouse and F. T. Ledig, 1987 The fitness consequences of multiple-locus heterozygosity: the relationship between heterozygosity and growth rate in pitch pine (*Pinus rigida* Mill.) Evolution 41: 787–798. [12, 16]
Byrne, I., and D. C. Rasmusson, 1974 Recurrent selection for mineral content in wheat and barley. Euphytica 23: 241–249. [17]

Cannings, C., and E. A. Thompson, 1981 *Genealogical and Genetic Structure*. Cambridge University Press, Cambridge. [11]
Carey, K., and F. R. Ganders, 1987 Patterns of isoenzyme variation in *Plectritis* (Valerianaceae). Syst. Bot. 12: 125–132. [3]
Carlquist, S., 1974 *Island Biology*. Columbia Univ. Press, New York. [15]
Carmon, J. L., H. A. Stewart, C. C. Cockerham and R. E. Comstock, 1956 Prediction equations for rotational cross-breeding. J. Ann. Sci. 15: 930–936. [20]
Carson, H. L., and A. R. Templeton, 1984 Genetic revolutions in relation to speciation phenomena: the founding of new populations. Ann. Rev. Ecol. Syst. 15: 97–131. [15]
Casey, R., C. Domoney and N. Ellis, 1986 Legume storage proteins and their genes. Oxford

Surveys Plant Mol. Cell Biol. 3: 1–95. [4]

Cavallini, A., C. Zolfino, G. Cionini, R. Cremonini, L. Natali, O. Sassoli and P. G. Cionini, 1986 Nuclear DNA changes within *Helianthus annus* L.: Cytophotometric, karyological and biochemical analyses. Theor. Appl. Genet. 73: 20–26. [6]

Cavers, P. B., and J. L. Harper, 1967 Studies in the dynamics of plant populations. I. The fate of seed and transplants introduced into various habitats. J. Ecol. 55: 59–71. [12]

Chapman, C. G. D., 1989 Collection strategies for the wild relatives of field crops. pp. 263–279. In: *The Use of Plant Genetic Resources*, Edited by A. H. D. Brown, O. H. Frankel, D. R. Marshall, and J. T. Williams. Cambridge University Press, Cambridge. [21]

Charlesworth, B., 1985 Genetic constraints on the evolution of plant reproductive systems. pp. 155–179. In: *Population Genetics in Forestry*, Edited by H.–R. Gregorius. Springer-Verlag, Berlin. [12]

Charlesworth, D., and B. Charlesworth, 1987 Inbreeding depression and its evolutionary consequences. Annu. Rev. Ecol. Syst. 18: 237–268. [11]

Chase, S. S., 1952 Production of homozygous diploids of maize from monoploids. Agron. J. 44: 263–267. [17]

Cheliak, W. M., B. P. Dancik, K. Morgan, F. C. H. Yeh and C. Strobeck, 1985a Temporal variation of the mating system in a natural population of jack pine. Genetics 109: 569–584. [12, 16]

Cheliak, W. M., J. A. Pitel and G. Murray, 1985b Population structure and mating system of white spruce. Can. J. For. Res. 15: 301–308. [16]

Cherry, J. P., 1975 Comparative studies of seed protein and enzymes of species and collections of *Arachis* by gel electrophoresis. Peanut Sci. 2: 57–65. [4]

Choo, T. M., H. R. Klinck and C. A. St. Pierre, 1980 The effect of location on natural selection in bulk populations of barley (*Hordeum vulgare* L.) II. Quantitative traits. Can. J. Plant Sci. 60: 41–47. [12]

Christiansen, F. B., and M. W. Feldman, 1983 Selection in complex genetic systems. V. Some properties of mixed selfing and random mating with two loci. Theor. Pop. Biol. 23: 257–272. [10]

Chung, M. S., 1981 Biochemical methods for determining population structure in *Pinus sylvestris*. L. Acta For. Fenn. 173: 1–28. [16]

Clausen, J., D. D. Keck and W. M. Hiesey, 1948 Experimental studies on the nature of species. III. Environmental responses of climatic races of *Achillea*. Carnegie Institute Washington Publ. 581. [12]

Clayton, G. A., and A. Robertson, 1955 Mutation and quantitative variation. Am. Nat. 89: 151–158. [20]

Clegg, M. T., 1978 Dynamics of correlated genetic systems. II. Simulation studies of chromosomal segments under selection. Theor. Pop. Biol. 13: 1–23. [12]

Clegg, M. T., 1980 Measuring plant mating systems. BioScience 30: 814–818. [9]

Clegg, M. T., and R. W. Allard, 1972 Patterns of genetic differentiation in the slender wild oat species *Avena barbata*. Proc. Natl. Acad. Sci., U.S.A. 69: 1820–1824. [19]

Clegg, M. T., and R. W. Allard, 1973 Viability vs. fecundity selection in the slender wild oat, *Avena barbata* L. Science 181: 667–668. [11, 12]

Clegg, M. T., and B. K. Epperson, 1985 Recent developments in population genetics. Adv. Genet. 23: 235–269. [6, 9, 14]

Clegg, M. T., R. W. Allard and A. L. Kahler, 1972 Is the gene the unit of selection? Evidence from two experimental plant populations. Proc. Natl. Acad. Sci. U.S.A. 69: 2474–2478. [10]

Clegg, M. T., A. L. Kahler and R. W. Allard, 1978a Estimation of life cycle components of selection in an experimental plant population. Genetics 89: 765–792. [9, 10, 11, 12, 15]

Clegg, M. T., A. L. Kahler and R. W. Allard, 1978b Genetic demography of plant populations. pp. 173–188. In: *Genetics and Ecology: The Interface*. Edited by P. F. Brussard, Springer-Verlag, Berlin. [11]

Clegg, M. T., A. H. D. Brown and P. R. Whitfield, 1984a Chloroplast DNA diversity in wild and cultivated barley: Implications for genetic conservation. Genet. Res. 43: 339–343. [6]

Clegg, M. T., J. R. Y. Rawson and K. Thomas, 1984b Chloroplast DNA evolution in pearl millet and related species. Genetics 106: 449–461. [6]

Cliff, A. D., and J. K. Ord, 1981 *Spatial Processes*. Pion, London. [14]

Cliff, A. D., R. L. Martin and J. K. Ord, 1975 A test for spatial autocorrelation in chloropleth maps based upon a modified x^2 statistic. Trans. Papers, Inst. Br. Geog. 65: 109–129. [14]
Cochran, W. G., 1951 Improvement by means of selection. Proc. Second Berkeley Symp. Math. Stat. Prob., pp. 449–470. [17]
Cockerham, C. C., 1956 Effects of linkage on the covariances between relatives. Genetics 41: 138–141. [11]
Cockerham, C. C., 1961 Implications of genetic variances in a hybrid breeding program. Crop Sci. 1: 47–52. [20]
Cockerham, C. C., 1967 Group inbreeding and coancestry. Genetics 58: 89–104. [2]
Cockerham, C. C., 1969 Variance of gene frequencies. Evolution 23: 72–84. [2]
Cockerham, C. C., 1971 Higher order probability functions of identity of alleles by descent. Genetics 69: 235–246. [2, 11]
Cockerham, C. C., 1973 Analyses of gene frequencies. Genetics 74:679–700. [2]
Cockerham, C. C., 1983 Covariances of relatives from self-fertilization. Crop Sci. 23: 1177–1180. [11]
Cockerham, C. C., and B. S. Weir, 1973 Descent measures for two loci with some applications. Theor. Pop. Biol. 4: 300–330. [11]
Cockerham, C. C., and B. S. Weir, 1984 Covariances of relatives stemming from a population undergoing mixed self and random mating. Biometrics 40: 157–164. [11]
Coen, E. S., and G. A. Dover, 1983 Unequal exchanges and the coevolution of X and Y rDNA arrays in *Drosophila melanogaster.* Cell 33: 849–855. [5]
Collinge, D. B., and A. J. Slusarenko, 1987 Plant gene expression in response to pathogens. Plant Mol. Biol. 9: 389–410. [20]
Cone, K. C., R. J. Schmidt, B. Burr and F. A. Burr, 1988 Advantages and limitations of using *Spm* as a transposon tag. pp. 149–159. In: *Plant Transposable Elements*, Edited by O. E. Nelson. Plenum, New York. [7]
Conkle, M. T., 1981 Isozyme variation and linkage in six conifer species. pp. 11–17. In: *Isozymes of North American Forest Trees and Forest Insects*, Edited by M. T. Conkle. PSW-48, USDA Gen. Tech. Rep. [19]
Cotterill, P. P., 1984 A plan for breeding radiata pine. Silv. Genet. 33: 84–90. [16]
Cournoyer, B. M., 1970 Crown rust epiphytology with emphasis on the quantity and periodicity of spore dispersal from heterogeneous oat cultivar—rust race populations. Ph.D. thesis, Iowa State University, Ames. [13]
Cox, T. S., D. J. Cox and K. J. Frey, 1987 Mutations for polygenic traits in barley under nutrient stress. Euphytica 36: 823–829. [20]
Crawford, T. J., 1984 The estimation of neighborhood parameters for plant populations. Heredity 52: 273–283. [14]
Crawford-Sidebotham, T. J., 1972 The role of slugs and snails in the maintenance of the cyanogenesis polymorphisms of *Lotus corniculatus* L. and *Trifolium repens* L. Heredity 28: 405–411. [11]
Crow, J. F., 1952 Dominance and over dominance. pp. 282–297. In: *Heterosis*, Edited by J. W. Gowen, Iowa State University Press, Ames. [20]
Crow, J. F., and C. Denniston, 1988 Inbreeding and variance effective population numbers. Evolution 42: 482–495. [15]
Crow, J. F., and M. Kimura, 1970 *An Introduction to Population Genetics Theory.* Harper & Row, New York. [6]
Crowe, L. K., 1971 The polygenic control of outbreeding in *Borago officinalis* L. Heredity 27: 111–118. [20]
Cullis, C. A., 1987 The generation of somatic and heritable variation in response to stress. Am. Natur. 130: S62–S73. [6]
Curtis, S. E., and M. T. Clegg, 1984 Molecular evolution of chloroplast DNA sequences. Mol. Biol. Evol. 1: 291–301. [6]
Curtis, S. E., and R. Haselkorn, 1984 Isolation, sequence and expression of two members of the 32 kd thylakoid membrand protein gene family from the cyanobacterium *Anabaena* 7120. Plant Mol. Biol. 3: 249–258. [8]
Cwynar, L. C., and G. M. MacDonald, 1987 Geographical variation of Lodgepole pine in relation to population history. Am. Nat. 123: 463–469. [15]

Dallas, J. F., 1988 Detection of DNA "fingerprints" of cultivated rice by hybridization with a human minisatellite DNA probe. Proc. Natl. Acad. Sci. U.S.A. 85: 6831–6835. [6]

Damian, R. T., 1964 Molecular mimicry: Antigen sharing by parasite and host and its consequences. Am. Nat. 98: 129–149. [8]

Daniellson, C. E., 1949 Seed globulins of the Gramineae and Leguminosae. Biochem. J. 44: 387–400. [4]

Darwin, C. R., 1876 The effects of cross- and self-fertilisation in the vegetable kingdom. John Murray, London. [20]

David, F. N., 1971 Measurement of diversity: Multiple cell contents. Proc. 6th Berkely Symp. Math. Stat. Prob. 4: 109–136. [14]

Davies, C. S., J. B. Coates and N. C. Nielsen, 1985 Inheritance and biochemical analysis of four electrophoretic variants of δ-conglycinin from soybean. Theor. Appl. Genet. 71: 351–358. [4]

Davies, C. S., T. J. Cho and N. C. Nielsen, 1987 Tests for linkage of seed protein genes Lx_1, Lx_3, and cgy_1 to markers on known linkage groups. Soybean Genet. Newslett. 14: 248–252. [4]

Davies, M. S., and R. W. Snaydon, 1973a Physiological differences among populations of *Anthoxanthum odoratum* L. collected from the Park Grass Experiment, Rothamsted I. Response to calcium. J. Appl. Ecol. 10: 33–45. [12]

Davies, M. S., and R. W. Snaydon, 1973b Physiological differences among populations of *Anthoxanthum odoratum* L. collected from the Park Grass Experiment, Rothamsted II. Response to aluminum. J. Appel. Ecol. 10: 47–55. [12]

Davies, M. S., and R. W. Snaydon 1974 Physiological differences among populations of *Anthoxanthum odoratum* L. collected from the Park Grass Experiment, Rothamsted III. Response to phosphate. J. Appl. Ecol. 11: 669–707. [12]

Davies, M. S., and R. W. Snaydon, 1976 Rapid population differentiation in a mosaic environment. III. Measures of selection pressures. Heredity 36: 59–66. [12]

Delaney, D. D., 1988 Improvement of percentage phaseolin in common bean (Phaseolus vulgaris L.) using marker-based and S_1 family recurrent selection. Ph.D. Thesis, University of Wisconsin, Madison. [18]

Delogu, G., C. Lorenzoni, A. Marocco, P. Martiniello, M. Odbardi and A. M. Stanca, 1988 A recurrent selection program for grain yield in winter barley. Euphytica 37: 105–110. [17]

Dennis, E. S., E. J. Finnegan, B. H. Taylor, T. A. Peterson, A. R. Walker and W. J. Peacock, 1988 Maize transposable elements: structure, function, and regulation. pp. 101–13. In: *Plant Transposable Elements*, Edited by O. E. Nelson. Plenum, New York. [7]

Devlin, B., K. Roeder and N. C. Ellstrand, 1988 Fractional paternity assignment: Theoretical development and comparison to other methods. Theor. Appl. Genet. 76: 369–380. [9, 11]

Dewey, S. E., and J. S. Heywood, 1988 Spatial genetic structure in a population of *Psychotria nervosa*. I. Distribution of genotypes. Evolution 42: 834–838. [14]

DiMichele, L., and D. A. Powers, 1982. Physiological basis for swimming endurance differences between LDH-B genotypes of *Fundulus heteroclitus*. Science 216: 1014–1016. [8]

Dinoor, A., 1970 Source of oat crown rust resistance in hexaploid and tetraploid wild oats in Israel. Can. J. Bot. 48: 153–161. [13]

Dinoor, A., and N. Eshed, 1987 The analysis of host and pathogen populations in natural ecosystems. pp. 75–88. In: *Populations of Plant Pathogens: Their Dynamics and Genetics*, Edited by M. S. Wolfe and C. E. Caten. Blackwell, Oxford. [13]

Dirzo, R., 1984 Herbivory: A phytocentric overview. pp. 141–165. In: *Perspectives on Plant Population Ecology*, Edited by R. Dirzo and J. Sarukhan. Sinauer, Sunderland, Mass. [12]

Dirzo, R., and J. L. Harper, 1982a Experimental studies on slug-plant interactions III. Differences in the acceptibility of individual plants of *Trifolium repens* to slugs and snails. J. Ecol. 70: 101–117. [12]

Dirzo, R., and J. L. Harper, 1982b Experimental studies on slug-plant interactions. IV. The performance of cyanogenic and acyanogenic morphs of *Trifolium repens* in the field. J. Ecol. 70: 119–138. [12]

Doebley, J., W. Renfroe and A. Blanton, 1987 Restriction site variation in the *Zea* chloroplast genome. Genetics 117: 139–147. [6]

Doering, H. P., and P. Starlinger, 1986 Molecular genetics of transposable elements in plants. Annu. Rev. Genet. 20: 175–200. [7]

Doggett, H., 1972 Recurrent selection in sorghum populations. Heredity 28: 9–29. [17]
Doll, H., and A. H. D. Brown, 1979 Hordein variation in wild (*Hordeum spontaneum*) and cultivated (*H. vulgare*) barley. Can. J. Genet. Cytol. 21: 391–404. [4]
Donald, C. M., 1968 The breeding of crop ideotypes. Euphytica 17: 385–403. [20]
Dooner, H. K., and O. E. Nelson, 1977a Genetic control of UDPglucose:flavonol 3-O-glucosyl transferase in the endosperm of maize. Biochem. Genet. 15: 509–519. [7]
Dooner, H. K., and O. E. Nelson, Jr., 1977b Controlling element-induced alterations in UDP glucose: flavonoid glucosyltransferase, the enzyme specified by the *bronze* locus in maize. Proc. Natl. Acad. Sci. U.S.A. 74: 5623–5627. [7]
Dooner, H. K., and O. E. Nelson, 1979 Heterogeneous flavonoid glucosyl transferases in purple derivatives from a controlling element-suppressed *bronze* mutant in maize. Proc. Natl. Acad. Sci. U.S.A. 76: 2369–2371. [7]
Dooner, H. K., E. Ralston and J. English, 1988 Deletions and breaks involving the borders of the *Ac* element in the *bz-m2(Ac)* allele of maize, pp. 214–226. In: *Plant Transposable Elements*, Edited by O. E. Nelson, Plenum, New York. [7]
Dover, G., 1982 Molecular drive: A cohesive mode of species evolution. Nature, (London) 299: 111–117. [5]
Dover, G. A., 1986 The spread and success of non-Darwinian novelties. pp. 199–238. In: *Evolutionary Processes and Theory*, Edited by S. Karlin and E. Nevo. Academic Press, New York. [6]
Doyle, J. J., and R. N. Beachy, 1985 Ribosomal gene variation in soybean (*Glycine*) and its relatives. Theor. Appl. Genet. 70: 369–376. [6]
Doyle, J. J., M. A. Schuler, W. D. Godette, V. Zenger, R. N. Beachy, and J. L. Slightom, 1985 The glycosylated seed storage protein of *Glycine max* and *Phaseolus vulgaris*. J. Biol. Chem. 261: 9228–9238. [4]
Drouin, G., and G. A. Dover, 1987 A plant processed pseudogene. Nature, (London) 328: 557–558. [20]
Durrant, A., 1962 The environmental induction of heritable changes in *Linum*. Heredity 17: 27–61. [6]
Duvick, D. N., 1984 Genetic diversity in major farm crops on the farm and in reserve. Econ. Bot. 38: 161–178. [18]
Dvořák, J., and R. Appels, 1982 Chromosome and nucleotide sequence differentiation in genomes of polyploid *Triticum* species. Theor. Appl. Genet. 63: 349–360. [5]
Dvořák, J., and R. Appels, 1986 Investigation of homologous crossing over and sister chromatid exchange in the wheat *Nor-B2* locus coding for rRNA and *Gli-B2* locus coding for gliadins. Genetics 115: 1037–1056. [5]
Dvořák, J., D. Jue and M. Lassner, 1987 Homogenization of tandemly repeated nucleotide sequences by distance-dependent nucleotide sequence conversion. Genetics 116: 487–498. [5, 6]
Dvořák, J., P. E. McGuire and B. Cassidy, 1988 Apparent sources of the A genomes of wheats inferred from polymorphism in abundance and restriction fragment length of repeated nucleotide sequences. Genome 30: 680–689. [5]
Dyck, P. L. and D. J. Samborski, 1968 Genetics of resistance to leaf rust in the common wheat varieties Webster, Loris, Brevit, Carina, Malakof and Centenario. Can. J. Genet. Cytol. 10: 7–17. [13]

Edmonds, J. M., and S. M. Glidewell, 1977 Acrylamide gel electrophoresis of seed proteins from some *Solanum* (sect. *Solanum*) species. Plant Syst. Evol. 127: 277–291. [4]
Edmunds, G. F., and D. N. Alstad, 1978 Coevolution in insect herbivores and conifers. Science 199: 941–945. [12]
Edwards, M. D., C. W. Stuber and J. F. Wendel, 1987 Molecular-marker-facilitated investigations of quantitative-trait loci in maize. I. Numbers, genomic distribution and types of gene action. Genetics 116: 113–125. [16, 19]
Efimov, V. A., A. V. Andreeva, S. V. Reverdatto, R. Jung and O. G. Chakhmakhcheva, 1988 Nucleotide sequence of the barley chloroplast *psbA* gene for the Q_B protein of photosystem II. Nucl. Acids Res. 16: 5685. [8]
Efron, B., 1982 Bootstrap methods: another look at the jackknife. Ann. Stat. 7: 1–26. [11]
Ehrendorfer, F., 1979 Reproductive biology in island plants. pp. 293–306. In: *Plants and Islands*,

Edited by D. Bramwell, Academic Press, London. [15]
Eldridge, K. G., and A. R. Griffin, 1983 Selfing effects in *Eucalyptus regnans*. Silv. Genet. 32: 216–221. [16]
El-Kassaby, Y., 1982 Associations between allozyme genotypes and quantitative traits in Douglas-fir (*Pseudotsuga menziesii* (Mirb.) Franco). Genetics 101: 103–115. [16]
El-Kassaby, Y. A., and K. Ritland, 1986a The relation of outcrossing and contamination of reproductive phenology and supplemental mass pollination in a Douglas-fir seed orchard. Silv. Genet. 35: 240–244. [16]
El-Kassaby, Y. A., and K. Ritland, 1986b Low levels of pollen contamination in a Douglas-Fir seed orchard as detected by allozyme markers. Silv. Genet. 35: 224–229. [16]
El-Kassaby, Y. A., F. C. Yeh and O. Sziklai, 1981 Estimation of the outcrossing rate of Douglas-fir (*Pseudotsuga menziesii* (Mirb.) Franco) using allozyme polymorphisms. Silv. Genet. 30: 182–184. [16]
El-Kassaby, Y. A., M. D. Meagher, J. Parkinson and F. T. Portlock, 1987 Allozyme inheritance, heterozygosity and outcrossing rate among *Pinus monticola* near Ladysmith, British Columbia. Heredity 58: 173–181. [9, 16]
Ellstrand, N. C., 1984 Multiple paternity within the fruits of the wild radish *Raphanus sativus*. Am. Nat. 123: 819–828. [9, 11]
Ellstrand, N. C., and J. Antonovics, 1985 Experimental studies of the evolutionary significance of sexual reproduction. II. A test of the density-dependent hypothesis. Evolution 39: 657–666. [12]
Ellstrand, N. C., and D. L. Marshall, 1985 Interpopulation gene flow by pollen in wild radish. Am. Nat. 126: 596–605. [3]
Ellstrand, N. C., and M. L. Roose, 1987 Patterns of genotypic diversity in clonal plant species. Am. J. Bot. 74: 123–131. [3]
Endler, J. A., 1977 *Geographic Variation, Speciation, and Clines*. Princeton University Press, Princeton, N.J. [14]
Endler, J. A., 1986 *Natural Selection in the Wild*. Princeton University Press, Princeton. [12]
Ennos, R. A., 1981 Detection of selection in populations of white clover (*Trifolium repens* L.) Biol. J. Linn. Soc. 15: 75–82. [12]
Ennos, R. A., 1982 Association of the cyanogenic loci in white clover. Genet. Res. 40: 65–72. [10]
Ennos, R. A., 1983 Maintenance of genetic variation in plant populations. Evol. Biol. 16: 129–155. [12]
Ennos, R. A., 1985a The mating system and genetic structure in a perennial grass *Cynosurus cristatus* L. Heredity 55: 121–126. [9, 10, 15]
Ennos, R. A., 1985b The significance of genetic variation for root growth within a natural population of white clover (*Trifolium repens*). J. Ecol. 73: 615–624. [12]
Ennos, R. A., and M. T. Clegg, 1982 Effect of population substructuring on estimates of outcrossing rate in plant populations. Heredity 48: 283–292. [14]
Ennos, R. A., and R. K. Dodson, 1987 Pollen success, functional gender and assortative mating in an experimental plant population. Heredity 58: 119–126. [9]
Ennos, R. A., and K. W. Swales, 1987 Estimation of the mating system in a fungal pathogen *Crumenulopsis sororia* (Karst.) Groves using isozyme markers. Heredity 59: 423–430. [9]
Ennos, R. A., and K. W. Swales, 1988 Genetic variation in tolerance of host monoterpenes in a population of the ascomycete canker pathogen *Crumenulopsis sororia*. Plant Pathol. 37: 407–416. [12]
Epperson, B. K., and R. W. Allard, 1984 Allozyme analysis of the mating system in lodgepole pine populations. Heredity 75: 212–214. [9, 14, 16]
Epperson, B. K., and R. W. Allard, 1987 Linkage disequilibrium between allozymes in natural populations of lodgepole pine. Genetics 115: 341–352. [10, 16]
Epperson, B. K., and R. W. Allard, 1989 Spatial autocorrelation analysis of the distribution of genotypes within populations of lodgepole pine. Genetics 121: 369–377. [14]
Epperson, B. K., and M. T. Clegg, 1986 Spatial autocorrelation analysis of flower color polymorphisms within substructured populations of morning glory (*Ipomoea purpurea*). Am. Nat. 128: 840–858. [14]
Epperson, B. K., and M. T. Clegg, 1987a Frequency-dependent variation for outcrossing rate among flower color morphs of *Ipomoea purpurea*. Evolution 41: 1302–1311. [14]

Epperson, B. K., and M. T. Clegg, 1987b Instability at a flower color locus in morning glory, *Ipomoea purpurea*. J. Hered. 78: 346–352. [14]
Erickson, J. M., M. Rahire, P. Bennoun, P. Delepelaire, B. Diner and J. D. Rochaix, 1984 Herbicide resistance in *Chlamydomonas reinhardtii* results from a mutation in the chloroplast gene for the 32-kilodalton protein of photosystem II. Proc. Natl. Acad. Sci. U.S.A. 81: 3617–3621. [8]
Eriksson, G., B. Schelander and V. Akebrand, 1973a Inbreeding depression in an old experimental plantation of *Picea abies*. Hereditas 73: 185–194. [16]
Eriksson, G., A. Jonsson and D. Lindgren, 1973b. Flowering in a clone trial of *Picea abies* Karst. Stud. For. Suecica 110: 1–44. [16]
Eriksson, G., S. Anderson, V. Eiche, J. Ifver and A. Persson, 1980 Severity index and transfer effects on survival and volume production of *Pinus sylvestris* in northern Sweden. Stud. For. Suecica 156: 1–32. [16]
Eshel, I., 1978 On the mutual maintenance of a polymorphism in two nonepistatic loci with partial selfing. Theor. Pop. Biol. 13: 99–111. [10]
Evans, D. R., J. Hill, T. A. Williams and I. Rhodes, 1985 Effects of coexistence on the performance of white clover–perennial ryegrass mixtures. Oecologia 66: 536–539. [12]
Ewens, W. J., 1968 A genetic model having complex linkage behavior. Theor. Appl. Genet. 38: 140–143. [10]
Ewens, W. J., 1979 *Mathematical Population Genetics*. Springer-Verlag, Berlin. [10]

Falconer, D. S. 1981 *Introduction to Quantitative Genetics*, 2nd Edition. Longman, London. [11, 12, 15, 17, 20]
Falkenhagen, E. R., 1985 Isozyme studies in provenance research of forest trees. Theor. Appl. Genet. 69: 335–347. [16]
Farabaugh, P. J., V. Schmeissner, M. Hofer and J. Miller, 1978 Genetic studies of the *lac* repressor. VII. On the molecular nature of spontaneous hotspots in the *lacI* gene of *Escherichia coli*. J. Mol. Biol. 126: 847–863. [5]
Farris, M. A. and J. B. Mitton, 1984 Population density, outcrossing rate and heterozygote superiority in ponderosa pine. Evolution 38: 1151–1154. [12, 16]
Farris, M. A., and B. A. Schaal, 1983 Morphological and genetic variation in ecologically central and marginal populations of *Rumex acetosella* L. (Polygonaceae). Am. J. Bot. 70: 246–255. [15]
Fasoulas, A. C., and R. W. Allard, 1962 Nonallelic gene interactions in the inheritance of quantitative characters in barley. Genetics 47: 899–907. [17]
Faulkner, R. (Ed.), 1975 *Seed Orchards*. Forestry commission bulletin No. 54. Her Majesty's Stationery Office, London. [16]
Fedoroff, N. V., 1983 Controlling elements in maize. pp. 1–63. In: *Mobile Genetic Elements*, Edited by J. A. Shapiro. Academic Press, New York. [7]
Fedoroff, N., D. B. Furtek and O. E. Nelson, 1984 Cloning of the *bronze* locus in maize by a simple and generalizable procedure using the transposable controlling element *Ac*. Proc. Natl. Acad. Sci. U.S.A. 81: 3825–3829. [7]
Feldman, M. W., and F. B. Christiansen, 1975 The effect of population subdivision on two loci without selection. Genet. Res. 24: 151–167. [14]
Felsenstein, J., 1985 Phylogenies and the comparative method. Am. Nat. 125: 1–15. [11]
Figdore, S. S., W. C. Kennard, K. M. Song, M. K. Slocum and T. C. Osborn, 1988 Assessment of the degree of restriction fragment length polymorphism in *Brassica*. Theor. Appl. Genet. 75: 833–840. [6]
Figueroa, F., E. Günther and J. Klein, 1988 MHC polymorphism predating speciation. Nature, (London) 355: 265–267. [8]
Finney, D. J., 1958 Plant selection for yield improvement. Euphytica 7: 83–106. [17]
Fins, L., and W. J. Libby, 1982 Population variation in *Sequoiadendron:* seed and seedling studies, vegetative propagation, and isozyme variation. Silv. Genet. 31: 102–110. [16]
Fisher, R. A., 1918 The correlation between relatives on the supposition of Mendelian inheritance. Trans. Roy. Soc. Edinburgh. 52: 399–433. [20]
Fisher, R. A., 1941 Average excess and average effect of a gene substitution. Ann. Eugen. 11: 53–63. [12]

Flavell, R. B., and M. O'Dell, 1979 Genetic control of nucleolus formation in wheat. Chromosoma 71: 135–152. [5]

Flavell, R. B., M. O'Dell, P. Sharp, E. Nevo and A. Beiles, 1986 Variation in the intergenic spacer of ribosomal DNA of wild wheat, *Tricum dicoccoides*, in Israel. Mol. Biol. Evol. 3: 547–558. [6]

Fogel, S., J. W. Welch and E. J. Louis, 1984 Meiotic gene conversion mediates gene amplification in yeast. Cold Spring Harbor Symp. Quant. Biol. 49: 55–65. [5]

Ford, E. B., 1971 *Ecological Genetics*. Wiley, New York. [10]

Ford-Lloyd B., and M. Jackson, 1986 *Plant Genetic Resources: An Introduction to Their Conservation and Use*. Edward Arnold, London. [21]

Fouilloux, G., 1980. Effectif et nombre de cycles de sélection á utiliser lors de l'emploi de la filiation unipaire (single seed descent method) ou de l'haplomethode pour la création de variétés lignées (pures) á partir d'une F_1. Ann. Amélior. Plantes 30: 17–38. [17]

Fowler, D. P., 1987 In defense of the polycross. Can. J. For. Res. 17: 1624–1627. [16]

Frankel, O. H., 1970 Genetic conservation in perspective. pp. 469–490. In: *Genetic Resources in Plants—Their Exploration and Conservation*, Edited by O. H. Frankel and E. Bennett. Blackwell, Oxford and Edinburgh. [21]

Frankel, O. H., 1974 Genetic conservation: Our evolutionary responsibility. Genetics 78: 53–65. [21]

Frankel, O. H., 1975 Genetic resources centres—a co-operative global network, pp. 473–481. In: *Crop Genetic Resources for Today and Tomorrow*, Edited by O. H. Frankel and J. G. Hawkes. Cambridge University Press, Cambridge. [21]

Frankel, O. H., 1984 Genetic perspectives of germplasm conservation. pp. 161–170. In: *Genetic Manipulation: Impact on Man and Society*, Edited by W. K. Arber, K. Llimensee, W. J. Peacock, and P. Starlinger. Cambridge University Press, Cambridge. [21]

Frankel, O. H., 1985a Genetic resources: The founding years. I. Early beginnings 1961–1966. Diversity 7: 26–29. [21]

Frankel, O. H., 1985b Genetic resources: The founding years. II. The movements constituent assembly. Diversity 8: 30–32. [21]

Frankel, O. H., 1986 Genetic resources: The founding years. III. The long road to the international board. Diversity 9: 30–33. [21]

Frankel, O. H., 1988 Genetic resources: Evolutionary and social responsibilities. pp. 19–46. In: *Seeds and Sovereignty: The Use and Control of Plant Genetic Resources*, Edited by J. R. Kloppenburg Jr. Duke University Press, Durham. [21]

Frankel, O. H., and E. Bennett, 1970 *Genetic Resources in Plants—Their Exploration and Conservation*. Blackwell, Oxford and Edinburgh. [21]

Frankel, O. H., and A. H. D. Brown, 1984 Current plant genetic resources—a critical appraisal. pp. 1–11. In: *Genetics: New Frontiers*, Vol. IV. Oxford and IBH, New Delhi. [21]

Frankel, O. H., and M. E. Soulé, 1981 *Conservation and Evolution*. Cambridge Univ. Press, Cambridge. [15, 21]

Frankel, O. H., W. L. Gerlach and W. J. Peacock, 1987 The ribosomal RNA genes in synthetic tetraploids of wheat. Theor. Appl. Genet. 75: 138–143. [5]

Franklin, I., and R. C. Lewontin, 1970 Is the gene the unit of selection? Genetics 65: 707–734. [10]

Frei, O. M., C. W. Stuber, and M. M. Goodman, 1986a Use of allozymes as genetic markers for predicting performance in maize single cross hybrids. Crop Sci. 26: 37–42. [19]

Frei, O. M., C. W. Stuber, and M. M. Goodman, 1986b Yield manipulation from selection on allozyme genotypes in a composite of elite corn lines. Crop Sci. 26: 917–921. [19]

Friedman, S. T., and W. T. Adams, 1985a Levels of outcrossing in two loblolly pine seed orchards. Silv. Genet. 34: 157–162. [16]

Friedman, S. T., and W. T. Adams, 1985b Estimation of gene flow into two seed orchards of loblolly pine (*Pinus taeda* L.). Theor. Appl. Genet. 69: 609–615. [16]

Fryxell, P. A., 1957 Mode of reproduction in higher plants. Bot. Rev. 23: 135–233. [9]

Furnier, G. R., and W. T. Adams, 1986 Mating system in natural populations of Jeffery pine. Am. J. Bot. 73: 1002–1008. [9, 16]

Furnier, G. R., P. Knowles, M. A. Clyde and B. P. Dancik, 1987 Effects of avian seed dispersal on the genetic structure of whitebark pine populations. Evolution 41: 607–612. [16]

Furtek, D. B., J. W. Schiefelbein, F. Johnston and O. E. Nelson, 1988 Sequence comparisons of

three wild-type *Bronze* alleles from *Zea mays*. Plant Mol. Biol. 11: 473–481. [7]
Fyfe, J. L., and N. T. J. Bailey, 1951 Plant breeding studies in leguminous forage crops. 1. Natural cross-breeding in winter beans. J. Agric. Sci. 41: 371–378. [9, 20]

Gabriel, K. R., and R. R. Sokal, 1969 A new statistical approach to geographic variation analysis. Syst. Zool. 18: 259–270. [14]
Ganders, F. R., K. Carey and A. J. F. Griffiths, 1977 Natural selection for a fruit dimorphism in *Plectritis congesta* (Valerianaceae). Evolution 31: 873–881. [12]
Gayler, K. R., and G. E. Sykes, 1985 Effects of nutritional stress on the storage proteins of soybeans. Plant Physiol. 78: 582–585. [4]
Gayler, K. R., B. G. Broadle, M. Snook and E. E. Johnson, 1984 Precursors of storage proteins in *Lupinus angustifolius*. Biochem. J. 221: 333–341. [4]
Gepts, P., 1988a Phaseolin as an evolutionary marker. pp. 215–241. In: *Genetic Resources of* Phaseolus *Beans*, Edited by P. Gepts. Kluwer, Dordrecht, The Netherlands. [4]
Gepts, P., 1988b A Middle American and an Andean common bean gene pool. pp. 375–390. In: *Genetic Resources of* Phaseolus *Beans*, Edited by P. Gepts. Kluwer, Dordrecht, The Netherlands. [4]
Gepts, P., and F. A. Bliss, 1984 Enhanced available methionine concentration associated with higher phaseolin levels in common bean seeds. Theor. Appl. Genet. 69: 47–53. [4]
Gepts, P., and F. A. Bliss, 1986 Phaseolin variability among wild and cultivated common beans (*Phaseolus vulgaris*) from Colombia. Econ. Bot. 40: 469–478. [4, 18]
Gepts, P., and F. A. Bliss, 1988 Dissemination pathways of common bean (*Phaseolus vulgaris*, Fabaceae) deduced from phaseolin electrophoretic variability. II. Europe and Africa. Econ. Bot. 42: 86–104. [4]
Gepts, P., and M. T. Clegg, 1989 Genetic diversity in pearl millet [*Pennisetum glaucum* (L.) R. Br.] at the DNA sequence level. J. Hered. 80: 203–208. [6]
Gepts, P., and D. G. Debouck, in press Origin, domestication, and evolution of the common bean (*Phaseolus vulgaris* L.). In: *Bean* (Phaseolus vulgaris *L.*) *Production and Improvement in the Tropics*, Edited by O. Voysest and A. Van Schoonhoven. Centro Internacional de Agricultura Tropical, Cali, Colombia. [4]
Gepts, P., T. C. Osborn, K. Rashka and F. A. Bliss, 1986 Phaseolin-protein variability in wild form and landraces of the common bean (*Phaseolus vulgaris*): evidence for multiple centers of domestication. Econ. Bot. 40: 451–468. [4, 18]
Gepts, P., K. Kmiecik, P. Pereira and F. A. Bliss, 1988 Dissemination pathways of common bean (*Phaseolus vulgaris*, Fabaceae) deduced from phaseolin electrophoretic variability. I. The Americas. Econ. Bot. 42: 73–85. [4]
Gerlach, W. L., and T. A. Dyer, 1980 Sequence organization of the repeated units in the nucleus of wheat which contain 5S rRNA genes. Nucleic Acid Res. 8: 4851–4865. [5]
Gerlach, W. L., D. Llewellyn and J. Haseloff, 1987 Construction of a plant disease resistance gene from the satellite RNA of tobacco ring spot virus. Nature, (London) 328: 802–805. [20]
Gieffers, W., and J. Heselbach, 1988 Disease incidence and yield of different cereal cultivars in pure stands and mixtures. I. Spring barley (*Hordeum vulgare* L.) J. Plant Dis. Prot. 95: 46–62. [13]
Giles, B. E., 1984 A comparison between quantitative and biochemical variation in the wild barley *Hordeum murinum*. Evolution 38: 34–41. [15]
Giles, B. E., and K. J. R. Edwards, 1983 Quantitative variation within and between populations of wild barley, *Hordeum murinum*. Heredity 51: 325–333. [21]
Gill, B. S., and R. Appels, 1988 Relationships between *Nor*-loci from different *Triticeae* species. Plant Syst. Evol. 160: 77–89. [5]
Glover, D. E., and S. C. H. Barrett, 1986 Variation in the mating system of *Eichhornia paniculata* (Spreng.) Solms. (Pontederiaceae). Evolution 40: 1122–1131. [9, 15]
Glover, D. E., and S. C. H. Barrett, 1987 Genetic variation in continental and island populations of *Eichhornia paniculata* (Pontederiaceae). Heredity 59: 7–17. [15]
Gojobori, T., and M. Nei, 1984 Concerted evolution of the immunoglobulin V_H gene family. Mol. Biol. Evol. 1: 195–212. [8]
Goldberg, R. G., G. Hoshek, G. S. Ditta and R. W. Breidenbach, 1981 Developmental regulation of cloned superabundant embryo mRNAs in soybean. Dev. Biol. 83: 218–231. [4]

Golden, S. S., and R. Haselkorn, 1985 Mutation to herbicide resistance maps within the *psbA* gene of *Anacystis nidulans* R2. Science 229: 1104–1107. [8]

Golding, G. B., and C. Strobeck, 1980 Linkage disequilibrium in a finite population that is partially selfing. Genetics 94: 777–789. [10]

Golding, G. B., C. F. Aquadro and C. H. Langley, 1986 Sequence evolution within populations under multiple types of mutation. Proc. Natl. Acad. Sci. U.S.A. 83: 427–431. [6]

Goloubinoff, P., M. Edelman and R. B. Hallick, 1984 Chloroplast-coded atrazine resistance in *Solanum nigrum: psbA* loci from susceptible and resistant biotypes are isogenic except for a single codon change. Nucl. Acids Res. 12: 9489–9496. [8]

Gonella, J. A., and P. A. Peterson, 1978 Isozyme relatedness of inbred lines of maize and performance of their hybrids. Maydica 23: 55–61. [19]

Goodman, M. M., and C. W. Stuber, 1983a Races of maize. VI. Isozyme variation among races of maize in Bolivia. Maydica XXVIII: 169–187. [3]

Goodman, M. M., and C. W. Stuber, 1983b Maize, pp. 1–33. In: Isozymes in Plant Genetics and Breeding, Part B, Edited by S. D. Tanksley and T. J. Orton. Elsevier, Amsterdam. [19]

Goodman, M. M., C. W. Stuber, K. Newton and H. H. Weissinger, 1980 Linkage relationships of 19 isozyme loci in maize. Genetics 96: 697–710. [19]

Goodnight, C. J., 1987 On the effect of founder events on epistatic genetic variance. Evolution 41: 80–91. [15]

Goodnight, C. J., 1988 Epistasis and the effect of founder events on the additive genetic variance. Evolution 42: 441–454. [15]

Gottlieb, L. D., 1981 Electrophoretic evidence and plant populations. Prog. Phytochem. 7: 1–46. [3]

Gottlieb, L. D., 1984 Genetics and morphological evolution in plants. Am.Nat. 123: 681–709. [15]

Gottschalk, W., and H. P. Müller, 1983 *Seed Proteins: Biochemistry, Genetics, Nutritive Value.* Martinus Nijhoff/Dr. W. Junk, The Hague. [4]

Gould, F., 1983 Genetics of plant herbivore systems: Interactions between applied and basic study. pp. 599–653. In: *Variable Plants and Herbivores in Natural and Managed Systems.* Edited by R. F. Denno and M. S. McClure, Academic Press, New York. [12]

Gould, S. J., and R. C. Lewontin, 1979 The spandrels of San Marco and the Panglossian paradigm: a critique of the adaptationist programme. Proc. R. Soc. Lond. Ser. B 205: 581–598. [12]

Govindaraju, D. R., 1988. Relationship between dispersal ability and levels of gene flow in plants. Oikos 52: 31–35. [16]

Graham, P. H., 1981 Some problems of nodulation and symbiotic nitrogen fixation in *Phaseolus vulgaris* L.: A review. Field Crops Res. 4: 92–112. [18]

Graur, D., and W. H. Li, 1989 Evolution of protein inhibitors of serine proteinases: Positive darwinian selection or compositional effects? J. Mol. Evol. 28: 286–298. [8]

Gray, A. J., 1987 Genetic change during succession in plants. pp. 273–293. In: *Colonization, Succession and Stability,* Edited by A. J. Gray, M. J. Crawley, and P. J. Edwards. Blackwell Scientific Publications, Oxford. [15]

Green, A. G., A. H. D. Brown and R. N. Oram, 1980 Determination of outcrossing rate in a breeding population of *Lupinus albus* L. Z. Pflanzenzüchtg. 84: 181–191. [9]

Green, T. R., and C. A. Ryan, 1972 Wound-induced proteinase inhibitor in plant leaves: A possible defense mechanism against insects. Science 175: 776–777. [8]

Gregorius, H. R., M. Ziehe and M. D. Ross, 1987 Selection caused by self-fertilization I. Four measures of self-fertilization and their effects on fitness. Theor. Pop. Biol. 31: 91–115. [16]

Griffing, B., 1975 Efficiency changes due to use of doubled-haploids in recurrent selection methods. Theo. Appl. Genet. 46: 367–386. [17]

Gullberg, U., R. Yazdani, D. Rudin and N. Ryman, 1985 Allozyme variation in Scots pine (*Pinus sylvestris* L.) in Sweden. Silv. Genet. 34: 193–201. [16]

Gupton, C. L., 1981 Phenotypic recurrent selection for increased leaf weight and decrease alkaloid content of burley tobacco. Crop Sci. 21: 921–925. [17]

Guries, R. P., and F. T. Ledig. 1982. Genetic diversity and population structure in pitch pine (*Pinus rigida* Mill). Evolution 36: 387–402. [16]

Hackett, W. E. R., and G. W. P. Dawson, 1958 The distribution of the ABO and simple 'Rhesus'

(d) blood groups in the Republic of Ireland from a sample of 1 in 37 of the adult population. Irish J. Med. Sci. 6(387): 99–109. [14]

Hadjinov, M. I., V. S. Scherbak, N. I. Benko, V. P. Gusev, T. B. Sukhorzheuskaya and L. P. Voronova, 1982 Interrelationships between isozyme diversity and combining ability in maize lines. Maydica 27: 135–149. [19]

Haining, R. P., 1977 Model specification in stationary random fields. Geog. Anal. 9: 107–129. [14]

Haining, R. P., 1978 The moving average model for spatial interaction. Trans. Inst. Br. Geog. 3: 202–225. [14]

Haining, R. P., 1979 Statistical tests and process generators for random field models. Geog. Anal. 11: 45–64. [14]

Haldane, J. B. S., 1924 The mathematical theory of natural and artificial selection. Part II. Proc. Cambridge Phil. Soc. Biol. Sci. 1: 158–163. [9]

Hallauer, A. R., 1981 Selection and breeding methods. pp. 3–55. In: *Plant Breeding* II. Edited by K. J. Frey. Iowa State Univ. Press, Ames. [17]

Hallauer, A. R., and J. B. Miranda, 1981 *Quantitative Genetics in Maize Breeding*. Iowa State Univ. Press, Ames. [17]

Hamrick, J. L., 1987 Gene flow and distribution of genetic variation in plant populations. pp. 53–67. In: *Differentiation Patterns in Higher Plants*, Edited by K. Urbanska. Academic Press, New York. [3]

Hamrick, J. L., In press Isozymes and the analyses of genetic structure of plant populations. In: *Isozymes in Plant Biology*, Edited by D. Soltis and P. Soltis, Dioscorides Press. [3]

Hamrick, J. L., and R. W. Allard, 1972 Microgeographical variation in allozyme frequencies in *Avena barbata*. Proc. Natl. Acad. Sci. U.S.A. 69: 2100–2104. [12, 14]

Hamrick, J. L., and R. A. Allard, 1975 Correlations between quantitative characters and enzyme genotypes in *Avena barbata*. Evolution 29: 438–442. [19]

Hamrick, J. L., and L. R. Holden, 1979 Influence of microhabitat heterogeneity on gene frequency distribution and gametic phase disequilibrium in *Avena barbata*. Evolution 33: 521–533. [3]

Hamrick, J. L., and M. D. Loveless, 1986 The influence of seed dispersal mechanisms on the genetic structure of plant populations, pp. 211–223. In: *Frugivores and Seed Dispersal*, Edited by A. Estrada and T. H. Fleming. Junk Publ., The Hague, Netherlands. [3]

Hamrick, J. L., and M. D. Loveless, 1986 Isozyme variation in tropical trees: procedures and preliminary results. Biotropica 18: 201–207. [16]

Hamrick, J. L., and M. D. Loveless, 1988 Isozyme variation in tropical trees. In: *Plant Evolutionary Ecology*, Edited by J. H. Bock and Y. B. Linhart. Westview Press, Boulder, CO. [16]

Hamrick, J. L., and M. D. Loveless, (in press) Associations between the breeding system and the genetic structure of tropical tree populations. In: *Evolutionary Ecology of Plants*, Edited by J. Bock and Y. B. Linhart. Westview Press, Boulder, Colo. [3, 16]

Hamrick, J. L., Y. B. Linhart and J. B. Mitton, 1979 Relationships between life history characteristics and electrophoretically-detectable genetic variation in plants. Annu. Rev. Ecol. Syst. 10: 173 200. [3, 16]

Hamrick, J. L., J. B. Mitton and Y. B. Linhart, 1981 Levels of genetic variation in trees: Influence of life history characteristics. pp. 35–41. In: *Isozymes of North American Forest Trees and Forest Insects*, Edited by M. T. Conkle, Pacific SW For. Range Expt. Sta. Tech. Report #48, Berkeley. [3, 16]

Hamrick, J. L., H. M. Blanton and K. J. Hamrick, (in press-a) Genetic structure of geographically marginal populations of ponderosa pine. Am. J. Bot. [3]

Hamrick, J. L., G. B. Griswold and M. J. Godt, (in press-b) Association between Slatkin's measure of gene flow and the dispersal ability of plant species. Am. Nat. [3]

Hanson, W. D., 1959 The theoretical distribution of lengths of parental gene blocks in the gametes of an F_1 individual. Genetics 44: 197–209. [19]

Harding, J., and R. W. Allard, 1969 Population studies in predominantly self-pollinated species. XII. Interactions between loci affecting fitness in a population of *Phaseolus lunatus*. Genetics 61: 721–736. [10]

Harding, J., and C. B. Mankinen, 1972 Genetics of *Lupinus*. IV. Colonization and genetic variability in *Lupinus succulentus*. Theor. Appl. Genet. 42: 267–271. [15]

Harding, J., and C. L. Tucker, 1964 Quantitative studies on mating systems. 1. Evidence for the non-randomness of outcrossing in *Phaseolus lunatus*. Heredity 19: 369–381. [9]

Harding, J., R. W. Allard and D. G. Smeltzer, 1966 Population studies in predominantly self-pollinated species. IX. Frequency dependent selection in *Phaseolus lunatus*. Proc. Natl. Acad. Sci. U.S.A. 56: 99–104. [12]

Harju, A., K. Kärkkäinen and O. Muona, 1987 Pollen migration into a Scots pine seed orchard. Hereditas 106: 3. [16]

Harlan, J. R., 1984 Evaluation of wild relatives of crop plants. pp. 212–224. In: *Crop Genetic Resources: Conservation and Evaluation*, Edited by J. H. W. Holden and J. T. Williams. Allen and Unwin, London. [21]

Harlan, J. R., and J. M. J. de Wet, 1971 Toward a rational classification of cultivated plants. Taxon 20: 509–517. [21]

Harmsen, R., F. A. Bliss, C. Cardona, C. E. Posso and T. C. Osborn, 1988 Transferring genes for arcelin protein from cultivated beans: Implications for bruchid resistance. Annu. Rpt. Bean Imp. Coop. 31: 54–55. [18]

Harpending, H. C., 1973 Discussion of relationship of conditional kinship to *apriori* kinship, following paper by N. E. Morton. pp. 78–79. In: *Genetic Structure of Populations*, Edited by N. E. Morton. University of Hawaii Press, Honolulu. [14]

Harper, J. L., 1977 *Population Biology of Plants*. Academic Press, New York. [11]

Harry, I. B., and D. D. Clarke, 1986 Race specific resistance in groundsel (*Senecio vulgaris*) to the powdery mildew *Erysiphe fischeri*. New Phytol. 103: 167–175. [12]

Hart, G. E., 1983 Hexaploid wheat (*Triticum aestivum* L. em Thell). pp. 35–56. In: *Isozymes in Plant Genetics and Breeding, Part. B*, Edited by S. D. Tanksley and T. J. Orton. Elsevier, Amsterdam. [19]

Hartana A., 1983 Genetic variability in seed protein levels associated with two phaseolin protein types in common bean (*Phaseolus vulgaris* L.). M.S. Thesis, University of Wisconsin, Madison. [18]

Hartana, A., 1986 Components of variability for seed protein of common bean (*Phaseolus vulgaris* L.). Ph.D. Thesis, University of Wisconsin, Madison. [18]

Hastings, A., 1981 Disequilibrium, selection and recombination: Limits in two-locus two-allele models. Genetics 98: 659–668. [10]

Hastings, A., 1984 Maintenance of high disequilibrium in the presence of partial selfing. Proc. Natl. Acad. Sci. U.S.A. 81: 4596–4598. [10]

Hastings, A., 1985a Four simultaneously stable polymorphic equilibria in two-locus two-allele models. Genetics 109: 225–261. [10]

Hastings, A., 1985b Stable equilibria at two loci in populations with large selfing rates. Genetics 109: 215–228. [10]

Hastings, A., 1986 Limits to the relationship among recombination, disequilibrium, and epistasis in two locus models. Genetics 113: 177–185. [10]

Haufler, C. H., and D. E. Soltis, 1984 Obligate outcrossings in a homosporous fern: field confirmation of a laboratory prediction. Amer. J. Bot. 71: 878–881. [9]

Hayward, M. D., 1970 Selection and survival in *Lolium perenne*. Heredity 25: 441–447. [12]

Hedge, P. J., and B. G. Spratt, 1985 Resistance to δ-lactam antibiotics by remodeling the active site of an *E. coli* penicillin-binding protein. Nature (London) 318: 478–480. [8]

Hedrick, P. W., 1980 Hitchhiking: A comparison of linkage and partial selfing. Genetics 94: 791–808. [10]

Hedrick, P. W., 1987 Population genetics of intragametophytic selfing. Evolution 41: 137–144. [9]

Hedrick, P. W., and L. Holden, 1979 Hitchhiking: An alternative to coadaptation for the barley and slender wild oat examples. Heredity 43: 79–86. [10]

Hedrick, P. W., S. Jain and L. Holden, 1978 Multilocus systems in evolution. Evol. Biol. 11: 104–184. [10]

Heidrich-Sobrinho, E., and A. R. Cordeiro, 1975 Codominant isoenzymic alleles as markers of genetic diversity correlated with heterosis in maize (*Zea mays* L.). Theor. Appl. Genet. 46: 197–199. [19]

Helbaeck, H., 1966 Commentary on the phylogenesis of *Triticum* and *Hordeum*. Econ. Bot. 20: 350–360. [5]

Helentjaris, T., G. King, M. Slocum, C. Siedenstang and S. Wegman, 1985 Restriction fragment polymorphisms as probes for plant diversity and their development as tools for applied plant breeding. Plant. Mol. Biol. 5: 109–118. [19]

Helentjaris, T., M. Slocum, S. Wright, A. Schaeffer and J. Nienhuis, 1986 Construction of genetic linkage maps in maize and tomato using restriction fragment length polymorphisms. Theor. Appl. Genet. 72: 761–769. [6]

Helentjaris, T., D. Weber and S. Wright, 1988 Identification of the genomic locations of duplicate nucleotide sequences in maize by analysis of restriction fragment length polymorphisms, Genetics 118: 353–363. [19]

Helenurm, K., and F. R. Ganders, 1985 Adaptive radiation and genetic differentiation in Hawaiian *Bidens*. Evolution 39: 753–765. [15]

Heuch, I., 1980 Loss of incompatibility types in finite populations of the heterostylous plant *Lythrum salicaria*. Hereditas 92: 53–57. [15]

Heywood, J. S., 1986 The effect of plant size variation on genetic drift in populations of annuals. Am. Nat. 127: 851–861. [15]

Hiebert, R. D., and J. L. Hamrick, 1983 Patterns and levels of genetic variation in Great Basin bristlecone pine, *Pinus longaeva*. Evolution 37: 302–311. [3, 16]

Higgins, T. J. V., 1984 Synthesis and regulation of major proteins in seeds. Annu. Rev. Plant Physiol. 35: 191–221. [4]

Hilder, V. A., A. M. R. Gatehouse, S. E. Sheerman, R. F. Barker and D. Boulte, 1987 A novel mechanism of insect resistance engineered into tobacco. Nature (London) 330: 160–163. [8]

Hill, R. E., and N. D. Hastie, 1987 Accelerated evolution in the reactive center regions of serine protease inhibitors. Nature (London) 326: 96–99. [8]

Hill, W. G., and P. D. Keightley, 1988 Interrelations of mutation, population size, artificial and natural selection. pp. 57–70. In: *Proceedings of the Second International Conference on Quantitative Genetics*, Edited by B. S. Weir, M. M. Goodman, E. J. Eisen and G. Namkoong. Sinauer, Sunderland, Mass. [20]

Hill, W. G., and A. Robertson, 1968 Linkage disequilibrium in finite populations. Theor. Appl. Genet. 38: 226–231. [10]

Hill, W. G., and B. S. Weir, 1988 Variances and covariances of squared linkage disequilbrium in finite populations. Theor. Pop. Biol. 33: 54–78. [6]

Hirschberg, J., and L. McIntosh, 1983 Molecular basis of herbicide resistance in *Amaranthus hybridus*. Science 222: 1346–1349. [8]

Holden, J. H. W., 1984 The second ten years, pp. 227–285. In: *Crop Genetic Resources: Conservation and Evaluation*, Edited by J. H. W. Holden and J. T. Williams. Allen and Unwin, London. [21]

Holden, J. H. W., and J. T. Williams, 1984 *Crop Genetic Resources: Conservation and Evaluation*. Allen and Unwin, London. [21]

Holden, L. R., 1979 New properties of the two-locus partial selfing model with selection. Genetics 93: 217–236. [10]

Holland, J., K. Spinkler, F. Horodyski, E. Grabau, S. Nichol and S. Vandepol, 1982 Rapid evolution of RNA genomes. Science 215: 1577–1585. [8]

Holsinger, K. E., 1987 Gametophytic self-fertilization in homosporous plants: Development, evaluation and application of a statistical method for evaluating its importance. Am. J. Bot. 74: 1173–1183. [9]

Holwerda, B. C., S. Jana and W. L. Crosby, 1986 Chloroplast and mitochondrial DNA variation in *Hordeum vulgare* and *Hordeum spontaneum*. Genetics 114: 1271–1291. [6]

Honma, S., and O. Heecht, 1959 Bean interspecific hybrid between *Phaseolus vulgaris* and *P. lunatus*. J. Hered. 50: 223–237. [18]

Horovitz, A., and J. Harding, 1972 The concept of male outcrossing in hermaphrodite higher plants. Heredity 29: 223–236. [9]

Hudspeth, R. L., and J. W. Grula, In press Sequence variation in the maize gene encoding the phospheonolpyruvate carboxylase isozyme involved in C_4 photosynthesis. Molec. Gen. Genet. [6]

Hughes, A. L., and M. Nei, 1988 Pattern of nucleotide substitution at major histocompatibility complex class I loci reveals overdominant selection. Nature (London) 335: 167–170. [8]

Hughes, A. L., and M. Nei, 1989 Nucleotide substitution at class II MHC loci: Evidence for overdominant selection. Proc. Natl. Acad. Sci. U.S.A. 86: 958–962. [8]

Hunter, R. B., and L. W. Kannenberg, 1971 Isozyme characterization of corn (*Zea mays* L.) inbreds and its relation to single cross hybrid performance. Can. J. Genet. Cytol. 13: 649–655. [19]

Hurcade, D., D. Dressler and J. Wolfson, 1973 The amplification of ribosomal RNA genes involves a rolling circle intermediate. Proc. Natl. Acad. Sci. U.S.A. 70: 2926–2930. [5]
Huston, M., 1979 A general hypothesis of species diversity. Am. Nat. 113: 81–101. [15]

Iltis, H. H., J. P. Doebley, R. M. Guzman and B. Pazy, 1979 *Zea diploperennis* (Gramineae): A new teosinte from Mexico. Science 203: 186–187. [21]
Imaizumi, Y., N. E. Morton, and D. E. Harris, 1970 Isolation by distance in artificial populations. Genetics 66: 569–582. [14]
Imam, A. G., and R. W. Allard, 1965 Population studies in predominantly self-pollinated species. VI. Genetic variability between and within natural populations of wild oats from differing habitats in California. Genetics 51: 49–62. [12]
Ingram, C. B., and J. T. Williams, 1984 *In situ* conservation of wild relatives of crops. pp. 163–179. In: *Crop Genetic Resources: Conservation and Evaluation*, Edited by J. H. W. Holden and J. T. Williams. Allen and Unwin, London. [21]
International Board For Plant Genetic Resources, 1985 *Biogeographical Surveying and* in situ *Conservation of Crop Relatives*. IBPGR, Rome. [21]
International Board For Plant Genetic Resources, 1987 *Annual Report*, 1986. IBPGR, Rome. [21]
International Board For Plant Genetic Resources, 1988 *Annual Report*, 1987. IBPGR, Rome. [21]

Jackson, L. F., A. L. Kahler, R. K. Webster and R. W. Allard, 1978 Conservation of scald resistance in barley cross populations. Phytopathology 68: 645–650. [13]
Jacquard, A., 1973 A distance between individuals whose genealogies are known. pp. 82–86. In: *Genetic Structure of Populations*, Edited by N. E. Morton. University of Hawaii Press, Honolulu. [14]
Jacquard, A., 1974 *The Genetic Structure of Populations*. Springer-Verlag, New York. [11]
Jain, S. K., 1975 Genetic reserves. pp. 379–396. In: *Crop Genetics Resources for Today & Tomorrow*, Edited by O. H. Frankel and J. G. Hawkes. Cambridge University Press, Cambridge. [21]
Jain, S. K., 1976 The evolution of inbreeding in plants. Annu. Rev. Ecol. Syst. 7: 469–495. [15]
Jain, S. K., 1984 Breeding systems and the dynamics of plant populations. Proc. Intl. Congr. Genetics (New Delhi) 4: 291–316. [9, 15]
Jain, S. K., and R. W. Allard, 1960 Population studies in predominantly self-pollinated species. I. Evidence for heterozygote advantage in a closed population of barley. Proc. Natl. Acad. Sci. U.S.A. 46: 1373–1377. [12]
Jain, S. K., and R. W. Allard, 1966 The effects of linkage, epistasis and inbreeding on population changes under selection. Genetics 53: 633–659. [10]
Jain, S. K., and A. D. Bradshaw, 1966 Evolutionary divergence among adjacent plant populations. I. The evidence and its theoretical analysis. Heredity 21: 407–441. [12]
Jain, S. K., and P. S. Martins, 1979 Ecological genetics of the colonizing ability of rose clover (*Trifolium hirtum* All.). Am. J. Bot. 66: 361–366. [9, 15]
Jain, S. K., and K. N. Rai, 1974 Population biology of *Avena*. IV. Polymorphism in small populations of *Avena fatua*. Theor. Appl. Genet. 44: 7–11. [15]
Jain, S. K., and C. A. Suneson, 1966. Increased recombination and selection in barley populations carrying a male sterility factor. I. Quantitative variability. Genetics 54: 1215–1224. [9]
Jana, S., and S. N. Acharya, 1981 The value of mass reservoirs in genetic conservation. pp. 51–57. In: *Barley Genetics IV. Proc. Fourth Int. Barley Genetics Symp.* Edinburgh, Scotland. [21]
Jana, S., and B. S. Khangura, 1986 Conservation of diversity in bulk populations of barley (*Hordeum vulgare* L.). Euphytica 35: 761–776. [21]
Jana, S., and L. N. Pietrzak, 1988 Comparative assessment of genetic diversity in wild and primitive cultivated barley in a center of diversity. Genetics 119: 981–990. [10]
Jarosz, A. M., and M. Levy, 1988 Effect of habitat and population structure on powdery mildew epidemics in experimental *Phlox* populations. Phytopathology 78: 358–352. [13]
Jayakar, S. D., 1970a A mathematical model for interaction of gene frequencies in a parasite and

its host. Theor. Pop. Biol. 1: 140–164. [13]

Jayakar, S. D., 1970b On the detection and estimation of linkage between a locus influencing a quantitative character and a marker locus. Biometrics 26: 451–464. [19]

Jeffreys, A. J., V. Wilson and S. L. Thein, 1985 Hypervariable "minisatellite" regions in human DNA. Nature (London) 314: 69–73. [6]

Jeffreys, A. J., V. Wilson and S. L. Thein, 1985 Individual-specific fingerprints of human DNA. Nature (London) 316: 76–79. [9, 11]

Jeger, M. J., G. D. Jones and E. Griffiths, 1981 Disease progress of non-specialised fungal pathogens in intraspecific mixed stands of cereal cultivars. II. Field experiments. Ann. Appl. Biol. 98: 199–210. [13]

Jensen J., J. H. Jorgensen, Y. P. Jensen, H. Gliese and H. Doll, 1980 Linkage of the hordein loci *Hor1 and Hor2* with the powdery mildew resistance loci *M1-k* and *M1-a* on barley chromosome 5. Theor. Appl. Genet. 58: 27–31. [4]

Jensen, N. F., 1988 *Plant Breeding Methodology*. Wiley, New York. [17]

Jensen, V., 1984 Legumin-like and vicilin-like storage proteins in *Nigella damascena* (Ranunculaceae) and six other dicotyledonous species. J. Plant Physiol. 115: 161–170. [4]

Jinks, J. L., H. S. Pooni and M. K. U. Chowdhury, 1985 Detection of linkage and pleiotropy between characters of *Nicotiana tabacum* using inbred lines produced by dihaploidy and single seed descent. Heredity 55: 327–333. [10]

Jinks-Robertson, S., and T. D. Petes, 1985 High-frequency meiotic gene conversion between repeated genes on nonhomologous chromosomes in yeast. Proc. Natl. Acad. Sci. U.S.A. 82: 3350–3354. [5]

Johanningmeier, U., and R. B. Hallick, 1987 The *psbA* gene of DCMU-resistant *Euglena gracilis* has an amino acid substitution at serine codon 265. Curr. Genet. 12: 465–470. [8]

Johns, M. A., J. N. Strommer and M. Freeling, 1983 Exceptionally high levels of restriction site polymorphism in DNA near the *Adh1* gene. Genetics 105: 733–743. [6]

Johnson, B. L., 1967 Confirmation of the genome donors of *Aegilops cylindrica*. Nature (London) 216: 859–852. [4]

Johnson, E. D., J. Knight and K. R. Gayler, 1985 Biosynthesis and processing of legumin-like storage proteins in *Lupinus angustifolius* (lupin). Biochem. J. 283: 673–679. [4]

Jones, D. A., 1962 Selection of the acyanogenic form of the plant *Lotus corniculatus* L. by various animals. Nature (London) 193: 1109–1110. [11]

Jones, R. N. and H. Rees, 1967 Genotypic control of chiasma behaviour in rye. XI. The influence of B chromosomes on meiosis. Heredity 22: 333–347. [12]

Jones, R. N. and H. Rees, 1982 *B Chromosomes*. Academic Press, London. [12]

Jonsson, A., I. Ekberg and G. Eriksson, 1976 Flowering in a seed orchard of *Pinus sylvestris* L. Stud. For. Suecica 135: 1–38. [16]

Jorde, L. B., A. W. Eriksson, K. Morgan and P. L. Workman, 1982 The genetic structure of Iceland. Hum. Hered. 32: 1–7. [14]

Kahler, A. L., 1985 Associations between enzyme marker loci and agronomic traits in maize. Proc. 40th Annu. Corn Sorghum Res. Conf., Am. Seed Trade Assoc. 40: 66–89. [19]

Kahler, A. L., and C. F. Wehrhahn, 1986 Associations between quantitative traits and enzyme loci in the F_2 population of a maize hybrid. Theor. Appl. Genet. 72: 15–26. [19]

Kahler, A. L., R. W. Allard, M. Krzakowa, C. F. Wehrhahn and E. Nevo, 1980 Associations between enzyme phenotypes and environment in the slender wild oat (*Avena barbata*) in Israel. Theor. Appl. Genet. 59: 101–111. [21]

Kalisz, S., 1986 Variable selection on the timing of germination in *Collinsia verna* (Scrophulariaceae). Evolution 40: 479–491. [12]

Kang, H., and H. Nienstaedt, 1987 Managing long-term tree breeding stock. Silv. Genet. 36: 30–39. [16]

Karlin, S., 1975 General two-locus selection models: Some objectives, results and interpretations. Theor. Pop. Biol. 7: 399–421. [10]

Karlin, S., and M. W. Feldman, 1970 Linkage and selection: Two locus symmetric viability model. Theor. Pop. Biol. 1: 39–71. [10]

Kasarda, D. D., J. E. Bernardin and C. O. Qualset, 1976 Relationship of gliadin protein components to chromosomes in hexaploid wheat (*Triticum aestivum* L.). Proc. Natl. Acad. Sci.

U.S.A. 73: 3646–3650. [4]
Kato, A., K. Yakura and S. Tanifuji, 1982 Organization of ribosomal DNA in the carrot. Plant Cell Physiol. 23: 151–154. [6]
Keddy, P. A., 1983 Shoreline vegetation in Axe Lake, Ontario: Effects of exposure on zonation patterns. Ecology 64: 331-344. [15]
Keeler, K. H., 1978 Intrapopulation differentiation in annual plants. II. Electrophoretic variation in *Veronica peregrina*. Evolution 32: 638–645. [15]
Keightley, P. D., and H. Kacser, 1987 Dominance, pleitrophy and metabolic structure. Genetics 117: 319–329. [20]
Kelly, J. F., 1971 Genetic variation in the methionine levels of mature seeds of common bean (*Phaseolus vulgaris* L.). J. Am. Soc. Hort. Sci. 96: 561–563. [18]
Kenrick, J., and R. B. Knox, 1982 Function of the polyad in reproduction of Acacia. Ann. Bot. 50: 721–727. [16]
Kenworthy, W. J., and C. A. Brim, 1979 Recurrent selection in soybeans. I. Seed yield. Crop Sci. 19: 315–318. [17]
Kesseli, R. W., and S. K. Jain, 1985 Breeding systems and population structure in *Limnathes*. Theor. Appl. Genet. 71: 292–299. [9, 11]
Kihara, H., 1944 Discovery of the DD-analyser, one of the ancestors of *Triticum vulgare* (Japanese). Agr. Hort. (Tokyo) 19: 13–14. [5]
Kihara, H., and N. Kondo, 1943 Studies on amphiploids of *Aegilops caudata* x *Ae. umbelluluta* induced by colchicine. Seiken Ziho 2: 24–42. [5]
Kihara, H., and S. Matsumura, 1941 Genomanalyse bei Triticum und Aegilops. VIII. Ruckkreuzung des Bastardes Ae. caudata x Ae. cylindrica zu Eltern und seine Nachkommen. Cytologia 11: 493–506. [5]
Kim, H. K., 1987 Molecular characterization of a functional mutation caused by insertion of a defective Suppressor-mutator transposable element in the *bzl* Locust of maize. Ph.D. Thesis, University of Wisconsin, Madison. [7]
Kim, H. Y., J. W. Schiefelbein, V. Raboy, D. B. Furtek and O. E. Nelson, 1987 RNA splicing permits expression of a maize gene with a defective Suppressor-mutator transposable element insertion in an exon. Proc. Natl. Acad. Sci. U.S.A. 84: 5863–5867. [7]
Kimura, M., 1968 Evolutionary rate at the molecular level. Nature (London) 217: 624–626. [8]
Kimura, M., 1983 *The Neutral Theory of Molecular Evolution*. Cambridge University Press, Cambridge. [8]
Kimura, M., and J. F. Crow, 1963 The measurement of effective population numbers. Evolution 17: 279–288. [15]
Kimura, M., and J. F. Crow, 1964 The number of alleles that can be maintained in a finite population. Genetics 49: 725–738. [8]
Kimura, M., and H. Kayano, 1961 The maintenance of supernumary chromosomes in wild populations of *Lilium callosum* by preferential segregation. Genetics 46: 1699–1712. [12]
Kimura, M., and T. Ohta, 1979 Population genetics of multigene family with special reference to decrease of genetic correlation with distance between gene members on a chromosome. Proc. Natl. Acad. Sci. U.S.A. 76: 4001–4005. [5]
Kimura, M., and G. H. Weiss, 1964 The stepping stone model of population structure and the decrease of genetic correlation with distance. Genetics 49: 561–576. [14, 15]
King, J. N., B. P. Dancik and N. K. Dhir, 1984 Genetic structure and mating system of white spruce (*Picea glauca*) in a seed production area. Can. J. For. Res. 14: 639–643. [16]
Kingman, J. F. C., 1978 A simple model for the balance between selection and mutation. J. Appl. Prob. 15: 1–12. [20]
Kirby, G. C., 1984 Breeding systems and heterozygosity in populations of tetrad forming fungi. Heredity 52: 35–41. [9]
Klein, A. S., M. Clancy, L. Paje-Manalo, D. B. Furtek, L. C. Hannah and O. E. Nelson, 1988 The mutation *bronze-mutable 4 Derivative* 6856 is caused by the insertion of a novel 6.7 kilobase pair transposonin the untranslated leader region of the *Bronze-1* gene. Genetics 120: 779–790. [7]
Klein, J., 1986 *Natural History of the Major Histocompatibility Complex*. Wiley, New York. [8]
Klein, J., and F. Figueroa, 1986 Evolution of the major histocompatibility complex. CRC Crit. Rev. Immunol. 6: 295–386. [8]
Klitz, W., and G. Thomson, 1987 Disequilibrium pattern analysis. II. Application to Danish

HLA*a* and *B* locus data. Genetics 116: 633–643. [10]
Kloppenburg, J. R., Jr., 1988 *Seeds and Sovereignity: The Use and Control of Plant Genetic Resources*. Duke University Press, Durham, N.C. [21]
Knapp, S. J., and T. S. Cox, 1988 S_1 family recurrent selection in autogamous crops based on dominant genetic male sterility. Crop Sci. 28: 277–231. [17]
Knight, S. E., and D. M. Waller, 1987 Genetic consequences of outcrossing in the cleistogamous annual, *Impatiens capensis*. I. Population-Genetic structure. Evolution 41: 969–978 [15]
Knowles, P., G. R. Furnier, M. A. Aleksuik and D. J. Perry, 1987 Significant levels of self-fertilization in natural populations of tamarack. Can. J. Bot. 65: 1087–1091. [9, 12, 16]
Koenig, R., S. P. Singh and P. Gepts, in press Novel phaseolin types in wild and cultivated common bean (*Phaseolus vulgaris*, Fabaceae). Econ. Bot. [4]
Koski, V., 1970 A study of pollen dispersal as a mechanism of gene flow. Commun. Inst. For. Fenn. 70: 1–78. [16]
Koski, V., 1971 Embryonic lethals of *Picea abies* and *Pinus sylvestris*. Commun. Inst. For. Fenn. 75: 1–30. [16]
Koski, V., 1980 On the variation of flowering and seed crop in mature stands of *Pinus sylvestris* L. Silv. Fenn. 14: 71–75. [16]
Koski, V., and O. Muona, 1986 Probability of inbreeding in relation to clonal differences in male flowering and embryonic lethals. pp. 391–400. In: *Proc. IUFRO Meeting of Working Groups on Breeding Theory, Progeny testing and Seed orchards*, Edited by A. V. Hatcher and R. J. Weir, Williamsburg, VA. [16]
Kreis, M., P. R. Shewry, B. G. Forde and B. J. Miflin, 1985 Structure and evolution of seed storage proteins and their genes with particular reference to those of wheat, barley and rye. Oxford Surv. Plant Mol. Cell Biol. 2: 253–317. [4]
Kreitman, M., 1983 Nucleotide polymorphism at the alcohol dehydrogenase locus of *Drosophila melanogaster.* Nature (London) 304: 412–417. [6, 8]
Krishna Iyer, P. V., 1949 The first and second moments of some probability distributions arising from points on a lattice and their applications. Biometrika 36: 135–141. [14]
Krishna Iyer, P. V., 1950 The theory of probability distributions of points on a lattice. Ann. Math. Stat. 21: 198–217. [14]
Kulshrestha, V. P., and S. Chowdhury, 1987 A new selection criterion for yield in wheat. Theor. Appl. Genet. 84: 275–279. [20]

Ladizinsky, G., 1979 The origin of lentil and its wild gene pool. Euphytica 18: 179–187. [4]
Ladizinsky, G., 1985 Founder-effect in crop-plant evolution. Econ. Bot. 39: 191–199. [4]
Ladizinsky, G., and A. Adler, 1975 The origin of chickpea *Cicer arietinum* as indicated by seed protein electrophoresis. Israel J. Bot. 24: 182–187. [4]
Ladizinsky, G., and A. Hamel, 1980 Seed protein profiles of pigeon pea (*Cajanus cajan*) and some *Atylosia* species. Euphytica 20: 313–317. [4]
Lagudah, E. S., and G. M. Halloran, 1988a Phylogenetic relationships of *Triticum tauschii*, the D genome donor to hexaploid wheat. I. Variation in HMW subunits of glutelin and Gliadin. Theor. Appl. Genet. 75: 592–598. [4]
Lagudah, E. S., and G. M. Halloran, 1988b Phylogenetic relationships of *Triticum tauschii*, the D genome donor to hexaploid wheat. 2. Inheritance and chromosomal mapping of the HMW subunits of glutenin and gliadin gene loci of *T. tauschii*. Theor. Appl. Genet. 75: 599–605. [4]
Lamb, E. M., and L. L. Hardman, 1985 Survey of bean varieties grown in Rwanda. US-AID/Univ. of Minnesota and Govt. of Rwanda Misc. Public Dept. of Agronomy and Plant Breeding, University of Minnesota, St. Paul. [18]
Lamkey, K. R., A. R. Hallauer, 1987 Allelic differences at enzyme loci and hybrid performance in maize. J. Hered. 78: 231–234. [19]
Lande, R., 1976 The maintenance of genetic variability by mutation in a polygenic character with linked loci. Genet. Res. 26: 221–235. [20]
Lande, R., 1979 Quantitative genetic analysis of multivariate evolution, applied to brain:body size allometry. Evolution 33: 402–416. [20]
Lande, R., 1980 Genetic variation and phenotypic evolution during allopatric speciation. Am. Nat. 116: 463–479. [15]
Lande, R., 1981 The minimum number of genes contributing to quantitative variation between

and within populations. Genetics 99: 541–553. [20]

Lande, R., 1982 A quantitative genetic theory of life history evolution. Ecology 63: 607–615. [11]

Lande, R., 1988 Genetics and demography in biological conservation. Science 241: 1455–1460. [16]

Lande, R., and S. J. Arnold, 1983 The measurement of selection on correlated characters. Evolution 37: 1210–1227. [11, 12]

Lande, R., and D. W. Schemske, 1985 The evolution of self-fertilization and inbreeding depression in plants. I. Genetic models. Evolution 39: 24–40. [11, 12]

Lander, E. S., and D. Botstein, 1989 Mapping Mendelian factors underlying quantitative traits using RFLP linkage maps. Genetics 121: 185–199. [11, 19]

Landry, B. S., R. Kesseli, H. Leung and R. W. Michelmore, 1987a Comparison of restriction endonucleases and sources of probes for their efficiency in detecting restriction fragment length polymorphisms in lettuce (*Lactuca sativa* L.). Theor. Appl. Genet. 74: 646–653. [6]

Landry, B. S., R. V. Kesseli, B. Farrara and R. W. Michelmore, 1987b A genetic map of lettuce (*Lactuca sativa* L.) with restriction fragment length polymorphism, isozyme, disease resistance and morphological markers. Genetics 116: 331–337. [6, 19, 20]

Landsmann, J., and H. Uhrig, 1985 Somaclonal variation in *Solanum tuberosum* detected at the molecular level. Theor. Appl. Genet. 71: 500–505. [5]

Langley, C. H., and C. F. Aquadro, 1987 Restriction-map variation in natural populations of *Drosophila melanogaster: White*-locust region. Mol. Biol. Evol. 4: 651–663. [6, 8]

Langley, C. H., A. E. Schrimpton, T. Yamazaki, N. Miyashita, Y. Matsuo and C. F. Aquadro, 1988 Naturally occurring variation in the restriction map of the *Amy* region of *Drosophila melanogaster.* Genetics 119: 619–629. [6]

Larkins, B. A., 1983 Genetic engineering of seed storage proteins. pp. 255–304. In: Genetic Engineering of Plants, Edited by T. Kosuge, C. P. Meredith, and A. Hollaender. Plenum, New York. [4]

Laroche-Raynal, M., and M. Delseny, 1986 Identification and characterization of the mRNA for major storage proteins from radish. Eur. J. Biochem. 157: 321–327. [4]

Larson, R. L., and E. H. Coe, 1977 Gene-dependent flavonoid glucosyltransferase in maize. Biochem. Genet. 15: 153–156. [7]

Laskowski, M., I. Kato, W. J. Kohr, S. J. Park, M. Tashiro and H. E. Whatley, 1987 Positive darwinian selection in evolution of protein inhibitors of serine proteinases. Cold Spring Harbor Symp. Quant. Biol. 52: 545–553. [8]

Lassner, M. W., and J. Dvořák, 1985 Organization of the 5S rRNA family in wheat. J. Cell. Biochem., Suppl. 9C p. 219. [5]

Lassner, M., and J. Dvořák, 1986 Preferential homogenization between adjacent and alternate subrepeats in wheat rDNA. Nucl. Acids Res. 14: 5499–5512. [5]

Lassner, M., O. Anderson and J. Dvořák, 1987 Hypervariation associated with a 12-nucleotide direct repeat and inferences in intergenomic homogenization of ribosomal RNA gene spacers based on the DNA sequence of a clone from the wheat *Nor-D3* locus. Genome 29: 770–781. [5]

Lawrence, G. J., and K. W. Shepherd, 1981a Inheritance of glutenin protein subunits of wheat. Theor. Appl. Genet. 60: 333–337. [4]

Lawrence, G. J., and K. W. Shepherd, 1981b Chromosomal location of genes controlling seed proteins in species related to wheat. Theor. Appl. Genet. 59: 25–31. [4]

Lawrence, M. J., 1984 The genetic analysis of ecological traits. pp. 27–63. In: *Evolutionary Ecology*, Edited by B. Shorrocks, Blackwell, Oxford. [12]

Lawrence, T., 1984 *Collection of Crop Germplasm: The First Ten Years 1974–84*. IBPGR, Rome. [21]

Learn, G. H., and B. A. Schaal, 1987 Population subdivision for ribosomal DNA repeat variants of *Clematis fremontii*. Evolution 41: 433–437. [6]

Lebeda, A., and T. Jendoulek, 1987 Cluster analysis as a method for evaluation of genetic similarity in specific host-parasite interation (*Lactuca sativa–Bremia lactucae*). Theor. Appl. Genet. 75: 194–199. [20]

Lebowitz, R. J., M. Soller and J. S. Beckmann, 1987 Trait-based analyses for the detection of linkage between marker loci and quantitative trait loci in crosses between inbred lines. Theor. Appl. Genet. 73: 556–562. [19]

Ledig, F. T., 1986 Conservation strategies for forest gene resources. For. Ecol. Managem. 14: 77–90. [16]
Ledig, F. T., and M. T. Conkle, 1983 Gene diversity and genetic structure in a narrow endemic, Torrey pine (*Pinus torreyana* Parry ex Carr). Evolution 37: 70–85. [15, 16]
Lefort, P. L., and N. Legisle, 1977 Quantitative stock-scion relationships in vine. Preliminary investigations by the analysis of reciprocal graftings. Vitis 16: 149–161. [20]
Leonard, K. J., 1969a Factors affecting rates of stem rust increase in mixed plantings of susceptible and resistant oat varieties. Phytopathology 59: 1845–1850. [13]
Leonard, K. J., 1969b Selection in heterogeneous populations of *Puccinia graminis* f. sp. *Avenae*. Phytopathology 59: 1851–1857. [13]
Leonard, K. J., 1977 Selection pressures and plant pathogens. Ann. N.Y. Acad. Sci. 287: 207–222. [13]
Leonard, K. J., 1984 Population genetics of gene-for-gene interactions between plant host resistance and pathogen virulence. pp. 131–148. In: *Proc.XV Int. Cong. Genetics*, New Delhi, 1983. Oxford and IBH, New Delhi. [13]
Leonard, K. J., and C. C. Mundt, 1984 Methods for estimating epidemiological effects of quantitative resistance to plant disease. Theor. Appl. Genet. 67: 219–230. [20]
Levin, D. A., 1975 Genic heterozygosity and protein polymorphism among local populations of *Oenothera biennis*. Genetics 79: 477–491. [15]
Levin, D. A., 1977 The organization of genetic variability in *Phlox drummondii*. Evolution 38: 477–494. [15]
Levin, D. A., 1988 Consequences of stochastic elements in plant migration. Am. Nat. 132: 643–651. [15]
Levin, D. A., and H. W. Kerster, 1974 Gene flow in seed plants. Evol. Biol. 7: 139–220. [16]
Levy, A. A., and M. Feldman, 1988 Biogeographical distribution of HMW glutenin alleles in populations of the wild tetraploid wheat *Triticum turgidum* var. *dicoccoides*. Theor. Appl. Genet. 75: 651–658. [4]
Lewontin, R. C., 1965 Selection for colonizing ability. pp. 79–92. In: *The Genetics of Colonizing Species*, Edited by H. G. Baker and G. L. Stebbins. Academic Press, New York. [15]
Lewontin, R. C., 1974 *The Genetic Basis of Evolutionary Change*. Columbia University Press, New York. [3]
Lewontin, R. C., and J. L. Hubby, 1966 A molecular approach to the study of genic heterozygosity in natural populations. II. Amount of variation and degree of heterozygosity in natural populations of *Drosophila pseudoobscura*. Genetics 54: 595–609. [6]
Lewontin, R. C., and K. Kojima, 1960 The evolutionary dynamics of complex polymorphisms. Evolution 14: 458–472. [10]
Li, W. H., and M. Nei, 1975 Drift variances of heterozygosity and genetic distance in transient states. Genet. Res. 25: 229–248. [2]
Lillis, M., and M. Freeling, 1986 *Mu* transposons in maize. Trends Genet. 2: 183-188. [7]
Lin, L. S., T. H. D. Ho and J. R. Harlan, 1985 Rapid amplification and fixation of new restriction sites in the ribosomal DNA repeats in the derivatives of a cross between maize and *Tripsacum dactyloides*. Dev. Genet. 6: 101–112. [5]
Linhart, Y. B., 1974 Intra-population differentiation in annual plants. I. *Veronica peregrina* L. raised under non-competitive conditions. Evolution 28: 232–243. [15]
Linhart, Y. B., J. B. Mitton, D. M. Bowman and K. B. Sturgeon, 1979 Genetic aspects of fertility differentials in ponderosa pine. Genet. Res. 33: 237–242. [16]
Linhart, Y. B., J. B. Mitton, K. B. Sturgeon and M. L. Davis, 1981 Genetic variation in space and time in a population of ponderosa pine. Heredity 46: 407–426. [3]
Link, G., and U. Langridge, 1984 Structure of the chloroplast gene for the precursor of the M_r 32,000 photosystem II protein from mustard (*Sinapsis alba* L.). Nucl. Acids Res. 12: 945–958. [8]
Lioi, L., and R. Bollini, 1984 Contribution of processing events to the molecular heterogeneity of four banding types of phaseolin, the major storage protein of *Phaseolus vulgaris* L. Plant Mol. Biol. 3: 345–353. [4]
Lloyd, D. G., 1980 Benefits and handicaps of sexual reproduction. Evol. Biol. 15: 69–111. [12]
Loveless, M. D., and J. L. Hamrick, 1984 Ecological determinants of genetic structure in plant populations. Annu. Rev. Ecol. Syst. 15: 65–95. [3, 16]
Lovett Doust, L., 1981 Population dynamics and local specialisation in a clonal perennial (*Ra-*

nunculus repens). II. The dynamics of leaves and a reciprocal transplant-replant experiment. J. Ecol. 69: 757–768. [12]

Lowry, T. K., and D. J. Crawford, 1983 Allozyme divergence and evolution in *Tetramolopium* (Compositae: Astereae) on the Hawaiian Islands. Sys. Bot. 10: 64–72. [15]

Luckett, D. D., and K. J. R. Edwards, 1986 Esterase genes in parallel composite cross barley populations. Genetics 114: 289–302. [10]

Luedders, V. D., 1978 Effect of planting date on natural selection in soybean populations. Crop Sci. 18: 943–944. [12]

Luig, N. H., and I. A. Watson, 1970 The role of barley, rye and grasses in the 1973–74 wheat stem rust epiphytotic and southern and eastern Australia. Proc. Linn. Soc. New South Wales 101: 65–76. [13]

Lundkvist, K., 1979 Allozyme frequency distributions in four Swedish populations of Norway spruce (*Picea abies* K.). 1. Estimations of genetic variation within and among populations, genetic linkage and a mating system parameter. Hereditas 90: 127–143. [16]

Lundkvist, K., and D. Rudin, 1977 Genetic variation in eleven populations of *Picea abies* as determined by isozyme analysis. Hereditas 85: 67–74. [3, 15]

Lycett, G. W., A. J. Delanney, W. Zhao, J. A. Gatehouse, R. R. D. Croy and D. Boulter, 1984 Two cDNA clones coding for the legumins protein of *Pisum sativum* L. contain sequence repeats. Plant Mol. Biol. 3: 91–96. [4]

Lynch, M., and W. G. Hill, 1986 Phenotypic evolution by neutral mutation. Evolution 40: 915–935. [20]

Mackay, T. F. C., 1987 Transposable element-induced polygenic mutations in *Drosophila melanogaster.* Genet. Res. 49: 225–233. [7]

MacKey, J., and C. O. Qualset, 1986 Conventional methods of wheat breeding. pp. 7–24. In: *Genetic Improvement of Yield in Wheat,* Edited by E. L. Smith. Crop Science Society of America, Madison Wisc. [17]

Mahmoud, S. H., and J. A. Gatehouse, 1984 Inheritance and mapping of vicilin storage protein genes in *Pisum sativum* L. Heredity 53: 185–191. [4]

Malécot, G., 1948 *Les mathématiques de l'hérédité.* Masson, Paris. [14]

Malécot, G., 1955 Remarks on decrease of relationship with distance, following paper by M. Kimura. Cold Spring Harbor Symp. Quant. Biol. 20: 52–53. [14]

Malécot, G., 1973 Isolation by distance. pp. 72–75. In: *Genetic Structure of Populations,* Edited by N. E. Morton. University of Hawaii Press, Honolulu. [14]

Manos, P. S., and D. E. Fairbrothers, 1987 Allozyme variation in populations of six northeastern American red oaks (Fagaceae: *Quercus* subg. Erythrobalanus). Syst. Bot. 12: 365–373. [16]

Mansur, Vergara, L., C. F. Konzak, A. K. Gerechter-Amital, A. Grama and A. Blum, 1984 A computer-assisted examination of the storage protein genetic variation in 841 accessions of *Triticum dicoccoides.* Theor. Appl. Genet. 69: 79–86. [4]

Mantel, N., 1967 The detection of disease clustering and a generalized regression approach. Cancer Res. 27: 209–220. [14]

Marshall, D. R., 1977 The advantages and hazards of genetic homogeneity. pp. 1–20. In: *The Genetic Basis of Epidemics in Agriculture,* Edited by P. R. Day. New York Academy of Science, N.Y. [13, 21]

Marshall, D. R., and R. W. Allard. 1970 Isozyme polymorphisms in natural populations of *Avena fatua* and *A. barbata.* Heredity 25: 373–382. [2]

Marshall, D. R., and R. W. Allard, 1970 Maintenance of isozyme polymorphisms in natural populations of *Avena barbata.* Genetics 66: 393–399. [9, 12]

Marshall, D. R., and A. H. D. Brown, 1974 Estimation of the level of apomixis in plant populations. Heredity 32: 321–333. [9]

Marshall, D. R., and A. H. D. Brown, 1975 Optimum sampling strategies in genetic conservation. pp. 53–80. In: *Crop Genetic Resources for Today and Tomorrow,* Edited by O. H. Frankel and J. G. Hawkes. Cambridge University Press, Cambridge. [14, 21]

Marshall, D. R., and A. H. D. Brown, 1981 Wheat genetic resources, pp. 21–40. In: *Wheat Science—Today and Tomorrow,* Edited by L. T. Evans and W. J. Peacock. Cambridge University Press, Cambridge. [21]

Marshall, D. R., and A. H. D. Brown, 1983 Theory of forage plant collection. pp. 135–148. In:

Genetic Resources of Forage Plants, Edited by J. G. McIvor and R. A. Bray. CSIRO, Melbourne. [21]
Marshall, D. R., and A. J. Pryor, 1978 Multiline varieties and disease control. I. The "dirty crop" approach with each component carrying a unique single resistance gene. Theor. Appl. Genet. 51: 177–184. [13]
Marshall, D. R., and A. J. Pryor, 1979 Multiline varieties and disease control. II. The "dirty crop" approach with components carrying two or more genes for resistance. Euphytica 28: 145–159. [13]
Marshall, D. R., and B. S. Weir, 1979 Maintenance of genetic variation in apomictic plant populations. 1. Single locus models. Heredity 42: 159–172. [9]
Marshall, D. R., and P. W. Weiss, 1982 Isozyme variation within and among Australian populations of *Emex spinosa* (L.) Campd. Aust. J. Biol. Sci. 35: 327–332. [3]
Marshall, D. R., J. J. Burdon and A. H. D. Brown, 1980 Optimal sampling strategies in the biological control of weeds. Proc. V Int. Symp. Biol. Contr. Weeds, Brisbane, Australia, 1980: 103–111. [15]
Marshall, D. R., J. J. Burdon and W. J. Müller, 1986 Multiline varieties and disease control. VI. Effects of selection at different stages of the pathogen life cycle on the evolution of virulence. Theor. Appl. Genet. 71: 801–809. [13]
Marshall, H. G., 1976 Genetic changes in oak bulk populations under winter survival stress. Crop Sci. 16:9–15. [12]
Martens, J. W., R. I. H. McKenzie and G. J. Green, 1970 Gene-for-gene relationships in the *Avena: Puccinia graminis* host-parasite system in Canada. Can. J. Bot. 48: 969–975. [13]
Martin, C., S. MacKay and R. Carpenter, 1988 Large-scale chromosomal restructuring is induced by the transposable element Tam3 at the *nivea* locus of *Antirrhinum majus*. Genetics 119: 171–184. [6]
Martins, P. S., and S. K. Jain, 1979 Role of genetic variation in the colonizing ability of rose clover (*Trifolium hirtum* All.). Am. Nat. 114: 591–595. [15]
Maruyama, T., 1973 Isolation by distance, genetic variability, the time required for a gene substitution, and local differentiation in a finite, geographically structured population. pp. 80–81. In: *Genetic Structure of Populations*, Edited by N. E. Morton. University of Hawaii Press, Honolulu. [14]
Maruyama, T., and M. Nei, 1981 Genetic variability maintained by mutation and overdominant selection in finite populations. Genetics 98: 441–459. [8]
Mather, K., and J. L. Jinks, 1971 *Biometrical Genetics*. Cornell University Press, Ithaca, N.Y. [19]
Matta, N. K., and J. A. Gatehouse, 1982 Inheritance and mapping of storage protein genes in *Pisum sativum*. Heredity 48: 383–392. [4]
Matzinger, D. F., C. C. Cockerham, and E. A. Wernsman, 1977 Single character and index mass selection with random mating in a naturally self-fertilizating species. pp. 503–518. In: *Proceedings of the International Conference on Quantitative Genetics*, Edited by E. Pollack, O. Kempthorne, and T. B. Bailey. Iowa State Univ. Press, Ames. [17]
May, C. E., and R. Appels, 1987 Variability and genetics of spacer DNA sequences between the ribosomal-RNA genes of hexaploid wheat (*Triticum aestivum*). Theor. Appl. Genet. 74: 617–624. [5]
Mayer, S. S., 1987 Breeding systems in Hawaiian *Wikstroemia* Endl. (Thymelaeaceae). Bot. Soc. Am. Abstr. 74: 743. [15]
Maynard Smith, J., 1978 *The Evolution of Sex*. Cambridge University Press, Cambridge. [12]
Mayo, O., 1966 On the problem of self-incompatibility alleles. Biometrics 22: 111–120. [20]
Mayo, O., 1971 Rates of change in gene frequency in tetrasomic organisms. Genetica 42: 329–337. [20]
Mayo, O., 1986 *The Theory of Plant Breeding*. Clarendon Press, Oxford. [17, 20]
Mayo, O., In press Conventional plant breeding and the ' new genetics.' *Proceedings of the SABRAO Symposium on Genetic Engineering in Developing Countries*, University Kebangsan Malaysia, Kuala Lumpur. [20]
Mayo, O., and A. M. Hopkins, 1985 Problems of estimating the minimum number of genes contributing to quantitative variation. Biomet, J. 27: 181–187. [20]
Mayo, O., and C. R. Leach, 1987 Stability of self-incompatibility systems. Theor. Appl. Genet. 74: 788–792. [20]

Mayo, O., and C. R. Leach, In press Quantitatively determined self-incompatibility. I. Theoretical considerations. Theor. Appl. Genet. [20]
Mayo, O., G. R. Fraser and G. Stamatoyannopoulos, 1969 Genetic influences on serum cholesterol in two Greek villages. Hum. Hered. 19: 86–99. [20]
Mayr, E., 1970 *Populations, Species and Evolution*. Harvard University Press, Cambridge, Mass. [14, 15]
Mazer, S. J., 1987 The quantitative genetics of life history and fitness components in *Raphanus raphanistrum* L. (Brassicaceae): ecological and evolutionary consequences of seed-weight variation. Am. Nat. 130: 891–894. [12]
McCauley, D. E., D. P. Whittier and L. M. Reilly, 1985 Inbreeding and the rate of self-fertilization in a grape fern. *Botrychium dissectum*. Am. J. Bot. 72: 1978–1981. [9]
McClintock, B., 1950 The origin and behavior of mutable loci in maize. Proc. Natl. Acad. Sci. U.S.A. 36: 344–355. [7]
McClintock, B., 1954 Mutations in maize and chromosomal aberrations in Neurospora. Carnegie Inst. Wash. Year Book 53: 254–260. [7]
McClintock, B., 1962 Topographical relations between elements of control systems in maize. Carnegie Inst. Wash. Year Book. 61: 448–461. [7]
McClintock, B., 1965 The control of gene action in maize. Brookhaven Symp. Biol. 18: 162–184. [7]
McClintock, B., 1978 Mechanisms that rapidly reorganize the genome. Stadler Symp. 10: 25-48. [7]
McConnell, T. J., W. S. Talbot, R. A. McIndoe and E. K. Wakeland, 1988 The origin of MHC class II gene polymorphism within the genus *Mus*. Nature (London) 332: 651–654. [8]
McFadden, E. S., and E. R. Sears, 1946 The origin of *Triticum spelta* and its free-threshing hexaploid relatives. J. Hered. 37: 81–89. [5]
McFerson, J. R., 1983 Genetic and breeding studies of dinitrogen fixation in common bean (*Phaseolus vulgaris* L.). Ph.D. Thesis, University of Wisconsin, Madison. [18]
McGraw, J. B., and J. Antonovics, 1983 Experimental ecology of *Dryas octopetala* ecotypes I. Ecotypic differentiation and life-cycle stages of selection. J. Ecol. 71: 879–897. [12]
McIntyre, S., and S. C. H. Barrett, 1986 A comparison of weed communities of rice in Australia and California. Proc. Ecol. Soc. Austr. 14: 237–250. [15]
McIvor, J. G., and R. A. Bray, 1983 *Genetic Resources of Forage Plants*. CSIRO, Melbourne. [21]
McMillan, I., and A. Robertson, 1974 The power of methods for detection of major genes affecting quantitative characters. Heredity 32: 349–356. [19, 20]
McNeal, F. H., C. F. McGuire, and M. A. Berg, 1978 Recurrent selection for grain protein content in spring wheat. Crop Sci. 18: 779–782. [17]
McNeill, C. T., and S. K. Jain, 1983 Genetic differentiation and phylogenetic inference in the plant genus *Limnanthes* (section Inflexae). Theor. Appl. Genet. 66: 257–269. [3]
Meagher, T. R., 1986 Analysis of paternity within a natural population of *Chamaelirium leuteum*. I. Identification of most-likely male parents. Am. Nat. 128:P 129–215. [3, 9, 11]
Meagher, T. R., and E. Thompson, 1987 Analysis of parentage for naturally established seedlings of *Chamaelirium luteum* (Liliaceae). Ecology 68: 803–812. [9, 11]
Medina-Filho, H. P., 1980 Linkage of *Aps-1*, *Mi* and other markers on chromosome 6. Rep. Tomato Genet. Coop. 30: 26–28. [19]
Meredith, C. P., 1984 Selecting better crops from cultured cells. pp. 503–528. In: *Gene Manipulation in Plant Improvement*, Edited by J. P. Gustafson. Plenum, New York. [17]
Merkle, S. A., and W. T. Adams, 1987 Patterns of allozyme variation within and among Douglas-fir breeding zones in southwest Oregon. Can. J. For. Res. 17: 402–407. [16]
Mertz, E. T., L. S. Bates and O. E. Nelson, 1964 Mutant gene that changes protein composition and increases lysine content in maize endosperm. Science 145: 279–280. [4]
Messing, J., D. Gerachty, G. Heidecker, N. T. Hu, J. Kridl and I. Rubenstein, 1983 Plant gene structure. pp. 211–227. In: *Genetic Engineering of Plants: An Agricultural Perspective*, Edited by T. Kosuge, C. P. Meredity, and A. Hollaender. Plenum, New York. [4]
Miège, M. N., 1982 Protein types and distribution. pp. 219–345. In: *Nucleic Acids and Proteins in Plants. I. Structure, Biochemistry, and Physiology of Proteins*. Encyclopedia of Plant Physiol., Vol. 14A, Edited by D. Boulter and B. Parthier. Springer, Berlin. [4]
Miflin, B. J., B. G. Forde, P. R. Shewry, M. Kreis and J. Forde, 1984 Repeated sequences in

cereal storage proteins. Oxford Surv. Plant Mol. Cell Biol. 1: 231–234. [4]
Mikola, J., 1982 Bud-set phenology as an indicator of climatic adaption of Scots pine in Finland. Silv. Fenn. 16: 178–184. [16]
Miller, P. A,., and J. O. Rawlings, 1967 Selection for increased lint yield and correlated responses in upland cotton, *Gossypium hirsutum* L. Crop Sci. 7: 637–640. [17]
Miller, T. E., W. L. Gerlach and R. B. Flavell, 1980 Nucleolus organizer variation in wheat and rye revealed by in situ hybridization. Theor. Appl. Genet. 61: 285–288. [5]
Miller, T. E., J. Hutchinson and S. M. Reader, 1983 The identification of the nucleolus organizer chromosomes of diploid wheat. Theor. Appl. Genet. 65: 145–147. [5]
Minvielle, F., 1987 Dominance is not necessary for heterosis: A two locus model. Genet. Res. 49: 245–247. [20]
Miranda, B. D., 1987 Breeding value, inheritance and response to indirect selection for N_2 fixation in common bean (*Phaseolus vulgaris* L.). Ph.D. Thesis, University of Wisconsin, Madison. [18]
Mitchell-Olds, T., 1986 Quantitative genetics of survival and growth in *Impatiens capensis*. Evolution 40: 107–116. [12, 15]
Mitchell-Olds, T., and J. J. Rutledge, 1986 Quantitative genetics in natural plant populations: a review of the theory. Am. Nat. 127: 379–402. [11, 12]
Mitchell-Olds, T., and R. G. Shaw, 1987 Regression analysis of natural selection: Statistical inference and biological interpretation. Evolution 41: 1149–1161. [12, 20]
Mitchell-Olds, T., and D. M. Waller, 1985 Relative performance of selfed and outcrossed progeny in *Impatiens capensis*. Evolution 39: 533–544. [11]
Mitton, J. B., 1983 Conifers. pp. 443–472. In: *Isozymes in Plant Genetics and Breeding*, Edited by S. D. Tanksley and T. J. Orton. Elsevier, Amsterdam. [16]
Mitton, J. B., and M. C. Grant, 1984 Associations among protein heterozygosity, growth rate and developmental homeostasis. Annu. Rev. Ecol. Syst. 15: 479–499. [11, 12, 16, 20]
Mitton, J. B., Y. B. Linhart, J. L. Hamrick and J. S. Beckman, 1977 Observations on the genetic structure and mating system of ponderosa pine in the Colorado Front Range, Theor. Appl. Genet. 51: 5–13. [16]
Mitton, J. B., K. B. Sturgeon and M. C. Davis, 1980 Genetic difference in ponderosa pine along a steep elevational transect. Silvae Genet. 29: 100–103. [10]
Mitton, J. B., Y. B. Linhart, M. L. Davis and K. B. Sturgeon, 1981 Estimation of outcrossing in ponderosa pine, *Pinus ponderosa* Laws., from patterns of segregation of protein polymorphisms and from frequencies of albino seedings. Silv. Genet. 30: 117–121. [16]
Moll, R. H., J. H. Lonnquist, J. Véler Fortuno and E. C. Johnson, 1965 The relationship of heterosis and genetic divergence in maize. Genetics 52: 139–144. [20]
Mooney, P. R., 1983 The law of the seed. Dev. Dialogue 1–2: 1–72. [21]
Moran, G. F., and J. C. Bell 1987 The origin and genetic diversity of *Pinus radiata* in Australia. Theor. Appl. Genet. 73: 616–622. [16]
Moran, G. F., and A. H. D. Brown, 1980 Temporal heterogeneity of outcrossing rates in alpine ash (Eucalyptus delegatensis R. T. Bak.). Theor. Appl. Genet. 57: 101–105. [12, 16]
Moran, G. F., and A. R. Griffin, 1985 Non-random contribution of pollen in polycrosses of *Pinus radiata* D. Don Silv. Genet. 34: 117–121. [16]
Moran, G. F., and S. D. Hopper, 1987 Conservation of genetic resources of rare and widespread eucalypts in remnant vegetation. pp. 151–162. In: *Nature Conservation: the Role of Remnants of Native Vegetation*, Edited by D. A. Saunders, G. W. Arnold, A. A. Burbidge, and A. J. M. Hopkins. Surrey Beatty, Sydney. [14, 16]
Moran, G. F., J. C. Bell and A. C. Matheson, 1980 The genetic structure and levels of inbreeding in a *Pinus radiata* D. Don seed orchard. Silv. Genet. 29: 190–193. [16]
Moran, G. F., J. C. Bell and A. R. Griffin, 1989c Reduction in levels of inbreeding in a seed orchard of *Eucalyptus regnans* compared with natural populations. Silv. Genet. 38: 32–36. [16]
Moran, G. F., O. Muona and J. C. Bell 1989a *Acacia mangium*: a tropical forest tree of the coastal lowlands with low genetic diversity. Evolution 43: 231–235. [16]
Moran, G. F., O. Muona and J. C. Bell 1989b Breeding systems and genetic diversity in *Acacia auriculiformis* and *A. crassicarpa*. Biotropica 21. [9, 16]
Moran, P. A. P., 1973a A gaussian Markovian process on a square lattice. J. Appl. Prob. 10: 54–62. [14]

Moran, P. A. P. 1973b Necessary conditions for Markovian processes on a lattice. J. Appl. Prob. 10: 605–612. [14]

Mori, T. S., Utsumi, H. Inaba, K. Kitamura and K. Harada, 1981 Differences in subunit composition of glycinin among soybean cultivars. J. Agric. Food Chem. 29: 20–23. [4]

Mortlock, R. P. (ed.), 1984 *Microorganisms as Model Systems for Studying Evolution*. Plenum, New York. [8]

Morton, N. E., 1973a Kinship and population structure. pp. 66–69. In: *Genetic Structure of Populations*, Edited by N. E. Morton. University of Hawaii Press, Honolulu. [14]

Morton, N. E., 1973b Isolation by distance. pp. 76–79. In: *Genetic Structure of Populations*, Edited by N. E. Morton. University of Hawaii Press, Honolulu. [14]

Morton, N. E., 1973c Prediction of kinship from a migration matrix. pp. 119–123. In: *Genetic Structure of Populations*, Edited by N. E. Morton. University of Hawaii Press, Honolulu. [14]

Morton, N. E., 1973d Kinship bioassay. pp. 158–163. In: *Genetic Structure of Populations*, Edited by N. E. Morton. University of Hawaii Press, Honolulu. [14]

Morton, N. E., 1973e Prediction of kinship from genealogies. pp. 89–91. In: *Genetic Structure of Populations*. Edited by N. E. Morton. University of Hawaii, Honolulu. [14]

Morton, N. E., 1982 Estimation of demographic parameters from isolation by distance. Hum. Hered. 32: 37–41. [14]

Morton, N. E., C. Miki and S. Yee, 1968 Bioassay of population structure under isolation by distance. Am. J. Hum. Genet. 20: 411–419. [14]

Morton, N. E., S. Yee, D. E. Harris and R. Lew, 1971 Bioassay of kinship. Theor. Pop. Biol. 2: 507–524. [14]

Morzycka-Wroblewska, E., E. V. Selker, J. N. Stevens and R. L. Metzenberg, 1985 Concerted evolution of dispersed *Neurospora crassa* 5S RNA genes: Pattern of sequence conservation between allelic and nonallelic genes. Mol. Cell. Biol. 5: 46–51. [5]

Mousseau, T. A., and D. A. Roff, 1987 Natural selection and the heritability of fitness components. Heredity 59: 181–197. [11]

Mulitze, D. K., and R. J. Baker, 1985 Evaluation of biometrical methods for estimating number of genes. Theor. Appl. Genet. 69: 553–558. [18]

Müller, G., 1977 Cross-fertilization in a conifer stand inferred from enzyme gene-markers in seeds. Silv. Genet. 26: 223–226. [3]

Müller, G., 1977 Untersuchungen über die natürliche Selbstbefruchtung in Beständen der Fichte (*Picea abies* (1.) Karst.) und Kiefer (*Pinus sylvestris* L.). Silv. Genet. 26: 207–217. [16]

Müller-Starck, G., 1982 Reproductive systems in conifer seed orchards. I. Mating probabilities in a seed orchard of *Pinus sylvestris*. L. Silv. Genet. 31: 188–197. [16]

Müller-Starck, G., and M. Ziehe, 1984 Reproductive systems in conifer seed orchards. 3. Female and male fitnesses of individual clones realized in seeds of *Pinus sylvestris* L. Theor. Appl. Genet. 69: 173–177. [16]

Mulligan, B., N. Schultes, L. Chen and L. Bogorad, 1984 Nucleotide sequence of a multi-copy gene for the B protein of photosystem II of a cyanobacterium. Proc. Natl. Acad. Sci. U.S.A. 81: 2693–2697. [8]

Mundt, C. C., and K. J. Leonard, 1985 Effect of host genotype unit area on epidemic development of crown rust following focal and general inoculations of mixtures of susceptible and immune oat plants. Phytopathology 75: 1141–1145. [13]

Mundt, C. C., and K. J. Leonard, 1986 Effect of host genotype unit area on development of focal epidemics of bean rust and common maize rust in mixtures of resistant and susceptible plants. Phytopathology 76: 895–900. [13]

Muntzing, A., 1943 Genetical effects of duplicated fragment chromosomes in rye. Hereditas 29: 91–112. [12]

Muona, O. A., 1982 Multilocus study of an experimental barley population. Hereditas 96: 247–254. [10]

Muona, O., and A. Harju, In press Effective population sizes, mating system and genetic variability in Scots pine natural populations and seed orchards. Silv. Genet. [16]

Muona, O., and A. E. Szmidt, 1985 A multilocus study of natural populations of *Pinus sylvestris*. pp. 226–240. In: *Population Genetics in Forestry*. Lecture notes in Biomathematics 60, Edited by H. R. Gregorius. Springer-Verlag, Berlin. [16]

Muona, O., R. W. Allard and R. K. Webster, 1982 Evolution of resistance to *Rhynchosporium secalis* (Oud.) Davis in barley composite cross II. Theor. Appl. Genet. 61: 209–214. [12, 13]

Muona, O., R. Yazdani and D. Rudin, 1987 Genetic change between life stages in *Pinus sylvestris:* Allozyme variation in seeds and planted seedlings. Silv. Genet. 36: 39–42. [16]

Muona, O., A. Harju and K. Kärkkäinen, 1988 Genetic comparison of natural and nursery grown seedlings of *Pinus sylvestris* using allozymes. Scand. J. For. Res. 3: 37–46. [16]

Murai, N., D. W. Sutton, M. G. Murray, J. L. Slightom, D. J. Merlo, N. A. Reichert, C. Sengupta-Gopalan, C. A. Stock, R. F. Barker, J. D. Kemp and T. C. Hall, 1983 Phaseolin gene from bean is expressed after transfer to sunflower via tumor-inducing plasmid vectors. Science 222: 477–482. [20]

Murphy, J. P., D. B. Helsel, A. Elliott, A. M. Thro and K. J. Frey, 1982 Compositional stability of an oat multiline. Euphytica 31: 33–40. [13]

Murray, B. E., I. L. Craig and T. Rajhathy, 1970 A protein electrophoresis study of three amphidiploids and eight species of *Avena*. Can. J. Genet. Cytol. 12: 651–665. [4]

Nagylaki, T., 1976 The evolution of one- and two-locus systems. Genetics 83: 583–600. [10]

Nagylaki, T., 1977 The evolution of one- and two-locus systems II. Genetics 85: 347–354. [10]

Nagylaki, T., 1984 The evolution of multigene families under intrachromosomal gene conversion. Genetics 106: 529–548. [8]

Nakamura, Y., M. Leppert, P. O'Connell, R. Wolfe, T. Holm, M. Culver, C. Martin, E. Fujimoto, M. Hoff, E. Kumlin and R. White, 1987 Variable number of tandem repeat (VNTR) markers for human gene mapping. Science 235: 1616–1622. [9]

Namkoong, G., 1984 A control concept of gene conservation. Silv. Genet. 33: 160–163. [16]

National Academy of Sciences, 1972 Soybeans and Other Edible Legumes. pp. 207–252. *Genetic Vulnerability of Major Crops*. National Academy of Science, Washington, D.C. [18]

Navashin, M., 1934 Chromosomal alterations caused by hybridization and their bearing upon certain general genetic problems. Cytologia 5: 159–203. [5]

Navot, N., and D. Zamir, 1987 Isozyme and seed protein phylogeny of the genus *Citrullus* (*Cucurbitaceae*). Plant Syst. Evol. 156: 61–67. [4]

Neale, D. B., and W. T. Adams, 1985a The mating system in natural and shelterwood stands of Douglas-fir. Theor. Appl. Genet. 71: 201–207. [3, 9, 16]

Neale, D. B., and W. T. Adams, 1985b Allozyme and mating-system variation in balsam fir (*Abies balsamea*) across a continuous elevational transect. Can. J. Bot. 63: 2448–2453. [9, 12, 16]

Neale, D. B., N. C. Wheeler and R. W. Allard, 1986 Paternal inheritance of chloroplast DNA in Douglas-fir. Can. J. For. Res. 16: 1152–1154. [6]

Neale, D. B., M. A. Saghai-Maroof, R. W. Allard, Q. Zang and R. A. Jorgensen, 1988 Chloroplast DNA diversity in populations of wild and cultivated barley. Genetics 120: 1105–1110. [6]

Nei, M. 1973a Analysis of gene diversity in subdivided populations. Proc. Natl. Acad. Sci. U.S.A. 70: 3321–3323. [2, 3, 14, 16]

Nei, M., 1973b The theory and estimation of genetic distances. pp. 45–51. In: *Genetic Structure of Populations*, Edited by N. E. Morton. University of Hawaii Press, Honolulu. [14]

Nei, M., 1987 *Molecular Evolutionary Genetics*. Columbia University Press, New York [8]

Nei, M., and T. Gojobori, 1986 Simple methods for estimating the numbers of synonymous and nonsynonymous nucleotide substitutions. Mol. Biol. Evol. 3: 418–426. [8]

Nei, M., and M. Syakudo, 1958 The estimation of outcrossing in natural populations. Jpn. J. Genet. 33: 46–51. [11]

Nei, M., and F. Tajima, 1981 Genetic drift and estimation of effective population size. Genetics 98: 625–640. [15]

Nei, M., T. Maruyama and R. Chakraborty, 1975 The bottleneck effect and genetic variability in populations. Evolution 29: 1–10. [15]

Nelson, O. E., 1968 The *waxy* locus in maize. II. The location of the controlling element alleles. Genetics 60: 507–524. [7]

Nelson, O. E., and A. S. Klein, 1984 Characterization of an *Spm-controlled bronze-mutable allele in maize. Genetics 106: 769–779. [7]*

Nevers, P., and H. Saedler, 1977 Transposable genetic elements as agents of gene instability and chomosomal rearrangements. Nature (London) 268: 109–115. [7]

Nevo, E., and P. I. Payne, 1987 Wheat storage protein diversity of HMW glutenin subunits in wild emmer from Israel. I. Biogeographical patterns and ecological predictability. Theor. Appl. Genet. 74: 827–836. [4]

Nevo, E., D. Zohary, A. H. D. Brown and M. Haber, 1979 Genetic diversity and environment associations in wild barley, *Hordeum spontaneum*. Evolution 33: 815–833. [4]

Nevo, E., A. H. D. Brown and D. Zohary, 1979 Genetic diversity in the wild progenitor of barley in Israel. Experientia 35: 1027–1029. [21]

Nevo, E., E. Golenberg, A. Beiles, A. H. D. Brown and D. Zohary, 1982 Genetic diversity and environmental associations of wild wheat, *Triticum dicoccoides* in Israel. Theor. Appl. Genet. 62: 241–254. [3, 4]

Nevo, E., A. Beiles, N. Storch, H. Doll and B. Andersen, 1983 Microgeographic edaphic differentiation in hordein polymorphisms of wild barley. Theor. Appl. Genet. 64: 123–132. [4]

Nevo, E., A. Beiles and R. Ben-Shlomi, 1984 The evolutionary significance of genetic diversity: Ecological, demographic and life history correlates. pp. 13–213. In: *Evolutionary Dynamics of Genetic Diversity.* Lecture Notes in Biomathematics Vol. 53, Edited by G. S. Mani. Springer-Verlag, New York. [3]

Nikkanen, T., and P. Velling, 1987 Correlations between flowering and some vegetative characteristics of grafts of *Pinus sylvestris*. For. Ecol. Manage. 19: 35–40. [16]

Nishikawa, K., 1983 Species relationship of wheat and its putative ancestors as viewed from isozyme variation. Proc. 6th Int. Wheat Genet. Symp., Kyoto, pp. 59–63. [5]

Nitzsche, W., and J. Hesselbach, 1983 Sortenmischungen statt viellinien-sorten. 1. Sommergerste (*Hordeum vulgare* L.). Z. Planzenzüchtg 90: 68–74. [13]

Nitzsche, W., and J. Hesselbach, 1984 Sortenmischungen statt viellinien-sorten. 2. Wintergerste (*Hordeum vulgare* L.) and winterweizen (*Triticum aestivum* L.). Z. Planzenzüchtg 92: 151–158. [13]

Oden, N. L., 1984 Assessing the significance of a spatial correlogram. Geog. Anal. 16: 1–16. [14]

Oden, N. L., and R. R. Sokal, 1986 Directional autocorrelation: An extension of spatial correlograms to two dimensions. Syst. Zool. 35: 608–617. [14]

Ohta, T., 1983 On the evolution of multigene families. Theor. Pop. Biol. 23: 216–240. [8]

Ohyama, K., H. Fukuzawa, T. Kohchi, H. Shirai, S. Sano, T. Sano, K. Umesono, Y. Shiki, M. Takeuchi, Z. Chang, S. Aota, H. Inokuchi and H. Ozeki, 1986 Chloroplast gene organization deducted from complete sequence of liverwort *Marchantia polymorpha* chloroplast DNA. Nature (London) 322: 572–574. [8]

Oishi, K. K., D. R. Shapiro and K. K. Tewari, 1984 Sequence organization of a pea chloroplast DNA gene coding for a 34,500-Dalton protein. Mol. Cell. Biol. 4: 2556–2563. [8]

Okagaki, R. J., and S. R. Wessler, 1988 Comparison of non-mutant and mutant *waxy* genes in rice and maize. Genetics 120: 1137–1143. [7]

Oka, H. I., 1975 Consideration of the population size necessary for conservation of crop germplasms. pp. 57–63. In: *Gene Conservation—Exploration, Collection, Preservation and Utilization of Genetic Resources*, Edited by T. Matsuo. J. I. B. P. Synthesis 5, University of Tokyo Press, Tokyo [21]

Oldfield, M. W., and J. B. Alcorn, 1987 Conservation of traditional agroecosystems. Bioscience 37: 197–208. [21]

O'Malley, D. M., and K. S. Bawa, 1987 Mating system of a tropical rain forest tree species. Am. J. Bot. 74: 1143–1149. [9, 16]

Oono, K., and M. Sugiura, 1980 Heterogeneity of the RNA gene clusters in rice. Chromosoma 76: 85–87. [6]

Oram, R. N., H. Doll and B. Koie, 1975 Genetics of two storage protein variants in barley. Hereditas 80: 53–58. [4]

Ord, J. K., 1980 Tests of significance using non-normal data. Geog. Anal. 12: 387–392. [14]

O'Reilly, C., N. S. Shepherd, A. Pereira, Z. S. Schwarz-Sommer, I. Bertram, D. S. Robertson, P. A. Peterson and H. Saedler, 1985 Molecular cloning of the *a1* locus of *Zea mays* using the transposable elements *En* and *Mu*. EMBO J. 4:877–882. [7]

Orton, T. J., 1984 Somaclonal variation: Theoretical and practical considerations. pp. 427–468. In: *Gene Manipulation in Plant Improvement*, Edited by J. P. Gustafson. Plenum, New York. [17]

Osborn, T. C., 1988 Genetic control of bean seed protein. CRC Crit. Rev. Plant Sci. 7: 93–116. [4]
Osborn, T. C., and F. A. Bliss, 1985 Effect of genetically removing lectin seed protein on horticultural and seed characteristics of common bean. J. Am. Soc. Hort. Sci. 110: 484–488. [4]
Osborn, T. C., J. W. S. Brown and F. A. Bliss, 1985 Bean lectins 5. Quantitative genetic variation in seed lectins of *Phaseolus vulgaris* L. and its relationships to qualitative lectin variation. Theor. Appl. Genet. 70: 22–31. [18]
Osborn, T. C., T. Blake, P. Gepts and F. A. Bliss, 1986 Bean arcelin. 2. Genetic variation, inheritance and linkage relationships of a novel seed protein of *Phaseolus vulgaris* L. Theor. Appl. Genet. 71: 847–855. [4, 18]
Osborn, T. C., D. C. Alexander and J. F. Fobes, 1987 Identification of restriction fragment length polymorphisms linked to genes controlling soluble solids content in tomato fruit. Theor. Appl. Genet. 73: 350–356. [18]
Osborn, T. C., D. C. Alexander, S. S. M. Sun, C. Cardona and F. A. Bliss, 1988 Insecticidal activity and lectin homology of arcelin seed protein. Science 240: 207–210. [4, 18]
Osborne, T. B., 1907 *The Vegetable Proteins*. Longmans & Green, London. [4]
Owens, K. W., F. A. Bliss and C. E. Peterson, 1985 Genetic variation within and between two cucumber populations derived via the inbred backcross line method. J. Am. Soc. Hort. Sci. 110: 437–441. [18]

Palmer, J. D., 1987 Chloroplast DNA evolution and biosystematic uses of chloroplast DNA variation. Am. Natur. 130: S6–S29. [6]
Palmer, J. D., 1988 Intraspecific variation and multicircularity in *Brassica* mitochondrial DNAs. Genetics 118: 341–351. [6]
Park, Y. S., D. P. Fowler and J. F. Coles, 1984 Population studies of white spruce. II. Natural inbreeding and relatedness among neighboring trees. Can. J. For. Res. 14: 909–913. [16]
Parker, M. A., 1986 Individual variation in pathogen attack and differential reproductive success in the annual legume *Amphicarpaea bracteata*. Oecologia 69: 253–259. [12]
Parker, M. A., 1988a Polymorphism for disease resistance in the annual legume *Amphicarpaea bracteata*. Heredity 60: 27–31. [12, 13]
Parker, M. A., 1988b Disequilibrium between disease-resistance variation and allozyme loci in an annual legume. Evolution 42: 239–247. [13]
Patel, J. D., E. Reisbergs and S. O. Fejer, 1985 Recurrent selection in doubled-haploid populations in barley (*Hordeum vulgare* L.). Can. J. Genet. Cytol. 27: 172–177. [17]
Paulis, J. W., and J. S. Wall, 1977 Comparison of the protein compositions of selected corns and their wild relative teosinte and *Tripsacum*. J. Agric. Food Chem. 85: 265–270. [4]
Pay, A., M. A. Smith, F. Nagy and L. Marton, 1988 Sequence of the *psbA* gene from wild type and triazin-resistant *Nicotiana plumbaginifolia*. Nucl. Acids Res. 16: 8176. [8]
Payne, P. E., 1987 Genetics of wheat storage proteins and the effect of allelic variation on bread-making quality. Annu. Rev. Plant Physiol. 38: 141–153. [4]
Payne, P. E., and G. J. Lawrence, 1983 Catalogue of alleles for the complex gene loci, *GLU-A1*, *GLU-B1*, *GLU-D1*, which code for high-molecular weight subunits of glutenin in hexaploid wheat. Cereal Res. Commun. 11: 2–35. [4]
Payne, P. E., C. N. Law and E. E. Mudd, 1980 Control by homoeologous group 1 chromosomes of the high molecular weight subunits of glutenin, a major protein of wheat endosperm. Theor. Appl. Genet. 58: 113–120. [4]
Payne, P. E., L. M. Holt, A. J. Worland and C. N. Law, 1982 Structural and genetical studies on the high molecular weight subunits of glutelin. III. Telocentric mapping of the subunit genes on the long arms of the homoeologous group 1 chromosomes. Theor. Appl. Genet. 63: 129–138. [4]
Payne, P. I., K. G. Corfield and J. A. Blackman, 1979 Identification of a high molecular weight subunit of glutenin whose presence correlates with bread-making quality in wheat of related pedigree. Theor. Appl. Genet. 55: 153–157. [17]
Payne, T. S., D. D. Stuthman, R. L. McGraw and P. P. Bregitzer, 1986 Physiological changes associated with three cycles of recurrent selection for grain yield improvement in oats. Crop Sci. 26: 734–736. [17]
Peng, J. Y., J. C. Glazmann and S. S. Virmani, 1988 Heterosis and isozyme divergence in indica

rice. Crop Sci. 28: 561–563. [19]

Perry, D. J., and B. P. Dancik, 1986 Mating system dynamics of Lodgepole pine in Alberta, Canada. Silv. Genet 35: 190–195. [16]

Person, C., 1966 Genetic polymorphism in parasitic systems. Nature (London) 212: 266–267. [13]

Perutz, M. F., C. Bauer, G. Gros, F. Leclercq, C. Vandecasserie, A. G. Schnek, G. Braunitzer, A. E. Friday, and K. A. Joysey, 1981 Allosteric regulation of crocodilian haemoglobin. Nature (London) 291: 682–684. [8]

Peterson, P. A., 1953 A mutable pale green locus in maize. Genetics 38: 682–683. [7]

Peterson, P. A., 1988 The mobile element systems in maize. pp. 43–68. In: *Plant Transposable Elements*, Edited by O. E. Nelson. Plenum, New York. [7]

Phillip, M., 1980 Reproductive biology of *Stellaria longipes* Goldie as revealed by a cultivation experiment. New Phytol. 85: 557–569. [12]

Phillips, M. A., and A. H. D. Brown, 1977 Mating system and hybridity in *Eucalyptus pauciflora*. Aust. J. Biol.Sci. 30: 337–344. [10]

Pielou, E. C., 1977 *Mathematical Ecology*, 2nd edition. Wiley, New York. [14]

Piper, J. G., B. Charlesworth and D. Charlesworth, 1984 A high rate of self-fertilization and increased seed fertility of homostyle primroses. Nature (London) 310: 50–51. [9]

Piper, J. G., B. Charlesworth and D. Charlesworth, 1986 Breeding system evolution in *Primula vulgaris* and the role of reproductive assurance. Heredity 56: 207–217. [12]

Plessas, M. E., and S. H. Strauss, 1986 Allozyme differentiation among populations, stands and cohorts in Monterey pine. Can. J. For. Res. 16: 1155–1164. [12]

Plucknett, D. L., N. H. J. Smith, J. T. Williams and N. Murthi Anishetty, 1983 Crop germplasm conservation and developing countries. Science 220: 163–169. [21]

Plucknett, D. L., N. J. H. Smith, J. T. Williams and N. Murthi Anishetty, 1987 *Gene Banks and the World's Food*. Princeton University Press, Princeton. [21]

Pollak, E., 1983 A new method for estimating the effective population size from allele frequency changes. Genetics 99: 531–548. [15]

Pollak, L. M., C. O. Gardner and A. M. Parkhurst, 1984 Relationships between enzyme marker loci and morphological traits in two mass selected maize populations. Crop Sci. 24: 1174–1179. [19]

Pont, G., F. Degroote and G. Picard, 1988 Illegitimate recombination in the histone multigenic family generates circular DNAs in Drosophila embryos. Nucl. Acids Res. 16: 8817–8833. [5]

Powers, H. R., 1980 Pathogenic variation among single aeciospore isolates of *Cronartium quercuum* f. sp. *fusiforme*. For. Sci. 26: 280–282. [12]

Price, H. J., K. L. Chambers, K. Bachmann and J. Riggs, 1983 Inheritance of nuclear 2C content variation in intraspecific and interspecific hybrids of *Microseris* (Asteraceae). Am. J. Bot. 70: 1133–1138. [6]

Price, S. C., K. N. Schumaker, A. L. Kahler, R. W. Allard and J. E. Hill, 1984 Estimates of population differentiation obtained from enzyme polymorphisms and quantitative characters. J. Hered. 75: 141–142. [3]

Price, S. C., R. W. Allard, J. E. Hill and J. Naylor, 1985 Associations between discrete genetic loci and genetic variability for herbicide reaction in plant populations. Weed Sci. 33: 650–643. [19]

Prout, T., 1965 The estimation of fitness from genotypic frequencies. Evolution 19: 546–551. [11]

Prout, T., 1969 The estimation of fitness from population data. Genetics 63: 949–967. [12]

Prout, T., 1973 Appendix to J. B. Mitton and R. K. Koehn. Population genetics of marine pelecypods. III. Epistasis between functionally related isoenzymes in *Mytilus edulus*. Genetics 73: 487–496. [10, 14]

Qualset, C. O., 1979 Mendelian genetics of quantitative characters with reference to adaptation and breeding in wheat. Proc. Fifth Int. Wheat Genet. Symp. 2: 577–590. Indian Soc. Genet. and Plant Breeding, New Delhi. [17]

Qualset, C. O., and C. W. Schaller, 1966 Aleurone color and quantitative genetic variability in isogenic lines of barley. Can. J. Genet. Cytol. 8: 584–591. [17]

Qualset, C. O., and H. E. Vogt, 1980 Efficient methods of population management and utilization in breeding wheat for Mediterranean-type climates. pp. 166–188. Proc. Third Int.

Wheat Conf., Madrid. [17]
Qualset, C. O., C. W. Schaller and J. C. Williams, 1965 Performance of isogenic lines of barley as influenced by awn length, linkage blocks, and environment. Crop Sci. 5: 489–494. [17, 19]
Quinn, T. W., J. S. Quinn, F. Cooke and B. N. White, 1987 DNA marker analysis detects multiple maternity and paternity in single broods of the lesser snow goose. Nature (London) 326: 392–394. [11]
Quisenberry, J. E., B. Roark, J. D. Bilbro and L. L. Ray, 1978 Natural selection in a bulked hybrid population of upland cotton. Crop Sci. 18: 799–801. [12]

Ralston, E. J., J. J. English and H. K. Dooner, 1988 Sequence of three *bronze* alleles of maize and correlation with fine structure. Genetics 119: 185–197. [6, 7]
Rao, C. R., 1973 *Linear Statistical Inference and Its Applications.* Wiley, New York. [14]
Rapp, A., and H. Mandery, 1986 Wine aroma. Experientia 42: 873–884. [20]
Reith, M., and N. A., Straus, 1987 Nucleotide sequence of the chloroplast gene responsible for triazine resistance in canola. Theor. Appl. Genet. 73: 357–363. [8]
Reynolds, J., B. S. Weir and C. C. Cockerham, 1983 Estimation of the coancestry coefficient: basis for a short-term genetic distance. Genetics 105: 767–779. [11]
Rhoades, D. F., 1983 Herbivore population dynamics and plant chemistry. pp. 155–220. In: *Variable Plants and Herbivores in Natural and Managed Systems.* Edited by R. F. Denno and M. S. McClure. Academic Press, New York. [12]
Rice, K., and S. K. Jain, 1985 Plant population genetics and evolution in disturbed environments. pp. 287–303. In: *The Ecology of Natural Disturbance and Patch Dynamics,* Edited by S. T. A Pickett and P. S. White. Academic Press, New York. [15]
Richards, A. J., 1986 *Plant Breeding Systems.* George Allen & Unwin, London. [20]
Rick, C. M., and J. F. Fobes, 1975 Allozymes of Galapagos tomatoes: Polymorphism, geographic distribution and affinities. Evolution 29: 443–457. [15]
Rick, C. M., and S. D. Tanksley, 1981 Genetic variability in *Solanum pennellii:* Comparisons with two other sympatric tomato species. Syst. Evol. 139: 11–45. [3]
Rick, C. M., and J. F. Fobes, 1974 Association of an allozyme with nematode resistance. Rep. Tomato Genet. Coop. No. 24: 25. [18, 19]
Rick, C. M., J. F. Fobes and M. Holle, 1977 Genetic variation in *Lycopersicon pimpinellifolium:* Evidence of evolutionary change in mating systems. Plant Syst. Evol. 127: 139–170. [12, 21]
Ripley, B. D., 1981 *Spatial Statistics.* Wiley, New York. [14]
Ritland, K., 1984 The effective proportion of self-fertilization with consanguineous matings in inbred populations. Genetics 106: 139–152. [9]
Ritland, K., 1985 The genetic-mating structure of subdivided populations. I. Open-mating model. Theor. Pop. Biol. 27: 51–74. [11, 14]
Ritland, K., 1986 Joint maximum likelihood estimation of genetic and mating structure using open-pollinated progenies. Biometrics 42: 25–43. [9, 11]
Ritland, K., 1987 Definition and estimation of higher-order gene fixation indices. Genetics 117: 783–793. [11]
Ritland, K., In press Correlated matings in the partial selfer *Mimulus guttatus.* Evolution. [9, 11]
Ritland, K., In press The genetic-mating structure of subdivided populations. II. Correlated mating models. Theor. Pop. Biol. [11]
Ritland, K., and Y. A. El-Kassaby, 1985 The nature of inbreeding in a seed orchard of Douglas-fir as shown by an efficient multilocus model. Theor. Appl. Genet. 71: 375–384. [16]
Ritland, K., and F. R. Ganders, 1985 Variation in the mating system of *Bidens menziesii* (Asteraceae) in relation to population substructure. Heredity 55: 235–244. [9]
Ritland, K., and F. R. Ganders, 1987a Covariation of selfing rates with parental gene fixation indices within populations of *Mimulus guttatus.* Evolution 41: 760–771. [9, 11]
Ritland, K., and F. R. Ganders, 1987b Crossability of *Mimulus guttatus* in relation to components of gene fixation. Evolution 41: 772–786. [11]
Ritland, K., and S. Jain, 1981 A model for the estimation of outcrossing rate and gene frequencies using n independent loci. Heredity 47: 35–52. [3, 9, 11]
Roberts, E. H., and R. H. Ellis, 1984 The implications of the deterioration of orthodox seeds during storage for genetic resources conservation. pp. 18–37. In: *Crop Genetic Resources:*

Conservation and Evaluation, Edited by J. H. W. Holden and J. T. Williams. Allen and Unwin, London. [21]

Roberts, E. H., M. W. King and R. H. Ellis, 1984 Recalcitrant seeds: Their recognition and storage. pp. 38–52. In: *Crop Genetic Resources: Conservation and Evaluation*, Edited by J. H. W. Holden and J. T. Williams. Allen and Unwin, London. [21]

Rode, A., C. Hartmann, A. Benslimane, E. Picard and F. Quetier, 1987 Gametoclonal variation detected in the nuclear ribosomal DNA from doubled haploid lines of a spring wheat (*Triticum aestivum* L., cv. Cesar). Theor. Appl. Genet. 74: 31–37. [5]

Roeder, G. S., R. L. Keil and K. A. Voelkel-Meiman, 1986 A recombination-stimulating sequences in the ribosomal RNA gene cluster of yeast. pp. 29–33. In: *Mechanisms of yeast recombination*. Cold Spring Harbor Laborator, Cold Spring Harbor, N.Y. [5]

Rogers, S., S. Honda and A. J. Bendich, 1986 Variation in the ribosomal RNA genes among individuals of *Vicia faba*. Plant Mol. Biol. 6: 339–345. [5]

Rogstad, S. H., J. C. Patton and B. A. Schaal, 1988 M13 repeat probe detects DNA minisatellite-like sequences in gymnosperms and angiosperms. Proc. Natl. Acad. Sci. U.S.A. 85: 9176–9178. [6]

Rohlf, F. J., and G. D. Schnell, 1971 An investigation of the isolation-by-distance model. Am. Nat. 105: 295–324. [14]

Romero Andreas, J., and F. A. Bliss, 1985 Heritable variation in the phaseolin protein of non-domesticated common bean, *Phaseolus vulgaris* L. Theor. Appl. Genet. 71: 478–480. [18]

Romero Andreas, J., B. S. Yandell and F. A. Bliss, 1986 Bean arcelin 1. Inheritance of a novel seed protein of *Phaseolus vulgaris* L. and its effect on seed composition. Theor. Appl. Genet. 72: 123–128. [18]

Rosahl, S., J. Schell and L. Willnitzer, 1987 Expression of a tuber-specific storage protein in transgenic tobacco plants: demonstration of an esterase activity. EMBO J. 6: 1155–1159. [20]

Rose, M. R., 1984 Genetic co-variation in *Drosophila* life history: Untangling the data. Am. Nat. 123: 565–569. [15]

Ross, M. D., 1984 Frequency-dependent selection in hermaphrodites: the rule rather than the exception. Biol. J. Linn. Soc. 23: 145–155. [16]

Ross, M. D., and B. S. Weir, 1975 Maintenance of male sterility in plant populations. III Mixed selfing and random mating. Heredity 35: 21–29. [12]

Rudin, D., G. Eriksson and M. Rasmuson, 1977 Inbreeding in a seed tree stand of *Pinus sylvestris* L., in northern Sweden. A study by the aid of the isozyme technique. Dept. For. Genet. Res. Notes 25: 1–45. [16]

Rudin, D., O. Muona and R. Yazdani, 1986 Comparison of the mating system of *Pinus sylvestris* in natural stands and seed orchards. Hereditas 104: 15–19. [16]

Sachs, M. M., E. S. Dennis, W. L. Gerlach and W. J. Peacock, 1986 Two alleles of maize *alcohol dehydrogenase 1* have 3' structural and poly(A) addition polymorphisms. Genetics 113: 449–467. [6]

Saedler, H., and P. Nevers, 1985 Transposition in plants: a molecular model. EMBO J. 4: 585–590. [7]

Saghai-Maroof, M. A., R. K. Webster and R. W. Allard, 1983 Evolution of resistance to scald, powdery mildew and net blotch in barley composite cross II populations. Theor. Appl. Genet. 66: 279–283. [12, 13]

Saghai-Maroof, M. A., K. M. Soliman, R. A. Jorgensen and R. W. Allard, 1984 Ribosomal DNA spacer-length polymorphisms in barley: Mendelian inheritance, chromosomal location, and population dynamics. Proc. Natl. Acad. Sci. U.S.A. 81: 8014–8018. [5, 6]

Saitou, N., and M. Nei, 1986 Polymorphism and evolution of influenza A virus genes. Mol. Biol. Evol. 3: 57–74. [8]

Sarkar, P., and G. L. Stebbins, 1956 Morphological evidence concerning the origin of the B genome in wheat. Am. J. Bot. 43: 297–304. [5]

Sarvas, R., 1962 Investigations on the flowering and seed crop of *Pinus sylvestris*. Commun. Inst. For. Fenn. 53: 1–98. [16]

Sarvas, R., 1970 Establishment and registration of seed orchards. Folia For. 89: 1–24. [16]

SAS Institute Inc. SAS/STAT™ Guide for Personal Computers, Version 6 Edition. Cary, NC, SAS Institute Inc., 1987, 1028 pp. [3]

Schaal, B. A., 1980 Measurement of gene flow in *Lupinus texensis*. Nature (London) 284: 450–451. [3]
Schaal, B. A., and D. A. Levin, 1976 The demographic genetics of *Liatris cylindracaea* Michx. (Compositae). Am. Nat. 110: 191–206. [12]
Schaal, B. A., W. J. Leverich and J. Nieto-Sotelo, 1987 Ribosomal DNA diversity in the native plant *Phlox divaricata*. Mol. Biol. Evol. 4: 611–621. [6]
Schaeffer, S. W., and C. F. Aquadro, 1987 Nucleotide sequence of the *Adh* gene region of *Drosophila pseudoobscura*: Evolutionary change and evidence for an ancient gene duplication. Genetics 117: 61–73. [6]
Schaeffer, S. W., C. F. Aquadro and W. W. Anderson, 1987 Restriction-map variation in the alcohol dehydrogenase region of *Drosophila pseudoobscura*. Mol. Biol. Evol. 4: 254–265. [6]
Schaeffer, S. W., C. F. Aquadro and C. H. Langley, 1988 Restriction-map variation in the *Notch* region of *Drosophial melanogaster.* Mol. Biol. Evol. 5: 3–40. [6, 8]
Scheiner, S. M., and C. J. Goodnight, 1984 The comparison of phenotypic plasticity and genetic variation in populations of the grass *Danthonia spicata*. Evolution 38: 845–855. [15]
Schemske, D. W., 1984 Population structure and local selection in *Impatiens pallida* (Balsaminaceae), a selfing annual. Evolution 38: 817–832. [12, 15]
Schemske, D. W., and R. L. Lande, 1985 The evolution of self-fertilization and inbreeding depression in plants. II. Empirical observations. Evolutin 39: 41–52. [11]
Schemske, D. W., and R. Lande, 1987 On the evolution of plant mating systems: A reply to Waller. Am. Nat. 130: 804–806. [20]
Schiefelbein, J. W., 1987 Molecular characterization of mutations at the *bronze-1* locus caused by the WIAc-Ds and *Spm-dSpm* transposable element families. Ph.D. Thesis, University of Wisconsin—Madison. [7]
Schiefelbein, J. W., D. B. Furtek, V. Raboy, J. A. Banks, N. V. Fedoroff and O. E. Nelson, 1985 Exploiting transposable elements to study the expression of a maize gene. pp. 445–449. In: *Plant Genetics*, Edited by M. Freeling. Liss, New York. [7]
Schiefelbein, J. W., D. B. Furtek, H. K. Dooner and O. E. Nelson, 1988 Two mutations in a maize *bronze-1* allele caused by transposable elements of the *Ac-Ds* family alter the quantity and quality of the gene product. Genetics 120: 767–777. [7]
Schiller, G., M. T. Conkle and C. Grunwald, 1986 Local differentiation among Mediterranean populations of Aleppo pine in their isoenzymes. Silv. Genet. 35: 11–19. [16]
Schinkel, C., and P. Gepts, 1988 Phaseolin diversity in the tepary bean, *Phaseolus acutifolius* A. Gray. Plant Breeding 101: 292–301. [4]
Schmidt, K. P., and D. A. Levin, 1985 The comparative demography of reciprocally sown populations of *Phlox drummondii* Hook. I. Survivorships, fecundities and finite rates of increase. Evolution 39: 396–404. [12]
Schmidt, R. J., F. A. Burr and B. Burr, 1987 Transposon tagging and molecular analysis of the maize regulatory locus *opaque-2*. Science 238: 960–963. [7]
Schnell, F. W., 1982 A synoptic study of the methods and categories of plant breeding. Z. Pflanzenzuchtg. 89: 1–18. [17]
Schoen, D. J., 1983 Relative fitnesses of selfed and outcrossed progeny in *Gilia achilleifolia* (Polemoniaceae). Evolution 37: 292–301. [11, 12]
Schoen, D. J., 1985 Correlation between classes of mating events in two experimental plant populations. Heredity 55: 381–385. [16]
Schoen, D. J., 1988 Mating system estimation via the one pollen parent model with the progeny array as the unit of observation. Heredity 60: 439–444. [9]
Schoen, D. J., and W. M. Cheliak, 1987 Genetics of the polycross. 2. Male fertility variation in Norway spruce, *Picea abies* (L.) Karst. Theor. Appl. Genet. 74: 554–559. [16]
Schoen, D. J., and M. T. Clegg, 1984 Estimation of mating system parameters when outcrossing events are correlated. Proc. Natl. Acad. Sci. U.S.A. 81: 5258–5262. [16]
Schoen, D. J., and S. C. Stewart, 1986 Variation in male reproductive investment and male reproductive success in white spruce. Evolution 40: 1109–1120. [16]
Schoen, D. J., and S. C. Stewart, 1987 Variation in male fertilities and pairwise mating probabilities in *Picea glauca*. Genetics 116: 141–152. [9, 16]
Schoenfeld, M., T. Yacoby, A. Ben-Yehuda, B. Rubin and J. Hirschberg, 1987 Triazine resistance in *Phalaris paradoxa*: Physiological and molecular analysis. Z. Naturforsch. 42c: 779–782. [8]

Schoonhoven, A. V., C. Cardona and J. Valor, 1983 Resistance to the bean weevil and the Mexican bean weevil (Coleoptera: Brunchidae) in non-cultivated common bean accessions. J. Econ. Entomol. 76: 1255–1259. [18]

Schumaker, K. M., and G. R. Babbel, 1980 Patterns of allozymic similarity in ecologically central and marginal populations of *Hordeum jubatum* in Utah. Evolution 34: 110–116. [15]

Schwarz-Sommer, Z. S., A. Gierl, H. Cuypers, P. A. Peterson and H. Saedler, 1985 Plant transposable elements generate the DNA sequence diversity needed in evolution.EMBO J. 4: 591–597. [7]

Schwenke, K. D., W. Heinze, M. Schultz, K. J Linow, L. Prahl, J. Behlke, R. Reichelt, E. E. Braudo and L. P. Sologub, 1979 Helianthinin—the main storage protein in sunflower seeds (*Helianthus anuus* L.). pp. 45–62. In: *Seed Proteins of Dicotyledonous Plants,* Edited by K. Müntz. Akademie Verlag, Berlin. [4]

Scoles, G. J., B. S. Gill, Z. Y. Xin, B. C. Clarke, C. L. McIntyre, C. Chapman and R. Appels, 1988 Frequent duplications and deletion events in the 5S RNA genes and the associated spacer regions of the Triticeae. Plant Syst. Evol. 160: 105–122. [5]

Sears, B. B., 1983 Genetics and evolution of the chloroplast. Stadler Symp. 15: 119–139. [6]

Second, G., 1985 Geographic origins, genetic diversity and the molecular clock hypothesis in the Oryzae. NATO ASI Series 65: 41–56. [3]

Sederhoff, R. R., 1987 Molecular mechanisms of mitochondrial-genome evolution in higher plants. Am. Natur. 130: S30–S45. [6]

Sgarbieri, V. C., and J. R. Whitaker, 1982 Physical, chemical, and nutritional properties of common bean (IPhaseolus) proteins. Adv. Food Res. 28: 93–166. [4]

Shaw, D. V., and R. W. Allard, 1982a Estimation of outcrossing rates in Douglas-fir using isozyme markers. Theor. Appl. Genet. 62: 113–120. [9, 14, 16]

Shaw, D. V., and R. W. Allard, 1982b Isozyme heterozygosity in adult and open-pollinated embryo samples of Douglas-fir. Silva. Fenn. 16: 115–121. [16]

Shaw, D. V., A. L. Kahler and R. W. Allard, 1981 A multilocus estimator of mating system parameters in plant populations. Proc. Natl. Acad. Sci. U.S.A. 78: 1298–1302. [3, 14]

Shaw, R. G., 1987 Maximum-likelihood approaches applied to quantitative genetics of natural populations. Evolution 41: 812–826. [20]

Shea, K. L., 1987 Effects of population structure and cone production on outcrossing rates in Engelmann spruce and subalpine fir. Evolution 41: 124–136. [16]

Shebeski, L. H., 1967 Wheat and breeding. pp. 249–272. In: *Proceedings of the Canadian Centennial Wheats Symposium,* Edited by K. F. Nielson. Modern Press, Saskatoon. [21]

Shen, H. H., D. Rudin and D. Lindgren, 1981 Study of the pollination pattern in a Scots pine seed orchard by means of isozyme analysis. Silv. Genet. 30: 7–15. [16]

Shewry, P. R., and B. J. Miflin, 1985 Seed storage proteins of economically important cereals. Adv. Cereal Sci. Technol. 7: 1–83. [4]

Shewry, P. R., H. M. Pratt, A. J. Faulks, M. J. Charlton and B. J. Miflin, 1977 The storage protein (hordein) polypeptide pattern of barley (*Hordeum vulgare* L.) in relation to varital identification and disease resistance. J. Natl. Inst. Agric. Bot. 15: 34–50. [4]

Shewry, P. R., H. M. Pratt, R. A. Fink and B. J. Miflin, 1978 Genetic analysis of hordein polypeptides from single seeds of barley. Heredity 40: 463–466. [4]

Shewry, P. R., A. J. Faulks, S. Parmer and B. J. Miflin, 1980 The genetic analysis of barley storage proteins. Heredity 44: 383–389. [4]

Shinozaki, K., M. Ohme, M. Tanaka, T. Wakasugi, N. Hayashida, T. Matsubayashi, N. Zaita, J. Chunwongse, J. Obokata, K. Yamaguchi-Shinozake, C. Ohto, K. Torazawa, B. Y. Meng, M. Sugita, H. Deno, T. Kamogashira, K. Yamada, J. Kusuda, F. Takaiwa, A. Kato, N. Tohdoh, H. Shimada and M. Sugiura, 1986 The complete nucleotide sequence of tobacco chloroplast genome: its gene organization and expression. EMBO J 5: 2043–2050. [6]

Sigurbjörnsson, B., 1983 Induced mutations. pp. 153–176. In: *Crop Breeding,* Edited by D. R. Wood. American Society of Agronomy, Madison, Wisc. [17]

Silander, J. A., Jr., 1984 The genetic basis of the ecological amplitude of *Spartina patens.* III. Allozymic variation. Bot. Gaz. 145: 569–577. [15]

Silander, J. A., Jr., 1985 The genetic basis of the ecological amplitude of *Spartina patens.* II. Variance and correlation analysis. Evolution 39: 1034–1052. [15]

Silvela, L., and R. Diez-Barra, 1985 Recurrent selection in autogamous species under forced random mating. Euphytica 34: 817–832. [17]

Simmonds, N. W., 1962 Variability in crop plants, its use and conservation. Biol. Rev. 37: 442–465. [21]

Simon, J. P., Y. Bergeron and D. Gagnon, 1986 Isozyme uniformity in populations of red pine (*Pinus resinosa*) in the Abitibi Region, Quebec. Can. J. For. Res. 16: 1133–1135. [3]

Singh, L. N., and L. P. V. Johnson, 1969 Natural selection in a composite cross of barley. Can. J. Geneet. Cytol. 11: 34–42. [12]

Singh, R. B., and J. T. Williams, 1984 Maintenance and multiplication of plant genetic resources. pp. 120–130. In: *Crop Genetic Resources: Conservation and Evaluation*, Edited by J. H. W. Holden and J. T. Williams. Allen and Unwin, London. [21]

Sirkkomaa, S., 1983 Calculations on the decrease of genetic variation due to the founder effect. Hereditas 99: 11–20. [15]

Slatkin, M., 1977 Gene flow and genetic drift in a species subject to frequent local extinctions. Theor. Pop. Biol. 12: 253–262. [15]

Slatkin, M., 1981 Estimating levels of gene flow in natural populations. Genetics 99: 323–335. [3]

Slatkin, M., 1985 Rare alleles as indicators of gene flow. Evolution 39: 53–65. [3, 15]

Slightom, J. L., and P. P. Chee, 1987 Advances in the molecular biology of plant seed storage proteins. Biotech. Adv. 5: 29–45. [4]

Slightom, J. L., A. E. Blechl and O. Smithies, 1980 Human fetal G- and A-globin genes: Complete nucleotide sequences suggests that DNA can be exchanged between these duplicated genes. Cell 21: 627–638. [4]

Slightom, J. L., R. F. Drong,R. C. Klassy and L. M. Hoffman, major storage proteins form *Phaseolus vulgaris* are encoded by two unique gene families. Nucl. Acid Res. 13: 6483–6498. [4]

Smith, D. B., and W. T. Adams, 1983 Measuring pollen contamination in clonal seed orchards with the aid of genetic markers. pp. 64–73. In: *Proc. 17th South. Forest Tree Improv. Conf.* [16]

Smith, G. P., 1976 Evolution of repeated DNA sequences by unequal crossover. Science 191: 528–535. [5]

Smith-Huerta, N. L., 1986 Isozymic diversity in three allotetraploid *Clarkia* and their putative diploid progenitors. J. Hered. 77: 349–354. [3]

Smith, J. S. C., and O. S. Smith, 1986 Environmental effects on zein chromatograms of maize inbred lines revealed by reversed-phase high-performance liquid chromatography. Theor. Appl. Genet. 71: 607–612. [4]

Smith, J. S. C., M. M. Goodman and C. W. Stuber, 1984 Variation within teosinte. III. Numerical analysis of allozyme data. Econ. Bot. 38: 97–113. [21]

Smyth, C. A., and J. L. Hamrick, 1986 Realized gene flow via pollen in artificial populations of musk thistle, *Carduus nutans*. Evolution 41: 613–619. [3]

Snape, J. W., R. B. Flavell, M. O'Dell, W. G. Hughes and P. E. Payne, 1985 Intrachromosomal mapping of the nucleolar organizer region relative to three marker loci on chromosome 1B of wheat (*Triticum aestivum*). Theor. Appl. Genet. 69: 263–270. [5]

Snape, J W , J de Buyser, Y. Henry and E. Simpson, 1986 A comparison of methods of haploid production in a cross of wheat, *Triticum aestivum*. Z Pflanzenzuchtg. 96: 320–329. [17]

Snape, J. W., L. A. Sitch, E. Simpson and B. B. Parke, 1988 Tests for the presence of gametoclonal variation in barley and wheat doubled haploids produced using *Hordeum bulbosum* system. Theor. Appl. Genet. 75: 509–513. [17]

Snaydon, R. W., and A. D. Bradshaw, 1962 The performance and survival of contrasting natural populations of white clover when planted into an upland *Festuca-Agrostis* sward. J. Br. Grassl. Soc. 17: 113–118. [12]

Snell, G. D., 1968 The H-2 locus of the mouse: Observations and speculations concerning its comparative genetics and its polymorphism. Folia Biol. (Prague) 14: 335–358. [8]

Snyder, T. P., D. A. Stewart and A. F. Strickler, 1985 Temporal analysis of breeding structure in jack pins (*Pinus banksiana* Lamb.). Can. J. For. Res. 15: 1159–1166. [16]

Soave, C., R. Reggiani, N. Di Fonzo and F. Salamini, 1981 Clustering of genes for 20 kD zein subunits in the short arm for maize chromosome 7. Genetics 97: 363–377. [4]

Soave, C., R. Reggiani, N. Di Fonzo and F. Salamini, 1982 Genes for zein subunits on maize chromosome 4. Biochem. Genet. 20: 1027–1038. [4]

Sokal, R. R., 1979 Ecological parameters inferred from spatial correlograms. pp. 167–196. In:

Contemporary Quantitative Ecology and Related Ecometrics, Edited by G. P. Patil and M. L. Rosenzweig. International Cooperative Publishing House, Fairland, Md. [14]

Sokal, R. R., 1988 Genetic, geographic, and linguistic distances in Europe. Proc. Natl. Acad. Sci. U.S.A. 85: 1722–1726. [14]

Sokal, R. R., and P. Menozzi, 1982 Spatial autocorrelations of HLA frequencies in Europe support demic diffusion of early farmers. Am. Nat. 119: 1–17. [14]

Sokal, R. R., and N. L. Oden, 1978a Spatial autocorrelation in biology. 1. Methodology. Biol. J. Linn. Soc. 10: 199–228. [14]

Sokal, R. R., and N. L. Oden, 1978b Spatial autocorrelation in biology. 2. Some biological implications and four applications of evolutionary and ecological interest. Biol. J. Linn. Soc. 10: 229–249. [14]

Sokal, R. R., and D. E. Wartenberg, 1983 A test of spatial autocorrelation analysis using an isolation-by-distance model. Genetics 105: 219–237. [14]

Sokal, R. R., and E. M. Winkler, 1987 Spatial variation among Kenyan tribes and subtribes. Human Biol. 59: 147–164. [14]

Sokal, R. R., P. E. Smouse and J. V. Neel, 1986 The genetic structure of a tribal population, the Yanomama Indians. XV. Patterns inferred by autocorrelation analysis. Genetics 114: 259–287. [14]

Sokal, R. R., N. L. Oden and J. S. F. Barker, 1987 Spatial structure in *Drosophila buzzatii* populations: Simple and directional spatial autocorrelation. Am. Nat. 129: 122–142. [14]

Sokal, R. R., G. M. Jacquez and M. C. Wooten, 1989 Spatial autocorrelation analysis of migration and selection. Genetics 121: 845-855. [14]

Soller, M., and J. S. Beckmann, 1983 Genetic polymorphism in varietal identification and genetic improvement. Theor. Appl. Genet. 67: 25–33. [19]

Soller, M., and J. S. Beckmann, 1987 Cloning quantitative trait loci by insertional mutagenesis. Theor. Appl. Genet. 74: 369–378. [20]

Soller, M., and J. Plotkin-Hazan, 1977 The use of marker alleles for the introgression of linked quantitative alleles. Theor. Appl. Genet. 51: 133–137. [19]

Soltis, D. E., and P. S. Soltis, 1987 Breeding system of the fern *Dryopteris expansa:* Evidence for mixed mating. Am. J. Bot. 74: 504–509. [9]

Soltis, P. S., and W. L. Bloom, 1986 Genetic variation and estimates of gene flow in *Clarkia speciosa* subsp. *polyantha* (Onagraceae). Am. J. Bot. 73: 1677–1682. [15]

Sorensen, F. C., and R. S. Miles, 1982 Inbreeding depression in height, height growth, and survival of Douglas-fir, ponderosa pine and noble fir to 10 years of age. For. Sci. 28: 283–292. [16]

Sorensen, F. C., and T. L. White, 1988. Effect of natural inbreeding on variance structure in tests of wind-pollination Douglas-fir progenies. For. Sci. 34: 102–118. [16]

Sousa, W. P., 1979 Disturbance in marine interidal boulder fields: The nonequilibrium maintenance of species diversity. Ecology 60: 1225–1239. [15]

Spagnoletti Zeuli, P. L., and C. O. Qualset, 1987 Geographical diversity for quantitative spike characters in a world collection of durum wheat. Crop Sci. 27: 235–241. [17]

Spielmann, A., and E. Stutz, 1983 Nucleotide sequence of soybean chloroplast DNA regions which contain the *psbA* and *trnH* genes and cover the ends of the large single copy region and one end of the inverted repeats. Nucl. Acids Res. 11: 7157–7167. [8]

Squillace, A. E., 1974 Average genetic correlations among offspring from open-pollinated forest trees. Silv. Genet. 23: 149–156. [16]

St. Clair, D. A., 1986 Segregation, selection and population improvement for ^{15}N- determined dinitrogen fixation in common bean (*Phaseolus vulgaris* L.). Ph.D. Thesis, University of Wisconsin, Madison. [18]

St. Clair, D. A., D. J. Wolyn, J. Dubois, R. H. Burris and F. A. Bliss, 1988 A field comparison of N_2-fixation determined with ^{15}N-depleted ammonium sulfate and ^{15} N-enriched ammonium sulfate in selected inbred backcross lines of common bean. Crop Sci. 28: 733–778. [18]

Stanton, M. L., A. A. Snow and S. N. Handel, 1986 Floral evolution: Attractiveness to pollinators increases male fitness. Science 232: 1625–1627. [12]

Staub, J. E., L. Fredrick and T. L. Marty, 1987 Electrophoretic variation in cross-compatible wild diploid species of *Cucumis*. Can. J. Bot. 65: 792–798. [3]

Stearns, S. C., 1976 Life history tactics: A review of the ideas. Rev. Biol. 51: 3–47. [11]

Stebbins, G. L., 1957 Self fertilization and population variability in the higher plants. Am. Nat. 91: 337–354. [9]
Stegemann, H., and G. Pietsch, 1983 Methods for quantitative and qualitative characterization of seed proteins of cereals and legumes. pp. 45–75. In: *Seed Proteins: Biochemistry, Genetics, Nutritive Value*, Edited by W. Gottschalk and H. P Müller. Martinus Nijhoff/Dr. W. Junk, The Hague. [4]
Stephens, J. C., 1985 Statistical methods of DNA sequence analysis: detection of intragenic recombination or gene conversion. Mol. Biol. Evol. 2: 539–556. [6]
Stephens, J. C., and M. Nei, 1985. Phylogenetic analysis of polymorphic DNA sequences at the Adh locust in *Drosophila melanogaster* and its sibling species. J. Mol. Evol. 22: 289–300. [6]
Stephenson, A. G., and R. I. Bertin, 1983 Male competition, female choice, and sexual selection in plants. pp. 109–149. In: *Pollination Biology*, Edited by L. Real. Academic Press, New York. [16]
Stern, K., and H. R. Gregorius, 1972 Schätzungen der effektiven Populationsgrösse bei *Pinus sylvestris*. Theor. Appl. Genet. 42: 107–110. [16]
Stern, K., and L. Roche, 1974 *Genetics of Forest Ecosystems*. Springer-Verlag, Berlin. [16]
Stevenson, N. D., A. H. D. Brown and B. D. H. Latter, 1972 Quantitative genetics of sugarcane IV. Genetics of Fiji disease resistance. Theor. Appl. Genet. 42: 262–266. [20]
Stewart, C. B., J. W. Schilling and A. C. Wilson, 1987 Adaptive evolution in the stomach lysozymes of foregut fermenters. Nature (London) 33: 401–404. [8]
Stoddard, F. L., 1986a Auto-fertility and bee visitation in winter/spring genotypes of faba beans (*Vicia faba*). Plant Breed. 97: 171–182. [20]
Stoddard, F. L., 1986b Pollination fertilisation and seed development in inbred lines and F_1 hybrids of spring faba beans (*Vicia faba* L.) Plant Breed. 97: 210:–211. [20]
Stoddard, F. L., 1986c Pollination and fertilization in commercial crops of field beans (*Vicia faba* L.) J. Agric. Sci. 106: 89–97. [20]
Strauss, S. H., 1986 Heterosis at allozyme loci under inbreeding and crossbreeding in *Pinus attenuata*. Genetics 113: 115–134. [12, 16]
Strauss, S. H., and W. J. Libby, 1987 Allozyme heterosis in radiata pine is poorly explained by overdominance. Am. Nat. 130: 879–890. [12, 16, 20]
Strobeck, C., 1979 Partial selfing and linkage: the effect of a heterotic locus on a neutral locus. Genetics 92: 305–315. [10]
Stuber, C. W., In press Isozymes as markers for studying and manipulating quantitative traits. In: Isozymes in Plant Biology, Edited by D. Soltis and P. Soltis, Dioscorides Press, Portland, Oregon. [19]
Stuber, C. W., and M. D. Edwards, 1986 Genotypic selection for improvement of quantitative traits in corn using molecular marker loci. Proc. 41st Annu. Corn Sorghum Res. Conf. Am. Seed Trade Assoc. 41: 70–83. [19]
Stuber, C. W., and M. M. Goodman, 1983 Allozyme genotypes for popular and historically important inbred lines of corn. U.S. Dept. of Agriculture. Agric. Res. Serv., Southern Series No. 16, 29 pp. [19]
Stuber, C. W., and R. H. Moll, 1972 Frequency changes of isozyme alleles in a selection experiment for grain yield in maize (*Zea mays* L.). Crop Sci. 12: 337–340. [19]
Stuber, C. W., R. H. Moll, M. M. Goodman, H. E. Schaffer and B. S. Weir, 1980 Allozyme frequency changes associated with selection for increased grain yield in maize (*Zea mays* L.). Genetics 95: 225–236. [19]
Stuber, C. W., M. M. Goodman and R. H. Moll, 1982 Improvement of yield and ear number resulting from selection at allozyme loci in a maize population. Crop Sci. 22: 737–740. [19]
Stuber, C. W., M. D. Edwards and J. F. Wendel, 1987 Molecular marker-facilitated investigations of quantitative trait loci in maize. II. Factors influencing yield and its component traits. Crop Sci. 27: 639–648. [17, 19]
Sullivan, J. G., and F. A. Bliss, 1983a Genetic control of quantitative variation in phaseolin seed protein of common bean. J. Am. Soc. Hort. Sci. 108: 782–787. [18]
Sullivan, J. G., and F. A. Bliss, 1983b Expression of enhanced seed protein content in inbred backcross lines of common bean. J. Am. Soc. Hort. Sci. 108: 787–791. [18]
Sumarno, and W R. Fehr, 1982 Response to recurrent selection for yield in soybeans. Crop Sci. 22: 295–299. [17]

Sun, M., 1987 Genetics of gynodioecy in Hawaiian *Bidens*. Heredity 59: 327–336. [15]
Sun, M., and F. R. Ganders, 1986 Female frequencies in gynodioecious populations correlated with selfing rates in hermaphrodites. Am. J. Bot. 73: 1645–1648. [15]
Sun, M., and F. R. Ganders, 1987 Microsporogenesis in male-sterile and hermaphroditic plants of nine gynodioecious taxa. Am. J. Bot 74:209–217. [15]
Sun, M., and F. R. Ganders, 1988 Mixed mating systems in Hawaiian *Bidens* (Asteraceae). Evolution 42: 516–527. [9, 11, 15]
Sun, S. M., J. L. Slightom and T. C. Hall, 1981 Intervening sequences in a plant gene-comparison of the partial sequence of cDNA and genomic DNA of French bean phaseolin. Nature (London) 289: 37–41. [4]
Suneson, C. A., 1956 An evolutionary plant breeding method. Agron. J. 48: 188–191. [21]
Suneson, C. A., 1968 Harland barley. Calif. Agric. 22: 9. [17]
Sutton, W. D., W. L. Gerlach, D. Schwartz and W. J. Peacock, 1984 Molecular analysis of *Ds* controlling element mutations at the *Adh1* locus of maize. Science 223: 1265–1268. [7]
Swaminathan, M. S., 1983 Relevance of protein improvement in plant breeding. pp. 1–44. In: Seed Proteins: Biochemistry, Genetics, Nutritive Value, Edited by W. Gottschalk and H. P. Müller. Martinus Nijhoff/Dr. W. Junk, The Hague. [4]
Swofford, D. L., and R. B. Selander. 1981 BIOSYS-1: A FORTRAN program for the comprehensive analysis of electrophoretic data in population genetics and systematics. J. Hered. 72: 281–283. [2]
Systma, K., and L. D. Gottlieb, 1986 Chloroplast evolution and phylogenetic relationships in *Clarkia* sect. Peripetasma (Onograceae). Evolution 40: 1248–1261. [11]
Szmidt, A. E., and O. Muona, 1985 Genetic effects of Scots pine (*Pinus sylvestris* L.) domestication. pp. 241–252. In: *Population genetics in Forestry*. Lecture Notes in Biomathematics 60, Edited by H. R. Gregorius. Springer-Verlag, Berlin. [16]
Szmidt, A. E., Y. A. El-Kassaby, A. Sigurgeirsson, T. Aldèn, D. Lindgren and J. E. Hällgren, 1988 Classifying seedlots of *Picea sitchensis* and *Picea clauca* in zones of introgression, using restriction analysis of chloroplast DNA. Theor. Appl. Genet. 76: 841–845. [16]

Tachida, H., and C. C. Cockerham, 1988 Variance components of fitness under stabilizing selection. Genet. Res. 51: 47–53. [20]
Tanaka, T., and M. Nei, 1988 Positive darwinian selection at the variable region genes of immunoglobulins. Genome 30, Suppl. 1: 367. [8]
Tanksley, S. D., 1983 Molecular markers in plant breeding. Plant Mol. Biol. Rep. 1: 3–8. [11, 19]
Tanksley, S. D., and J. Hewitt, 1988 Use of molecular markers in breeding for soluble solids content in tomato—a re-examination. Theor. Appl. Genet. 75: 811–823. [18, 19]
Tanksley, S. D., and C. M. Rick, 1980 Isozyme linkage map of the tomato: Applications in genetics and breeding. Theor. Appl. Genet. 57: 161–170. [19]
Tanksley, S. D., H. Medina-Filho and C. M. Rick, 1982 Use of naturally-occurring enzyme variation to detect map genes controlling quantitative traits in an interspecific backcross of tomato. Heredity 49: 11–25. [18, 19]
Tartof, K. D., 1975 Redundant genes. Annu. Rev. Genet. 9: 355–385. [5]
Technical Advisory Committee (TAC), 1986 *Report of the Second External Program and Management Review of the International Board for Plant Genetic Resources*. TAC Secretariat, FAO, Rome. [21]
Templeton, A. R., 1980 The theory of speciation via the founder principle. Genetics 94: 1011–1038. [15]
Thoday, J. M., 1961 Location of polygenes. Nature (London) 191: 368–370. [17]
Thompson, E. A., S. Deeb, D. Walker and A. G. Motulsky, 1988 The detection of linkage disequilibrium between closely linked markers; RFLPs at the AI-CIII apolipoprotein genes. Am. J. Hum. Genet. 42: 113–124. [10]
Thomson, G., 1977 The effect of a selected locus on linked neutral loci. Genetics 85: 753–788. [10]
Thomson, G., and W. Klitz, 1987 Disequilibrium pattern analysis. I. Theory. Genetics 116: 625–632. [10]
Tigerstedt, P. M. A., 1973 Studies on isozyme variation in marginal and central populations of

Picea abies. Hereditas 75: 47–60. [15]

Tigerstedt, P. M. A., D. Rudin, T. Niemelä and J. Tammisola, 1982 Competition and neighbouring effect in a naturally regenerating population of Scots pine. Silv. Fenn. 16: 122–129. [16]

Timothy, D. H., C. S. Levings III, D. R. Pring, M. F. Conde and J. L. Kermicle, 1979 Organgelle DNA variation and systematic relationships in the genus *Zea*: Teosinte. Proc. Natl. Acad. Sci. U.S.A. 76: 4220–4224. [6]

Tobler, W. R., 1975 Linear operators applied to areal data. pp. 14–37. In: *Display and Analysis of Spatial Data*, Edited by J. C. Davis and M. J. McCullagh. Wiley, London. [14]

Tombs, M. P., and M. Low, 1967 A determination of the sub-units of arachin by osmometry. Biochem. J. 105: 181–187. [4]

Turelli, M., 1984 Heritable genetic variation via mutation-selection balance: Lerch's zeta versus the abdominal bristle. Theor. Pop. Biol. 25: 138–193. [20]

Turelli, M., 1985 Effects of pleiotropy on predictions concerning mutation-selection balance for polygenic traits. Genetics 111: 165–195. [20]

Turkington, R., and J. L. Harper, 1979 The growth, distribution and neighbour relationships of *Trifolium repens* in a permanent pature. IV. Fine scale biotic differentiation. J. Ecol. 67: 245–254. [12]

Turner, M. E., J. C. Stephens and W. W. Anderson, 1982 Homozygosity and patch structure in plant populations as a result of nearest-neighbor pollination. Proc. Natl. Acad. Sci. U.S.A. 79: 203–207. [14]

Tuschall, D. M., and L. C. Hannah, 1982 Altered maize endosperm ADP-glucose pyrophosphorylases from revertants of a *shrunken-2-Dissociation* allele. Genetics 100: 105–111. [7]

Upton, G. J. G., and B. Fingleton, 1985 *Spatial Data Analysis by Example. Vol. 1. Point Pattern and Quantitative Data*. Wiley, New York. [14]

Usberti, J. A., and S. K. Jain, 1978 Variation in *Panicum maximum*: A comparison of sexual and asexual populations. Bot. Gaz. 139: 112–116. [9]

Utsumi, S., Z. Yokoyama and T. Mori, 1980 Comparative studies of subunit compositions of legumins from various cultivars of *Vicia faba* L. seeds. Agric. Biol. Chem. 44: 595–601. [4]

Utz, H. F., 1981 Comparison of three methods of line selection. pp. 119–126. In: *Quantitative Genetics and Breeding Methods*. Edited by A. Gallais. INRA Lusignan, France. [17]

Utz, H. F., and F. W. Schnell, 1973 Optimum number of places and optimum intensity of selection in three-stage selection processes. pp. 63–68. In: *Proc. 1st Meeting Section Biometrics in Plant Breeding*, Edited by H. Rundfelt and G. Wricke, Hanover. [17]

Vaeck, M., A. Reynaerts, H. Höfte, S. Jansens, M. De Beuckeleer, C. Dean, M. Zabeau, M. Van Montagu and J. Leemans, 1987 Transgenic plants protected from insect attack. Nature (London) 328: 33–37. [20]

Valentini, G., S. Soave and E. Ottaviano, 1979 Chromosomal location of zein genes in *Zea mays*. Heredity 42: 32–40. [4]

Vallega, V., and J. G. Waines, 1987 High molecular weight glutenin subunit variation in *Triticum turgidum* var. *dicoccum*. Theor. Appl. Genet. 75: 706–710. [4]

Vallejos, C. E., and S. D. Tanksley, 1983 Segregation of isozyme markers and cold tolerance in an interspecific backcross of tomato. Theor. Appl. Genet. 66: 241–247. [19]

Van Damme, J. M. M., and Van Delden, W., 1984 Gynodioecy in *Plantago lanceolata* L. IV. Fitness components of sex types in different life cycle stages. Evolution 38: 1326–1336. [12]

Van der Plank, J. E., 1963 *Plant Diseases: Epidemics and Control*. Academic Press, New York. [20]

Vanselow, M., In press Recurrent selection in oat breeding. *Proc. 3rd Int. Oat Conf.* Lund, Sweden. [17]

Van Sloten, D. H., 1984 The genetic resources of leafy vegetables. pp. 240–248. In: *Crop Genetic Resources: Conservation and Evaluation*, Edited by J. H. W. Holden and J. T. Williams. Allen and Unwin, London. [21]

Viotti, A., N. E. Pogna, C. Balducci and M. Durante, 1980 Chromosomal location of zein genes by *in situ* hybridization in *Zea mays*. Mol. Gen. Genet. 178: 34–41. [4]

Viotti, A., D. Abildsten, N. Pogna, E. Sala and V. Pirotta, 1982 Multiplicity and diversity of clones zein cDNA sequences and their chromosomal localization. EMBO J. 1: 53–58. [4]
Vitale, A., C. Soave and E. Galante, 1980 Peptide mapping of IEF zein components from maize. Plant Sci. Lett. 18: 57–64. [4]
Vogelmann, J. E., and G. J. Gastony, 1987 Electrophoretic enzyme analysis of North American and Eastern Asian populations of *Agastache* sect. Agastache (Labiatae). Am. J. Bot. 74: 385–393. [3]

Wade, M. J., and D. E. McCauley, 1988 Extinction and recolonization: Their effects on the genetic differentiation of local populations. Evolution 42: 995–1005. [15]
Wadley, F. M., 1954 Limitations of the "zero method" of population counts. Science 119: 689–690. [14]
Wagner, D. B., G. R. Furnier, M. A. Saghai-Maroof, S. M. Williams, B. P. Dancik and R. W. Allard, 1987 Chloroplast DNA polymorphisms in lodgepole and jack pines and there hybrids. Proc. Natl. Acad. Sci. U.S.A. 84: 2097–2100. [6, 16]
Wagner, D. B., S. Zhong-Xu, D. R. Govindaraju and B. P. Dancik, In press The population genetic structure of chloroplast DNA polymorphism in a *Pinus banksiana–P. contorta* sympatric region: Spatial autocorrelation. Studia Forestalia Suecica. [14]
Waines, J. G., R. M. Manshardt and W. C. Wells, 1988 Interspecific hybridization between *Phaseolus vulgaris* and *P. acutifolius*. pp. 485–502. In: *Genetic Resources of Phaseolus Beans*, Edited by P. L. Gepts. Kluwer Academic Publ., Boston. [18]
Walbot, V., and C. A. Cullis, 1985 Rapid genomic change in higher plants. Annu. Rev. Plant Physiol. 36: 367–396. [6]
Waller, D. M., and S. E. Knight, In press Genetic consequences of outcrossing in the cleistogamous annual, *Impatiens capensis*. III. Outcrossing rates and genetic correlations. Evolution. [9]
Warwick, S. I., B. K. Thompson and L. D. Black, 1987 Genetic variation in Canadian and European populations of the colonizing weed species *Apera spica-venti*. New Phytol. 106: 301–317. [15]
Weber, M. F., and R. F. Stettler, 1981 Isoenzyme variation among ten populations of *Populus trichocarpa* Torr. et Gray in the Pacific Northwest. Silv. Genet. 30: 82–87. [15]
Weber, W. E., 1976 A 10-parent diallel for quantitative genetic studies in peas (*Pisum sativum* L.). Z. Pflanzenzuchtg. 77: 30–42. [17]
Weber, W. E., 1979 Number and size of cross progenies from a constant total number of plants manageable in a breeding program. Euphytica 28: 453–456. [17]
Weber, W. E., 1980. Quantitativ genetische Untersuchungsmethoden und ihre Beduetung fur die Selektion in der Selbstbefruchterzuchtung. Habilitationsschrift, Univ. Hannover. [17]
Weber, W. E., 1981. Two stage selection in segregating generations of autogamous species. pp. 127–134. In: *Quantitative Genetics and Breeding Methods*, Edited by A. Gallais. INRA, Lusignan, France. [17]
Weber, W. E., 1982 Selection in segregating generations of autogamous species. I. Selection response for combined selection. Euphytica 31: 493–502. [17]
Weber, W. E., 1983 The efficience of repeated crosses within a population of self-fertilizing species z. Pflanzenzuchtg. 91: 129–139. [17]
Weber, W. E., 1984 Selection in segregating generations of autogamous species. II. Response to selection in two stages. Euphytica 33: 447–454. [17]
Weber, W. E., and G. Wricke, 1981 Genetische Varianzkomponenten verschiedener Kreuzungen bei Hafer. Deutsche Landwirtschaftsgesellschaft. Ausschuss fur Biometrie und Versuchswesen. [17]
Webster, R. K., M. A. Saghai-Maroof and R. W. Allard, 1986 Evolutionary response of barley composite cross II to *Rhynchosporium secalis* analyzed by pathogenic complexity and by gene-by-race relationships. Phytopathology 76: 661–668. [13]
Wehrhahn, C., and R. W. Allard, 1965 The detection and measurement of the effects of individual genes involved in the inheritance of a quantitative character in wheat. Genetics 51: 109–119. [17, 18]
Weir, B. S., 1979 Inferences about linage disequilibrium. Biometrics 35: 235–254. [2, 10]
Weir, B. S., and L. D. Brooks, 1986 Disequilibrium on human chromosome 11p. Genet. Epi-

demol. Suppl. 1: 177–183. [6]
Weir, B. S., and C. C. Cockerham, 1969 Group inbreeding with two linked loci. Genetics 63: 711–742. [11]
Weir, B. S., and C. C. Cockerham, 1973 Mixed self and random mating at two loci. Genet. Res. 21: 247–262. [2, 10, 11]
Weir, B. S., and C. C. Cockerham, 1984 Estimating *F*-statistics for the analysis of population structure. Evolution 38: 1358–1370. [11, 14]
Weir, B. S., and C. C. Cockerham. 1988 Complete characterization of disequilibrium at two loci. pp. 88–110. In: *Mathematical Evolutionary Theory*, Edited by M. W. Feldman. Princeton University Pres, Princeton. [2]
Weir, B. S., R. W. Allard and A. L. Kahler, 1972 Analysis of complex allozyme polymorphisms in a barley population. Genetics 72: 55–523. [10]
Weir, B. S., R. W. Allard and A. L. Kahler, 1974 Further analysis of complex allozyme polymorphisms in a barley population. Genetics 78: 911–919. [10]
Weir, B. S., E. J. Eisen, M. M. Goodman and G. Namkoong, 1988 *Proceedings of the Second International Conference on Quantitative Genetics*. Sinauer, Sunderland, Mass. [20]
Weller, J. I., M. Soller and T. Brody, 1988 Linkage analysis of quantitative traits in an interspecific cross of tomato (*Lycopersicon esculentum* x *Lycopersicon pimpinellifolium*) by means of genetic markers. Genetics 118: 329–339. [19]
Wendel, J. F., C. W. Stuber, M. D. Edwards and M. M. Goodman, 1986 Duplicated chromosome segments in maize (*Zea mays* L.): Further evidence from hexokinase isozymes. Theor. Appl. Genet. 72: 178–185. [19]
Wendel, J. F., M. M. Goodman, C. W. Stuber and J. B. Beckett, 1988 New isozyme systems for maize (*Zea mays* L.): Aconitate hydratase, adenylate kinase. NADH dehydrogenase, and shikimate dehydrogenase. Biochem. Genet. 26: 421–445. [19]
Werr, W., W. B. Frommer, C. Maas and P. Starlinger, 1985 Structure of the sucrose synthase gene on chromosome 9 of *Zea mays* L. EMBO J. 4: 1373–1380. [6, 7]
Wessler, S., and M. J. Varagona, 1985 Molecular basis of mutations at the *waxy* locus of maize: Correlation with the fine structure map. Proc. Natl. Acad. Sci. U.S.A. 82: 4177–4181. [7]
Wessler, S., G. Baran, M. Varagona and S. L. Dellaporta, 1986 Excision of *Ds* produces *waxy* proteins with a range of enzymatic activities. EMBO J. 5: 2427–2432. [7]
Wessler, S. R., G. Baran and M. Varagona, 1987 The maize transposable element *Ds* is spliced from RNA. Science 237: 916–918. [7]
Wheeler, N. C., and R. P. Guries, 1982 Population structure, genic diversity, and morphological variation in *Pinus contorta* Dougl. Can. J. For. Res. 12: 595–606. [16]
Wheeler, N. C., and R. P. Guries, 1987 A quantitative measure of introgression between lodgepole and jack pines. Can. J. Bot. 65: 1876–1885. [16]
Whittle, P., 1954 On stationary processes in the plane. Biometrika 41: 434–449. [14]
Whyte, R. O., 1958 *Plant Exploration, Collection and Introduction*. FAO Agriculture Studies 41, Rome. [21]
Wilcox, M. D., 1983 Inbreeding depression and genetic variances estimated from self- and cross-pollinated families of *Pinus radiata*. Silv. Genet. 32: 89–96. [16]
Wilkes, H. G., 1983 Current status of crop plant germplasm. CRC Crit. Rev. Plant Sci. 1: 133–181. [21]
Williams, J. T., 1984 A decade of crop genetic resources research. pp. 1–17. In: *Crop Genetic Resources: Conservation and Evaluation*, Edited by J. H. W. Holden and J. T. Williams. Allen and Unwin, London. [21]
Williams, P. H., and C. B. Hill, 1986 Rapid-cycling populations of *Brassica*. Science 232: 1385–1389. [13]
Williams, S. M., and C. Strobeck, 1985 Sister chromatid exchange and the evolution of rDNA spacer length. J. Theor. Biol. 116: 625–636. [5]
Wilson, A. C., 1975 Evolutionary importance of gene regulation. Stadler Symp. 7: 117–134. University of Missouri, Columbia, Mo. [8]
Wilson, C. M., 1985 Mapping of zein polypeptides after isoelectric focusing on agarose gels. Biochem. Genet. 23: 115–124. [4]
Witcombe, J. R., and M. M. Gilani, 1979 Variation in cereals from the Himalayas and the optimum strategy for sampling plant germplasm. J. Appl. Ecol. 16: 633–640. [21]
Withers, L. A., 1989 *In vitro* conservation and germplasm utilisation. pp. 309–334. In: *The Use*

of Plant Genetic Resources, Edited by A. H. D. Brown, O. H. Frankel, D. R. Marshall, and J. T. Williams. Cambridge University Press, Cambridge. [21]

Witter, M. S., and G. D. Carr, 1988 Adaptive radiation and genetic differentiation in the Hawaiian Silversword alliance (Compositae; Madiinae). Evolution 42: 1278–1287. [15]

Wolfe, K. H., W. H Li and P. M. Sharpe, 1987 Rates of nucleotide substitution vary greatly among plant mitochondrial, chloroplast, and nuclear DNAs. Proc. Natl. Acad. Sci. U.S.A. 84: 9054–9058. [6]

Wolfe, M. S., 1984 Trying to understand and control powdery mildew. Plant Pathol. 33: 451–466. [13]

Wolfe, M. S., P. N. Minchin and S. E. Wright, 1976 Effects of variety mixtures. Annual Report, Plant Breeding Institute (1975). [13]

Wolff, K., 1988 Genetic analysis of ecologically relevant morphological variability in *Plantago lanceolata* L. III. Natural selection in an F_2 population. Theor. Appl. Genet 75: 772–778. [12]

Woods, T. A., 1987 The preservation and interpretation of historical germplasm: Living history forms and agricultural museums. Diversity 10: 35–38. [21]

Workman, P. L., and R. W. Allard, 1964 Population studies in predominantly self-pollinated species. V. Analysis of differential and random viabilities in mixtures of competing pure lines. Heredity 19: 181–189. [12]

Workman, P. L., and S. K. Jain, 1966 Zygotic selection under mixed random mating and self-fertilisation: theory and problems of estimation. Genetics 54: 159–171. [12]

Wricke, G., and W. E. Weber, 1986 *Quantitative Genetics and Selection in Plant Breeding*. de Gruyter, Berlin. [17, 20]

Wricke, G., and W. E. Weber, 1988. Application of quantitative genetics in plant breeding. Proc. Int. Conf. Population Mathematics, Edited by D. Rasch, F. Pirchner and J. Adam. Prob. Angewand. Stat. 25: 134–147. [17]

Wright, A. J., 1977 Inbreeding in synthetic varieties of field beans (*Vicia faba* L.) J. Agric. Sci. 89: 495–501. [20]

Wright, S., 1931 Evolution in mendelian populations. Genetics 16: 97–159. [15]

Wright, S., 1940 Breeding structure of populations in relation to speciation. Am. Nat. 74: 232–248. [15]

Wright, S., 1943 Isolation by distance. Genetics 28: 114–138. [14]

Wright, S., 1946 Isolation by distance under diverse systems of mating. Genetics 31: 39–59. [14]

Wright, S., 1951 The genetical structure of populations. Ann. Eugen. 15: 313–354. [15]

Wright, S., 1965 The interpretation of population structure by F-statistics with special regard to systems of mating. Evolution 19: 395–420. [14]

Wright, S., 1966 Polyallelic random drift in relation to evolution. Proc. Natl. Acad. Sci. U.S.A. 55: 1074–1081. [8]

Wright, S., 1969 *Evolution and the Genetics of Populations*. Volume 2, *The Theory of Gene Frequencies*. University of Chicago Press, Chicago. [11, 15]

Wright, S., 1978 *Evolution and the Genetics of Populations*. Volume 4. *Variability Within and Among Natural Populations*. University of Chicago Press, Chicago. [14]

Wrigley, C. W., and K. W. Shepherd, 1973 Electrofocusing of grain proteins from wheat genotypes. Ann N.Y. Acad. Sci. 109: 154–162. [4]

Wu, L., A. D. Bradshaw and D. A. Thurman, 1975 The potential for evolution of heavy metal tolerance in plants. III. The rapid evolution of copper tolerance in *Agrostis stolonifera*. Heredity 34: 165–187. [12]

Wu, N. H., J. C. Cote and R. Wu, 1987 Structure of the chloroplast *psbA* gene encoding the Q_B protein from *Oryza sativa* L. Dev. Genet. 8: 339–350. [8]

Yang, J-Ch., T. M. Ching and K. K. Ching, 1977 Isozyme variation of coastal Douglas fir. I. A study of geographic variation in 3 enzyme systems. Silv. Genet. 26: 10–18. [15]

Yasuda, N., 1968 An extension of Wahland's principle to evaluate mating type frequency. Am. J. Hum. Genet. 29: 1–23. [14]

Yasuda, N., 1973 Mating type frequency in terms of the gene frequency moments with random genetic drift. In: *Genetic Structure of Population*, Edited by N. E. Morton. University of Hawaii Press, Honolulu. [14]

Yazdani, R., D. Lingren and D. Rudin, 1985a Gene dispersion and selfing frequency in a seed tree stand of *Pinus sylvestris* (L.). pp. 1139–154. In: *Population Genetics in Forestry*. Lecture Notes in Biomathematics 60, Edited by H. R. Gregorius. Springer-Verlag, Berlin. [16]

Yazdani, R., O. Muona, D. Rudin and A. E. Szmidt, 1985b Genetic structure of a *Pinus sylvestris* L. seed-tree stand and naturally regenerated understory. For. Sci. 31: 430–436. [12, 16]

Yeh, F. C., and J. T. Arnott, 1986 Electrophoretic and morphological differentiation of *Picea sitchensis, Picea glauca*, and their hybrids. Can. J. For. Res. 16: 791–798. [16]

Yeh, F. C., and K. Morgan, 1987 Mating system and multilocus associations in a natural population of *Pseudotsuga menziesii* (Mirb). Franco. Theor. Appl. Genet. 73: 799–808. [9]

Yeh, F., and D. O'Malley, 1980 Enzyme variation in natural populations of Douglas fir, *Pseudotsuga menziesii* (Mub.) Franco, from British Columbia. I. Genetic variation patterns in Coastal populations. Silv. Genet. 29: 83–92. [15]

Yeh, F. C., W. M. Cheliak, B. P. Dancik, K. Illingworth, D. C. Trust and B. A. Pryhitka, 1985 Population differentiation in lodgepole pine, *Pinus contorta* ssp. *latifolia*: A discriminant analysis of allozyme variation. Can. J. Genet. Cytol. 27: 210–218. [16]

Yonezawa, K., 1985 A definition of the optimal allocation of effort in conservation of plant genetic resources, with application to sample size determination for field collection. Euphytica 34: 345–354. [21]

Yonezawa, K., and H. Yamagata, 1978 On the number and size of cross combinations in a breeding programme of self-fertilizing crops. Euphytica 27: 113–116. [17]

Yu, Q., H. V. Colot, C. P. Kyriacou, J. C. Hall and M. Rosbash, 1987 Behavior modification by *in vitro* mutagenesis of a variable region within the period gene of *Drosophilia*. Nature (London) 326: 765–769. [8]

Yunushkanov, S., and A. P. Ibragimov, 1984 Identification of seed protein markers of cotton species *Gossypium hirsutum* L. and *Gossypium barbadense* L., and study of the relationship of their inheritance with some characters. Sov. Genet. 10: 789–796. [4]

Zack, C. D., R. J. Ferl and L. C. Hannah, 1986 DNA sequence of a *Shrunken* allele of maize: Evidence for visitation by insertional sequences. Maydica 31: 5–16. [7]

Zamir, D., N. Navot and J. Rudich, 1984 Enzyme polymorphism in *Citrullus lanatus* and *C. colycynthis* in Israel and Sinai. Plant Syst. Evol. 146: 163–170. [3]

Zhang, Q., and R. W. Allard. 1986 Sampling variance of the genetic diversity index. J. Hered. 77: 54–55. [2]

Zobel, B. J., and J. Talbert, 1984 *Applied Forest Tree Improvement*. Wiley, New York. [16]

Zouros, E., 1987 The use of allelic isozyme variation for the study of heterosis. pp. 1–59. In: *Isozymes: Current Topics in Biological and Medical Research*, Vol. 13, Edited by M. C. Rattazi, J. G. Scandalios, and G. S. Whitt, Liss, New York. [11]

Zurawski, G., and M. T. Clegg, 1987 Evolution of higher-plant DNA-encoded genes: Implications for structure, function and phylogenetic studies. Annu. Rev. Plant Physiol. 38: 391–418. [6, 8]

Zurawski, G., H. J. Bohnert, P. R. Whitfield and W. Bottomley, 1982 Nucleotide sequence of the gene for the M_r 32,000 thylakoid membrane protein from *Spinacea oleracea* and *Nicotiana debneyi* predicts a totally conserved primary translation product of M_r 38,950. Proc. Natl. Acad. Sci. U.S.A. 79: 7699–7703. [8]

Index